AF838017

Mechanics of Advanced Materials and Structures

The 4th International Conference on Mechanics of Advanced Materials and Structures (ICMAMS 2024) was held from 11–13 December 2024 at Ramaiah Institute of Technology, Bengaluru, India. The event was jointly organized by Ramaiah Institute of Technology (India), Texas A&M University (USA), and Politecnico di Torino (Italy). The conference featured plenary lectures, keynote addresses, mini-symposia, and oral presentations, showcasing contributions from faculty members, research scholars, and industry experts. Discussions spanned a broad spectrum of themes, including: Composite and architected materials, Material design and manufacturing, Multifunctional, nano, and smart materials and structures, Energy harvesting technologies, Damage, fracture, and fatigue, Dynamics and vibrations, Metamaterials, Fluid–solid interactions, Micro- and nano-scale phenomena. More than 120 delegates presented their research, representing prestigious institutions worldwide, including the University of Alabama, Johns Hopkins University, Texas A&M University, Virginia Tech, Politecnico di Torino, and leading Indian institutions such as the IITs and NITs.

Dr. N V R Naidu, currently serving as Principal at M S Ramaiah Institute of Technology, Bangalore, India. With over 44 years of experience in academia and Research. Dr. Naidu academic journey includes B.Tech. in Mechanical Engineering, M.Tech and Ph.D in Industrial Engineering from Sri Venkateshwara University. He has received Dr. J Mahajan Award (2008-09) from the Indian Institution of Industrial Engineering . Throughout his illustrious career, he has held pivotal roles such as Dean Faculty of Engineering and Executive council Member at Visvesvaraya Technological University, reflecting his dedication to the advancement of academic excellence. He is serving on various academic boards and contributing his expertise to prestigious institutions in the field of Engineering. Dr. Naidu's expertise extends to research, evident in his 140+ published papers in prestigious national and international journals and conferences. He contributes as a reviewer for reputable journals and has chaired sessions at various conferences. Dr. Naidu has authored nine textbooks spanning Management, Total Quality Management, and Operations Research. Dr. Naidu has significantly impacted academic research, guiding seven doctoral candidates in areas like robust design and supply chain networks and he is adjudicator for Ph.D. theses across India. His international engagements in countries like the USA, Japan, and Canada have enhanced industry-institution interactions. He has executed numerous funded programs and projects, advancing academia and industry collaboration.

Dr. Subramanian K.R.V is a distinguished Professor and HOD of Mechanical Engineering and Head R & D with Ramaiah institute of Technology, with a robust background in nanotechnology and metallurgy. With a Ph.D. from the University of Cambridge and extensive experience in academia and industry, he leads groundbreaking research in nanocomposites, solar energy, and nanofluids. His pioneering work includes the development of nanostructured solar cells and supercapacitors, earning recognition globally. As a mentor, he has guided numerous Ph.D. and M.Tech students, fostering innovation in energy storage and materials science. Dr. Subramanian's contributions epitomize excellence at the intersection of nanotechnology and sustainable technology. Dr. Subramanian K.R.V pursued his Ph.D. in Nanotechnology from the University of Cambridge, England, focusing on spin-coatable oxide resists for electron beam nanolithography. He Completed his post doctoral study from University of Illinois at Urbana Champaign, USA from 2006 to 2008. His master's degree in Metallurgy from the Indian Institute of Science and bachelor's degree in Metallurgical Engineering from the National Institute of Technology, Tiruchirapalli, laid the foundation for his illustrious career in materials science. Dr. Subramanian has held significant roles in academia and industry, including positions at Ramaiah Institute of Technology, GITAM University, and Shriram Institute for Industrial Research. His expertise spans nanocomposites, solar energy, nanofluids, and nanotechnology applications in various fields. Dr. Subramanian is a fellow of prestigious organizations like the Cambridge Commonwealth Society and the Royal Society of Chemistry. His contributions have earned him recognition in the Marquis Who's Who of the World and as one of the Top 100 Professionals by the International Biographical Centre. He has over 31 years of academic and industrial experience and has published over 150 journals and conference papers. He has been a principal investigator for many government-funded research projects.

Dr Nagaraju Kottam is currently working as an Associate Professor and HOD in the Department of Chemistry, Ramaiah Institute of Technology, Bengaluru. He has obtained his PhD in Physical Chemistry from Central College, Bangalore University. His research interests include the design and development of advanced nano-materials and their composites for wastewater treatment, hydrogen generation, electrochemical, and biosensors applications. He has to his credit more than 100 research publications in refereed international journals, 06 Patents, 03 book chapters. He has guided 09 Ph.D's for their doctoral degrees and another 03 are on-going in which one is DST-Inspire fellow. He has mentored more than 15 UG/PG projects and the majority have received best proj-ect awards. Recently, received as the **best innovative project award of the year 2024** and 2020 from KSTA and KSCST, Bengaluru. He earned some government/institutional-funded projects and received several pres-tigious awards including the **"Award of Excellence in Teaching" from MSRIT for the year 2018-19** for his contribution to research and teaching. There is on-going institute level DST-FIST program. He had reviewed more than hundred articles in international journals and was awarded the Outstanding Reviewer award from an Elsevier journal and also served as guest editor for a few journals. Dr. Nagaraju was a recipient of the INSA-IAS summer research fellowship in 2014. He was also nominated as an expert member for evaluating Switzerland's government-funded project in 2020.

Dr G N Anil Kumar is working as an Assistant Professor in the Department of Physics, Ramaiah Institute of Technology, Bengaluru. He obtained his PhD in chemical crystallography from Bangalore University in 2012. His Research interests include single-crystal and powder X-ray diffraction studies on organic & inorganic materials, Co-crystal design and ab-initio studies of solids. He has published 60 research papers in international peer-reviewed journals and completed a VGST-sponsored funded project. He is also served as Guest editor and Reviewer for few peer review journals. He is guided one PhD and currently guiding two scholars for PhD. He is currently working on project related to design of MOF and spinel metal oxides for Li-ion batteries funded by RIT.

Mechanics of Advanced Materials and Structures

Edited by
Dr N V R Naidu
Dr K R V Subramanian
Dr Nagaraju Kottam
Dr G N Anil Kumar

CRC Press
Taylor & Francis Group
Boca Raton London New York

CRC Press is an imprint of the
Taylor & Francis Group, an **informa** business

First edition published 2026
by CRC Press
44 Park Square, Milton Park, Abingdon, Oxon, OX14 4RN

and by CRC Press
2385 NW Executive Center Drive, Suite 320, Boca Raton FL 33431

© 2026 selection and editorial matter, N V R Naidu, K R V Subramanian, Nagaraju Kottam and G N Anil Kumar, individual chapters, the contributors

CRC Press is an imprint of Informa UK Limited

The right of N V R Naidu, K R V Subramanian, Nagaraju Kottam and G N Anil Kumar to be identified as the authors of the editorial material, and of the authors for their individual chapters, has been asserted in accordance with sections 77 and 78 of the Copyright, Designs and Patents Act 1988.

All rights reserved. No part of this book may be reprinted or reproduced or utilised in any form or by any electronic, mechanical, or other means, now known or hereafter invented, including photocopying and recording, or in any information storage or retrieval system, without permission in writing from the publishers.

For permission to photocopy or use material electronically from this work, access www. copyright.com or contact the Copyright Clearance Center, Inc. (CCC), 222 Rosewood Drive, Danvers, MA 01923, 978-750-8400. For works that are not available on CCC please contact mpkbookspermissions@tandf.co.uk

Trademark notice: Product or corporate names may be trademarks or registered trademarks, and are used only for identification and explanation without intent to infringe.

British Library Cataloguing-in-Publication Data
A catalogue record for this book is available from the British Library

ISBN: 9781041296256 (pbk)
ISBN: 9781041295068 (hbk)
ISBN: 9781003770442 (ebk)

DOI: 10.1201/9781003770442

Typeset in Times New Roman
by HBK Digital

Contents

List of Figures

List of Tables

Preface

The 4th edition of ICMAMS (ICMAMS 2024) was held in Bangalore, India, from 11th to 13th of December 2024, hosted by the Ramaiah Institute of Technology (RIT), Karnataka, India, a leading academic institution known for its excellence in engineering and applied sciences. The organizing committee was led by Prof. J. N. Reddy, Texas A&M University, Honorary Chair, Professor N.V.R. Naidu, Principal of Ramaiah Institute of Technology, Chair, Prof. Erasmo Erasmo Carrera Politecnico di Torino, Co-Chair, alongside Dr. Subramanian K R V, Professor Nagaraju Kottam and many others who contributed significantly to the success of the event. More than 200 participants, including eminent researchers, scientists and industry professionals, gathered to present their work and exchange ideas on the latest advances in the field.

This edition was privileged to have Professor J.N. Reddy, founder of the journal Mechanics of Advanced Materials and Structures, as both chairman and keynote speaker. His keynote lecture, On Non-local Mechanics Models for Architected Materials and Fracture in Solids, explored the role of non-local theories in predicting the mechanical behavior of architected materials, providing new insights into fracture mechanics and structural reliability.

ICMAMS 2024 also featured several distinguished speakers, including Jagadeesh Gopalan from the Indian Institute of Science (IISc), Karnataka, India, Erasmo Carrera from Politecnico di Torino, Italy, Bruno Castanie from Institute Clément Ader, INSA, France, Somnath Ghosh from Johns Hopkins University, Maryland, USA, Arun Srinivasa from Texas A&M University, Texas, USA, and Weiqiu Chen from Zhejiang University, China.

The conference covered a wide range of topics, reflecting the diversity of contemporary research in the mechanics of advanced materials and structures. Discussions covered Integrated Computational Materials Engineering (ICME), composites and architected materials, multifunctional nanomaterials and their structural property relationships. Special attention was given to the role of machine learning in structural and materials engineering, process optimization, smart systems and robust design methodologies for composite structures. In addition, topics such as fracture mechanics, fatigue and damage analysis were explored in depth, as well as the emerging field of deployable structures for space exploration. The event provided a great opportunity for researchers at different stages of their careers, including PhD students and early career researchers, to present their findings and engage in high-level discussions with established experts in the field.

Professor J.N. Reddy awarded the J.N. Reddy Medal to Professor Somnath Ghosh (Hopkins University, USA) in recognition of his pioneering contributions to multiscale-multiphysics computational mechanics and materials science (). His extensive research has had a profound impact on the fields of computational mechanics, advanced numerical modeling and artificial intelligence applications in structural engineering. Professor Ghosh holds the Michael G. Callas Professorship at Johns Hopkins University, Maryland, USA, with appointments in Civil and Systems Engineering, Mechanical Engineering, and Materials Science and Engineering. The 4th edition of ICMAMS was a resounding success, increasing international collaboration and promoting state-of-the-art research in the mechanics of advanced materials and structures.

Honorary Chair: Prof J.N. Reddy
Conference Chair: Prof N.V.R. Naidu
Conference Co-chair: Prof Erasmo Carrera

1 Thermo-mechanical and structural characteristics of hybrid nanofiller and natural fibre strengthened biodegradable composite laminates

B. Deepanraj[1,a], *N. Senthilkumar*[2,b], *G. Perumal*[3,c] *and M. Asad*[1,d]

[1]Department of Mechanical Engineering, College of Engineering, Prince Mohammad Bin Fahd University, Al Khobar, Saudi Arabia

[2]Department of Mechanical Engineering, Saveetha School of Engineering, Saveetha Institute of Medical and Technical Sciences, SIMATS, Chennai, Tamil Nadu, India

[3]Department of Mechanical Engineering, V.R.S College of Engineering and Technology, Arasur, Villupuram, Tamil Nadu, India

Abstract

The development of light weight biodegradable hybrid nanofillers and hybrid natural fibre laminates constitute a major step forward in the field of sustainable materials research. This study focuses towards hybrid nanofillers made of graphene nanoplatelets and multi-walled carbon nanotubes in combined proportion of 3 wt.% is added in hybrid natural fibre (NF) of bamboo fibre (BF) and kenaf fibre (KF) stacked in the sequence of BF0+KF0+BF0+KF0+BF0 (BKF1), BF90+KF0+BF90+KF0+BF90 (BKF2), BF0+KF90+BF0+KF90+BF0 (BKF3), and BF90+KF90+BF90+KF90+BF90 (BKF4) using vacuum infusion technique, followed by compression moulding. The biodegradable polymer designated as poly-lactic acid (PLA) served as matrix and the mechanical testing were performed on manufactured specimens viz., tensile strength, % elongation, shore D hardness, impact, interlaminar shear, and flexural strength, and thermo-mechanical testing in accordance with ASTM recommendations. The results show that aligning the fibres seems to make the laminates' thermo-mechanical and mechanical properties better. The BKF3 sample beats the other laminates because it is more stable in heat and has better mechanical properties.

Keywords: Sustainable composite, bamboo fibre, kenaf fibre, mechanical strength, thermo-mechanical properties, nanofillers

Introduction

Aeronautical constructions may have their total mass decreased by using composites, including reflectors, optical elements, and satellite antenna components. For structural applications, composite materials offer increased specific stiffness and strength, longer fatigue cycles, less weight, and flexibility. Composites have made strides in a number of fields, including conductivity, acoustic soundproofing, resistance to corrosion, durability against wear, aesthetically pleasing, temperature-dependent behaviour, environmental endurance, and heat insulation [1]. The attributes of composite materials will alter due to service-related deterioration. The composite material becomes less rigid as a result of damage that appears at the micrometre scale [2]. Epoxy resins have numerous remarkable properties, which make them ideal for usage in composites; nonetheless, the deficiencies of epoxy resin regarding mechanical qualities and thermal stability restrict its usage to some degree. Epoxy resins can be reinforced using fibres viz., glass, carbon, basalt, and silk, to increase their mechanical properties resins [3]. Nanofillers, together with fibre reinforcements, have been widely used to make epoxy resins stronger and more stable in heat because they are stiffer, stronger, and more stable in heat. Nanofillers may be added to the polymer matrix at the nanoscale with greater proficiency than fibre reinforcements, which gives them reinforced composites with outstanding mechanical characteristics [4]. However, improving the functionality of polymers through the addition of either fibres or nanofillers is still not enough.

The orientation of natural fibres (NFs) within polymer laminates has a profound consequence on both

[a]babudeepan@gmail.com, [b]nskmfg@gmail.com, [c]perumal_harish@yahoo.com, [d]masad@pmu.edu.sa

DOI: 10.1201/9781003770442-1

mechanical and thermo-mechanical characteristics. The laminate has greater tensile strength, stiffness, and modulus when fibres line up in the same orientation with the force that is being applied, as the fibres efficiently transfer stress along their length. This unilateral alignment also helps with thermal stability by stopping deformation when heat is applied [5]. Although random or perpendicular fibre arrangements often result in lower tensile strength, they have the potential to increase impact resistance and flexibility in laminates by allowing them to absorb energy from various directions. For uses requiring more durability, this might be a good fit. By arranging the fibres at angles of 0° and 90°, a multi-axial design makes the laminate tougher in every direction and more temperature stable, allowing it to tolerate deformation across a broad temperature range [6]. For components that are subject to fluctuating temperatures, this reverse orientation makes the laminate more resistant to distortion and thermal fatigue. Polymer laminates may have their properties enhanced to meet specific mechanical and thermal performance requirements by adjusting the orientations of the fibres. Hybrid composites made with fibres and nanofillers are more robust and improved, according to recent studies [7].

Dehrooyeh et al. [8] conducted investigations into CNT-glass fibre hybrid composites, demonstrating that CNTs substantially enhance the modulus and tensile strength. As reported by Azhary et al. [9], adding 1 wt.% CNFs made the tensile strength and modulus of the glass fibre composites 9% and 10% more durable, and the bending modulus and strength 6% and 16% greater, respectively. Adding nanofillers might potentially make carbon fibre-thermoset composites more resilient when they are squeezed [10]. Using both fibre and nanofillers at the same time in the polymer composite hybridisation can considerably improve their mechanical attributes deprived of gaining weight. This quality is very important for the structural materials used in the military, aviation, and automotive industries. For instance, lowering a car's weight by 100 kg could lower its fuel use by 0.3–0.5 dm^3/100 km and its carbon dioxide emissions by 7.5–12.5 g/km [11]. NFs are currently at the top of the list in the sports, construction, and automotive industries since they are so strong and flexible. Some of them are NFs including flax, sisal, hemp, jute, coir, and kenaf [12]. Natural fibres can be very useful as strengthening phases in novel structural composites since they are low in density, cost-effective, long-lasting, produce less greenhouse gases, are easy to recycle, and are carbon neutral. These materials might be very good for the market, the ecosystem, and society [13]. The textile and garment sectors are where silk fibres are most commonly used. At this point, many experts agree that silk fibres have a lot of promise for usage in advanced technological materials that are designed to be more environmentally friendly while also being more resistant to impacts [14]. This is because silk fibres are very strong, flexible, and recyclable. Silk fibres may also be used in helmets and shields for cars when they are blended with other strong fibres like carbon and glass that have a high modulus. This integration lets them take in or let out a lot of impact energy, which greatly improves their ability to handle impacts. Unfortunately, silk fibres rigidity and strength have largely prevented their usage in structural composites [15].

This study is focused on investigating innovative strategies to tackle the challenges encountered in natural fibre composites, as well as their integration with other fibres that possess extreme strength and stiffness. As mentioned earlier, nanofillers have established themselves as a dependable method for enhancing the mechanical performance of composites made of NFs. However, no research has been reported on this issue. This investigation entailed the fabrication of two categories of NF/nanofiller fortified polymer composites utilising the vacuum infusion method. Their mechanical and thermo-mechanical characteristics were evaluated to govern their influence on their fortification in the matrix. Microscopic inspections were used to look at how the developed hybrid nanocomposites ruptured.

Materials and methods

Materials

The renewable poly-lactic acid (PLA) polymer matrix was fortified with hybrid nanofillers consisting of graphene nanoplatelets (GNP) and multi-walled carbon nanotubes (MWCNT), together with bamboo fibre (BF) and kenaf fibre (KF) weighing 220 gsm. Sisco Research Laboratories Pvt. Ltd Chennai, India supplied both MWCNTs and GNPs. The 98% pure MWCNTs have lengths of 0.5–3 μm, outer and inner diameters of 30–50 nm, and 5–15 nm. GNP sheets were supplied with lateral dimensions of 1–10 μm and thicknesses of 1–5 nm. Composite Mart supplied BF and KF mats. Hayael Aerospace India Pvt. Ltd., Chennai, India, offered biodegradable PLA polymer

Table 1.1 Design and material details of the composite's lamination.

Sample ID	KF mat (No. of layers)	BF mat (No. of layers)	MWCNT + GNP (wt.%)	Stacking sequence
BKF1	2	3	3	BF0+KF0+BF0+KF0+BF0
BKF2	2	3	3	BF90+KF0+BF90+KF0+BF90
BKF3	2	3	3	BF0+KF90+BF0+KF90+BF0
BKF4	2	3	3	BF90+KF90+BF90+KF90+BF90

Source: Authors

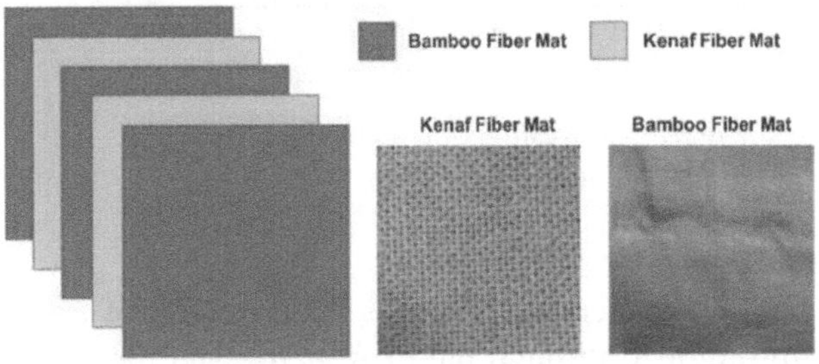

Figure 1.1 Laminate sequence
Source: Authors

matrix. Fibre reinforcing materials included woven BF and KF with a 300×300 mm plain size.

Fabrication of composites

Using vacuum infusion technique, 300 × 300 × 3.75 mm hybrid BF/KF/MWCNT/GNP-reinforced nanopolymer composite laminates were made. Ultrasonic sonicator was used to blend a specified amount of PLA polymer with MWCNT and GNP at a combined ratio of 3 wt. % for 60 minutes to create homogenous slurry. The composite laminates were sandwiched by layering KF and BF, a topcoat of BF/matrix over the KF, and mixed resin over the BF. To remove resin from the woven fabric, composite laminates were vacuum-bagged at ambient temperature. The hybrid laminates were removed from the mould after a day. This composite plate has a sandwich structure made up of a stiff BF layer and an interior layer of natural KFs that alternates. Table 1.1 shows the composite laminated sample for BKF1, BKF2, BKF3, and BKF4 and its material description, and Figure 1.1 illustrates the schematic of laminate formulation.

Characterisation

Composite materials' mechanical characteristics were measured using standard techniques. ADIS' Analogue Shore D Rubber Hardness Tester captures the polymer composite laminate hardness according to ASTM D2240. An evaluation of tensile test at a 2 mm/min strain rate was carried out using Dark System Inc., Model-TJ145 UTM in accordance with ASTM D638 [16]. ASTM D790 is used to estimate the flexural strength of samples 1 (20 × 25 × 3.75 mm). The flexural testing done to assess the flexural strength and modulus with three-point bending at 2 mm/min loading on 60 mm length sample positioned on wedges [17]. According to ASTM D256, a 65 × 10 × 3.75 mm specimen was tested with an impact tester (Hi Tech Calibration Services, EQ-10 Impact Tester). A 2.63 kg hammer with a 30.68 mm gravity distance and 150° lift angle was used [18]. The interlaminar shear strength (ILSS) of polymer laminates evaluates the resistance of a laminate to shear forces between its layers, essential for assessing structural integrity in demanding applications. ASTM D2344 is adopted with Instron 5967 under 30 kN load, a ±0.5% high-precision load cell, and 1–2 mm/min crosshead velocity [19].

Dynamic mechanical analysis

According to ASTM D 5023, the laminates were examined with a dynamic mechanical analyser (DMA 850). The specimen was 50 × 13 mm^2 in size, loss modulus, tan delta, and storage modulus [20]. It is used to look at how rheological properties change in real time, such as when stress, duration, temperature, frequency, environment, and other factors change. DMA behaviour was evaluated at temperatures fluctuating from ambient temperature to 180°C, at 1 Hz frequency, with 2°C/min rate of heating [26]. Liquid nitrogen was used as the coolant [21].

Thermo gravimetric analysis

PerkinElmer TGA equipment was adopted for thermo gravimetric analysis (TGA) conferring to

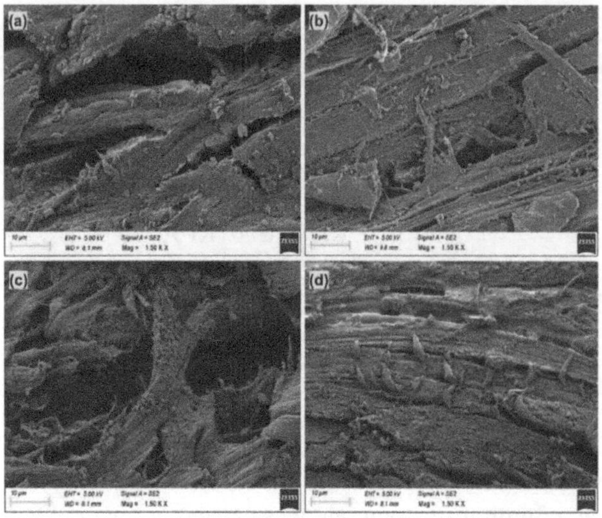

Figure 1.2 SEM of fabricated laminates (a) BKF1 (b) BKF2 (c) BKF3 (d) BKF4
Source: Authors

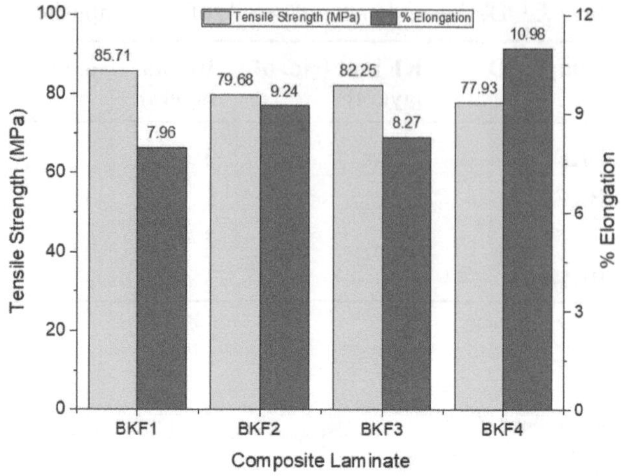

Figure 1.3 Results of tensile test
Source: Authors

ASTM E1131. TGA assess the stability in heat and degrading properties of materials by measuring the changes in substance weight in relation to time or heat [22], heating from 25°C (ambient temperature) to 750°C at 15°C per minute of constant rate. To keep the sample from getting into contact with air while it was heating, which could have affected the results, a 20 ml/min flow rate of N2 was utilised. During TGA analysis, the sample's mass is constantly measured and any changes in mass are documented. Nanocomposites' thermal endurance, degradation temperature, and characteristics may all be inferred from the collected data [23].

Results and discussion

The constructed composite laminates are subjected to SEM micrographs as illustrated in Figure 1.2. The MWCNTs and GNPs dispersed uniformly, create a robust network within the matrix (PLA), facilitating stress transmit and thermal conductivity. The incorporation of BF and KF further supports this structure by providing mechanical reinforcement. Due to differences in fibre morphology and surface characteristics, MWCNTs and GNPs may exhibit varied interactions with BFs and KFs, potentially influencing the laminate's overall distribution and bonding characteristics [24]. Ultra sonication ensures a homogeneous distribution, preventing agglomeration which makes the composite stronger. This uniform dispersion enhances effectiveness of transmitting load through the PLA matrix and the fibres, thus improving the material's mechanical strength, stiffness, and durability [25].

The tensile testing findings for the PLA composite laminate are shown in Figure 1.3. Observation shows that the composites containing 90° orientation has lower tensile strength than the unidirectional laminate. The BFs are stronger than the KFs, which means that the tensile strength is higher for all PLA laminates [26]. It's clear that the ductility goes down as the tensile strength goes up. Increasing tensile strength usually involves enhancing the stiffness and rigidity of the composite, often through higher fibre content or improved fibre-matrix bonding [27]. These alterations make it harder for the polymer chains to move around in the PLA matrix and make it harder for the laminate to deform plastically. The extra fibres make the laminate stronger, but they also make it less flexible, which makes it more likely to break when pulled [28]. Figure 1.4 depicts the fractography SEM images resulting out of tensile test, showing BF, and KF rupture, break down and pull out from the laminate.

The hardness and impact strength of the PLA laminate is presented in Figure 1.5. BFs have higher impact strength than KFs due to crystalline structure and cellulose content. When fibres are aligned at 0°, the laminate shows higher Shore D hardness along the fibre axis because BFs and KFs alignment resists indentation and distributes the applied load effectively, enhancing rigidity [29]. However, impact

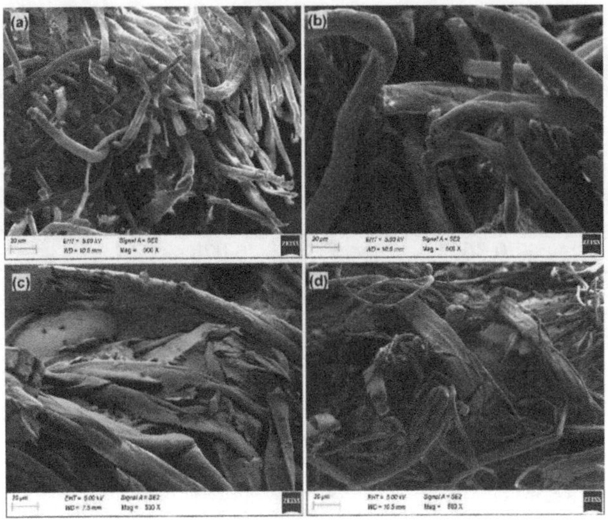

Figure 1.4 Fractography SEM (a) BKF1 (b) BKF2 (c) BKF3 (d) BKF4
Source: Authors

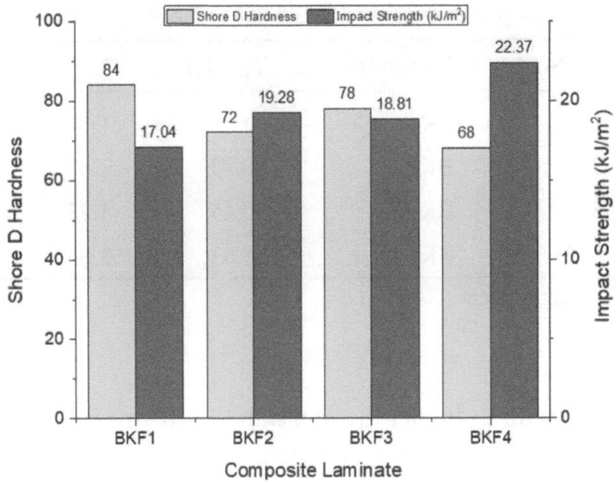

Figure 1.5 Hardness and impact strength of laminates
Source: Authors

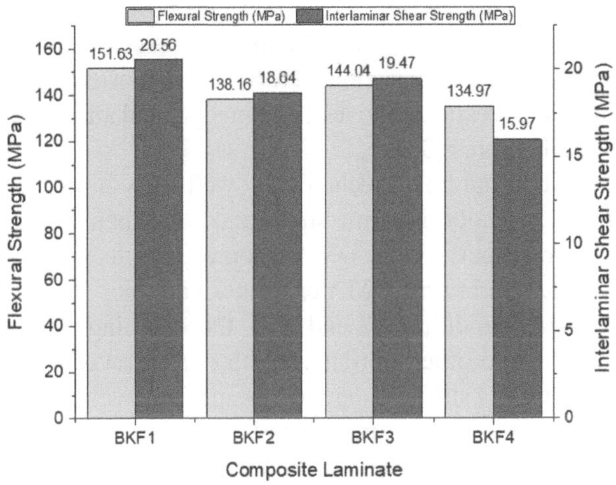

Figure 1.6 Flexural and ILSS of laminates
Source: Authors

strength may be lower in this direction, as aligned BFs and KFs provide less lateral resistance to forces applied perpendicular to the fibre direction, leading to possible delamination or brittle fracture under impact. Conversely, at 90° orientation, BFs/KFs are normal to the force, making the composite more capable of absorbing impact energy across planes and improving impact strength. This orientation allows the laminate to deform slightly under impact, dispersing energy and reducing likelihood of cracking [30].

Figure 1.6 shows the PLA laminates' flexural and ILSS. The alignment of BFs and KFs with respect to the applied loads affects the ILSS and flexural strength when the fibres are oriented at dissimilar angles (0° and 90°). When fibres are oriented at 0°, the laminate typically exhibits higher flexural strength, as fibres positioned parallel to the bending force effectively carry tensile and compressive stresses. This alignment maximises stress transfer along the fibre length, enhancing the laminate's ability to resist bending forces [31]. Conversely, a 90° fibre orientation tends to lower flexural strength, as the fibres offer less resistance to bending stress across the laminate, relying more on the PLA matrix, which is weaker in flexural capacity than the fibres. At 0°, fibres do not contribute significantly to interlaminar shear, as they are aligned with the force, leaving the matrix to bear most shear stress between layers [32]. This can reduce shear strength and potentially increase delamination risk under stress. On the contrary, when the fibres are at a 90° angle, they cross the interlaminar border, which makes the matrix stronger and prevents layers from slipping past each other. This cross-linking action makes it harder for layers to come apart or break, since the fibres provide the layers additional support against pressures that want to separate them [33].

When fibres are positioned in 0° and are parallel to the applied load, both fibres strengthen the composite against thermal deformation and enlargement under stress. This orientation allows for effective stress transfer along the fibre axis, reducing the laminate's thermal expansion along that direction, which is beneficial for applications requiring dimensional stability in response to temperature changes [34].

Table 1.2 DMA results of laminates.

Sl. No.	Laminate	Tg (°C)	E' (1 × 10^9 Pa)	E" (1 × 10^8 Pa)	Tan δ
1	BKF1	84.44	1.79	0.55	0.072
2	BKF2	86.72	2.47	0.81	0.086
3	BKF3	90.45	3.23	0.83	0.068
4	BKF4	85.63	2.86	0.86	0.077

Source: Authors

BFs, with its higher thermal stability and mechanical strength compared to KFs, tends to provide superior thermo-mechanical performance, making laminates with 0° BFs particularly stable under thermal loads. At a 90° position (normal to the load), the fibres resist dimensional changes across the thickness of the laminate, improving the composite's performance in resisting warping and thermal stresses that occur across multiple directions. However, this orientation has lower resilience to heat and mechanical attributes than the fibres [35].

To find out how mechanically well PLA laminates function under dynamic stress and what their visco-elastic properties are, DMA is a prerequisite. At various temperatures, DMA examines variables storage and loss modulus (E' and E"), the damping factor (tan δ), and other similar variables. This reveals the composite's stiffness, energy dissipation mechanism, and transformation properties [36]. Table 1.2 shows the results of the DMA of PLA laminates. The BKF3 composite outperforms similar laminates in terms of Tg, which stands at 90.45°C. E' is 3.23 × 10^6 kPa and E" is 0.83 × 10^5 kPa, both of which are larger in the BKF3 composite. Materials with a higher Tg are better able to withstand elevated temperatures, making them ideal for applications where resistance to heat is a key factor. In addition, a higher storage modulus indicates that the material is stiffer and so better able to withstand stress-induced deformation. This works well for uses where there must be some kind of load bearing. An increase in the loss modulus suggests that the material is better able to dissipate energy, which might lead to better damping and make it suitable for uses that need vibration or shock absorption [37].

The storage modulus goes up when compared to pure PLA. Apparently due to the synergistic effect of the BFs, KFs, and MWCNTs, the material is now stiffer and capable of supporting a greater load. Changes in the Tg may show that molecules are unable to move freely in the PLA, suggesting stronger interface bonding of the fibres, GNPs, and MWCNTs [38]. The tan δ also helps identify the capability of the material to disperse energy, with lower values implying better load transfer within the composite. Figures 1.7 and 1.8 depicts the E' and E" of PLA laminates and Figure 1.9 presents the tan δ of PLA laminates.

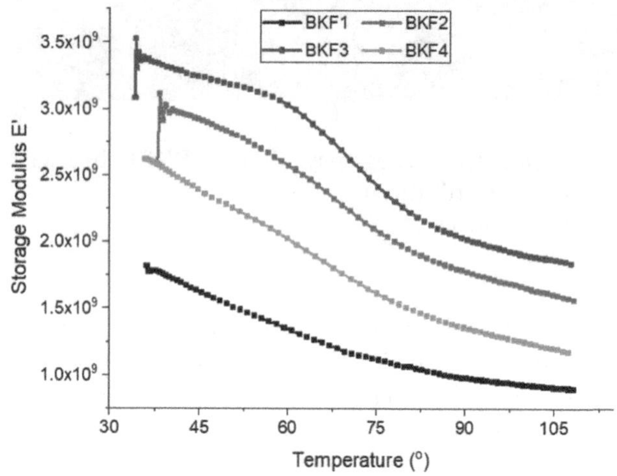

Figure 1.7 DMA storage modulus of laminates
Source: Authors

Table 1.3 shows the findings of the DSC analysis, and Figure 1.10 presents the DSC curve obtained for all the PLA laminate samples. For understanding the thermal transitions and crystallinity of PLA-BF-KF laminates with MWCNTs. DSC measures the heat flow associated with temperature changes in a material, revealing key thermal properties like Tg, melting and crystallisation temperature (Tm and Tc). Addition of BFs and KFs usually increases the Tg of PLA, as the fibres restrict the polymer chains' mobility, making the material more rigid and requiring more thermal energy to reach the glass transition

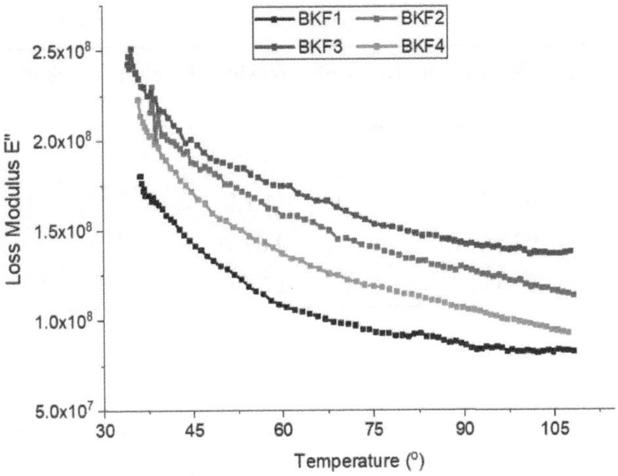

Figure 1.8 DMA loss modulus of laminates
Source: Authors

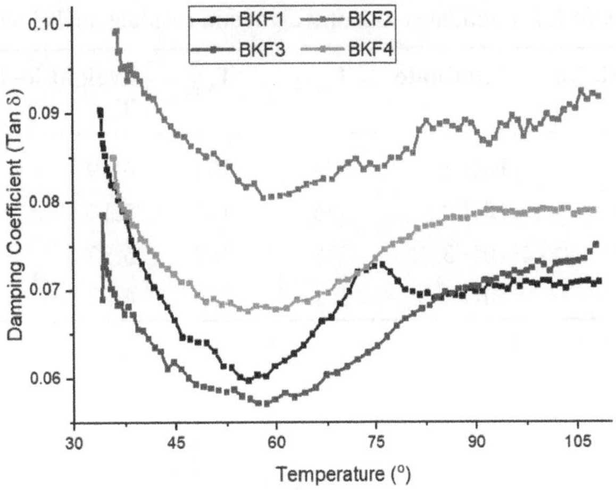

Figure 1.9 DMA tan delta of laminates
Source: Authors

Table 1.3 DSC results of laminates.

Sl. No.	Laminate	Low temp morphology (T_{G1}) °C	High temp morphology (T_{G2}) °C	Low temp crystalline (T_{C1}) °C	High temp crystalline (T_{C2}) °C	Melting temp (Tm)°C	Crystallinity %
1	BKF1	212	421	389	456	503	12.71
2	BKF2	224	435	414	463	489	10.58
3	BKF3	232	431	402	444	508	9.49
4	BKF4	230	427	392	439	494	10.71

Source: Authors

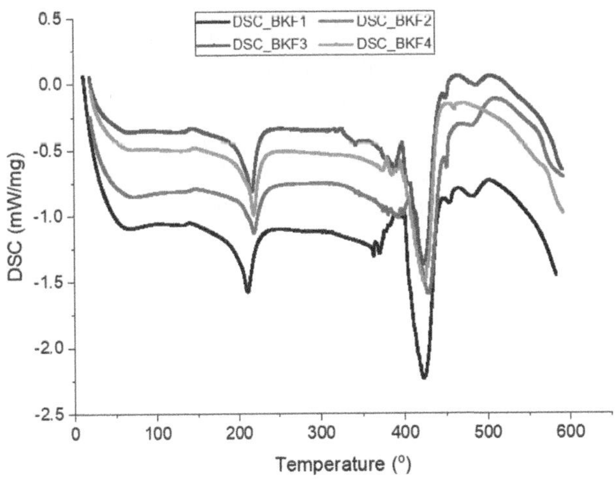

Figure 1.10 DSC results of laminates
Source: Authors

[39]. While the Tm of PLA generally remains relatively unchanged by the fibres, slight shifts may occur due to fibre-matrix interactions. However, BFs and KFs serve as nucleating agents, which often lower the Tc of PLA by providing sites for the PLA chains to organise and crystallise more readily [40].

The increased melting temperature suggests that this laminate can sustain greater temperatures before turning into a molten form, making it advantageous for applications requiring greater thermal stability. The lower crystallinity percentage, however, implies a more amorphous polymer structure, which can enhance flexibility and impact resistance but may reduce rigidity and tensile strength to some extent [41]. This balance of properties may result from the natural fibres disrupting the PLA's crystallisation process, leading to a less ordered structure. These characteristics suggest that the bamboo-kenaf reinforcement creates a material with a favourable combination of heat resistance and resilience, making it useful for usages where extreme thermal tolerance and moderate flexibility are needed [42].

Table 1.4 Laminates loss of weight and residues in TG analysis.

Sl. No.	Laminate	T_{Onset}	T_{Max}	Weight loss in % at T_{Onset}	Weight loss in % at T_{Max}	Residual weight in % at 600°C
1	BKF1	343	404	6.89	84.45	12.22
2	BKF2	356	439	7.14	95.33	11.12
3	BKF3	378	482	5.53	93.56	9.67
4	BKF4	361	454	6.82	81.11	17.49

Source: Authors

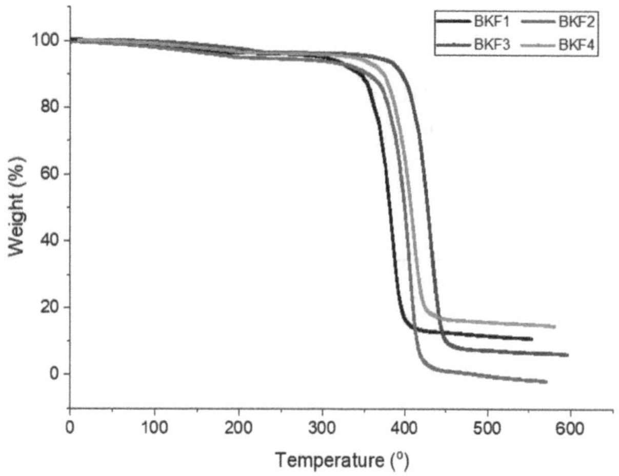

Figure 1.11 TG results of laminates
Source: Authors

TGA examines how much weight loss happens at different temperatures. This gives information about how the material disintegrates, how stable its constituents are, and how MWCNTs nanofillers affect the thermal characteristics of PLA laminates. The TGA curve may show the break down temperatures of PLA, BFs, and KFs break down, and the benefits achieved when they are blended. The incorporation of MWCNTs into the material may enhance its thermal endurance, perhaps leading to a higher breakdown temperature compared to the unreinforced version. This transformation demonstrates that MWCNTs are effective heat resistors and may potentially increase the composite's durability at elevated temperatures [43]. Table 1.4 displays the percentage of mass lost and total mass remaining for each specimen that underwent testing. This can be seen from the TGA curve for each PLA laminate in Figure 1.11.

Due to the reduced percentage of weight loss, the material degrades more slowly even when heated to high temperatures. The structure remains stable due to stronger interaction between the fibres and matrix. Decomposition with less residual weight also indicates a lower breakdown with less char residues. A more homogeneous composition and improved fibre embedding in the PLA matrix could be the result of this [44]. These findings point to evidence that the composite maintains its strength more effectively when heated, making it a good choice for places where durability in severe temperatures is important. The mix of BFs and KFs in the PLA matrix probably makes the composite stronger and less prone to break down when exposed to heat, which means it will work better and last longer [45].

Conclusion

In this study, vacuum infusion technique is adopted for fabricating PLA laminate made of different orientations of BFs and KFs which are tested for their structural and thermal-mechanical characteristics. The uniform spreading of nanofillers in the BF/KF reinforced PLA laminate, increases its effectiveness of load transfer, which makes the material stronger, stiffer, and more durable. In the PLA laminate, BFs has higher mechanical strength than the KFs due to its crystalline structure and cellulose content and the strength of tensile force is greater for all PLA laminates. However, impact strength of PLA laminates at a 90° orientation, BFs/KFs are normal to the force, making the composite more capable of absorbing impact energy across different planes and improving impact strength. At 0° orientation, fibres do not contribute significantly to interlaminar shear, as they are aligned with the force, leaving the matrix to bear the majority of shear stress between layers. This can reduce shear strength and potentially increase delamination risk under stress. When the fibres are lined up at 0° and parallel to the load, both BFs and KFs make the composite more resistant to thermal expansion and deformation but at 90° orientation, the fibres resist dimensional changes, resisting warping and thermal stresses. The Tg of BKF3 composite is higher (90.45°C) than the other PLA laminates, indicating that the material remains more stable at elevated temperatures. The increase in loss modulus suggests enhanced energy dissipation capacity. The increased melting temperature of BKF3 suggests that this laminate can sustain elevated temperatures before turning on to a molten form, making it advantageous for applications requiring greater thermal stability. The lower weight loss of BKF3 indicates that the material decomposes more slowly under high temperatures, likely due to stronger fibre-matrix interactions that help stabilise the structure.

References

[1] Sharma, S. Composite Materials: Mechanics, Manufacturing and Modeling. United Kingdom: CRC Press, 2021;1–30.

[2] Kanu, N. J. Modeling of stress wave propagation in matrix cracked laminates. *AIP Adv.*, 2021;11(8):085217.

[3] Hemnath, A., Anbuchezhiyan, G., NanthaKumar, P., Senthilkumar, N. Tensile and flexural behaviour of rice husk and sugarcane bagasse reinforced polyester composites. *Mater. Today Proc.*, 2021;46:3451–3454.

[4] Vinay, S. S., Sanjay, M. R., Siengchin, S., Venkatesh, C. V. Effect of Al2O3 nanofillers in basalt/epoxy composites: Mechanical and tribological properties. *Polymer Comp.*, 2021;42(4):1727–1740.

[5] Jamshaid, H., Mishra, R. K., Chandan, V., Nazari, S., Shoaib, M., Bizet, L., Ivanova, T. A., Muller, M., Valasek, P. Mechanical and thermo-mechanical performance of natural fibre-based single-Ply and 2-Ply woven prepregs. *Polymers*, 2023;15(4):994.

[6] Huang, S., Fu, Q., Yan, L., Kasal, B. Characterization of interfacial properties between fibre and polymer matrix in composite materials – A critical review. *J. Mater. Res. Technol.*, 2021;13:1441–1484.

[7] Mehdikhani, M., Gorbatikh, L., Verpoest, I., Lomov, S. V. Voids in fibre-reinforced polymer composites: A review on their formation, characteristics, and effects on mechanical performance. *J. Comp. Mater.*, 2019;53(12):1579–1669.

[8] Dehrooyeh, S., Vaseghi, M., Sohrabian, M., Sameezadeh, M. Glass fibre/carbon nanotube/epoxy hybrid composites: Achieving superior mechanical properties. *Mech. Mater.*, 2021;161:104025.

[9] Azhary, T., Kusmono, Wildan, M. W., Herianto. Mechanical, morphological, and thermal characteristics of epoxy/glass fibre/cellulose nanofiber hybrid composites. *Polymer Test.*, 2022;110:107560.

[10] Mechin, P.-Y., Keryvin, V., Grandidier, J.-C. Effect of the nano-filler content on the compressive strength of continuous carbon fibre/epoxy matrix composites. *Comp. Part B Engg.*, 2021;224:109223.

[11] Akampumuza, O., Wambua, Paul. M., Ahmed, A., Li, W., Qin, X. Review of the applications of biocomposites in the automotive industry. *Polymer Comp.*, 2017;38(11):2553–2569.

[12] Arun, K. V., Kumar, D. S., Murugesh, M. C. Drilling of TiO₂ and Zns filled Gfrp composites. *J. Minerals Mater. Characteriz. Engg.*, 2012;11(08):785–789.

[13] Rangappa, S. M., Siengchin, S., Parameswaranpillai, J., Jawaid, M., Ozbakkaloglu, T. Lignocellulosic fibre reinforced composites: Progress, performance, properties, applications, and future perspectives. *Polymer Comp.*, 2022;43(2):645–691.

[14] Yu, H., Zhang, J., Fang, M., Ma, T., Wang, B., Zhang, Z., Hu, Z., Li, H., Cao, X., Ding, C., Deng, H., Yang, K. Bio-inspired strip-shaped composite composed of glass fabric and waste selvedge from A. pernyi silk for lightweight and high-impact applications. *Comp. Part A Appl. Sci. Manufac.*, 2023;174:107715.

[15] Ho, M., Lau, K., Wang, H., Bhattacharyya, D. Characteristics of a silk fibre reinforced biodegradable plastic. *Comp. Part B Engg.*, 2011;42(2):117–122.

[16] Senthilkumar, N., Kalaichelvan, K., Elangovan, K. Mechanical behaviour of aluminum particulate epoxy composite - experimental study and numerical simulation. *Internat. J. Mech. Mater. Engg.*, 2012;7(3):214–221.

[17] Del Prete, C., Buratti, N., Mazzotti, C. Experimental analysis of time dependent phenomena and temperature effects on macro-synthetic fibre reinforced concretes in different loading conditions. *Constr. Build. Mater.*, 2022;326:126904.

[18] Rao, H. J., Singh, S., Singh, N., Ramulu, P. J., Santos, T. F., Santos, C. M., Senthamaraikannan, P., Suyambulingam, I., Gapsari, F., Raharjo, R., Mavinkere Rangappa, S., Siengchin, S. Enhancing mechanical performance and water resistance of Careya-Banana fibre epoxy hybrid composites through PLA coating and alkali treatment. *J. Mater. Res. Technol.*, 2024;32:4304–4315.

[19] Subbiah, A., Chockalingam, P., Munimathan, A., Mayakannan, K. Effect of fibre orientation on interlaminar shear stresses and thermal property of sisal fibre reinforced epoxy composites. *Matéria (Rio de Janeiro)*, 2024;29(4):e20240491.

[20] Haris, N. I. N., Hassan, M. Z., Ilyas, R. A., Suhot, M. A., Sapuan, S. M., Dolah, R., Mohammad, R., Asyraf, M. R. M. Dynamic mechanical properties

of natural fibre reinforced hybrid polymer composites: A review. *J. Mater. Res. Technol.*, 2022;19: 167–182.

[21] Vasanthkumar, P., Balasundaram, R., Senthilkumar, N., Palanikumar, K., Lenin, K., Deepanraj, B. Thermal and thermo-mechanical studies on seashell incorporated Nylon-6 polymer composites. *J. Mater. Res. Technol.*, 2022;21:3154–3168.

[22] Kumar, R., Ganguly, A., & Purohit, R. Thermogravimetric analysis of natural fiber reinforced hybrid composites – A review. Mater. Today Proc., 2023;1–4. https://doi.org/10.1016/j.matpr.2023.08.025.

[23] Dev, B., Rahman, Md. A., Repon, Md. R., Rahman, M. M., Haji, A., Nawab, Y. Recent progress in thermal and acoustic properties of natural fibre reinforced polymer composites: Preparation, characterization, and data analysis. *Polymer Comp.*, 2023;44(11):7235–7297.

[24] Prabhudass, J. M., Palanikumar, K., Natarajan, E., Markandan, K. Enhanced thermal stability, mechanical properties and structural integrity of MWCNT filled bamboo/kenaf hybrid polymer nanocomposites. *Materials*, 2022;15(2):506.

[25] Dejene, B. K. Advancing natural fibre-reinforced composites through incorporating ZnO nanofillers in the polymeric matrix: A review. *J. Natural Fibers*, 2024;21(1):2356015.

[26] Venkategowda, T., Manjunatha, L., Anilkumar, P. Dynamic mechanical behaviour of natural fibres reinforced polymer matrix composites – A review. *Mater. Today Proc.*, 2021;54:395–401.

[27] Jia, X., Madhu, S., Naveen, S., Vellingiri, S., Arun, J. Utilization of biosilica from Bermuda grass ash on silver-grey magnesium: influence of biosilica on its mechanical and tribological properties. *Biomass Conver. Biorefin.*, 2024;14(18):23015–23030.

[28] Ashothaman, A., Sudha, J., Senthilkumar, N., Palanikumar, K. Evaluation of polylactic acid polymer composites strengthened with chopped vetiver fibre and pearl millet-derived- nano silica towards environmental sustainability. *J. Polymer Res.*, 2024;31(11):335.

[29] Prashanth, M., Gouda, P. S. S., Manjunatha, T. S., Banapurmath, N. R., Edacheriane, A. Understanding the impact of fibre orientation on mechanical, interlaminar shear strength, and fracture properties of jute–banana hybrid composite laminates. *Polymer Comp.*, 2021;42(10):5475–5489.

[30] Hashim, M. K. R., Abdul Majid, M. S., Mohd Jamir, M. R., Kasim, F. H., Alshahrani, H. A., Deros, M. A. M., Hui, D. Effects of ply orientations and stacking sequences on impact response of pineapple leaf fibre (PALF)/carbon hybrid laminate composites. *Materials*, 2022;15(17):6121.

[31] Meravi, M. K., Panchore, V. Fiber Alignment's Effect on the Properties of Hybrid Glass/Flax Fiber-Reinforced Epoxy Composite Laminates. *Polymers Adv. Technol.*, 2024;35(10):e6618.

[32] Karthikeyan, R., Madhu, S. Impact of brown algae particles on the mechanical properties of jute-reinforced polymeric composites for sustainable development. *Results Engg.*, 2025;25:104548.

[33] Huang, Y., Lv, X., Huo, H., Zhang, B., Peng, G., Ge, J., Guo, H., Liu, Y. Interlaminar reinforced carbon fibre/epoxy composites by electrospun ultrafine hybrid fibres. *Comp. Part B Engg.*, 2024;281:111578.

[34] Hao, X., Zhou, H., Mu, B., Chen, L., Guo, Q., Yi, X., Sun, L., Wang, Q., Ou, R. Effects of fibre geometry and orientation distribution on the anisotropy of mechanical properties, creep behaviour, and thermal expansion of natural fibre/HDPE composites. *Comp. Part B Engg.*, 2020;185:107778.

[35] Nurazzi, N. M., Norrrahim, M. N. F., Sabaruddin, F. A., Shazleen, S. S., Ilyas, R. A., Lee, S. H., Padzil, F. N. M., Aizat, G., Aisyah, H. A., Mohidem, N. A., Asyraf, M. R. M., Abdullah, N., Sapuan, S. M., Abdan, K., Nor, N. M. Mechanical performance evaluation of bamboo fibre reinforced polymer composites and its applications: a review. *Function. Comp. Struct.*, 2022;4(1):015009.

[36] Gara, D. kumar, Raghavendra, G., Ojha, S., Prakash, M. O., Prasad, P. S. Rheology of epoxy/natural fibre composites. In Handbook of Epoxy/Fibre Composites. Springer Singapore. 2022:1–36.

[37] Maurya, A., Kumar, P., Sinha, S. Optimization of nanofiller compositions for enhancing thermo-mechanical properties of epoxy-based composites through the application of response surface methodology with central composite design. *J. Ind. Chem. Soc.*, 2024;101(11):101417.

[38] Safdar, M., Shafique, M. Z., Tahir, M. S., Mirza, M., Bajwa, S. Z., Khan, W. S. Thermal characterizations carbon nanotube-polymer composites. In Handbook of Carbon Nanotubes. Springer International Publishing. 2022:1147–1185.

[39] Binoj, J. S., Jaafar, M., Mansingh, B. B., Bharathiraja, G. Extraction and characterization of novel cellulosic biofiber from peduncle of Areca catechu L. biowaste for sustainable biocomposites. *Biomass Convers. Biorefin.*, 2024;14(17):20359–20367.

[40] Tripathi, N., Misra, M., Mohanty, A. K. Durable polylactic acid (PLA)-based sustainable engineered blends and biocomposites: Recent developments, challenges, and opportunities. *ACS Engg. Au*, 2021;1(1):7–38.

[41] Lin, Y., Bilotti, E., Bastiaansen, C. W. M., Peijs, T. Transparent semi-crystalline polymeric materials

and their nanocomposites: A review. *Polymer Engg. Sci.*, 2020;60(10):2351–2376.

[42] Mylsamy, B., Shanmugam, S. K. M., Aruchamy, K., Palanisamy, S., Nagarajan, R., Ayrilmis, N. A review on natural fibre composites: Polymer matrices, fibre surface treatments, fabrication methods, properties, and applications. *Polymer Engg. Sci.*, 2024;64(6):2345–2373.

[43] Manimaran, M., Norizan, M. N., Kassim, M. H. M., Adam, M. R., Norrahim, M. N. F., Knight, V. F. Critical assessment of the thermal stability and degradation of chemically functionalized nanocellulose-based polymer nanocomposites. *Nanotechnol. Rev.*, 2024;13(1):20240005.

[44] Momeni, S., Craplewe, K., Safder, M., Luz, S., Sauvageau, D., Elias, A. Accelerating the biodegradation of poly (lactic acid) through the inclusion of plant fibers: A review of recent advances. *ACS Sustain. Chem. Engg.*, 2023;11(42):15146–15170.

[45] Yang, Y., Wan, H., Wang, B., Wang, B., Chen, K., Tan, H., Sun, C., Zhang, Y. Preparation and properties of bamboo fibre/polylactic acid composite modified with polycarbodiimide. *Indus. Crops Prod.*, 2024;218:118829.

2 Numerical investigations on tool edge preparation effects on dry cutting of AA2024

Muhammad Asad[1,a], Muhammad Azhar Ali Khan[1,b], Muhammad Usman Rashid[2,c], Mushtaq Khan[1,d], Deepanraj Balakrishnan[1,e], Hassan Ijaz[3,f] and Faramarz Djavanroodi[1,g]

[1]Mechanical Engineering Department, Prince Mohammad Bin Fahd University, Saudi Arabia

[2]Civil Engineering Department, Punjab Tianjin University of Technology, Lahore, Pakistan

[3]School of Mechanical Engineering, Coventry University, United Kingdom

Abstract

This work investigates how honed and chamfered tool edges affect orthogonal machining performance of aerospace aluminium alloy AA2024. A finite element framework, coupling material` fracture energy and its ductile damage to model chip creation and residual stress development under finishing conditions. Thermo-mechanical simulations were executed to replicate real machining conditions, while chip morphology and cutting force trends were validated against previously reported experimental findings. The analysis highlights how tool edge design affects chip behaviour, force evolution, and subsurface stresses, offering guidance for selecting edge geometries in dry machining of AA2024.

Keywords: Finite element machining, cutting edge design, chamfer edge, honed edge, residual stresses, AA2024

Introduction

In machining processes, the term *tool edge preparation* refers to the micro-geometry present at the insert tip. Examples include chamfer edges, honed edges, combinations of chamfer and hone, wiper edges, or waterfall designs [1–3]. While tool macro-geometries such as rake angle and clearance angle are well recognised, it is the micro-scale edge modifications that significantly influence tool life, chip formation, and surface integrity [4–6].

Among these, chamfer and hone edges are frequently employed for both roughing and finishing operations, high-speed machining, and hard turning applications. Numerous investigations have linked these edges with differences in burr formation, surface roughness, wear progression, and residual stress development [7–9]. Both experimental and computational methods have been employed to assess their impact.

Within this research, Abaqus® was utilised to establish a finite element model aimed at examining how chamfered and honed edges affect finishing processes of AA2024 alloy. The tool and work piece geometries were constructed in line with actual machining configurations and cutting conditions reported in the literature [10]. A baseline orthogonal cutting case using a honed edge radius of 20 µm was validated against experimental data [11], after which alternative hone radii and chamfer geometries were explored.

Finite element and constitutive modelling

The FE setup was created in Abaqus/Explicit ver. 6.13 (Figure 2.1). In the model, the machining insert was assumed rigid and assigned a velocity of V_C = 800 m/min, while the work piece consisted of three meshed regions: (i) chip layer (thickness equal to feed), (ii) chip detachment region (thickness sized to insert` edge radius [4]), (iii) base work piece where residual stresses were analysed. These regions were connected through the built-in *tie constraint*.

Quadrilateral thermo-mechanically coupled elements (CPE4RT) were applied in the FE model, with mesh refinement of approximately 20–30 µm in the critical regions, including the rake face, tool edge, and chip zone. The material response of the

[a]masad@pmu.edu.sa, [b]mkhan6@pmu.edu.sa, [c]usman.rashid@ptut.edu.pk, [d]mkhan7@pmu.edu.sa, [e]dbalakrishnan@pmu.edu.sa, [f]hassan.ijaz@coventry.ac.uk, [g]fdjavanroodi@pmu.edu.sa

DOI: 10.1201/9781003770442-2

work piece was characterised using the Johnson–Cook constitutive relation (Equation 1) [1, 7, 12], which has been extensively employed for high-speed machining simulations.

Chip separation was represented using a combined damage and fracture energy approach [5]. The initiation of material failure was defined by the Johnson–Cook ductile damage model (Equations 2 and 3), while subsequent degradation and fracture were controlled through the damage evolution laws (Equations 4 and 5). The displacement at failure and associated fracture energy release rate were further evaluated using Equations 6 and 7, respectively. Work piece` thermo-mechanical and constitutive data are compiled in Tables 2.1 and 2.2. Corresponding properties of the machining insert are listed in Table 2.1.

$$\bar{\sigma}_{JC} = (A + B\bar{\epsilon}^n)\left[1 + C\, ln\left[\frac{\dot{\bar{\epsilon}}}{\dot{\bar{\epsilon}}_0}\right]\right]\left[1 - \left[\frac{T-T_r}{T_m-T_r}\right]^m\right] \quad (1)$$

$$\omega = \sum \frac{\Delta\bar{\varepsilon}}{\bar{\varepsilon}_{0i}} \quad (2)$$

$$\bar{\varepsilon}_{0i} = \left[D_1 + D_2\, exp\left(D_3 \frac{P}{\bar{\sigma}}\right)\right]\left[1 + D_4\, ln\left[\frac{\dot{\bar{\varepsilon}}}{\dot{\bar{\varepsilon}}_0}\right]\right] \left[1 - D_5\left[\frac{T-T_r}{T_m-T_r}\right]\right] \quad (3)$$

$$D = 1 - exp\left(-\int_0^{\bar{u}} \frac{\bar{\sigma}}{G_f}\, d\,\bar{u}\right) \quad (4)$$

$$D = \frac{\bar{u}}{\bar{u}_f} \quad (5)$$

$$\bar{u}_f = \frac{2\,G_f}{\sigma_y} \quad (6)$$

$$(G_f)_{I,II} = \frac{1-v^2}{E}(K_C^2)_{I,II} \quad (7)$$

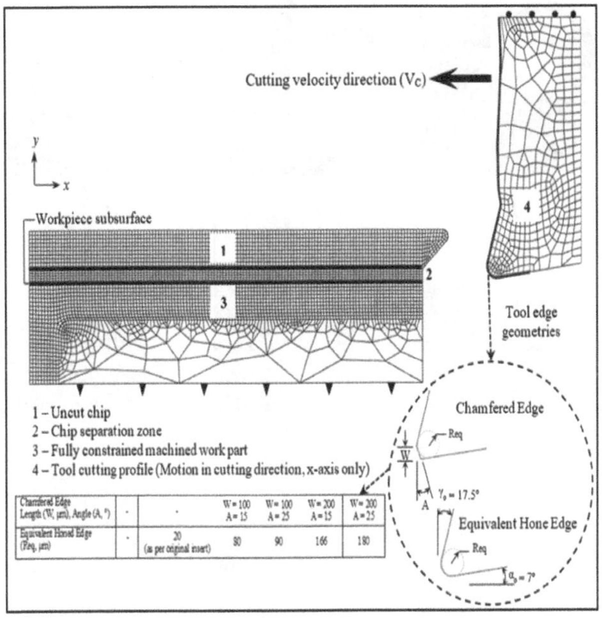

Figure 2.1 Two-dimensional orthogonal cutting setup showing tool and work piece geometry, applied boundary conditions, and the dimensions of chamfered and honed edges
Source: Authors

Table 2.1 Insert and work piece properties [11].

Parameters	Work piece (AA2024)	Tool (Tungsten Carbide)
Density, ρ	2700	11,900
Young's modulus, E	73,000	534,000
Poisson's ratio, v	0.33	0.22
Fracture energy, G_f	20×10^3	X
Specific heat, C_p	$0.557\,T + 877.6$	400
Expansion coefficient, α	$8.91^{-3}\,T$+22.2	5.5
Thermal conductivity, λ	$25 \leq T \leq 300$: $\lambda = 0.247T + 114.4$ $300 \leq T \leq T_m$: $\lambda = -0.125T + 226$	50
Meting temperature, T_m	520	2870
Room temperature, T_r	25	25
Fracture toughness (K_{IC} and K_{IIC})	26 and 37	X

Source: Authors

Table 2.2 Constitutive model parameters [11].

A	*B*	*N*	*C*	*m*	*D1*	*D2*	*D3*	*D4*	*D5*
352	440	0.42	0.0083	1	0.13	0.13	−1.5	0.011	0

Source: Authors

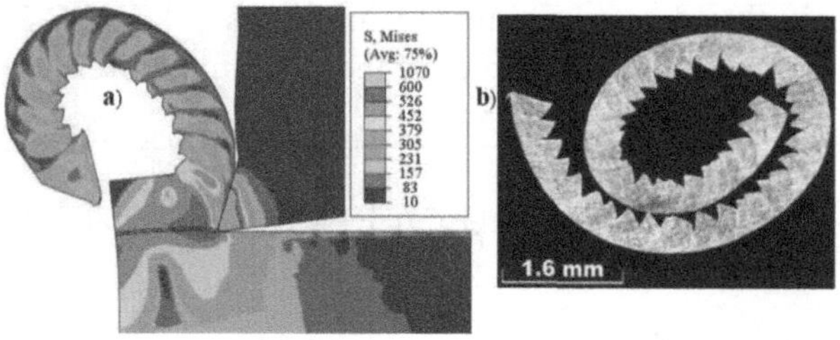

Figure 2.2 Comparison of segmented chip formation in AA2024 during rough cutting at V_C = 800 m/min and $f = 4 \times 10^{-4}$ m/rev (a) numerically predicted chip, (b) experimentally obtained chip [11]
Source: Authors

Results

Model validation

Preliminary simulations were carried out for rough cutting conditions (feeds of 3×10^{-4} and 4×10^{-4} m/rev at V_C = 400 and 800). Chip morphology and cutting forces obtained numerically showed close agreement with earlier experimental measurements [11] (Figure 2.2 and Table 2.3).

Chip morphology under finish cutting

The effect of the insert geometry was studied using the established model including chamfer edges, varying hone radii, etc. on chip morphology. Chip formation directly affects cutting forces, tool wear, and service life [1, 2], while subsurface residual stresses were also evaluated. To focus on finishing operations, the cutting speed was fixed at 800 m/min and $f = 1 \times 10^{-4}$ m/rev.

Simulated chip morphologies for finishing cuts with different hone and chamfer edges are depicted in Figure 2.3. The original tool with a 20 μm hone edge (Figure 2.3a) produced smooth, upward-curling chips, which may be difficult to fragment despite yielding a good surface finish [12]. Larger hone radii led to slightly segmented, downward-curling chips (Figures 2.3c, e, g, i). Chamfered edges, in contrast, generated highly segmented, serrated chips that fragment easily but can compromise work piece quality and increase the risk of tool fatigue.

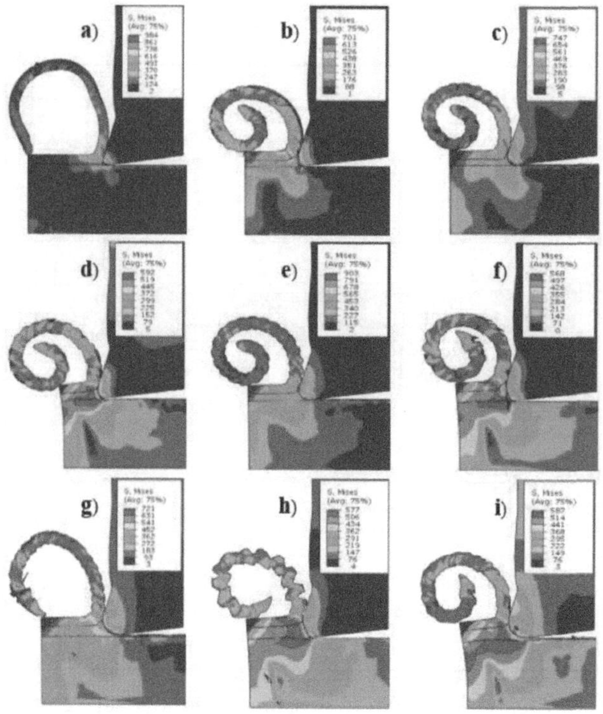

Figure 2.3 Simulated chip morphologies in finish cutting considering different tool-edge designs: (a) hone edge, Rn = 20 μm; (b) chamfer edge, W = 100 μm, A = 15°; (c) hone edge, Rn = 80 μm; (d) chamfer edge, W = 100 μm, A = 25°; (e) hone edge, Rn = 90 μm; (f) chamfer edge, W = 200 μm, A = 15°; (g) hone edge, Rn = 166 μm; (h) chamfer edge, W = 200 μm, A = 25°; (i) hone edge, Rn = 180 μm
Source: Authors

Residual stress distribution

Figure 2.4 illustrates the residual subsurface stresses in the insert` advance direction during finishing cuts ($f = 1 \times 10^{-4}$ m/rev) for different hone and chamfer edge geometries. The original tool with a 20 μm hone edge produced the highest tensile residual stresses. Chamfered edges slightly reduced these stresses, but they remained tensile. In contrast, larger hone radii (Rn > 90 μm) generally promoted a shift toward compressive residual stresses. Compressive stresses are typically preferred, as they help prevent crack initiation and propagation.

Table 2.3 Comparison of experimental and numerical results.

Process variables	Computed forces	Experimental registered forces
$V_C = 800, f = 3 \times 10^{-4}$	167	193
$V_C = 800, f = 4 \times 10^{-4}$	210	244
$V_C = 400, f = 4 \times 10^{-4}$	208	245
$V_C = 400, f = 3 \times 10^{-4}$	163	190

Source: Authors

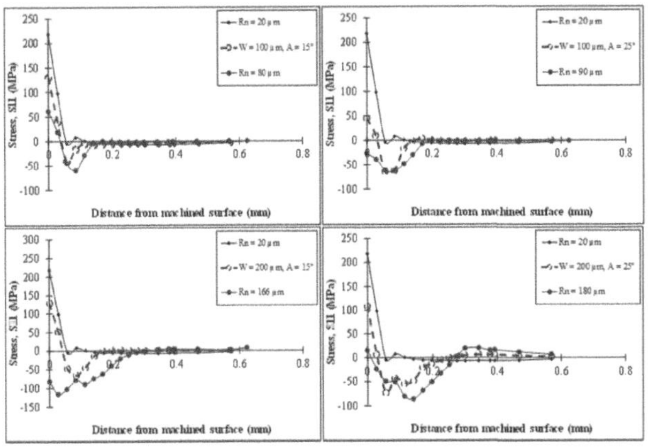

Figure 2.4 Distribution of residual stresses using different hone radii and chamfer geometries
Source: Authors

Conclusion

A numerical model for orthogonal cutting was created to investigate how tool geometry affects the machining. The study demonstrated that:

- Chamfered tools produce segmented chips that ease chip breakage but can reduce surface integrity and accelerate tool wear.
- Hone edge tools, especially with larger radii, are more effective in inducing compressive residual stresses, thereby improving fatigue resistance and surface quality.
- A properly designed hone edge with adequate radius, possibly in combination with optimised chip breaker geometry, is recommended for finishing operations on AA2024.

Acknowledgement

The authors gratefully acknowledge Prince Mohmmad Bin Fahd University for funding and cooperation to conduct the research.

References

[1] Özel, T. Computational modelling of 3D turning: Influence of edge micro-geometry on forces, stresses, friction and tool wear in PcBN tooling. *J. Mater. Proc. Technol.*, 2009;146:5167–5177.

[2] Fang, N., Wu, Q. The effects of chamfered and honed tool edge geometry in machining of three aluminium alloys. *Int. J. Mach. Tools Manuf.*, 2005;45:1178–1187.

[3] Asad, M. Tool-edge geometry effects in semi-to-finish cut machining operations on subsurface integrity and tool thermo-mechanical loadings: A coupled damage-fracture energy based finite element approach. *Mech. Adv. Mater. Struct.*, 2024:1–14.

[4] Subbiah, S., Melkote, S. N. Effect of finite edge radius on ductile fracture ahead of the cutting tool edge in micro-cutting of Al2024-T3. *Mater. Sci. Engg. A.*, 2008;474:283–300.

[5] Asad, M., Ijaz, H., Khan, M. A. A., Khan, M., et al. Comparative analyses and investigations of chamfered and honed-edge tool geometries on tool wear, chip morphology, residual stresses and end-burr formation. *J. Manuf. Proc.*, 2022;80:196–209.

[6] Weng, J., Zhuang, K., Cheng, H., Ding, H. A PSO-based semi-analytical force prediction model for chamfered carbide tools considering different material flow state caused by edge geometry. *Int. J. Mech. Sci.*, 2022;169:105329.

[7] Asad, M. Effects of tool edge geometry on chip segmentation and exit burr: A finite element approach. *Metals*, 2019;9:1234.

[8] Asad, M., Ijaz, H., Khan, M. A., Khan, Mabrouki, T. Saleem, W. Turning modeling and simulation of an aerospace grade aluminium alloy using two-dimensional and three-dimensional finite element method. *Proc. IMechE, Part B J. Engg. Manuf.*, 2014;228:367–375.

[9] Ren, H., Altintas, Y. Mechanics of machining with chamfered tools. *Trans. ASME J. Manuf. Sci. Engg.*, 2000;122:650–659.

[10] Ijaz, H., Mohd, D., Asad, M., Rubaiee, S. A three-dimensional finite element-approach to investigate the

optimum cutting parameters in machining AA2024. *Mech. Indus.*, 2020;21:615.

[11] Mabrouki, T., Girardin, F., Asad, M., Rigal, J. F. Numerical and experimental study of dry cutting for an aeronautic aluminium alloy (A2024-T351). *Int. J. Mach. Tools Manuf.*, 2008;48:1187–1197.

[12] Asad, M., Mabrouki, T. On the modelling of an aluminium alloy milling: 3D FEM approach. *Mechanics*, 2013;19:588–592.

3 Dynamic response of beam type structures subjected to bi-directional excitation

Mrinmoy Saha[a], B. Panigrahi[b] and Sabareesh G. R.[c]

Department of Mechanical Engineering, BITS Pilani, Hyderabad Campus, Hyderabad, Telangana, India

Abstract

Engineering structures are subjected to complex dynamic loadings in various practical operating environments. Conventionally, the research on the response analysis of such structures focuses on uni-directional excitation. However, such structures may experience bi-directional as well as multi-directional excitations, which significantly affect their dynamic response. In this regard, the present study investigates the effects of bi-directional excitation on beam type structures to explore and predict their dynamic behaviour under such conditions. In this work, a mathematical model based on the energy method is developed to calculate and predict a more accurate response of the beam under bi-directional excitation. The excitation is considered from axial as well as from base excitation in transverse direction. In the mathematical model, one excitation (axial excitation) is considered within the framework of time dependent stiffness. The model is validated against experimental data and applied to different beam configurations considering different sizes and material properties. The results reveal that bi-directional excitation leads to significantly different dynamic responses compared to uni-directional excitation, particularly in terms of complex vibrational modes. Limiting values for axial excitation parameters are reported after which the response is dominated by axial excitation. The study highlights the necessity of incorporating bi-directional excitation scenarios in the design and analysis of beams to ensure structural integrity and reliability.

Keywords: Beam dynamics, bi-directional excitation, multi-directional loading, dynamic response, energy method, mathematical modelling

Introduction

Engineering structures, particularly beams, play a crucial role in various engineering fields and are often subjected to dynamic loading conditions throughout their operational life. Conventionally, the analysis of dynamic responses in beams has focused on uni-directional excitations, which simplifies the complexities inherent in real-world situations that may overlook critical aspects of the structural response, especially when considering structures experienced bi-directional as well as multi-directional excitations, which significantly affect their dynamic response.

Recent studies emphasise the importance of considering multi-directional and bi-directional excitations in structures, as they significantly influence dynamic behaviour and create complex vibrational modes that conventional uni-directional models fail to capture [1]. Roy et al. [2] reviewed the impact of biaxial interaction and incidence angles on structural responses, highlighting their significant influence on system performance. Christos A. Zeris et al. [3] developed a finite-element model to understand the non-linear behaviour of reinforced concrete columns during biaxial excitation, highlighting the complexity of axial-flexural interactions and local damage concentrations and emphasising the importance of proper structural analysis. M. R. Salami et al. [4] studied reinforced concrete columns and revealed that bi-directional loading significantly increases the damage and decreases capacity, particularly in rectangular columns, using advanced modeling tools to assess the impact of various excitations on their structural integrity. H. H. Patil investigated the buckling and vibration properties of graded graphene platelet composite beams and revealed that grading pattern significantly impacts both buckling loads and mode forms [5]. E. Carrera et al. developed a higher-order precise formulation for linearised buckling analysis of beam-columns and compared it to current

[a]p20230090@hyderabad.bits-pilani.ac.in, [b]brajesh@hyderabad.bits-pilani.ac.in, [c]sabareesh@hyderabad.bits-pilani.ac.in

DOI: 10.1201/9781003770442-3

literature and finite element approaches [6]. Sanati et al. introduced an iso-geometric buckling analysis method that integrates CAD geometry with finite element analysis, reducing discretisation errors and avoiding concerns in complex geometries of laminated beams [7].

Various methods exist for analysing vibration, each offering different advantages depending on the complexity and requirements. Among these various techniques, the energy method is an effective technique in addressing complex structural dynamics by using kinetic and potential energy relationships. B. Panigrahi et al. developed a non-linear time domain model to study the dynamic response of cracked beams under various excitations, enhancing predictive maintenance strategies for fatigue-loaded structures [8]. Kumar G. D. developed a mathematical model to analyse free vibration characteristics of thin-walled anisotropic biconvex beams, emphasising structural health monitoring's importance in engineering applications [9]. B. Upendra used the Ritz method to optimise energy expressions for a piezo-patched beam, analysing vibrational response under axial loads and transverse vibration on a fixed-fixed beam [10].

The analysis of beam dynamic responses has evolved from uni-directional models to complex bi-directional and multi-directional excitations, which is crucial for accurately predicting real-world vibrational modes. Recent research has demonstrated the effect of biaxial interactions, non-linear material characteristics, and multi-frequency excitations on structural performance. However, there is a research gap in studying bi-directional excitations, which are less studied. As engineering applications become more complex, models are needed to accurately predict dynamic responses under bi-directional excitations. This study presents a mathematical model using the energy method to more accurately calculate and predict the response of a beam subjected to bi-directional excitation, considering both axial and transverse base excitations.

Methodology

Mathematical modelling

This study develops a mathematical model using the energy method to calculate the dynamic response of beams subjected to bi-directional excitations. The mathematical model accounts for the simultaneous effects of axial excitation and transverse base excitation. The formulation consists of an axially loaded beam with a boundary condition fixed at one end, and another end is free to move in axial direction only [10]. To accomplish the required conditions, axial excitation is considered within the framework of time dependent stiffness because when the axial force is time-dependent, the stiffness will also change with the axial force. Figure 3.1 demonstrates the experimental setup and schematic diagram of the beam, where the mid-plane of the beam is denoted as z=0, and the thickness of the beam is "t". The base excitation is represented as Wb = Y sin(ωt), and the transverse, axial and rotational displacements from the static equilibrium position are denoted as W, U, and Ψ, respectively.

The displacement field is considered in x and z direction (U and W) with space and time coordinates, which can be written using beam theory

$$U_0\,(x,z,t) = U\,(x,t) + z\Psi\,(x,t)$$
$$W_0\,(x,z,t) = W\,(x,t) \qquad (1)$$

where U (x, t), W (x, t), and Ψ (x, t) denote the axial displacements of the mid-plane, transverse displacements of the mid-plane, and the cross-sectional rotation, respectively.

By considering the mid-plane stretching resulting from large deformation (von Karman geometric non-linearity), the constitutive and kinematic expressions can be written in the following ways:

$$\varepsilon_{xx} = \frac{\partial U}{\partial x} + z\frac{\partial \Psi}{\partial x} + \frac{1}{2}\left(\frac{\partial W}{\partial x}\right)^2, \gamma_{xz} = \frac{\partial W_0}{\partial x} + \frac{\partial U_0}{\partial z} + \frac{\partial W}{\partial x} + \Psi$$

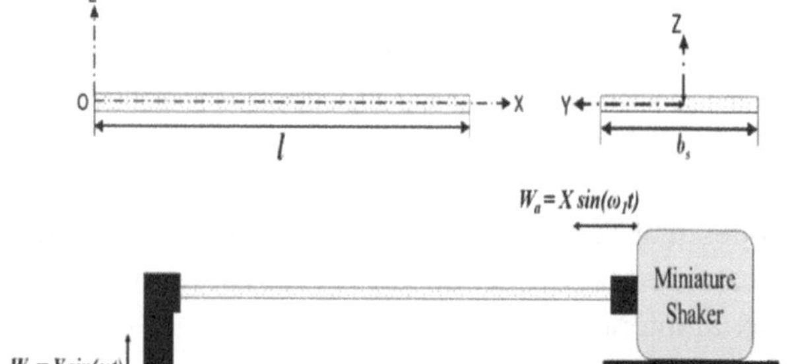

Figure 3.1 Block diagram of the experimental setup
Source: Authors

$$\sigma_{xx} = E\varepsilon_{xx} = E\left[\frac{\partial U}{\partial x} + z\frac{\partial \Psi}{\partial x} + \frac{1}{2}\left(\frac{\partial W}{\partial x}\right)^2\right]$$

$$\sigma_{xz} = \frac{kE}{2(1+v)}\gamma_{xz} = \frac{kE}{2(1+v)}\left(\frac{\partial W}{\partial x} + \Psi\right) \quad (2)$$

Now, the potential energy (E_p) and kinetic energy (E_k) of the system can be derived as Equations (3) and (4) by using Equation (2) as follows:

$$E_p = \frac{1}{2}\int_0^{l_s}\int_{-\frac{t_s}{2}}^{\frac{t_s}{2}}\int_{-\frac{b_s}{2}}^{\frac{b_s}{2}}\left[E\left\{\frac{\partial U}{\partial x} + z\frac{\partial \Psi}{\partial x} + \frac{1}{2}\left(\frac{\partial W}{\partial x}\right)^2\right\}^2 + \frac{kE}{2(1+v)}\left(\frac{\partial W}{\partial x} + \Psi\right)^2\right] dx\, dy\, dz \quad (3)$$

$$E_k = \frac{1}{2}\int_0^{l_s}\int_{-\frac{t_s}{2}}^{\frac{t_s}{2}}\int_{-\frac{b_s}{2}}^{\frac{b_s}{2}}\rho_s\left[\left\{\frac{\partial U}{\partial t} + z\frac{\partial \Psi}{\partial t}\right\}^2 + \left\{\frac{\partial W}{\partial t} + \frac{\partial W}{\partial t}\right\}^2\right] dx\, dy\, dz \quad (4)$$

As shown in Equation (2), the strain terms ε_{xx} and γ_{xz} are normal and shear strains, while the σ_{xx} and σ_{xz} are normal and shear stresses. By the application of axial harmonic excitation at the right-hand side of the beam, the boundary condition can be written at $x=L$ as $N(L, t) = -| P \times \sin \sin \omega t |$. By considering this boundary condition, one may derive the work performed by the external compressive force P (W_a) as

$$W_a = -P \times |\sin \omega t| \int_0^{l_s}\left[\frac{1}{2}\left(\frac{\partial W}{\partial x}\right)^2\right] dx \quad (5)$$

After converting E_p, E_k and W_a as non-dimensional term the Equations (3–5) can be written as,

$$E^*_{plinear} = \int_0^1 k_1\dot{u}^2 d\xi + \int_0^1 2k_2\dot{u}\dot{\psi} d\xi + \int_0^1 \dot{\psi}^2 d\xi + \int_0^1 k_3\dot{w}^2 d\xi + \int_0^1 k_4\psi^2 d\xi + 2\int_0^1 k_5\dot{w}\psi d\xi$$

$$E^*_{pnonlinear} = \int_0^1 k_9\dot{w}^4 d\xi + \int_0^1 k_{10}\dot{u}\dot{w}^2 d\xi + \int_0^1 k_{11}\dot{\psi}\dot{w}^2 d\xi$$

$$E^*_k = KE = -\omega^{*2}\left[\left\{\int_0^1 \psi^2 d\xi + m_1\int_0^1 (u^2 + w^2 + y^2 + 2wy) d\xi + 2\int_0^1 m_2 u\psi d\xi\right\} + m_3\left\{m_4\int_0^1 (u^2 + w^2 + y^2 + 2wy) d\xi + 2\int_0^1 m_5 u\psi d\xi\right\}\right]$$

$$W^*_a = -p_1|\sin \omega t|\int_0^1 \dot{w}^2 d\xi \quad (6)$$

The expression for the energy functional (π) is expressed as

$$\pi = E^*_{plinear} + E^*_{pnonlinear} + E^*_k + W^*_a \quad (7)$$

Considering the kinematically admissible function for the displacement fields, one may consider of as the linear combination of the modal functions as,

$$u = \sum_{i=1}^{n} a_i\varphi_{ui}, w = \sum_{i=1}^{n} b_i\varphi_{wi}, \psi = \sum_{i=1}^{n} c_i\varphi_{\psi i} \quad (8)$$

The algebraic non-linear equation can be expressed as,

$$[k_{linear}]\{q\} + [K(a_i, b_i, c_i)_{nonlinear}]\{q\} - \omega^{*2}[M]\{\ddot{q}\} = \{F_b(t)\} \quad (9)$$

The non-linear stiffness matrix in Equation (7) is a function of displacement fields, making linear algebraic solution approaches ineffective. An amplitude incremental-based iterative method is used to solve this, starting with static response and gradually increasing vibration amplitude. Unknown displacement field coefficients are scaled up, creating non-linear stiffness matrices and achieving convergence of response frequencies. This method is more effective than other algorithms as it captures stable and unstable non-linear system responses.

Experimental validation

Experimental validation is performed to ensure the reliability and accuracy of the proposed mathematical model. A beam sample made of galvanised iron is fixed at one end and the other end is attached to miniature shaker to apply axial excitation, as illustrated in Figure 3.1. Axial excitation is generated using a function generator connected to the shaker. Transverse vibration is produced by the Sdyn SEV 240 electrodynamic shaker, controlled by a vibration control system. Displacement and velocity data from the excitations are collected by Polytec VGO-200 (LDV) and data acquisition system, which includes a cDAQ-9178 chassis and NI-9232 input module. This data is then processed and visualised using LabVIEW software on a computer. The entire experimental setup is shown in Figure 3.2 [11–14].

Result and discussion

This section discusses a detailed comparison of the experimental data and the mathematical model. Increasing the number of polynomial terms in the

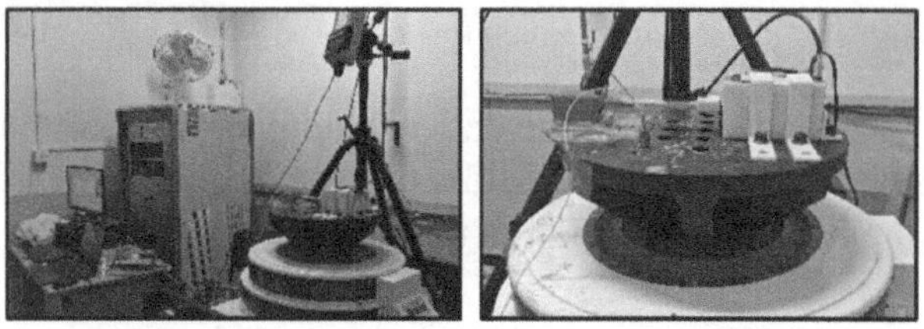

Figure 3.2 Experimental validation is performed to ensure the reliability and accuracy of the proposed mathematical model
Source: Authors

kinematically admissible function Equation (8), where the natural frequencies converge after the value of N=8, is the first step toward verifying the efficacy of the mathematical model. Thus, eight factors are taken into account in each analysis. Additionally, the outcomes of the experiments verify the answer that was derived from the mathematical model. Simultaneous application of distinct axial and transverse stimulation is detected. The material properties is GI steel beam is considered as $E = 210$ GPA, $\rho = 7850$ kg/m^3, $v = 0.26$ and the dimensions of the beam are length (L) = 0.132 m, width (b) = 0.029 m, thickness (h) = 0.0005 m.

Validation of mathematical model

To compare and validate the numerically simulated results with experimental responses, distinct bi-directional frequency ratios were imposed. These ratios represent the relationship between transverse to axial frequencies (transverse/axial) as follows: 0.25, 0.5. Figure 3.3 presents a comparison between the experimental and numerical displacement responses in the time domain; from Figure 3.3, it is evident that the numerical results match closely with the experimental results. The overall consistency between the experimental and numerical responses is significant, with very little variation observed, despite small differences in displacement values. This consistency shows that the numerical model is reliable and successfully describes the experimental behaviour with an average of 5.95% error at different frequency ratios.

Dynamic response of the beam

Dynamic displacement behaviour

The time versus displacement curves for the five experimental bi-directional frequency ratios, 0.25, 0.5, 1, 2, and 4, are shown in Figure 3.4. These

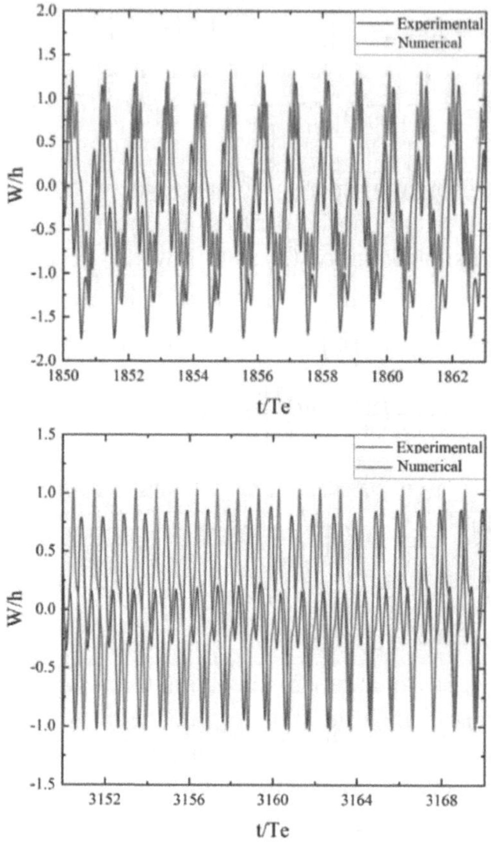

Figure 3.3 Comparison between experimental and numerical response of displacement in time domain. Transverse and axial frequency ratio: (a) 0.25; (b) 0.5
Source: Authors

plots represent the dynamic response of the beam to transverse and axial vibrations over time, where each frequency ratio reflects the relation between the two vibrational modes as the transverse and axial direction.

The time vs. displacement curve for the 0.25 ratio shows the highest displacement amplitudes and the most significant oscillatory behaviour. The transverse vibrational frequency is significantly lower than the axial frequency, causing the system to face more external energy in the axial direction leading to larger peak-to-peak oscillations in the transverse direction. This results in greater transverse displacement. It indicates that there is less resistance to transverse motion, which permits the displacement to increase, making the system more exposed to lateral disturbances. The displacements in the 0.5 ratio became lesser than that of 0.25 ratios. The system responds relatively stable periodically when the transverse frequency is closer to the axial frequency.

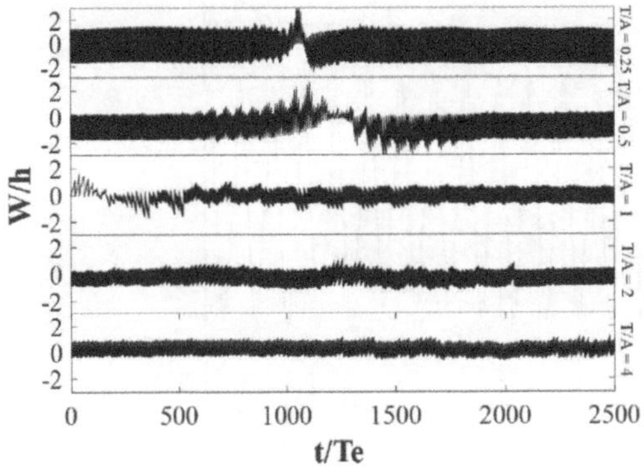

Figure 3.4 Displacement history of different bi-directional excitation ratios; (1) T/A = 0.25; (2) T/A = 0.5; (3) T/A = 1; (4) T/A = 2; (5) T/A = 4
Source: Authors

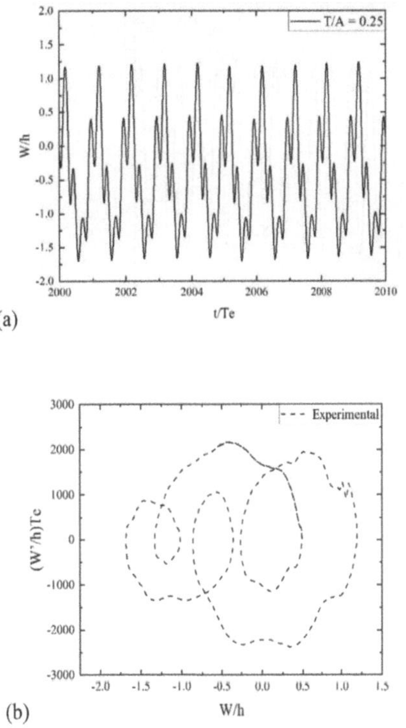

(a)

(b)

Figure 3.5 (a) Time-displacement plot; (b) Phase plane portrait of the beam mid-point under bi-directional excitation ratio (T/A) 0.25
Source: Authors

Although there is still a large lateral displacement, the total vibrational amplitude of the system starts to decrease due to the closer frequency in both directions than it achieved at the 0.25 ratio and is becoming less susceptible to transverse vibrations. The displacement curve shows a simpler oscillation like sine wave with moderate amplitude at a ratio of 1. When the transverse and axial frequencies are equal, the system experiences both vibrations in the same phase, resulting in an even distribution of energy and less displacement. As the transverse frequency increases, the system's ability to oscillate laterally is significantly dampened, limiting its displacement. This makes the system more stable and less responsive to vibrational forces, resulting in lower oscillation amplitudes. At the maximum ratio, the transverse frequency is four times the axial frequency, causing a significant decrease in displacement. The system's vibrational response is more sensitive to transverse motion at lower frequency ratios, but at higher ratios, the response is reduced in the transverse direction, resulting in fewer lateral oscillation amplitude.

Trajectory phase analysis

For the ratio of 0.25, the transverse excitation progresses relatively slower than the axial excitation, causing multiple orbits in the phase plane plot, where these two frequencies create complex, periodic trajectories that reflect the interaction of two modes of vibration, as shown in Figure 3.5 (b).

Figure 3.6 (b) represents the response for the frequency ratio of 0.5, where both frequencies are becoming closer, causing a more stable condition than the 0.25 frequency ratio and creating lesser

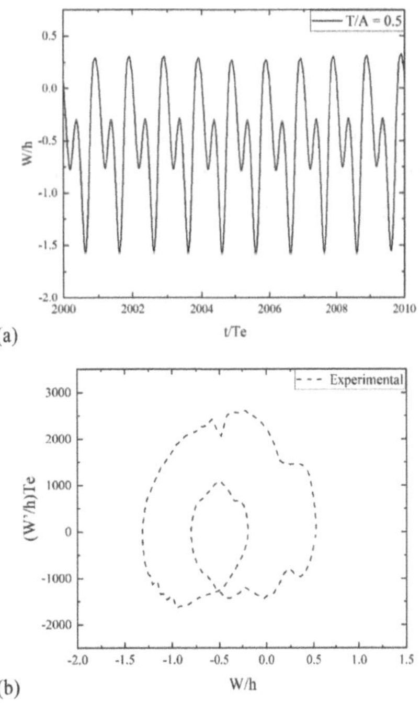

(a)

(b)

Figure 3.6 (a) Time-displacement plot; (b) Phase plane portrait of the beam mid-point under bi-directional excitation ratio (T/A) 0.5
Source: Authors

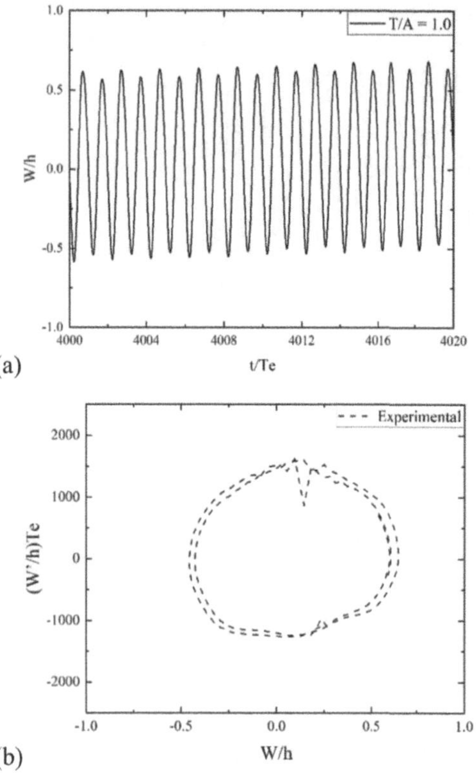

(a)

(b)

Figure 3.7 (a) Time-displacement plot; (b) Phase plane portrait of the beam mid-point under bi-directional excitation ratio (T/A) 1

Source: Authors

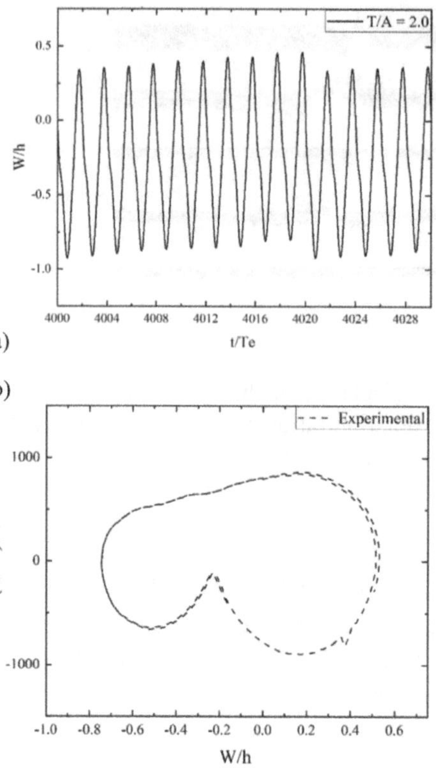

(a)

(b)

Figure 3.8 (a) Time-displacement plot; (b) Phase plane portrait of the beam mid-point under bi-directional excitation ratio (T/A) 2

Source: Authors

orbits and a less complex response in the phase plane plot.

Figure 3.7 (b) shows a simpler response indicating synchronisation between these two excitations in ratio 1. In this case, the displacement vs. velocity plot is likely to form a closed orbit, which indicates a stable periodic harmonic motion because of the same energy distribution in both directions. For the frequency ratio of 2, the transverse motion introduces faster motion while the axial motion produces slower and broader oscillations, resulting in a more intricate pattern than a simple harmonic motion, as shown in Figure 3.8, which indicates the interaction of two different modes.

In the case of frequency ratio 4 in Figure 3.9(a), the time vs. displacement plot shows a quasi-periodic response, and in Figure 3.9(b), the phase plane plot indicates a complex loop that contains multiple orbits due to the large difference between two excitation frequencies where transverse excitation induces rapid oscillation and slower oscillation in the axial direction. The faster transverse oscillations dominate the beam, leading to frequent changes in displacement and velocity. The axial motion with slower oscillations contributes to the overall complexity of the system's response. This complex quasi-periodic response reflects the non-linear dynamics of the beam under multi-frequency excitation, showcasing the influence of frequency ratio on the vibrational behaviour.

The phase plane portrait for all frequency ratios reveals distinct interactions between the transverse and axial vibrational modes. At lower frequency ratio, such as 0.25 and 0.5, the system exhibits multiple orbits, indicating the presence of non-linearity which is driven by the beam's large deformations and non-linear stiffness properties. The phase plane portrait for higher ratios tends to be dominated by the transverse motion, leading to more regular patterns than the lower ratio cases.

Frequency spectrum analysis

The frequency spectrum analysis for the transverse to axial excitation ratios demonstrates distinct

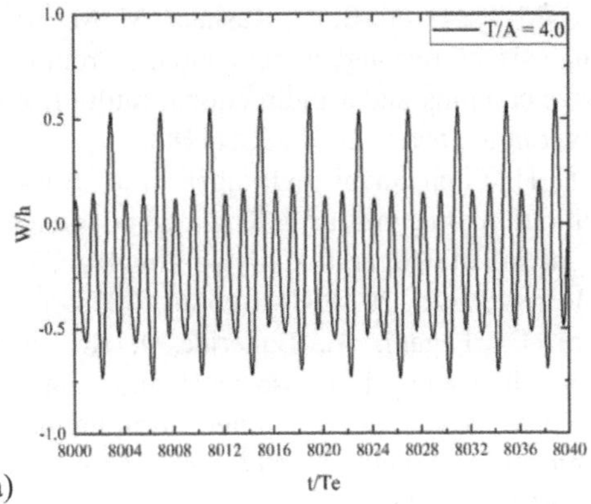

(a)

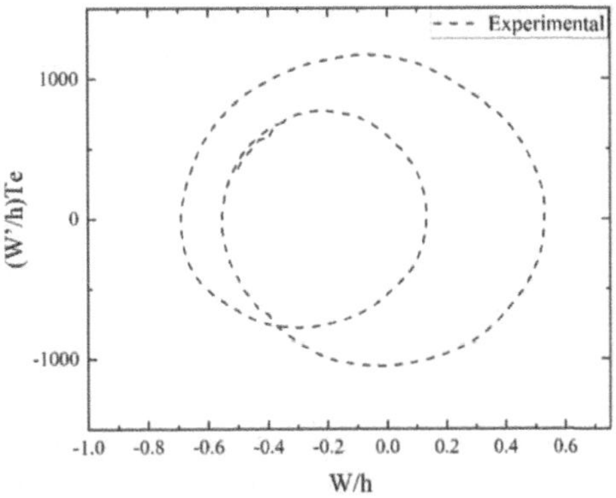

(b)

Figure 3.9 (a) Time-displacement plot; (b) Phase plane portrait of the beam mid-point under bi-directional excitation ratio (T/A) 4
Source: Authors

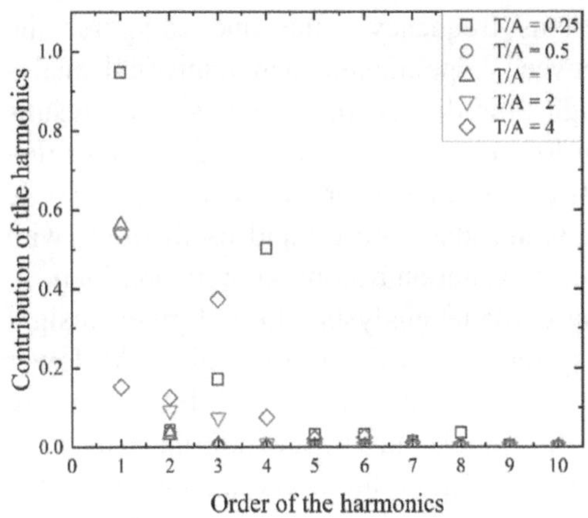

Figure 3.10 Frequency spectrum – Harmonic breakdown
Source: Authors

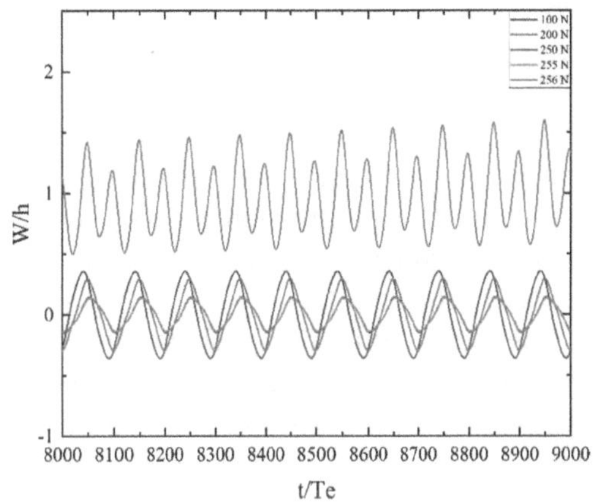

Figure 3.11 Change in response amplitude with increasing axial load under bi-directional excitation
Source: Authors

harmonic responses that evolve as the frequency ratio increases. As shown in Figure 3.10, for the lower ratios (e.g., 0.25, 0.5), the spectrum is dominated by the lower order harmonics, suggests that the energy concentration is more at lower order harmonic. As the frequency ratio increased to 1, the energy distribution becomes more uniform across the harmonic orders, reflecting a more balanced interaction between transverse and axial frequency. In this configuration, neither mode is dominant, and the beam exhibits a more symmetric response in terms of harmonic content.

For the higher frequency ratios (e.g., 2, 4), the domination is shifted towards higher order harmonic, as the rapid transverse oscillations excite higher frequency vibrational modes within the beam. The transition from lower order harmonic dominance to higher order harmonic dominance with increasing frequency ratio highlights the non-linear nature of the beam's response, with excitation frequency strongly influencing the modal interactions and structure.

Influence of axial load

The system's response shows an interesting change in behaviour under bi-directional excitation. The amplitude of the response keeps increasing, indicating a reaction to the applied load. After reaching 256 N load, the pattern of the response shifts so that the response no longer oscillates symmetrically around zero; it shifts to oscillate only in the positive domain. This suggests that the system is limiting the negative side of the motion (Figure 3.11).

Conclusion

This study presents a comprehensive investigation into the non-linear vibrational behaviour of a beam subjected to simultaneous transverse and axial excitations across various frequency ratios and compares the results between experimental and numerical analysis. Through time-displacement analysis, the results reveal that lower frequency ratios lead to larger displacements due to the slower axial excitations, while higher ratios introduce more rapid oscillations, with the transverse excitation becoming more dominant.

The phase plane analysis offers further insight into the system's non-linear dynamics. At lower frequency ratios, complex multiple orbit figures are indicating the modal coupling and non-linearity, with multiple orbits forming due to the interaction of the transverse and axial vibrations. As the frequency ratio increases, the system exhibits more regular, predictable orbits with the influence of axial excitation decreasing in comparison to the transverse motion.

In the frequency spectrum analysis, the harmonic content evolves as the frequency ratio increases. Lower order harmonics are more dominated in lower frequency ratio and higher order harmonics are more dominated in higher order ratio due to the decreasing dominance of the axial excitation frequency into the beam structure. This progression underscores the critical role of the frequency ratio in shaping the system's vibrational behaviour.

Overall, the study provides insights into the dynamic response of beams under multi-frequency excitations and varying axial load, highlighting the importance of frequency ratio, load variation, modal interactions, and non-linear effects. These findings have practical implications for vibration control, structure health monitoring, and energy harvesting applications.

Acknowledgment

This research work is not funded by any funding agency.

References

[1] Tang, Y., Ding, Q. Nonlinear vibration analysis of a bi-directional functionally graded beam under hygro-thermal loads. *Comp. Struct.*, 2019;225:111076.

[2] Roy, R., Acharjya, A., Roy, A., Santra, A., Bhattacharya, G. Response of structures to bidirectional seismic loading: Research progress and future directions. *J. Struct. Engg.*, 2022;148(7):03122002.

[3] Zeris, C. A., Mahin, S. A. Behavior of reinforced concrete structures subjected to biaxial excitation. *J. Struct. Engg.*, 1991;117(9):2657–2673.

[4] Salami, M. R., Dizaj, E. A., Kashani, M. M. Fragility analysis of rectangular and circular reinforced concrete columns under bidirectional multiple excitations. *Engg. Struct.*, 2021;233:111887.

[5] Patil, H. H., Pitchaimani, J., Eltaher, M. A. Buckling and vibration of beams using Ritz method: effects of axial grading of GPL and axially varying load. *Mech. Adv. Mater. Struct.*, 2024;31(16):3861–3874.

[6] Carrera, E., Pagani, A., Banerjee, J. R. Linearized buckling analysis of isotropic and composite beam-columns by Carrera Unified Formulation and dynamic stiffness method. *Mech. Adv. Mater. Struct.*, 2016;23(9):1092–1103.

[7] Sanati, S., Daraei, B., Shojaee, S., Hamzehei-Javaran, S. Isogeometric buckling analysis of orthotropic thin-walled laminated beams using the Carrera unified formulation. *Mech. Adv. Mater. Struct.*, 2023;30(14):2926–2940.

[8] Panigrahi, B., Pohit, G. Nonlinear dynamic response of open and breathing cracked functionally graded beam under single and multi-frequency excitation. *Engg. Struct.*, 2021;242:112437.

[9] Kumar, G. D., Panigrahi, B. Influence of flexural-torsional couplings and rotational effects on vibration behaviour of tapered thin-walled anisotropic biconvex beam. *J. Sound Vibr.*, 2024;570:118100.

[10] Upendra, B., Panigrahi, B., Sabareesh, G. R. On exploiting the transition of nonlinear characteristics of a broadband piezoelectric energy harvester for tuning it to low and high-frequency applications. *Mech. Adv. Mater. Struct.*, 2023;31:1–14.

[11] Carrera, E., Giunta, G., Petrolo, M. Beam structures: classical and advanced theories. John Wiley & Sons. 2011; Chapter 2: 9–22.

[12] Neves, A. M. A., Ferreira, A. J. M., Carrera, E., Cinefra, M., Roque, C. M. C., Jorge, R. M. N., Soares, C. M. Static, free vibration and buckling analysis of isotropic and sandwich functionally graded plates using a quasi-3D higher-order shear deformation theory and a meshless technique. *Comp. Part B Engg.*, 2013;44(1):657–674.

[13] Tian, R. L., Yang, X. W., Ren, L. F. Study on midspan deflection of beam bridge under moving loads by the recently proposed oscillator with time-dependent stiffness. *Adv. Mater. Res.*, 2011;179:1096–1101.

[14] Nguyen, Q. H., Hjiaj, M., Uy, B. Time-dependent analysis of composite beams with continuous shear connection based on a space-exact stiffness matrix. *Engg. Struct.*, 2010;32(9):2902–2911.

4 Digital twin framework for structural health monitoring using PWT-excited Lamb waves

Vishnu Harikumar[a] *and Bijudas C. R.*

Department of Aerospace Engineering, Indian Institute of Space Science and Technology, Trivandrum, Kerala, India

Abstract

Non-destructive structural health monitoring (SHM) employs techniques that allow for the continuous evaluation of structural integrity without causing harm to the structure itself. By utilising methods such as Lamb wave propagation and detection using piezoelectric wafer transducer (PWT), these systems enable the early detection of defects or damage, ensuring the safety and longevity of critical infrastructure. PWTs are employed to generate and detect guided waves, such as Lamb waves (LW), within the structures. These waves propagate through the structure, and the responses can be analysed to identify the presence of damage. Nonetheless, damage identification via damage-sensitive features derived from diverse signal processing techniques may prove ineffective in real-world settings characterised by several uncertainties. To address these challenges, a digital twin model that accommodates these uncertainties in SHM using PWAS is presented. The bonding between the sensor and the structure can significantly impact SHM applications where partial contact loss may lead to incorrect conclusions, potentially being misinterpreted as damage within the structure. A data-driven approach is implemented to develop a digital twin (DT) model that distinguishes between healthy and damaged states. Experimental studies are conducted with a wafer-type sensor coupled to an elastic plate, and a laser vibrometer is used as a receiver to analyse the A0 mode behaviour at different points within a 30 cm radius surrounding the sensor. Mode tuning curves for a frequency range from 13 kHz to 260 kHz are obtained, with the amplitude found to vary with bonding conditions. The obtained response is then used to train the digital twin model, which can be further tested with damage scenarios.

Keywords: Digital twin, structural health monitoring, Lamb waves, transfer learning

Introduction

Mechanical structures are continuously exposed to operating stresses and varying environmental conditions, leading to gradual deterioration over time. Traditionally, the inspection and maintenance of these structures have relied on preventive, time-based approaches. While this method offers some level of protection, it often results in unnecessary maintenance or failure to detect critical damage. In response to these limitations, structural health monitoring (SHM) techniques have gained significant importance in recent years. SHM facilitates continuous, real-time health assessment, enabling accurate prediction of remaining useful life (RUL) and supporting condition-based maintenance (CBM) strategies [1]. By monitoring structural degradation in real time, SHM improves safety, optimises maintenance schedules, and reduces costs by targeting interventions based on the actual condition of the structure rather than fixed intervals.

The SHM approach relies on data gathered from various sensors installed on structures, with the findings analysed to support decision-making under different operating conditions. A variety of non-destructive evaluation (NDE) methods, including bulk wave ultrasonic, X-ray, infrared thermography, and eddy current testing, are frequently employed for structural health inspections. However, these methods are often slow and cumbersome, particularly for large-scale structures. Lamb waves (LW) present a more efficient alternative due to their ability to cover large areas and detect small-scale damage [2]. By integrating permanently installed sensors, LW-based methods enable continuous monitoring, facilitating targeted inspections and early damage detection. Additionally, the real-time monitoring capabilities of LW-based SHM allow for tracking the damage

[a]vishnuharikumar92@gmail.com

DOI: 10.1201/9781003770442-4

as it develops over time, providing a more dynamic approach to the structural assessment.

Piezoelectric actuators and sensors are mounted on the surface to detect and localise damage in a plate structure. As symmetric (*S0*) and anti-symmetric (*A0*) modes of LW propagate through the plate, they interact with any existing damage, leading to changes in wave characteristics such as amplitude variation or phase shift in both modes [3]. The damage identification process leverages these changes by analysing the time of flight of the guided wave modes transmitted and received between transducers at specific observation points or by evaluating the amplitude and phase alterations of the transmitted and received signals at these points.

In addition to structural damage, several other parameters can affect the characteristics of LW, such as bonding layer degradation, debonded transducers, operational conditions, and the evolution of material properties in both the structure and transducers over time [4, 5]. As a result, individuals applying SHM techniques to these structures may mistakenly interpret a healthy structure as damaged due to the influence of these factors. Careful analysis is essential to avoid such misinterpretations and ensure accurate assessments of structural integrity.

Lamb wave-based damage identification generally requires the interpretation of damage features in guided wave signals. Effectively identifying these traits is essential for the successful detection of structural degradation. Typically, appropriate features are selected from raw wave signals based on expertise and professional knowledge, rendering the feature selection process both challenging and time-consuming. To reduce this burden, researchers have been investigating machine learning algorithms that allow for automated damage identification without the need for prior explicit feature definitions [6–8]. This method enhances both the identification procedure and the precision and efficiency of damage assessment.

This study is focused on proposing a digital twin framework for the SHM of plate structures. Experiments were conducted using piezoelectric wafer transducers (PWT) mounted on an aluminium plate, with the A0 mode amplitude measured using a laser vibrometer. The experiments were carried out across a frequency range of 70–350 kHz. Mode-tuning curves were plotted from the experimental data and compared with both analytical results and simulations performed using ABAQUS. By integrating experimental data, simulations, and machine learning, this work aims to develop a comprehensive DT framework that enhances real-time damage detection and structural assessment.

Lamb waves in plates

Lamb waves are a kind of directed elastic waves that go through thin, plate-like materials, making them especially advantageous for SHM. These waves exist in two fundamental modes: symmetric (S) and anti-symmetric (A), each with unique propagation characteristics that interact with structural defects, making them highly sensitive to damage such as cracks, delamination, and corrosion. One of the key advantages of Lamb waves is their ability to travel long distances with minimal energy loss, allowing them to cover large areas with relatively few sensors. Additionally, Lamb waves can detect even small defects, making them ideal for early damage detection. Lamb waves in structures can be generated using a piezoelectric actuator attached to the surface of the structure. When an alternating voltage is applied to the terminals of a PWT, it induces oscillatory expansion and contraction of the PWT material. This oscillatory motion, produced through the piezoelectric effect, is coupled by the PWT to the in-plane strain on the material's surface, effectively exciting Lamb waves within the structure. There are conditions under which the PWT will preferentially couple with only one of the LW modes, either the symmetric or the anti-symmetric mode. This phenomenon is known as the mode tuning of the PWT. Mode tuning allows for selective excitation of specific wave modes, optimising the detection of particular types of structural damage based on the sensitivity of the chosen mode.

Consider a surface-mounted PWT, as shown in the Figure 4.1.

The PWT is excited with a time-harmonic voltage, and as a result, time-harmonic interfacial shear stress is developed [9]. Applying this boundary condition in the wave equation, the response to the PWT excitation is obtained as;

$$U_x(x,t) = \frac{-i}{2\mu}\left[\sum \frac{\bar{\tau}(\xi^S)N_S(\xi^S)}{\xi^S D_S'(\xi^S)} + \sum \frac{\bar{\tau}(\xi^A)N_S(\xi^A)}{\xi^A D_A'(\xi^A)}\right]$$

where N_S, D_S, N_A, and D_A are the functions of wave number (ξ). This expression, for the S_0 and A_0 modes

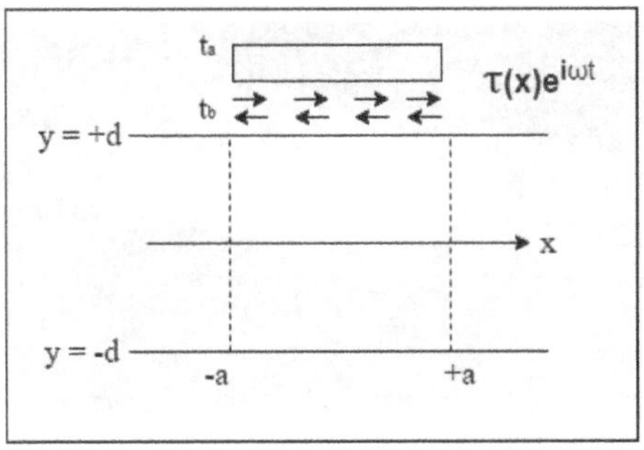

Figure 4.1 Schematic diagram of a PWT bonded to a surface
Source: Authors

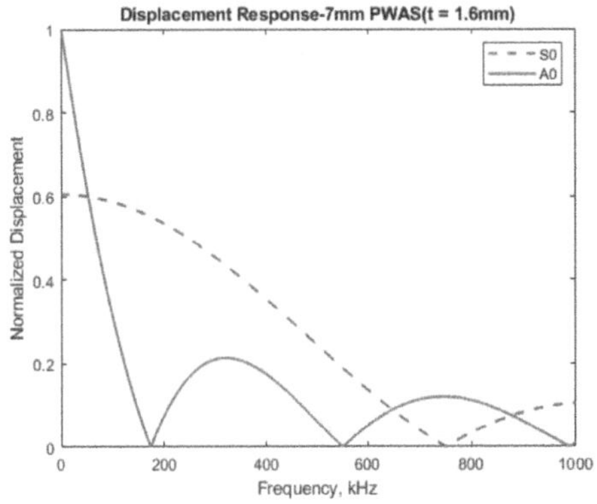

Figure 4.2 Predicted displacement response for a PWT length of 7 mm with a plate thickness of 1.6 mm
Source: Authors

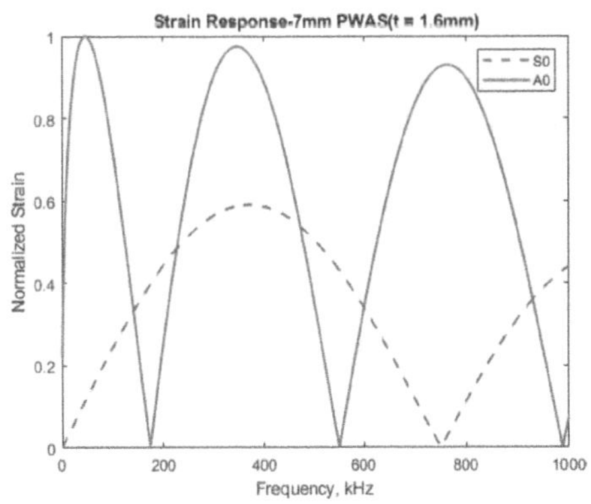

Figure 4.3 Predicted strain response for a PWT length of 7 mm with a plate thickness of 1.6 mm
Source: Authors

in a 7 mm length PWT for a plate of thickness of 1.6 mm, is plotted as shown in Figures 4.2 and 4.3.

When the Lamb wave encounters damage in the structure, changes occur in both the amplitude and phase of the A_0 mode, leading to a deviation in the response curve from the nominal, undamaged condition. This change in wave characteristics serves as a key indicator of damage presence and severity. It is this fundamental concept that underpins the development of the DT framework for SHM.

Figure 4.4 Structure assembly
Source: Authors

Table 4.1 Material properties.

	Density (kg/m3)	Youngs's Modulus (GPa)	Poisson's ratio
Aluminium	2680	60	0.33
PZT 5A	7600		
Adhesive	1100	3.5	0.3

Source: Authors

Numerical simulation

In the simulation, finite element analysis is performed using ABAQUS. The adhesive layer, bonding the PWT to the plate, is modelled with dimensions of 1 cm × 1 cm and a thickness of 0.03 mm. This ensures an accurate representation of the adhesive interface between the PWT and the structure for simulating the Lamb wave propagation. The model assembly is shown in Figure 4.4.

The material properties of the aluminium plate are obtained through mechanical testing to accurately replicate the experimental conditions in the simulation. The properties of the PWT and the adhesive layer are sourced from the literature, ensuring that the simulation closely matches the behaviour of the materials used in the physical experiment. The material properties used are given in the Table 4.1.

The properties of the PZT-5A are provided in Table 4.2. The polarisation direction of the PWT is defined as perpendicular to the thickness of the

Table 4.2 Parameters of the PZT materials.

Dielectric constant (F/m)	Elastic constant (GPa)	Piezoelectric constant (C/m2)
D11 8.11 × 10^{-9}	D1111 121	e2 23 12.3
D22 8.11 × 10^{-9}	D1122 75.4	e3 11−5.4
D33 7.35 × 10^{-9}	D2222 121	e1 13 12.3
	D1133 75.2	e3 22–5.4
	D2233 75.2	e3 33 15.8
	D3333 111	
	D1212 22.6	
	D1313 21.1	
	D2323 21.1	

Source: Authors

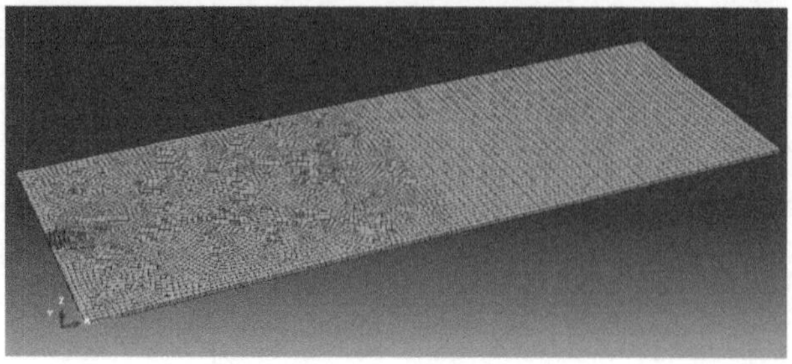

Figure 4.5 Meshed assembly
Source: Authors

plate, extending from the upper surface to the lower surface.

A tie constraint is imposed between the lower surface of the PWT and the upper surface of the adhesive. In this configuration, the PWT surface is identified as the master surface, whereas the adhesive surface is regarded as the slave surface. Similarly, a tie constraint is also applied at the interface between the adhesive and the plate, where the plate surface serves as the master surface, and the adhesive surface acts as the slave surface. A zero potential boundary condition is imposed on the lower surface of the PWT, while a 13-count sine burst signal, modulated by a Hanning window, is applied as the electric potential on the top surface of the PWT.

The element size and time increment in the finite element method significantly affect the convergence and accuracy of the numerical solution. Typically, smaller time increments and finer grid divisions lead to higher accuracy in the model, ensuring that the numerical solution more closely approximates the actual physical behaviour. According to Leckey et al. [10], the maximum mesh size for finite element modelling (FEM) must meet the criterion

$$S = \frac{\lambda_{min}}{20}$$

where S is the maximum mesh size. λ_{min} is the minimum wavelength of the Lamb wave modes. This criterion ensures convergence by maintaining sufficient resolution to capture wave propagation phenomena accurately. Based on the criterion, the minimum wavelength corresponds to the highest excitation frequency of 350 kHz. Accordingly, the maximum allowable mesh size is determined to be 0.5 mm. To ensure accuracy in simulating the Lamb wave propagation and interaction with the structure, a finer mesh size of 0.1 mm is selected near the PWT, where higher precision is needed, and a 0.3 mm mesh size is applied at the edges of the plate. The meshed assembly is shown in Figure 4.5.

The time increment in finite element analysis is crucial, especially for fine meshes, as smaller time steps are required to ensure stability and accuracy. A fine mesh demands a smaller time increment, significantly increasing the computational cost. The rule for determining the appropriate time increment is typically based on the highest frequency present in the model. A small enough time increment must be chosen to accurately capture the wave propagation.

$$\Delta t = \frac{1}{20 * f_{max}}$$

For the maximum frequency of 350 kHz, the corresponding time increment is calculated to be 0.1 μs, and this increment is implemented in the simulation. The total simulation time is set to 0.4 ms. The mesh for the PWT is assigned using a piezoelectric element type to account for the electro-mechanical coupling, while the plate is meshed with 8-noded linear brick elements to model the structural response accurately.

Experimental analysis

The experiment is conducted on an aluminium plate of dimensions length of 40 cm, width of 15 cm, and thickness of 2 mm. A PZT-5A piezoelectric transducer, with dimensions of 1 cm × 1 cm and a thickness of 1.5 mm, is used as the actuator to generate Lamb waves. The PZT-5A transducer is mounted at the midpoint along one of the shorter edges of the plate. The schematic diagram for the experiment setup is shown in Figure 4.6.

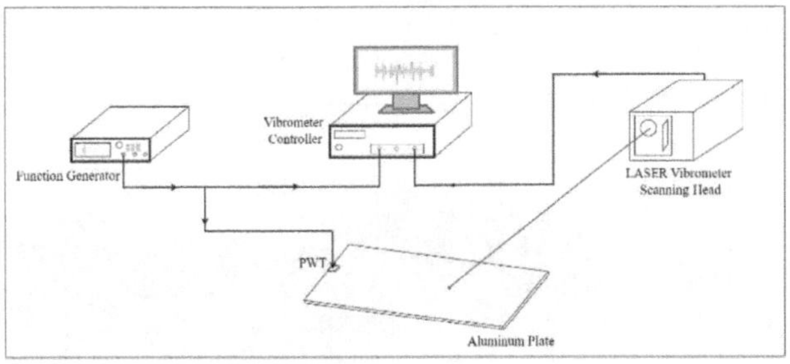

Figure 4.6 Schematic diagram for the experiment setup
Source: Authors

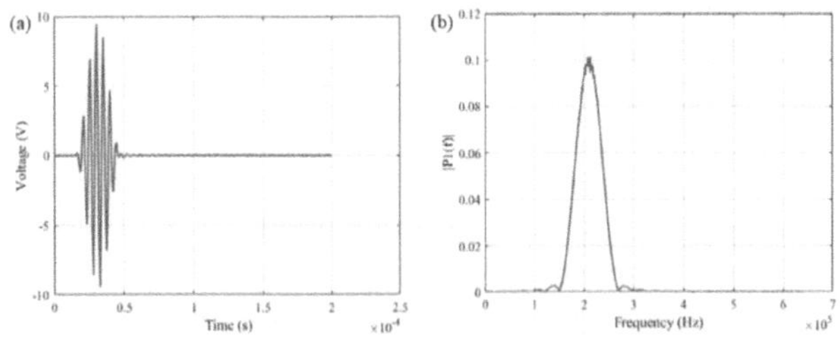

Figure 4.7 (a) 7 Count Hanning window signal with central frequency of 210 kHz. (b) FFT of the signal
Source: Authors

A 13-count sine burst, modulated by a Hanning window, serves as the LW excitation signal, with peak-to-peak amplitude of 20 V. The central frequency of the LW excitation is swept from 65 kHz to 325 kHz, with increments of 13 kHz.

$$V(t) = \frac{1}{2}\sin(2\pi f_c t)\left(1 - cos\left(\frac{2\pi t}{T}\right)\right)$$

where f_c is the central frequency, T is the total duration of the burst signal (Figure 4.7).

The signal is generated using a key sight waveform generator and then supplied to the PWT. The out-of-plane displacement is measured at points 20 cm and 25 cm away from the PWT using a Polytec PSV 500 Laser Doppler Vibrometer. The vibrometer is situated 600 mm from the specimens under examination, guaranteeing that the laser beam is perpendicular to the scanned surface for precise displacement readings. The excitation signal from the waveform generator is employed to initiate the data acquisition process, guaranteeing synchronisation between Lamb wave excitation and the collection of displacement responses. Data is acquired with a sampling frequency of 1.28 MHz over a sample time of 1.6 ms. A band-pass filter is applied with limits set at 50 kHz and 500 kHz to isolate the relevant frequency components of the measured Lamb waves.

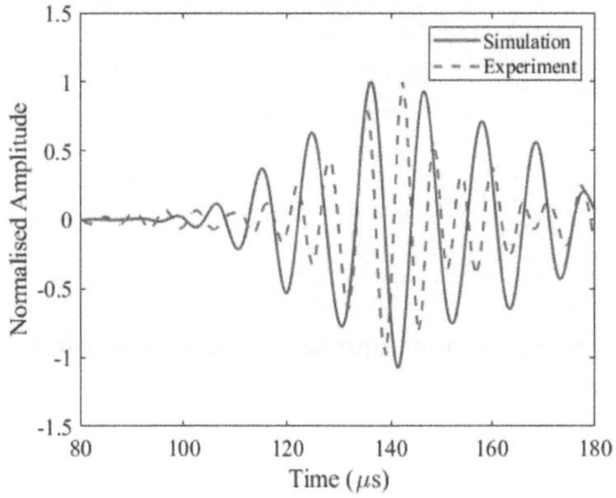

Figure 4.8 Normalised displacement curves for *S0* mode corresponding to a 5 count 150 kHz excitation
Source: Authors

Experimental verification

The simulation outcomes are compared with empirical data to verify the model's authenticity. By analysing the out-of-plane displacement and amplitude response at various frequencies, the consistency between simulation and experimental findings can be assessed. The displacement plots from both simulation and experiment were compared by normalising their amplitudes, allowing a direct evaluation of their relative wave characteristics. Figure 4.8 shows the experimentally received signal corresponding to a 5-count, 150 kHz excitation alongside the simulated signal for the same input.

The experimental and simulation responses show strong alignment, with minor discrepancies likely arising from boundary conditions in the experiment or differences in the sampling frequency between the experiment and simulation. This close agreement supports the simulation model's accuracy in replicating experimental behaviour.

Digital twin-based damage detection

The simulation study reveals that displacement response curves exhibit amplitude shifts when there are changes in the bonding area or thickness of adhesive, etc. These shifts can resemble those caused by structural damage, potentially leading to misinterpretation during SHM. To accurately differentiate between structural damage and variations in bonding

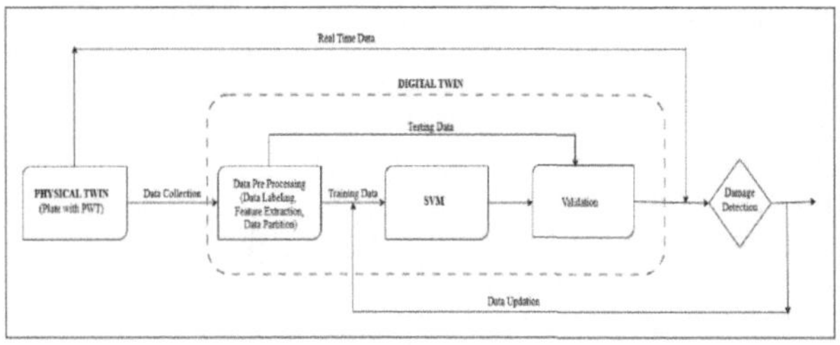

Figure 4.9 Schematic representation of the digital twin framework
Source: Authors

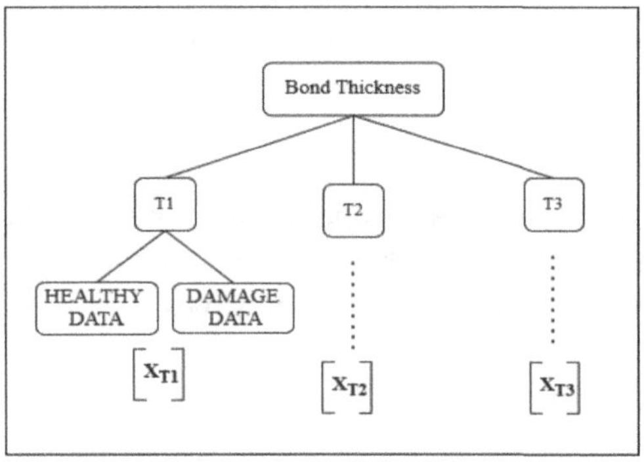

Figure 4.10 Flow chart representing the data collection
Source: Authors

conditions, a digital twin model is introduced. This model provides a reliable framework for monitoring the health of the structure by accounting for both damage-induced changes and bonding variations in the response curves. A schematic representation is provided in Figure 4.9, outlining the structure and construction of the digital twin framework for SHM.

Data preparation

As shown in the Figure 4.10, data generation forms a foundational component. This data is gathered from experimental responses as described in the Experiment section. Displacement responses are recorded at points, each located 20 cm and 25 cm from the PWT. For these measurements, the excitation frequency is swept from 65 kHz to 325 kHz. The resulting displacement response curves at each point serve as the primary dataset for constructing and validating the DT model. The data is collected for various conditions of the plate–PWT combinations, as shown in Figure 4.10.

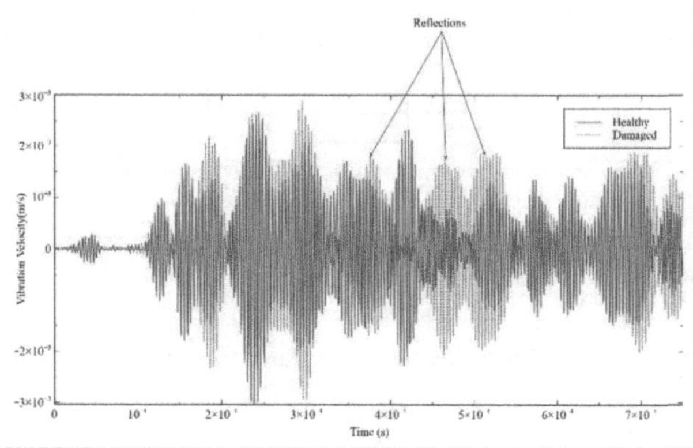

Figure 4.11 Flow chart representing the data collection
Source: Authors

The data collection process involved varying bond thickness levels to capture the effects of bonding on structural responses. Specifically, data was gathered at bond thicknesses of 0.03 mm, 0.04 mm, and 0.05 mm. For each bond thickness level, corresponding data for damaged conditions was also recorded to assess how damage impacts the system under different bonding scenarios. The damage was introduced as a notch, with dimensions of 2 cm in length and 0.5 mm in depth, positioned between the PWT and the measurement point. This setup enables a comprehensive understanding of how bond thickness variations and structural damage interact. Figure 4.11 illustrates the displacement response measured at a point 20 cm away from the source in response to a 260 kHz Hanning window input signal. Comparing the responses for both undamaged and damaged conditions reveals a noticeable increase in amplitude and a shift in the amplitude level of the *A0* mode due to the presence of damage. However, it is essential to note that this amplitude shift in the *A0* mode may also result from variations in bonding thickness, which can mimic the effects of damage. This overlap in response characteristics introduces potential ambiguity, as the system might incorrectly interpret amplitude changes as indicators of damage, reducing the accuracy of SHM system.

The collected data undergoes feature extraction, followed by a split into training and testing datasets. The training data is used to train a support vector machine (SVM) classifier, while the remaining data is reserved for testing the classifier's accuracy. Together, the data pre-processing unit and the validated SVM classifier form the core of the DT.

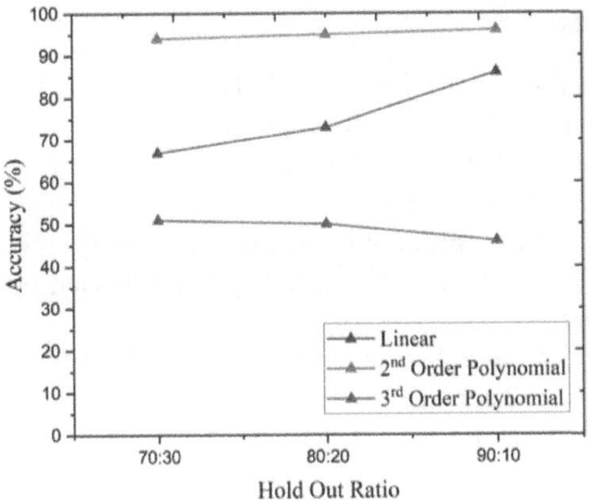

Figure 4.12 Accuracy plot for different SVM kernels with hold out ratio
Source: Authors

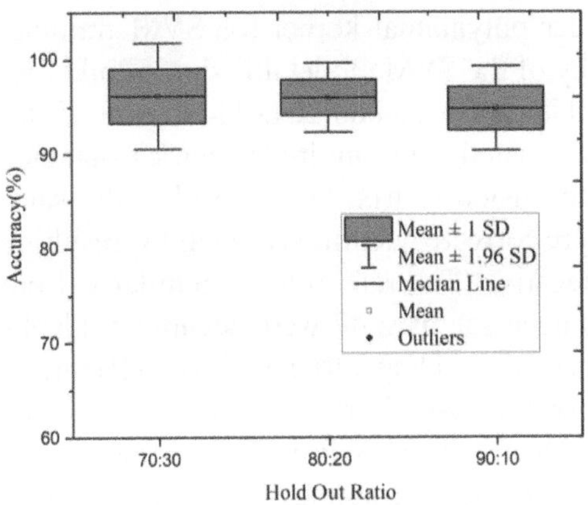

Figure 4.13 Variation in accuracies across hold out ratio for 2nd order polynomial kernel
Source: Authors

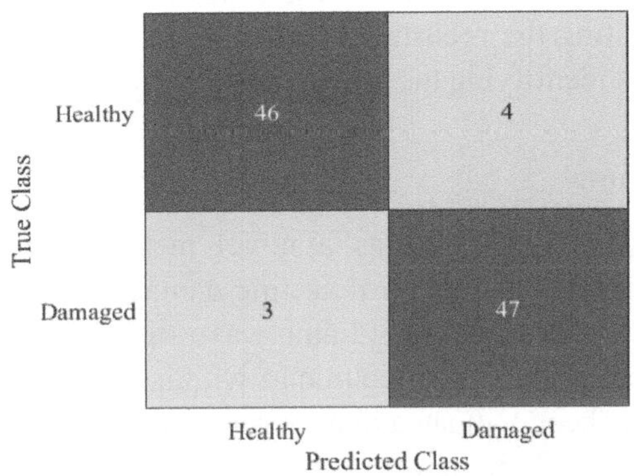

Figure 4.14 Confusion matrix corresponding to 80:20 hold out ratio and 2nd order polynomial kernel for SVM training
Source: Authors

Real-time data from the physical twin can be input to the DT for damage detection. Additionally, the predictions generated by the DT can be continuously added to the training dataset, enhancing accuracy and adaptability over time.

Results and discussion

In validating the developed DT framework, it is essential to assess its accuracy in distinguishing responses resulting from structural damage versus variations in bonding thickness. This validation process ensures that the DT can reliably separate true damage indicators from changes due to environmental or operational factors, thereby enhancing the precision of the SHM system.

Testing and evaluation

The DT model undergoes rigorous testing with a range of datasets, which are divided into training, and testing sets across three different training-to-testing ratios. To optimise model performance, various SVM kernel functions are explored within these different hold-out ratios. Additionally, stratified cross-validation is applied to maintain balanced class representation during both the training and testing phases, enhancing the model's robustness and reliability.

Figure 4.12 demonstrates that among the different kernels, the 2nd-order polynomial consistently exhibits superior performance across varied hold out ratios, while the 3rd-order polynomial demonstrates the lowest accuracy. This discrepancy may be attributed to overfitting, as the higher-order polynomial tends to capture noise and minor fluctuations in the training data, thereby impairing the model's ability to generalise effectively on new data.

Figure 4.13 shows a box plot depicting the fluctuations in accuracy for SVM training with the 2nd order polynomial kernel for different hold-out ratios. A 5-fold cross-validation method is utilised to evaluate the accuracy during testing, facilitating a rigorous assessment of the model's performance. From this chart, it is observed that there is a minimum change in the accuracy for the 80:20 hold-out ratio when compared to the mean accuracy value. The consistency demonstrates that the SVM model employing a 2nd-order polynomial kernel with an 80:20 hold-out ratio is proficient in accurately detecting structural damage.

Figure 4.14 displays the confusion matrix associated with the 80:20 hold out ratio utilising the second order polynomial kernel for SVM training. The efficacy of the SVM model utilising a 2nd order polynomial kernel and an 80:20 hold-out ratio is statistically evaluated using multiple metrics obtained from the confusion matrix. Out of 50 healthy samples, 46 were correctly identified as healthy, resulting in a true negative (TN) rate of 92%. Similarly, from the 50 damaged samples, 47 were accurately identified as damaged, yielding a true positive (TP) rate of 94%. The overall 5-fold cross-validation efficiency of the classification is 94%. Precision, defined as the ratio of true positive predictions to the total number of positive predictions, is computed at 91.5%, signifying a substantial accuracy in detecting damaged samples. Recall, or sensitivity, is reflected in the true positive rate, confirming the model's effectiveness in detecting damage. The F1-score is obtained as 92.2%, demonstrating the robustness of the SVM model in accurately identifying the structural condition.

Conclusion

The present study presents a novel methodology based on the DT model for detecting damage in thin metal plate structures using Lamb wave signals. The integration of the SVM technique within the SHM system has been validated and thoroughly examined to develop a DT framework for plate structures utilising Lamb waves. This study observes a variation in the amplitude of the A_0 mode resulting from changes in bond thickness as well as the presence of damage within the structure. Such variations pose a risk of misleading the SHM system, potentially interpreting changes in the A_0 amplitude of a healthy structure as indicative of damage. To address this issue, a DT framework is developed for damage detection. The DT model is trained to identify damage in a 2 mm thick aluminium plate using experimentally obtained wave signals. The accuracy of the proposed model is evaluated across various training-to-testing ratios and different kernels for the SVM classifier. Results demonstrate that the proposed DT model effectively identifies the presence of damage by discerning changes in operational variables, such as bonding thickness.

References

[1] Huang, C., Siqi, B., Hung, L. H., Hung, C. C., Kong, S. W., Y. W. K. C. Prognostics and health management for predictive maintenance: A review. *J. Manufac. Sys.*, 2024;75:78–101.

[2] V. Giurgiutiu, Structural health monitoring: with piezoelectric wafer active sensors, US: Elsevier, pp. 25, 2007.

[3] G. Santoni and V. Giurgiutiu, Lamb Wave-Mode Tuning of Piezoelectric Wafer Active Sensors for Structural Health Monitoring., *Journal of Vibration and Acoustics*. 129, 752–762, 2007.

[4] Park, G., Farrar, C., Scalea, F. d. Performance assessment and validation of piezoelectric active- sensors in structural health monitoring. *Smart Mater. Struct.*, 2006;15:1673–1683.

[5] Lanzara, G. Y., Kim, Y., Chang, Y., Fu-Kuo. Influence of interface degradation on the performance of piezoelectric actuators. *J. Intel. Mater. Sys. Struct.*, 2009;20:1699–1711.

[6] V. Nerlikar, R. Miorelli, A. Recoquillay and O. d'Almeida, Machine learning-based digital twin framework for realistic guided wave signal generation, applied to reliability assessment and global sensitivity analysis in SHM, in *50th Annual Review of Progress in Quantitative Nondestructive Evaluation*,, 001, 498–508, 2023.

[7] D. Wagg, K. Worden, R. Barthorpe and P. Gardner, Digital twins: state-of-the art and future directions for modeling and simulation in engineering dynamics applications, *Journal of Risk and Uncertainty in Engineering Systems*, 6, 907–918, 2020.

[8] A. Thelen, X. Zhang, O. Fink, Y. Lu, S. Ghosh, B. Youn, M. Todd, S. Mahadevan, C. Hu and Z. Hu, A comprehensive review of digital twin part 1: modeling and twinning enabling technologies, *Structural and MultiDisciplinary Optimization.*, 65, 354–374, 2022.

[9] V. Giurgiutiu, Tuend Lamb Wave Excitation and Detection with Piezoelectric Wafer Active Sensors for SHM, *Journal of Intelligent Material Systems and Structures*, 16, 291–305, 2005.

[10] Leckey, C. A. R., Parker, M. D., Raymond, F. Guided waves in anisotropic and quasi-isotropic aerospace composites: Three-dimensional simulation and experiment. *Ultrasonics*, 2014;54(1):385–394.

5 Element-free Galerkin method for static analysis of non-local elastic solids using differential and integral strain-driven models

Akhil S. L., Praveen Krishna I. R.[a] *and Gaurab Kumar Khanra*

Indian Institute of Space Science and Technology, Thiruvananthapuram, Kerala, India

Abstract

The element-free Galerkin (EFG) method is formulated and applied to analyse the bending mechanics of Euler-Bernoulli beams, utilising Eringen's strain-driven constitutive model. EFG, a meshless numerical method, is particularly well-suited for simulating non-local solids due to its inherent ability to handle non-local interactions. The formulation effectively incorporates long-range interactions within the entire domain through carefully constructed weighting functions, allowing for a more accurate representation of the physical behaviour of materials with non-local effects. This study examines the bending response of nano/micro beams subjected to standard boundary conditions and a set of non-local parameters. The analysis uses Eringen's differential and original integral forms of the non-local constitutive models. A discrete set of governing equations for the integral model is obtained by minimising the potential energy of the system, which captures the non-local effects through integral operators that account for interactions over finite distances. The boundary conditions, crucial for accurately capturing the beam's response, are implemented using a scaled transformation method and Lagrange multipliers. The study investigates key structural responses under various boundary conditions, including deflections, slopes, bending moments, and shear forces. These results are then compared against available analytical solutions and numerical results from other methods.

Keywords: Element free Galerkin, non-local elasticity, strain-driven model

Introduction

The meshless method is inherently non-local and provides a basis for interpolating field variables in a more accurate, reliable, consistent, and robust approach for solving integral or partial differential equations under varied boundary conditions using a set of scattered nodes. These methods have been shown to be particularly useful for problems involving complex geometries, moving boundaries, or large deformations and singularities [1]. Meshless methods can be grouped into three primary classes based on the construction of shape functions. Integral representation methods, series representation methods, and differential representation methods [2]. The EFG method is based on the global weak form framework and relies on MLS shape functions. The EFG method is successfully used to solve problems in science and engineering, like large strain rate problems, fracture mechanics, biomechanics, and dynamic problems that otherwise require remeshing if solved using FEM. Currently, numerous iterations of EFGM strategies are available to tackle specific problems, including stochastic meshless methods for random fields [3], fracture mechanics and contact problems [4], thermomechanical analysis [5], etc. Simulation of non-local elastic problems using EFGM is scarcely done in the literature. A non-local framework allows the inclusion of size effects into the constitutive relationships using one or more size-dependent parameters [6]. Classical continuum mechanics, the formalism used to simulate fundamental problems in engineering fields and physical science, cannot reflect the non-local/size effects and yields results that conflict with experimental evidence. The theory defines stress at a location depends solely on the strain at that point, thereby disregarding the effects of stress at other material points on the point of interest. Among these, Cosserat continuum theory is the earlier concept, stemming from non-Boltzmannian mechanics where the stress tensor is not required to be symmetric [7]. Building on Cosserat continuum theory, Mindlin and Tiersten [8] and Koiter [9] introduced the initial form of couple stress theory, which

[a]praveenkrishna@iist.ac.in

DOI: 10.1201/9781003770442-5

describes deformations entirely through a continuous displacement field. Later, Mindlin [10] proposed an enhanced version of couple stress theory aimed at modelling the linear elastic behaviour of isotropic materials while incorporating microstructural effects. Eringen's non-local elasticity theory [6] is the most widely recognised theory, which includes the effects of long-range forces in the formulation. According to the theory, the stress at a material location is influenced by strains at all material locations. This is managed by using a kernel function that decreases with the distance of a material point from the neighbouring point. The original integral constitutive law is less explored due to the difficulty in obtaining the solution of the integro-differential equation. For special types of kernel functions, the original integral non-local model can be recast into an equivalent differential form [11]. The simplicity of differential forms encouraged researchers to explore this model in studying micro/nanostructures. Analytic solutions for static and dynamic problems using the model have been extensively studied [12–17]. In 2006, Xu studied the free vibration of nano/micro beams and developed an analytic solution for natural frequency [18]. Aydogdu carried out a study on the dynamic characteristics of the nano rod model, which revealed size-dependency of frequency of free vibration [29], to mention a few.

On numerical simulation of micro/nanostructures, Plizzotto studied the non-local model using the thermodynamic and variational framework and developed NL-FEM (non-local FEM), which includes non-locality features in the global stiffness matrix [19–21]. NL-FEM is used along with differential non-local models and a two-phase non-local model for both static and dynamic analysis [22–24]. Attempts to use the meshless method in non-local analysis of micro/nanosystems in the literature include [25–28].

The quest to make smaller and smaller devices for aerospace, electronics, space systems, etc., in the form of micro/nano electro-mechanical devices (MEMS/NEMS) is ever-increasing. Analytical solutions for a few simple structures with standard boundary conditions are available. Analysis of complex nano/micro systems needs a reliable and robust numerical scheme. This paper presents the meshless scheme as an efficient and apt numerical scheme for non-local systems. The present study derives a 1D meshless formulation of a non-local Euler-Bernoulli beam using differential and integral models. The meshless formulation for the differential model is obtained by considering the equilibrium and integral model using the principle of minimum potential energy of the system. This paper is structured as follows: Section 2 contains the basics of Euler-Bernoulli beam theory, Section 3 briefly describes Eringen's non-local elasticity, and Section 4 explains the EFG method, including the shape function.

Euler Bernoulli beam theory

Revisiting the governing equations of the Euler-Bernoulli beam, the coordinate system selected is defined so that the represents the length, y represents the width of beam, and z-axes represent the height of the beam, respectively, as given in Figure 5.1. Based on the beam theory, the components of the displacements along the coordinate axes $\bar{u}_x$, $\bar{u}_y$, $\bar{u}_z$, depend only on the x and z coordinates. Only transverse loading is taken into consideration for analysis.

$$\begin{aligned} \bar{u}_{x(x)} &= u_{0(x,y,z)} - z\frac{\partial u_{z(x)}}{\partial x} \\ \bar{u}_{y(x,y,z)} &= 0 \\ \bar{u}_{z(x,y,z)} &= u_{z(x,t)} \end{aligned} \quad (1)$$

With the assumption of small displacement fields, the only strain component that remains non-zero is,

$$\epsilon_{xx} = \frac{\partial \bar{u}}{\partial x} = \frac{\partial u_0}{\partial x} - z\frac{\partial^2 u_z}{\partial x^2} \quad (2)$$

For Euler–Bernoulli beam theory, the rotation angle due to bending is directly derived from the transverse deflection as shown below,

$$\phi_y(x) = -\frac{du_{z(x,t)}}{dx} \quad (3)$$

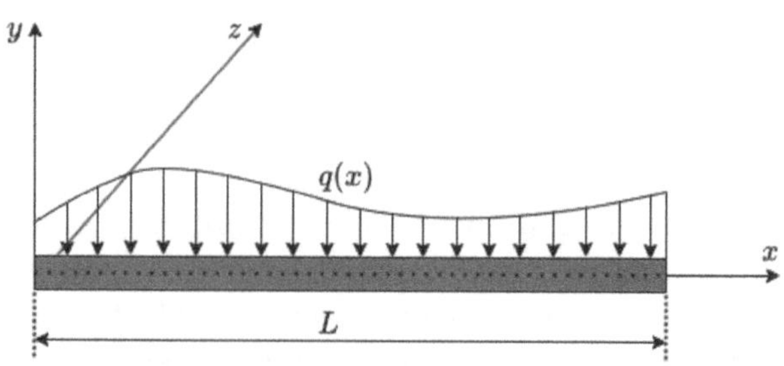

Figure 5.1 Euler Bernoulli beam of length L, arbitrary transverse loading $q(x)$
Source: Authors

Eringen's non-local elasticity

Eringen's non-local theory proposes that stress at a specific location is a function of the strain throughout the domain. This is achieved using a kernel function $k(|x-\bar{x}|,\kappa)$, denote the point of interest, and $\bar{x}$ denote the surrounding material point.

$$\sigma_{ij}(x) = \int_V k(|x-\bar{x}|,\kappa)t_{ij}(\bar{x})d\bar{V} \quad (4)$$

Here, σ_{ij} is the non-local stress tensor, $t_{ij}(\bar{x})$ is the classical stress tensor. Classical stress t_ij in its constitutive relation is, $\sigma_{ij} = C_{ijkl}\epsilon_{kl}$. Eq. (4) is Eringen's non-local constitutive equation in the integral form. which will be later used to derive the EFGM equations for static analysis. In this work, the exponential kernel function is used, which takes the form:

$$k(|x-\bar{x}|,\kappa) = \frac{1}{2\kappa}e^{-\frac{|x-\bar{x}|}{\kappa}} \quad (6)$$

The non-local parameter κ is defined as the product of a material constant (e_0) and internal length scale (a), expressed as $\kappa = e_0 a$. Under certain conditions, the equivalent differential form of the integral constitutive relation [11] takes the form,

$$(1-\mu\Delta^2)\sigma = t \quad (7)$$

where $\mu = (e_0 a)^2 = \kappa^2 = c^2$

Element free Galerkin method

Unlike traditional finite element-based methods, which rely on a mesh of elements to discretise the problem, the EFG method operates on nodes distributed in the domain. The moving least squares (MLS) is central to the EFG method. Polynomial fitting of the field variables at a point using values of neighbouring nodes weighted using a monotonic, smooth, and compactly supported weighing function. Along the length of the beam, the field variable $u^h(x)$ of $u(x)$ can be approximated as follows:

$$u^h(x) = \sum_{i=1}^{m} p_i(x)a_i(x) = \mathbf{p}^T(\mathbf{x})\,\mathbf{a}(\mathbf{x}) \quad (8)$$

Here, $\mathbf{p}^T(\mathbf{x}) = [p_1(x),\ldots\ldots, p_m(x)]$ represents the basis function of order m. These basis functions serve as the fundamental building blocks used to approximate the field variable. $a(x) = [a_1(x), a_2(x),\ldots, a_m(x)]$ is a vector whose entries are unknown parameters related to x. The unknown coefficients $\mathbf{a}(\mathbf{x})$, at any given point are computed by applying a weighted least-squares fitting. A support domain or compact support is a region adjacent to the point of interest. In fundamental bending analysis, the degree of freedom per node encompasses nodal displacements and rotations. Thus L_2 norm is,

$$J = \sum_{i=1}^{n}(w(x-x_i)\,[p^T(x_i)\,a(x) - \widehat{u_i}]^2) \quad (9)$$

Here, $w(x-x_i)$ is the smooth compactly supported weight function of the node i. Nodes near to the point x will have larger weights, while those farther away will have less weight. Out of the available weight functions, this study adopts the cubic spline weight function, defined piecewise, depending on the value of the normalised distance r,

$$\omega(r) = \begin{cases} \frac{2}{3} - 4r^2 + 4r^3 & for\, r \le \frac{1}{2} \\ \frac{4}{3} - 4r + 4r^2 - \frac{4}{3}r^3 & for\, \frac{1}{2} < r \le 1 \\ 0 & for\quad r > 1 \end{cases} \quad (10)$$

Here the variable $r = \frac{d_i}{d_{m_i}}$, d_i is the absolute value of the distance between the current node and another node in the domain. d_{mi}, is the product of non-dimensional scale d_{max} and. c_i is also a distance measure. In this study, it is taken as the nodal spacing. This will ensure the theoretical minimum of two nodes in the compact support to make the stiffness matrix invertible. To determine the unknown coefficients in the approximation, Equation 2, is minimised.

$$A(x)\,a(x) = C(x)\bar{d} \quad (11)$$

Here, $\mathbf{A}(\mathbf{x}) = \mathbf{W}\mathbf{P}\mathbf{P}^T$, $\mathbf{C}(\mathbf{x}) = \mathbf{W}\mathbf{P}$ and $\bar{d}$ is the output nodal coefficients corresponding to the nodal displacements, matrix $\mathbf{P}^T = [\mathbf{P}^T(\mathbf{x}_1), \mathbf{P}^T(\mathbf{x}_2),\ldots.,\mathbf{p}^T(\mathbf{x}_n)]$ collects the basis function at all nodal positions. $\mathbf{W}$ is the diagonal matrix containing the weights. Solving for $\mathbf{a}(\mathbf{x})$ and substituting back into Equation 1 to get the approximate field variable,

$$u^h(x) = \phi^T_{(x)}\bar{d} \quad (12)$$

ϕ is the moving least squares shape function, which is of the form:

$$\phi^T(x) = P^T(x)A^{-1}(x)C(x) \quad (13)$$

The MLS shape function lacks the Kronecker delta property. The consequence is that the displacement at a node is not guaranteed to be equal to nodal parameter obtained from solving the set of equations [29]. Over the years, numerous methods have been developed to deal with this issue. Lagrange multiplier method [30] introduced additional unknowns, called as multipliers help in enforcing boundary conditions. penalty factor method [31] imposes boundary conditions approximately by introducing a penalty term into the formulation, transformation method [32] the variables are transformed such that Kronecker delta property is satisfied, scaled transformation method [33], an improved version of transformation method, where displacement field along the essential boundary alone is scaled to match the boundary conditions. The present study uses a combination of scaled transformation method and Lagrange multiplier for simpler implementation of essential and constitutive boundary conditions [33].

4.1 Differential non-local model

The equation governing the bending of size, size-dependent beam according to the differential non-local model is given as [15],

$$EI\frac{\partial^4 w}{\partial x^4} - q(x) = (e_0 a)^2 \frac{d^2(-q(x))}{dx^2} \quad (14)$$

Multiplying throughout by δw and integrating along the length leads to Equation 15. Using meshless approximation $u^h(x) = \phi^T(x)d_s$, in Equation 15 to obtain Equation 16.

$$EI\frac{d^3w}{dx^3}\Big|_0^L \delta w - \left[\left[EI\frac{d\delta w}{dx}\frac{d^2w}{dx^2}dx\right]_0^L - \int EI\frac{d^2\delta w}{dx^2}\frac{d_w^2}{dx^2}dx\right] = \int_0^L q(x)\delta W\,dx - \int^L \mu\frac{d^2w}{dx^2}q(x)\delta w\,dx \quad (15)$$

$$= \int_0^L q(x)\delta(\phi^T(x)\,d_s)^T dx - \int_0^L \delta(\phi^T(x)\,ds)^T(e_0a)^2\frac{d^2}{dx^2}q(x)dx + \mathrm{EI}\frac{d}{dx}(\delta(\phi^T(x)d_s)^T)\frac{d^2(\phi^T ds)}{dx^2}^T\Big|_0^L - \mathrm{EI}\delta(\phi^T(x)ds)^T\frac{d^3(\phi^T(x)d_s)}{dx^3}\Big|_0^L = 0 \quad (16)$$

The stiffness matrix is:

$$K = \int_0^L \frac{d^2}{dx^2}((\phi^T(x)\,d_s))^T EI\frac{d^2}{dx^2}(\phi^T(x)\,d_s)dx \quad (17)$$

The force vector carries the non-local parameter and is given by:

$$f = \int_0^L q(x)(\phi^T(x)\,d_s)^T\,dx - \int_0^L(\phi^T(x)\,d_s)^T(e_0a)^2\frac{d^2}{dx^2}q(x)\,dx \quad (18)$$

4.2 Integral non-local model

For the integral non-local model, starting with the total potential and applying the principle of minimum potential energy yields governing differential equation. The formulation is identical to the finite element method given in ref [34]. According to this, in equilibrium, a system or structure attains a state of minimum total potential energy. The elastic structure is discretised into *N* nodes. The displacement field associated with these nodes can be determined using the relation.

$$\frac{\partial \Pi}{\partial u_1} = \frac{\partial \Pi}{\partial u_2} = \frac{\partial \Pi}{\partial u_3} = \cdots = \frac{\partial \Pi}{\partial u_N} = 0 \quad (19)$$

The total potential energy functional of the structure is given by:

$$\Pi[u] = \Pi[u_1] + \Pi[u_2] + \cdots + \Pi[u_N] \quad (20)$$

Subscripts 1,2, ... *N* represents the node number, and *u* the nodal degrees of freedom. In non-local elasticity, the terms $\frac{\partial \Pi_1}{\partial u_2}, \frac{\partial \Pi_1}{\partial u_3}, \ldots, \frac{\partial \Pi_1}{\partial u_N}$ .... in these derivatives persist, for simple beam bending, the total potential is of the form.

$$\Pi = \mathrm{U} + \mathrm{W} \quad (21)$$

Using Equation 21, total elastic strain energy takes the form:

$$U = \frac{1}{2}\int_V \epsilon_{ij}(x)\sigma_{ij}(x)dV = \frac{1}{2}\int_V \epsilon_{ij}(x)\int_V k(|-\bar{x}|,\kappa)\,C_{ijkl}(\bar{x})\,\epsilon_{kl}(\bar{x})\,dV(\bar{x})\,dV(x) \quad (22)$$

By using Equation 22, the total potential for a uniform Euler Bernoulli beam can be written as:

$$\Pi[w(x)] = \frac{EI}{2}\int_0^L$$

$$\int_0^L \frac{1}{2\kappa} e^{-\frac{|x-\bar{x}|}{\kappa}} \frac{d^2w(x)}{dx^2} \frac{d^2w(\bar{x})}{d\bar{x}^2} d\bar{x}dx - \int_0^L w(x)\, q(x)\, dx - \sum_{i=1}^{N} F_i w(x_i) + \sum_{i=1}^{N} C_i \frac{dw(x_i)}{dx} \tag{23}$$

$\int_A^{z^2} dA = I$ denotes the second moment of area of the beam's cross-section, $q(x)$ represents the distributed load in transverse direction, F_i the point load, C_i discrete bending moment. A background discretisation is used for numerical integration. Force vector F_i takes the form $Q = \{F_1, -C_1, \ldots, F_N, -C_N\}^T$. Differentiating the Equation 22 with respect to each nodal parameter yields:

$$\frac{\partial \Pi}{\partial d_i} = \frac{EI}{2\kappa}\left[\left(\int_0^L \int_0^L e^{-\frac{|x-\bar{x}|}{\kappa}} \phi_i''(x)\phi_i''(\bar{x}) d\bar{x}\, dx\right) d_i\right] + \frac{EI}{4\kappa}\left[\left(\int_0^L \int_0^L e^{-\frac{|x-\bar{x}|}{\kappa}} \phi_i''(x)\phi_i''(\bar{x}) d\bar{x}\, dx\right)\right] + \frac{EI}{4\kappa}\left[\left(\int_0^L \int_0^L e^{-\frac{|x-\bar{x}|}{\kappa}} \phi_i''(x)\phi_i''(\bar{x}) d\bar{x}\, dx\right)\right] - \int_0^L \phi^T(x) q(x) dx - Q_i = 0 \tag{24}$$

The first term represents the effect of node i on itself, the second term represents the influence of other nodes on i^{th} node and third term represents the effect of ith node on other nodes. For static beam problems,

$$K_{ij} d_j = F_i \tag{25}$$

$$K_{ij} = \frac{EI}{2\kappa}\left[\left(\int_0^L \int_0^L e^{-\frac{|x-\bar{x}|}{\kappa}} \phi_i''(x)\phi_j''(\bar{x}) d\bar{x}\, dx\right)\right] \tag{26}$$

The force vector corresponding to node i is given by the expression:

$$F_i = \int_0^L \phi^T(x) q(x) dx - Q_i \tag{27}$$

Numerical results of case studies

Numerical case studies using the developed EFG method for the differential and integral Eringen's non-local model are carried out for standard boundary conditions. Uniform node distribution is used for simplicity. Corresponding results are non-dimensionalised using classical maximum values. The beam properties used in the numerical results are taken from a recent paper [35].

The results are non-dimensionalised using classical results given in Table 5.1.

Table 5.1 Non-dimensional factor for different boundary conditions, C stands for cantilever beam, PL for point load, UDL for uniformly distributed loading, and SSB means simply supported beam.

Case	u_z	$\theta_y(x)$	$M(x)$	$V(x)$
C-PL	$\frac{3EI}{\lvert P\rvert L^3}$	$\frac{2EI}{\lvert P\rvert L^2}$	$\frac{1}{\lvert P\rvert L}$	$\frac{1}{\lvert P\rvert}$
C-UDL	$\frac{8EI}{\lvert q\rvert L^4}$	$\frac{6EI}{\lvert q\rvert L^3}$	$\frac{1}{\lvert q\rvert L^2}$	$\frac{1}{\lvert q\rvert L}$
SSB-UDL	$\frac{384EI}{5\lvert q\rvert L^4}$	$\frac{24EI}{\lvert q\rvert L^3}$	$\frac{8}{\lvert q\rvert L^2}$	$\frac{2}{\lvert q\rvert L}$

Source: Authors

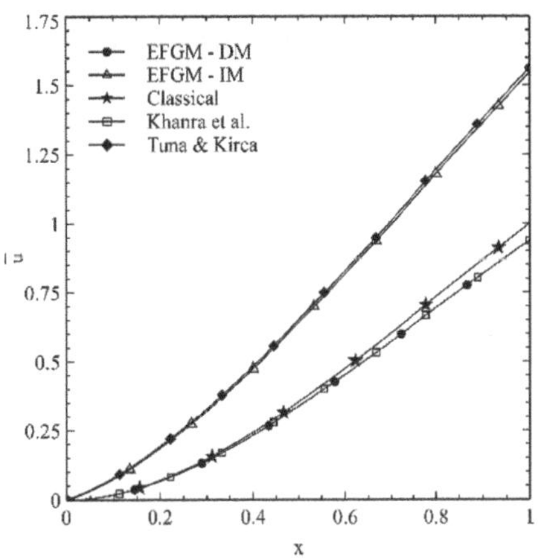

Figure 5.2 Comparison of normalised displacement for a cantilever nano beam with UDL for non-local parameter c = 1.125

Source: Authors

Cantilever nanobeam with UDL

Non-dimensional displacements of Euler Bernoulli nanobeam under UDL are compared with analytical solutions given in Khanra et al., and Tuna & Kirca (Figure 5.2). The displacements obtained using EFGM–DM are exactly the same as the analytical solutions by Khanra et al., and EFGM-IM is in exact agreement with the analytical solutions of Tuna & Kirca. When compared with classical elasticity solution integral model shows a strong non-local effect with softening behaviour and differential model shows a negligible stiffening effect.

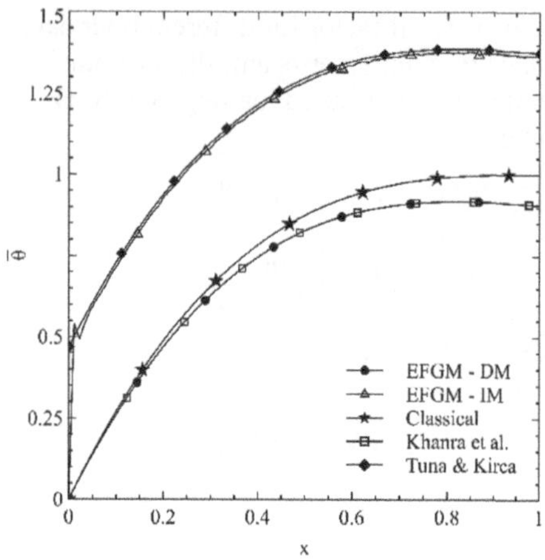

Figure 5.3 Comparison of the normalised slope of a cantilever beam with UDL for non-local parameter c = 1.125
Source: Authors

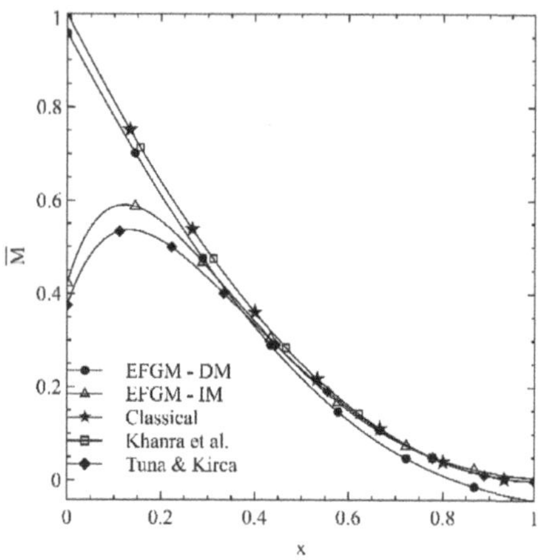

Figure 5.4 Normalised bending moments of cantilever nanobeam with UDL and for non-local parameter c = 1.125
Source: Authors

Figure 5.3 shows the slopes obtained using EFGM-DM matches with analytical solutions given by Khanra et al., and EFGM-IM agrees with Tuna & Kirca. Non-dimensionalised bending moment for EFGM-DM is like Khanra et al., and that of EFGM-IM is similar to that of bending moment obtained from Tuna & Kirca's results using convolution definition of stress resultants. Figures 5.4 and 5.5 depicts bending moment and shear force also follow parallel trend. In Figure 5.5, the shear force obtained using EFGM-DM near the boundary deviates from analytical solutions. This is due to the issues related to enforcing boundary conditions in the EFGM setting.

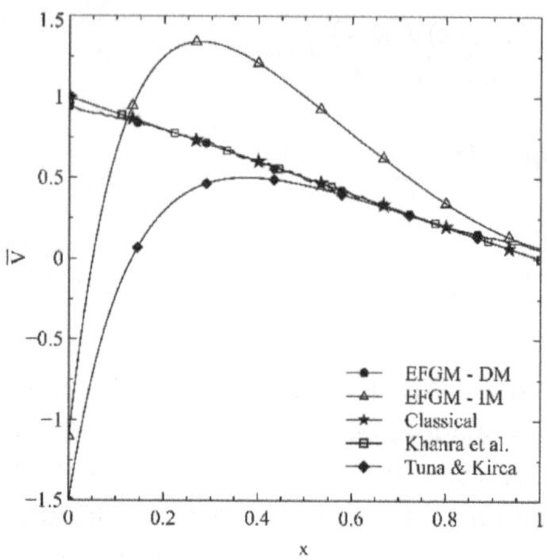

Figure 5.5 Non-dimensionalised shear force of cantilever nanobeam with UDL for non-local parameter c = 1.125
Source: Authors

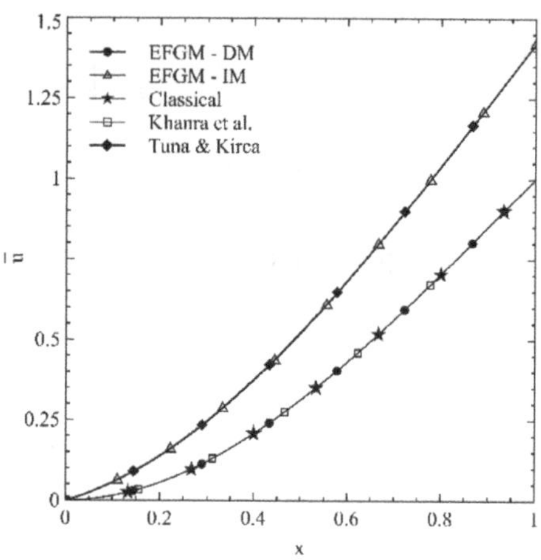

Figure 5.6 Normalised displacements for cantilever beam with a tip load and for non-local parameter c = 1.125
Source: Authors

Cantilever nanobeam with tip load

Non-dimensional displacements of Euler Bernoulli nanobeam under tip load are compared with analytical solutions given in Khanra et al. and Tuna & Kirca (Figure 5.6). The displacements obtained using EFGM–DM are the same as the analytical solutions by Khanra et al., and EFGM-IM is in exact agreement with the analytical solutions of Tuna & Kirca.

Figure 5.7 shows the slopes obtained using EFGM-DM, which matches the analytical solutions given by Khanra et al. and EFGM-IM, which agrees with Tuna & Kirca. The non-dimensionalised bending moment for EFGM-DM is like that of Khanra

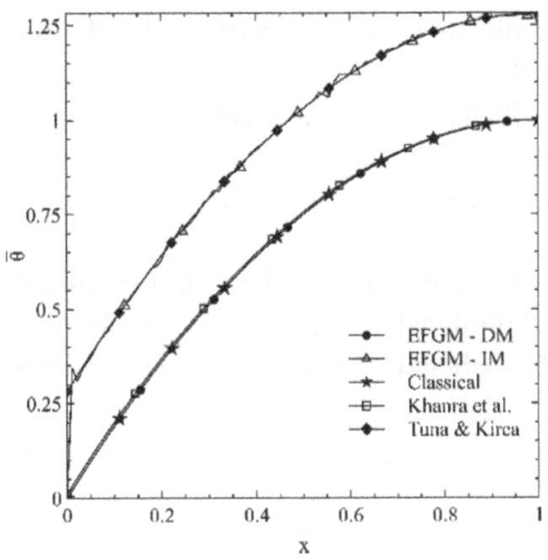

Figure 5.7 Non-dimensionalised slopes for cantilever beam with a tip load and for non-local parameter c = 1.125
Source: Authors

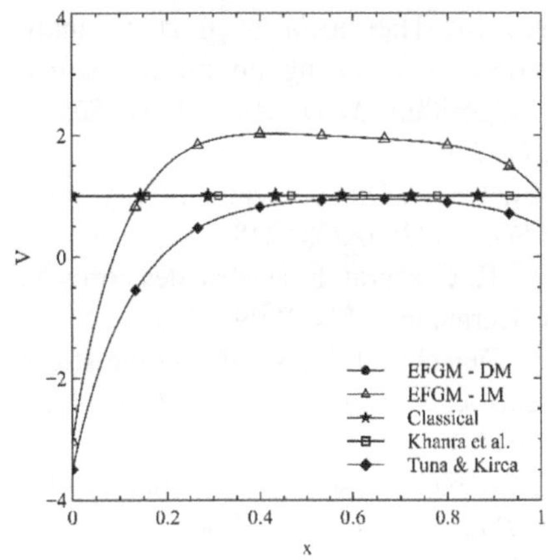

Figure 5.9 Non-dimensionalised shear force for cantilever beam with tip load and for non-local parameter c = 1.125
Source: Authors

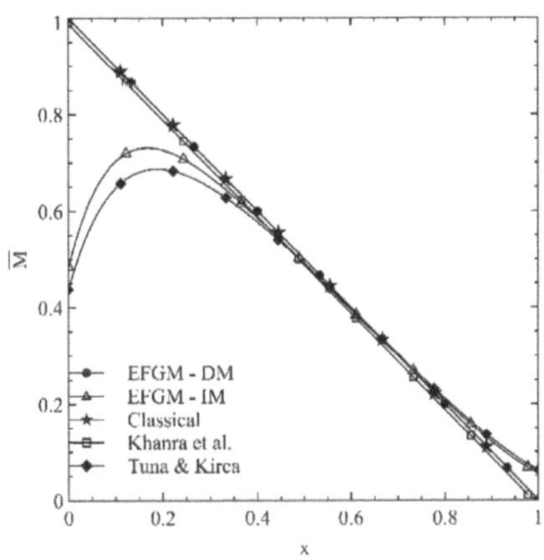

Figure 5.8 Non-dimensionalised bending moment for cantilever beam with tip load and for non-local parameter c = 1.125
Source: Authors

et al., and that of EFGM-IM is similar to that of the bending moment obtained from Tuna & Kirca's results using the convolution definition of stress resultants. Figures 5.8 and 5.9 depict the normalised bending moment and normalised shear force, which also follow the parallel trends. In Figure 5.9, the shear force obtained using EFGM-DM near the boundary deviates from analytical solutions. This is due to the issues related to enforcing boundary conditions in the EFGM setting.

Conclusion

Based on Eringen's strain-driven non-local constitutive model, the non-local EFG method is developed for both differential and integral forms. The method is much simpler than FEM formulation using integral model. Due to the non-local nature of the formulation, EFG method can be extended to other non-local models like the stress-driven non-local model, the local-nonlocal model, etc.

The results of numerical simulations for standard boundary conditions match with the solutions available in the literature for respective models. There are deviations of the normalised shear force and bending moment at the boundary while using EFG. This can be reduced by considering a greater number of nodes in local support. But complete satisfaction of essential equilibrium conditions cannot be obtained. Careful choice of weight function, polynomial order, integration scheme, and local support is crucial for EFG to get accurate and reliable solutions.

References

[1] Liu, G.-R., Gu, Y.-T., An introduction to meshfree methods and their programming, 1st ed. Dordrecht: Springer, 2005. 480 pp, 2005.
[2] Liu, G., Gu, Y. A local radial point interpolation method (LRPIM) for free vibration analyses of 2-D solids. *J. Sound Vibr.*, 2001;246(1):29–46.
[3] Aswathy, M., Arun, C. Stochastic analysis of thin beams and plates incorporating von-Kármán nonlinear strains using meshless method. *Thin-Walled Struct.*, 2024;196:111478.
[4] Lone, G. H., Jameel, A. Enriched element free Galerkin method for solving frictional contact between solid bodies. *Mech. Adv. Mater. Struct.*, 2023;30(20):4227–4245.

[5] Awasthi, Pant, M. Thermoelastic fracture analysis in orthotropic media using optimized element free Galerkin algorithm. *Mech. Adv. Mater. Struct.*, 2024;31(2):271–296.

[6] Eringen, C., Edelen, D. On nonlocal elasticity. *Internat. J. Engg. Sci.*, 1972;10:233–248.

[7] Cosserat, E. M. P., Cosserat, F. Theorie des corps deformables. A. Hermann et fils, 1909.

[8] Mindlin, R., Tiersten. Effects of couple-stresses in linear elasticity. *Arch. Rational Mech. Anal.*, 1962;11:415–448.

[9] Koiter, W. T. Couple-stresses in the theory of elasticity, I & II. *Philos. Trans. Royal Soc. London B*, 1969;67:17–44.

[10] Mindlin, R. On the equations of elastic materials with micro-structure. *Internat. J. Solids Struct.*, 1965;1:73–78.

[11] Eringen, C. On differential equations of nonlocal elasticity and solutions of screw dislocation and surface waves. *J. Appl. Phy.*, 1983;54:4703–4710.

[12] Peddieson, J., Buchanan, G. R., McNitt, R. P. Application of nonlocal continuum models to nanotechnology. *Internat. J. Engg. Sci.*, 2003;41:305–312.

[13] Wang, Q. Wave propagation in carbon nanotubes via nonlocal continuum mechanics. *J. Appl. Phy.*, 2005:98.

[14] Wang, Q., Liew, K. Application of nonlocal continuum mechanics to static analysis of micro-and nano-structures. *Phy. Lett. A*, 2007;363(3):236–242.

[15] Reddy, J. Nonlocal theories for bending, buckling and vibration of beams. *Internat. J. Engg. Sci.*, 2007;45(2):288–307.

[16] Lu, P., Lee, H., Lu, C., Zhang. Application of nonlocal beam models for carbon nanotubes. *Internat. J. Solids Struct.*, 2007;44(16):5289–5300.

[17] Reddy, J., Pang, S. Nonlocal continuum theories of beams for the analysis of carbon nanotubes. *J. Appl. Phy.*, 2008;103:23511.

[18] Xu, M. Free transverse vibrations of nano-to-micron scale beams. *Proc. Royal Soc. A Mathemat. Phy. Engg. Sci.*, 2006;462:2977–2995.

[19] Polizzotto. Nonlocal elasticity and related variational principles. *Internat. J. Solids Struct.*, 2001;38:7359–7380.

[20] Polizzotto. Unified thermodynamic framework for nonlocal/gradient continuum theories. *Eur. J. Mech.-A/Solids*, 2003;22(5):651–668.

[21] Polizzotto. Strain-gradient elastic-plastic material models and assessment of the higher order boundary conditions. *Eur. J. Mech.-A/Solids*, 2007;26(2):189–211.

[22] Numanoglu, H. M., Civalek, O. On the torsional vibration of nanorods surrounded by elastic matrix via nonlocal FEM. *Internat. J. Mech. Sci.*, 2019;161:105076.

[23] Lim, M. I., Zhang, G. A nonlocal finite element method for torsional statics and dynamics of circular nanostructures. *Internat. J. Mech. Sci.*, 2015;94:232–243.

[24] Numanoglu, H., Ersoy, H., Civalek, O., Ferreira, A. Derivation of nonlocal FEM formulation for thermo-elastic Timoshenko beams on elastic matrix. *Comp. Struct.*, 2021;273:114292.

[25] Tian, Y., Zhang, C., Sun, Y. The application of meshfree method in the numerical simulation of beams with the size effect. *Mathemat. Problems Engg.*, 2014;2014:590271.

[26] Ahmadi. Vibration analysis of 2D-functionally graded nanobeams using the nonlocal theory and meshless method. *Engg. Anal. Bound. Elem.*, 2021;124:142–154.

[27] Kiani. A meshless approach for free transverse vibration of embedded single-walled nanotubes with arbitrary boundary conditions accounting for nonlocal effect. *Internat. J. Mech. Sci.*, 2010;52:1343–1356.

[28] Rezaiee-Pajand, M., Mokhtari, M. A novel meshless particle method for nonlocal analysis of two-directional functionally graded nanobeams. *J. Brazil. Soc. Mech. Sci. Engg.*, 2019;41:1–23.

[29] Belytschko, T., Lu, Y. Y., Gu, L. Element-free Galerkin methods. *Internat. J. Num. Methods Engg.*, 1994;37(2):229–256.

[30] Dolbow, Belytschko, T. An introduction to programming the meshless element free Galerkin method. *Arch. Comput. Methods Engg.*, 1998;5:207–241.

[31] Liu, G.-R., Karamanlidis, D. Mesh free methods: moving beyond the finite element method. *Appl. Mech. Rev.*, 2003;56:B17–B18.

[32] Krongauz, Y., Belytschko, T. Enforcement of essential boundary conditions in meshless approximations using finite elements. *Comp. Methods Appl. Mech. Engg.*, 1996;131:133–145.

[33] Arun, B. R., Sivakumar, S. On material discontinuity modeling in element free Galerkin method. *J. Struct. Engg.*, 2009;36:182–194.

[34] Tuna, Kirca, M. Bending, buckling and free vibration analysis of Euler-Bernoulli nanobeams using Eringen's nonlocal integral model via finite element method. *Comp. Struct.*, 2017;179:269–284.

[35] Khanra, G. K., Krishna, I. R. P., Raveendranath, P. A re-look into the modeling aspects of Eringen's strain-driven nonlocal Euler-Bernoulli nanobeam bending problems. *Mech. Adv. Mater. Struct.*, 2023:1–14.

6 Re-analysis of Euler-Bernoulli microbeam bending problems modelled after Eringen's strain-driven non-local theory, strain gradient theory, and the stress-driven theory: A comparative study

Gaurab Kumar Khanra[a], *I. R. Praveen Krishna and Raveendranath P.*

Department of Aerospace Engineering, Indian Institute of Space Science and Technology, Thiruvananthapuram, Kerala, India

Abstract

Micro/nano dimensioned structures such as carbon nanotubes (CNTs) and graphene have received widespread attention among researchers worldwide. Due to the superior material properties exhibited by such structures of very low dimensions, these structures have been identified as potential building blocks for next-generation advanced sensors, actuators, and transducers. At such a small scale of physical dimensions, a phenomenon called "size effects" starts to prevail and exerts an influence upon the constitutive behaviour of the micro/nanostructure as a consequence of inter-atomic interactions. Classical continuum mechanics fail to incorporate these effects, so non-classical continuum mechanics, essentially "size-dependent," were developed. A plethora of non-classical continuum theories exist, out of which the present work lays its focus on three distinct theories, namely Eringen's strain-driven differential non-local model, strain gradient theory, and the recently proposed stress-driven non-local model. Analysis of Euler-Bernoulli microbeam bending behaviour forms the core objective of this study. Here the numerical results of fundamental and derived quantities emanating from the conventional FEM, Strain gradient FEM (SG-FEM), and analytical solutions based on the stress-driven non-local theory are compared against each other, following which some interesting results are discussed.

Keywords: Micro/nanostructures, size-effects, non-classical continuum mechanics, Eringen's non-local model, strain gradient theory, stress-driven model

Introduction

Micro- and nano-sized structures are the essential components for creating micro and nanoscale devices like micro electro mechanical systems (MEMS) and nano electro mechanical systems (NEMS). Such systems find application in developing sophisticated instrumentation components like sensors, actuators, and transducers. Hence, a set of methodical design principles is necessary to understand such low-dimensional structures' overall mechanics. At such a scale of minuscule dimensions, classical continuum mechanics becomes inadequate due to the emergence of "size effects", attributed to the predominance of interatomic and intermolecular forces. Consequently, the constitutive behaviour of the materials that make up these structures' deviates from the generalised Hooke's law. Within a micro/nanostructure, the dimensions of the bulk material are comparable to those of its internal structures. Therefore, the particles that make up these structures cannot be treated as point masses, as their rotation and deformation cannot be ignored. Therefore, a specialised class of updated "size-dependent" continuum theories must be developed. In the available literature, the works by the Cosserat brothers [1] serve as the earliest reference where such "size-dependent" continuum theories were discussed for the first time. Their theory is also known as the micropolar continuum theory. According to this theory, each material point can have additional degrees of freedom, including rotational motion. It introduces couple stresses (torques per unit area) and considers the asymmetric stress tensor, which is not present in classical continuum mechanics. The ideas developed in the works as mentioned earlier by the Cosserat brothers were later carried forward to develop theories such as gradient elasticity theory (GET), couple stress theory (CST), consistent couple stress theory (CCST), modified couple stress theory (MCST), peridynamics, Eringen's

[a]gaurabkhanra@gmail.com

DOI: 10.1201/9781003770442-6

non-local elasticity theory, strain gradient theory (SGT), stress-driven theory and many more [2–11]. Thereafter, numerous studies were conducted using these non-local continuum theories to create mathematical models designed to understand the static and dynamic characteristics of structures at the micro and nanoscale [12–21]. Studying the static bending behaviour of micro/nanostructures is a basic consideration in the design of the structural components of any sensor. It is, therefore, of interest in the present study. It was often observed that some theoretical models developed utilising the previously mentioned theories led to complicated mathematical expressions, which were difficult to evaluate. These complications resulted from considering several length scale parameters used to represent the "size effects". Therefore, the present work considers three different theories, viz., Eringen's strain-driven theory [6, 7, 22], strain gradient theory [23–26], and stress-driven theory [9–11] to study the micro/nano beam elastostatics. Out of these, Eringen's strain-driven model is of three types: the integral model, the differential model, and the (local/non-local) mixture model. The integral model leads to integro differential equations (IDEs), which brings forth ill-posedness [19, 27] to the governing IDEs and leads to solutions laden with inconsistencies in the satisfaction of essential and natural boundary conditions. As such, IDEs do not always guarantee the existence of unique solutions. On the contrary, the differential model manages to circumvent these difficulties associated with IDEs and lead to mathematical computations that require less effort than the integral non-local model-based formulations. Meanwhile, the two-phase non-local model combines the local phase of nearby atomic or molecular interactions with a non-local phase incorporating more distant, non-local effects within the material. Theoretically, the two-phase non-local models for micro/nanostructures often led to well-posed solutions to the associated governing IDEs. The present work, however, considers Eringen's differential non-local model. Regarding the strain gradient theory, unlike conventional theories, which consider only strain and stress at a point, strain gradient theory includes higher-order strain terms. These terms account for the effect of strain gradients, making it possible to model phenomena such as size-dependent stiffness and strength observed in micro- and nano-scale structures. This framework relies on intrinsic material parameters determined by experiments [23–26], making it a powerful tool for capturing "size effects". Besides, the recently proposed stress-driven theory has also received widespread acceptance amongst the research community. Unlike other non-local models, which often display boundary condition incompatibilities, the stress-driven model employs a convolution-based integral approach to define non-local stress with a bi-exponential kernel or error kernel function, capturing the physical response more accurately for low-dimensioned structures. This model has shown promising results in predicting the mechanical behaviour of nanostructures, showing size-dependent stiffness effects. With this, the present research effort proposes to investigate the bending characteristics of microbeams modelled after these three theories. The numerical value of the length scale parameter "l", (i.e., its magnitude without the unit of "l") which is used to denote the "size-effects" of microbeam as per strain gradient theory, is assumed to be equal to the value of the non-local parameter "" arising in the context of Eringen's strain-driven theory and the stress-driven theory. This assumption makes it possible to directly compare the results of the fundamental and derived quantities, which are the solutions to the governing differential equations for static microbeam bending problems arising out of the three different theories.

Solution methodologies

Analytical solutions based on Eringen's strain-driven non-local theory and the stress-driven non-local theory

Eringen's strain-driven non-local theory considers that within a micro/nano dimensioned domain, if one considers a point "x" within this domain, the non-local stress (σ_{ij}) at a given point depends on the strains that have developed at all other points within the material. Mathematically, this is represented as the volume integral (over v) written below:

$$\sigma_{ij} = \int \psi(|x-\bar{x}|, L_c)\sigma_{ij}^c(\bar{x})\, dv, \ \forall x \in v \tag{1}$$

In the above Equation (1), $\psi(|x-\bar{x}|, L_c)$ is the Kernel function which incorporates the non-local effects into the above constitutive equation. Here, $|x-\bar{x}|$ implies the Euclidean distance. L_c represents the characteristic length which is the product of λ (non-local parameter) and the physical length (L) of the microbeam (i.e., $L_c = \lambda L$). Also, according to generalised

Hooke's law, the local/classical stress tensor σ_{ij}^{c} is related to the corresponding strain-tensor $\epsilon_{kl}(\bar{x})$ as:

$$\sigma_{ij}^{c}(\bar{x}) = C_{ijkl}\epsilon_{kl}(\bar{x}) \tag{2}$$

where C_{ijkl} is the elastic modulus tensor. Employing Equation (1) for deriving the governing equations for the static/dynamic behaviour of micro/nano structures results in IDEs which leads to mathematical complexities but are of strongly non-local form [28, 29]. Therefore, with the aim of alleviating the computational efforts required to solve IDEs, Eringen proposed a differential counterpart to his originally proposed integral model. Consequently, after a series of transformations, Eringen successfully reduced the IDEs to their equivalent singular partial differential equations (PDEs) format valid for a specific type of physically admissible Kernel functions [22]. As a direct result of this transformation, Equation (1) now took the weakly non-local form [12, 13] as given by:

$$1 - L_c^2\nabla^2\left(\sigma_{ij}(x)\right) = C_{ijkl}\epsilon_{kl}(x) \tag{3}$$

For one-dimensional problems and especially for micro/nanostructures subjected to a uni-axial state of stress, the above Equation (3) reduces to:

$$\sigma_{xx} - L_c^2\frac{d^2\sigma_{xx}}{dx^2} = E.\epsilon_{xx} \tag{4}$$

where E is the elastic modulus of the microbeam material. The equilibrium equations for forces and moments are given by:

$$\frac{dV}{dx} + f(x) \quad = 0 \tag{5}$$

$$V - \frac{dM}{dx} \quad = 0 \tag{6}$$

where $M(x)$ and $V(x)$ are the bending moment and shear force resultants, while $f(x)$ is the externally applied load. For Euler-Bernoulli beams, the strain-displacement relationship is given as

$$\epsilon_{xx} = -z\frac{d^2w}{dx^2} \tag{7}$$

where "z" is the transverse distance of a point in the microbeam from the neutral axis and "w" is the transverse displacement of the microbeam. Also, from the definition of stress-resultant for bending-moment, one has $M = \int_A z\sigma_{xx}\, dA$. Hence, substituting Equation (7) into Equation (4) followed by multiplication of the resulting expression by "z" and subsequent integration along the microbeam's cross-section, yields the non-local moment-curvature relationship:

$$M - L_c^2\frac{d^2M}{dx^2} = -EI\frac{d^2w}{dx^2} \tag{8}$$

where I is 2nd moment of area of the microbeam. Furthermore, upon considering Equation (8) and the equilibrium Equations (5) and (6) the moments and shear forces can be obtained as:

$$M(x) \quad = -EI\frac{d^2w(x)}{dx^2} - L_c^2 f(x) \tag{9}$$

$$V(x) \quad = -EI\frac{d^3w(x)}{dx^3} - L_c^2\frac{df(x)}{dx} \tag{10}$$

Finally, the governing differential equation which dictates the bending behaviour of the microbeam can be obtained from Equations (5, 6, 10) as:

$$EI\frac{d^4w(x)}{dx^4} - f(x) + L_c^2\frac{d^2f(x)}{dx^2} = 0 \tag{11}$$

The above Equation (11) can be re-written in a more general format, taking into account a series of concentrated forces F_i at $x = A_i$, i∈{1,2,3,…, n} and externally applied concentrated moments, M_j at $x = B_j$, j∈{1,2,3,…, m} and in terms of the Dirac-delta function (δ) as [12, 13]:

$$EI\frac{d^4w(x)}{dx^4} = f(x) - L_c^2\frac{d^2f(x)}{dx^2} + \sum_{i=1}^{n} F_i\left(\delta(x - A_i) - L_c^2\delta^{(2)}(x - A_i)\right) - \sum_{j=1}^{m} M_j\left(\delta^{(1)}(x - B_j) - L_c^2\delta^{(3)}(x - B_j)\right) \tag{12}$$

Where, $\delta(x - a) = \begin{cases} \infty\,, & if\ x = a \\ 0, & otherwise \end{cases}$

and "a" is any arbitrary location within the microbeam, while the superscripts (1), (2) and (3) respectively indicate the 1st, 2nd and 3rd derivative of the Dirac-delta function. Subjecting Equation (12) to relevant boundary conditions according to the microbeam configuration, the analytical solutions for the microbeam bending problems can be obtained.

Interestingly, Eringen's originally proposed strain-driven non-local model had a fundamental flaw due to which the analytical solutions derived from the integral model [18, 19] and the differential model [12, 13, 17] displayed paradoxical behaviour and were inconsistent in capturing "size-effects" for certain micro/nano beam loading and boundary conditions. The root cause of such discrepancies was investigated rigorously by Barretta et al. Based on their investigations, it was found that certain additional constraint conditions (Constitutive Boundary Conditions (CBCs)) were required to be imposed to endow closure to the formulations and thereby generate unique solutions to the governing IDEs or PDEs emanating out of Eringen's non-local models [30, 31]. These extra boundary conditions were, however, in conflict with the equilibrium equations and thereby induced ill-posedness to the formulations related to Eringen's non-local models [27]. Therefore, an alternate well-posed stress-driven constitutive model was developed [9–11]. As per this novel constitutive model, the local stress tensor is viewed as an output of the convolution integral between the kernel function and the non-local stress tensor and is expressed as:

$$\sigma_{ij}^{c}(x) = \int \psi(|x-\bar{x}|, L_c)\sigma_{ij}(\bar{x})dv, \ \forall x \in v \quad (13)$$

From the perspective of Euler-Bernoulli beam theory, the above Equation (13) in terms of its equivalent moment-curvature form equates to [14]:

$$\frac{d^2w}{dx^2} = \int_0^L \psi\,(|x-\bar{x}|, L_c)\frac{1}{EI}M(\bar{x})\ d\bar{x} \quad (14)$$

The above Equation (14) presents an advantage over the equivalent strain-driven counterpart as the equilibrated bending moments become the input to the above convolution integral with the Kernel function and produces the elastic curvature $\left(\kappa = \frac{d^2w}{dx^2}\right)$ as the output.

A linear 2nd order differential equation results, bypassing the conflicting requirement of simultaneous satisfaction of the CBCs and the equilibrium equations. Subsequently, the corresponding analytical solutions can be derived upon imposing the boundary conditions for various microbeam configurations to the above Equation (14).

FEM and SG-FEM solutions based on Eringen's strain-driven non-local theory and the strain-gradient elasticity theory

Finite element formulations were contributed by Phadikar and Pradhan [16]. Applying Galerkin's method of weighted residuals to Equation (11), the weighted residual statement takes the form:

$$\int_{x_1^e}^{x_2^e}\left(EI\frac{d^4w(x)}{dx^4} - f(x) + L_c^2\frac{d^2f(x)}{dx^2}\right) N_k\ dx = 0,\ k = 1-4 \quad (15)$$

in which the transverse displacement, w of the microbeam is approximated as:

$$w(x) = N_1(x)w_1 + N_2(x)\theta_1 + N_3(x)w_2 + N_4(x)\theta_2 \quad (16)$$

Here, (w_1, w_2) represents the nodal Degree of Freedoms (DOFs) associated with the transverse displacement, whereas (θ_1, θ_2) denotes the nodal DOFs related to the slopes. Also, (x_1^e and x_2^e are the physical "x" coordinates of node 1 and node 2 of the eth element.

For microbeams modelled using FEM, the shape functions based upon the cubic-hermite polynomials for beam elements in terms of the Cartesian coordinates, "x" are given as in the Appendix section. These shape functions were used by Phadikar and Pradhan [16] to interpolate the deflection, (w) and the slopes, (θ). Shape functions $N_1(x)$ and $N_3(x)$ correspond to the deflection DOF, where as $N_2(x)$ and $N_4(x)$ relates to the slope DOF. Two DOFs per node are associated with one finite element. These shape functions are as shown in Figure 6.1.

Henceforth, continuing with the formulation and subsequent evaluation of Equation (15) leads to the final discretised set of equations as follows:

$$[\mathbf{K}^{FEM}]\{\mathbf{w}\} = \{\mathbf{Q_{dist}}\} + \{\mathbf{Q_{dist}^{NL}}\} + \{\mathbf{P}\} \quad (17)$$

Here, $\{\mathbf{Q_{dist}}\} = \begin{Bmatrix} \frac{fL_e}{2} \\ \frac{fL_e^2}{12} \\ \frac{fL_e}{2} \\ -\frac{fL_e^2}{12} \end{Bmatrix}$, $\{\mathbf{Q_{dist}^{NL}}\} = L_c^2\begin{Bmatrix} 0 \\ f \\ 0 \\ -f \end{Bmatrix}$ and $\{\mathbf{P}\} = \begin{Bmatrix} -V_1 \\ M_1 \\ V_2 \\ M_2 \end{Bmatrix}$

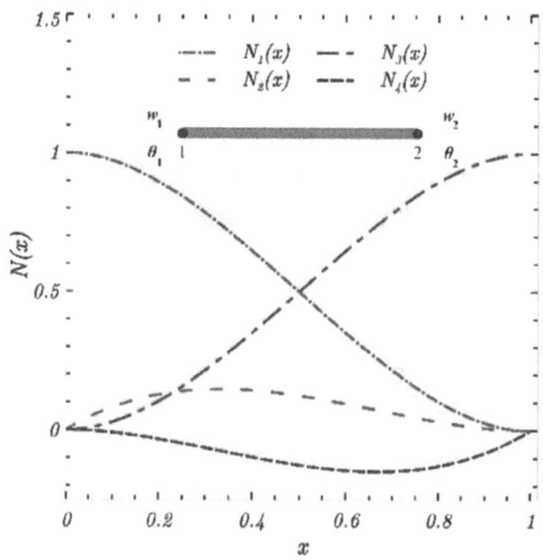

Figure 6.1 FEM shape functions
Source: Authors

The above Equation (17) is then solved for the nodal values of deflections and slopes upon imposition of relevant boundary conditions depending upon the microbeam configuration.

When the SG-FEM formulations are considered, Kahrobaiyan et al. [32] deduced the governing 6th order differential equation utilising Hamilton's principle. Contributing terms related to the kinetic energy (T), potential energy (U), and the work due to external forces (W) were first computed. From the perspective of strain-gradient theory, the potential energy (U) depends on constituent strain measures whose components are built upon the strain tensor (ϵ), the deviatoric stretch gradient tensor ($\eta^{(1)}$) and also the rotation gradient tensor χ^s which were represented by the symbols ϵ_{ij}, $\eta^{(1)}_{ijk}$ and χ^s_{ij} in [32]. These quantities are functions of displacement vector, u, dialation gradient vector γ and the infinitesimal rotation vector ϕ. Altogether, these terms generate complicated expressions relating the stress measures to their corresponding strain tensors and their gradients, which are not discussed in this article.

One may, however, refer to [32] for further intricate details. Therefore, following the procedure outlined in [32], the potential energy stored in the strain-gradient Euler-Bernoulli beam U is given as:

$$U = \frac{1}{2}\int_0^L \left\{ D_1 \left(\frac{\partial^2 w}{\partial x^2}\right)^2 + D_2 \left(\frac{\partial^3 w}{\partial x^3}\right)^2 \right\} dx \tag{18}$$

Where, $D_1 = EI + \mu A\left(2l_0^2 + \frac{8}{15}l_1^2 + l_2^2\right)$ and

$$D_2 = \mu I\left(2l_0^2 + \frac{4}{5}l_1^2\right)$$

In the above two equations l_0, l_1 and l_2 represent the supplementary length scale parameters of the material related to l as $l_0 = l_1 = l_2 = 1$, [32] which appear in the constitutive equations of higher order stresses. Also, μ represents the Lame's constant defined as $\mu = \frac{E}{2(1+v)}$, with v being the Poisson's ratio of the nanobeam material and the area of cross-section of the microbeam is given by "A".

The kinetic energy was determined as:

$$T = \frac{1}{2}\int_A \int_0^L \rho \left(\frac{\partial w}{\partial t}\right)^2 dxdA = \frac{1}{2}\int_0^L \rho A \left(\frac{\partial w}{\partial t}\right)^2 dx \tag{19}$$

where ρ stands for the microbeam's density. Also, due to the external forces acting upon the beam, the expression for the variation of the work W done is expressed as:

$$\delta W = \int_0^L f(x,t)\delta w dx + \hat{V}\,\delta w|_{x=0,L} + \hat{M}\,\delta\left(\frac{\partial w}{\partial x}\right)|_{x=0,L} + \hat{Q}\,\delta\left(\frac{\partial^2 w}{\partial x^2}\right)|_{x=0,L} \tag{20}$$

With $f(x,t)$, $\hat{V}$, $\hat{M}$ and $\hat{Q}$ respectively representing the distributed lateral load, externally applied lateral shear load, externally applied bending moment and externally applied higher order load at the end-sections of the microbeam. Furthermore, applying Hamilton's principle given as $\int_{t_1}^{t_2} (\delta T - \delta U + \delta W)\, dt = 0$, the associated governing 6th order differential equation along with the boundary conditions for the strain-gradient Euler-Bernoulli microbeam is therefore determined as:

$$D_1 \frac{\partial^4 w}{\partial x^4} - D_2 \frac{\partial^6 w}{\partial x^6} + \rho A \frac{\partial^2 w}{\partial t^2} = f(x,t) \tag{21}$$

$$(V(x,t) + \hat{V})|_{x=0,L} = 0 \text{ or}$$
$$\text{or } (w(x,t) - \hat{w})|_{x=0,L} = 0 \tag{22}$$

$$(M(x,t) - \hat{M})|_{x=0,L} = 0 \text{ or}$$
$$(\theta(x,t) - \hat{\theta})|_{x=0,L} = 0 \tag{23}$$

$$(Q(x,t) - \hat{Q})|_{x=0,L} = 0 \text{ or}$$
$$\text{or } (\kappa(x,t) - \hat{\kappa})|_{x=0,L} = 0 \tag{24}$$

It is worth noting that this formulation results in an additional DOF, in the form of the curvature, $\left(\kappa = \frac{\partial^2 w}{\partial x^2}\right)$. Additionally, Equations (22) and (23) refer to the classical boundary conditions, while Equation (24) is a non-classical boundary condition. Kahrobaiyan et al. [20] considered the case of static microbeam bending. Therefore, the inertial and temporal terms in Equation (21) were neglected, and it was expressed in terms of the natural coordinate "s" as:

$$\frac{d^6 w}{ds^6} - q^2 \frac{d^4 w}{ds^4} = 0 \tag{25}$$

Where, $q = L_e\sqrt{\frac{D_1}{D_2}}$. with L_e being the effective length of one finite element, i.e., $L_e = x_2^e - x_1^e$, which also serves as the Jacobian enabling the transformation between the physical Cartesian coordinate "x" and the natural coordinate "s." Therefore, "x" and "s" are related by the relation, $dx = L_e ds$. For the above linear homogeneous differential Equation (25), the general solution of this equation can be written as:

$$w = c_1 + c_2 s + c_3 s^2 + c_4 s^3 + c_5 \cosh qs + c_6 \sinh qs \tag{26}$$

in which the unknown constants $c_1, c_2, \ldots, c_6$ is determined by imposing the boundary conditions:

$$\text{At } s = 0:\ w = w_1,\ \frac{dw}{ds} = L_e\theta_1,\ \frac{d^2w}{ds^2} = L_e^2\kappa_1 \tag{27}$$

$$\text{At } s = 1:\ w = w_2,\ \frac{dw}{ds} = L_e\theta_2,\ \frac{d^2w}{ds^2} = L_e^2\kappa_2 \tag{28}$$

After that, a systematic series of matrix manipulations were carried out, following which the shape functions in terms of the natural coordinates were obtained. These shape functions approximate the nodal DOFs (w, θ and κ) at each node within one SG-FEM finite element and are mentioned in the Appendix section. With this, the displacement field can now be linked to the nodal displacement vector δ as:

$$\boldsymbol{w(s,t) = N(s)\delta(t)} \tag{29}$$

where, $\mathbf{N(s)} = [N_1(s)\ N_2(s)\ N_3(s)\ N_4(s)\ N_5(s)\ N_6(s)]$ and $\delta = [w_1\ \theta_1\ \kappa_1\ w_2\ \theta_2\ \kappa_W]^T$. The shape functions $N_1(s), N_2(s), \ldots, N_6(s)$ are mentioned in the Appendix section.

The resulting size-dependent strain-gradient beam element differs from FEM-based shape functions. As previously indicated, SG-FEM beam elements have an additional DOF for curvature, (κ). Thus, it produces a 3 DOF per node finite element instead of 2 DOF per node as in the case of standard FEM-based beam finite elements.

Shape functions corresponding to SG-FEM in terms of the natural coordinates, "s", is given in the Appendix section. Unlike the conventional cubic-hermite polynomial based shape functions used in FEM, the key feature that makes the SG-FEM shape functions distinct from FEM based shape functions is that these are hyperbolic transcendental functions. Additionally, these functions are also dependent upon the ratio, $q = L_e\sqrt{\frac{D_1}{D_2}}$. The SG-FEM based shape functions are as shown in Figures 6.2 and 6.3.

Therefore, moving ahead with the formulation, Equation (21) was then reconsidered in terms of the natural coordinate system, following which Equation (29) was substituted into it. In the next step, Galerkin's weighted residual procedure was implemented, followed by the method of integration

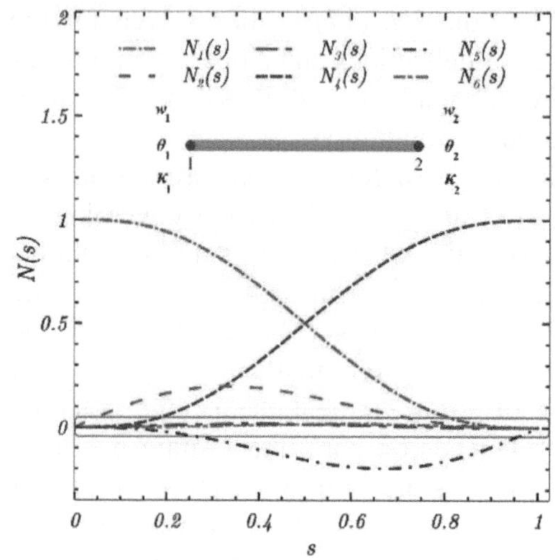

Figure 6.2 SG-FEM shape functions
Source: Authord

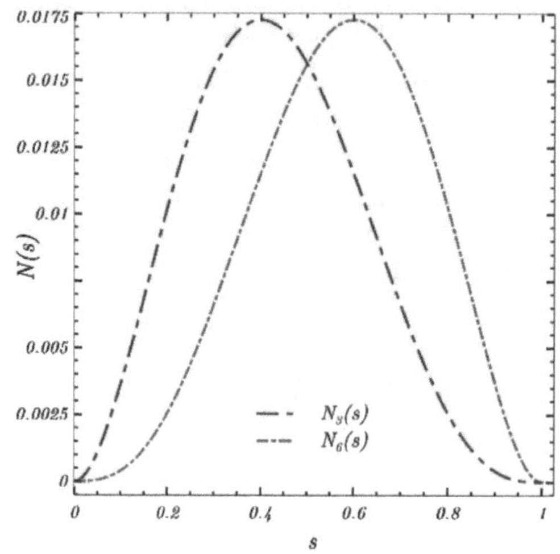

Figure 6.3 SG-FEM shape functions $N_3(s)$ and $N_6(s)$ corresponding to the curvature DOF
Source: Authors

by parts to derive the corresponding weak form of Equation (21).

Finally, the discretised set of equations took the form:

$$\mathbf{M}\,\ddot{\boldsymbol{\delta}} + \mathbf{K}\,\boldsymbol{\delta} = \mathbf{F_n} + \mathbf{F_e} \tag{30}$$

were,

$$\mathbf{M} = \rho A L_e \int_0^1 \mathbf{N}^T\,\mathbf{N}\,ds \tag{31}$$

$$\mathbf{K} = \mathbf{K}^1 + \mathbf{K}^2 \tag{32}$$

Table 6.1 List of geometric and material properties used as adopted from Kahrobaiyan et al. [32].

Parameter	Description	Value
F	Applied point load	10 μN
l	Length scale parameter as per SGT	0.25 μm
H	Height of the microbeam	21 μm
B	Width of the microbeam	2h μm
L	Length of the microbeam	20h μm
I	Area moment of inertia of the microbeam	$\frac{1}{12}$ bh³ μm⁴
A	Area of cross-section of the microbeam	bh μm²
E	Modulus of Elasticity of the microbeam	1.44 GPa
ν	Poisson's ratio	0.38
μ	Second Lame's constant	$\frac{E}{2(1+\nu)}$

Source: Authors

Table 6.2 List of case studies conducted.

Beam type	Description	Boundary conditions
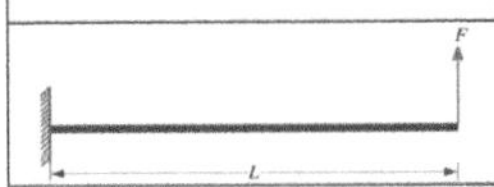	C-F PL @ $x = L$	$w(0) = 0,$ $\theta(0) = 0$
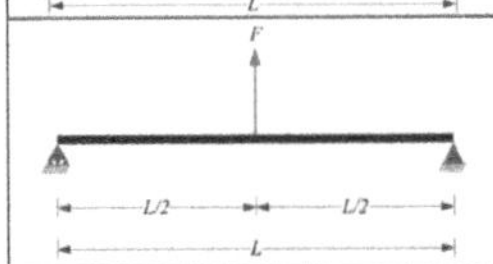	S-S PL @ $x = \frac{L}{2}$	$w(0) = 0,$ $w(L) = 0,$ $M(0) = 0,$ $M(L) = 0$
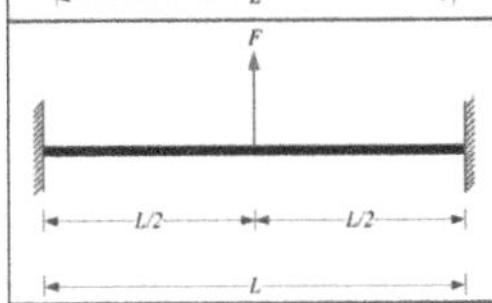	C-C PL @ $x = \frac{L}{2}$	$w(0) = 0,$ $w(L) = 0,$ $\theta(0) = 0,$ $\theta(L) = 0$

Source: Authors

In which,

$$\mathbf{K}^1 = \frac{D_1}{L_e^3}\int_0^1 \left(\frac{d^2\mathbf{N}}{ds^2}\right)^T \left(\frac{d^2\mathbf{N}}{ds^2}\right) ds \tag{33}$$

$$\mathbf{K}^2 = \frac{D_2}{L_e^5}\int_0^1 \left(\frac{d^3\mathbf{N}}{ds^3}\right)^T \left(\frac{d^3\mathbf{N}}{ds^3}\right) ds \tag{34}$$

$$\mathbf{F_n} = [\bar{V}(0,t)\ \ -\bar{M}(0,t)\ \ -\bar{Q}(0,t)\ \ -\bar{V}(1,t)\ \bar{M}(1,t)\ \bar{Q}(1,t)]^T \tag{35}$$

$$\mathbf{F_e} = L_e \int_0^1 \mathbf{N}^T f(s,t)\, ds \tag{36}$$

Since the present work concerns static microbeam bending problems, the mass matrix (**M**) is not considered. Hence, only the stiffness matrix (**K**), the natural boundary condition vector ($\mathbf{F_n}$), and the nodal force vector due to distributed lateral load ($\mathbf{F_e}$) are relevant.

Case studies

Microbeams in Cantilevered (C-F), simply-supported (S-S), and clamped-clamped (C-C) configurations when subjected to point load are considered. Deflections, slopes, bending moments, shear forces, and elastic curvatures evaluated as per all the discussed methods [14, 16, 21, 32] are compared. Input data regarding the microbeam's geometric and material properties are as tabulated in Table 6.1 below.

The case studies conducted along with corresponding boundary conditions which are implemented are shown in Table 6.2.

Results for a cantilevered microbeam subjected to tip load

Figure 6.4a shows the deflections estimated by the listed methods. Here, the calculated microbeam deflections, according to Wang et al. and FEM, entirely agree with the corresponding results obtained from classical theory. The evaluated deflections as per the analytical solutions of the stress-driven model and SG-FEM yield a stiffened response. Figure 6.4b shows the slopes as predicted by the different methods. As expected, the slopes which correspond to the FEM and Wang et al.'s analytical solutions coincide with that of the equivalent classical microbeam. Slopes predicted by the stress-driven analytical solutions and SG-FEM differ from those of the classical solutions. The impact of the stiffening effect is evident in the slope calculations derived from both the SG-FEM and stress-driven models. Portrayed in Figure 6.4c are the bending moments. In this figure, except for the bending moments evaluated as per Wang et al.'s analytical solution, the remaining methods satisfy the moment equilibrium equations. This is due to the fundamental definition of the stress resultants used to calculate the bending moments. Since Wang et al.'s solutions arise from Eringen's integral non-local model, the stress resultants are evaluated according to the convolution integral-based definition given by, $M(x) = \int_0^L \frac{1}{2L_c} e^{-\frac{|x-t|}{L_c}} * M^E(t)\, dt$, with $M^E(t)$ representing the classical bending moment

Cantilevered microbeam subjected to tip load

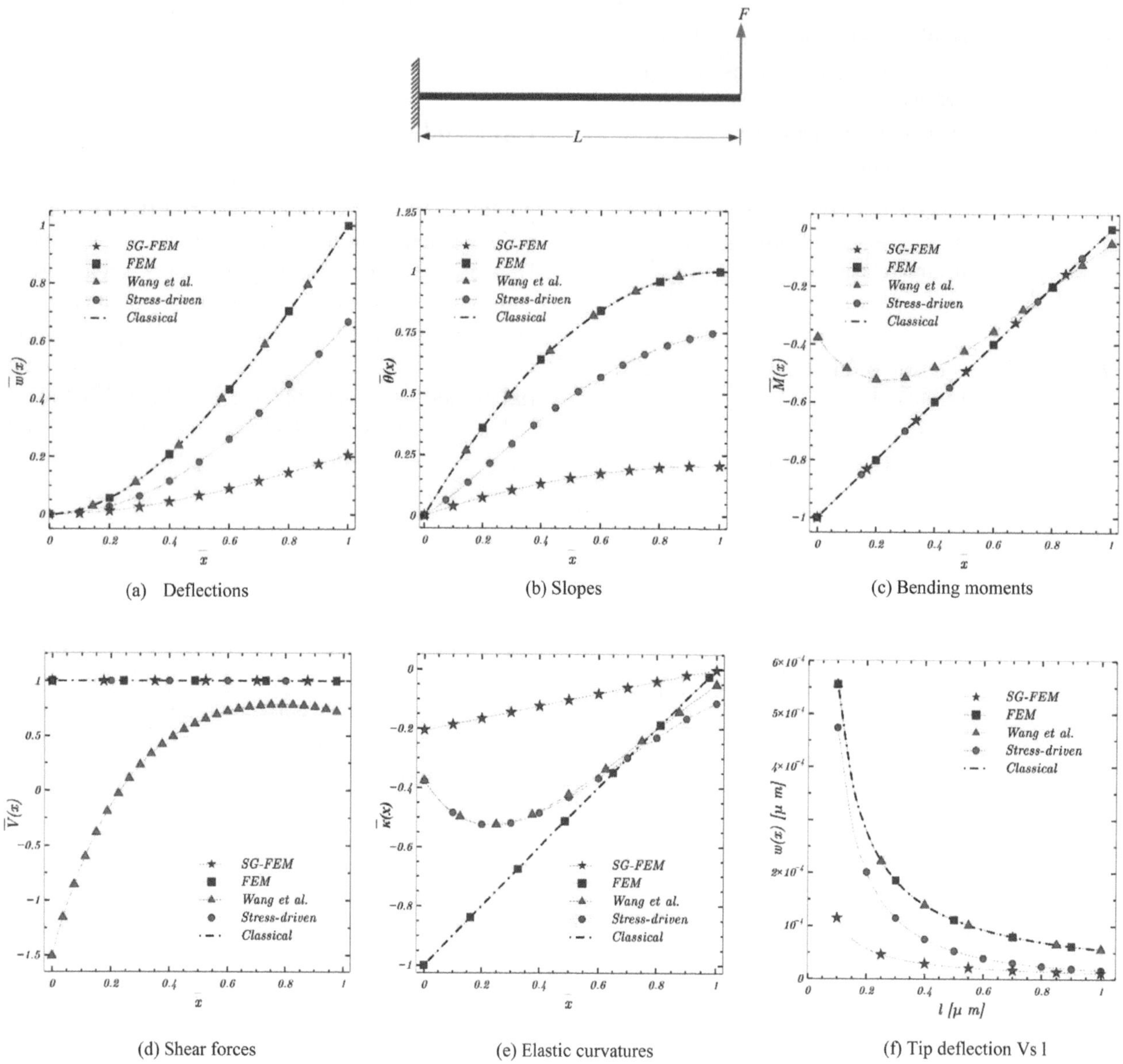

Figure 6.4 Comparison of non-dimensionalised quantities for a cantilevered microbeam subjected to concentrated load at its tip
Source: Authors

definition. This causes the bending moment to deviate from the moment equilibrium equations. FEM and SG-FEM solutions for the bending moments are, however, not based on this definition. These are calculated by multiplying the element-level stiffness matrices with the element-level nodal DOF vector. Shear forces are represented in Figure 6.4d. Here too, for the shear forces evaluated according to Wang et al., the convolution integral-based definition of the shear forces causes it to deviate from the force equilibrium conditions. Meanwhile, the remaining methods satisfy the same. The elastic curvatures are shown in Figure 6.4e. FEM solutions for elastic curvatures are seen to match with that of the classical theory. However, since Wang et al.'s and Barretta et al.'s solutions for elastic curvatures are computed based on the convolution integral definitions, a disparity between the elastic curvatures as per these two methods and the classical theory is visible. Elastic curvatures according to SG-FEM are evaluated directly from the system of algebraic equations Equation (30).

The obtained elastic curvature from SG-FEM computations suggests that it is influenced by the length scale parameter "l." Tip deflections are plotted against the length scale parameter "l" in Figure 6.4f. As generally discussed in the available literature, especially as per Wang et al.'s analytical solution and FEM solutions, the deflections of microbeams in the cantilevered configurations, when subjected to tip load, are independent of any size-effects. This situation is called the "cantilever beam paradox" problem in non-local elasticity. However, in this investigation, the microbeam's physical length "L" depends on its height "h," which in turn is dependent on the length scale parameter "l" as assumed in [32]. This assumption facilitates a direct comparison between the various methods and helps understand the softening or stiffening behaviour that "l" exerts upon the microbeam according to the different techniques. Consequently, the tip deflections, according to Wang et al., FEM, and classical solutions now become functions of the length scale parameter "l" via the geometric parameter "L" of the microbeam and display a consistent stiffened response with increasing values of "l." One may also note that the stress-driven model-based tip deflection and SG-FEM display a stiffening effect as "l" increases.

Results for a simply-supported microbeam subjected to concentrated load at its mid-span

A microbeam in a simply supported configuration is now being investigated. The deflection curves for this case are plotted in Figure 6.5a. Unlike the previously discussed case of the cantilevered microbeam, in the present case, the deflections, as per Wang et al.'s solutions, now show a significant dependency on the length scale parameter, indicating a strong influence from the "size effects". Meanwhile, FEM solutions match with classical theory-based solutions. Stress-driven analytical and SG-FEM solutions yield significantly higher stiffened microbeam bending responses. With regards to Wang et al.'s solution, one may note that at the point where the concentrated load is applied, a sharp change in the value of the deflection is registered. This is due to the presence of the Heaviside function in the analytical expression of the solution of the microbeam deflections. Figure 6.5b shows the slopes according to the various methods. A sudden discontinuity in the slopes, as per Wang et al.'s solution, is expected here as one evaluates the slope by taking the first derivative of the deflections. In a mathematical sense, the derivative of the Heaviside function now becomes a Dirac-delta function at the location where the concentrated load is applied on the microbeam, causing this sudden jump discontinuity. The slopes evaluated based on the remaining methods do not exhibit such behaviour and maintain slope continuity at the point of application of the concentrated load, "F". Depicted in Figure 6.5c is the bending moments. As mentioned previously, and due to the convolution integral based definition utilised to estimate the bending moments, Wang et al.'s solutions fail to satisfy the moment equilibrium equations as opposed to the remaining methods which satisfy the same. Shown in Figure 6.5d are the shear forces. Exempting for Wang et al.'s solution, the remaining methods satisfy the force equilibrium equations. Elastic curvatures are depicted in Figure 6.5e. The elastic-curvatures as per FEM coincide with the classical elasticity solutions. However, the solutions for elastic-curvatures according to SG-FEM, stress-driven and Wang et al.'s analytical solution for elastic-curvatures differ from the classical elasticity solutions. In Figure 6.5f the influence which "l" exerts on the mid-span deflections as per the different methods are shown. One may infer that Wang et al.'s analytical solution for mid-span deflections displays a fluctuating behaviour as it can be observed that when l [0.1–0.3 μm], initially the mid-span deflections appear to stiffen the microbeam. But as l ≥0.2 μm, a gradual increase in the microbeam's flexibility is observed. The remaining methods however do display a consistent stiffening effect with increase in the value of "l". For l ≥0.7 μm, the mid-span deflections as per SG-FEM and the stress-driven model appear to agree well with each other. Meanwhile, FEM solutions match with the classical elasticity theory-based solutions.

Results for a clamped-clamped microbeam subjected to concentrated load at its mid-span

A statically-indeterminate problem, in the form of a doubly-clamped microbeam subjected to point load at its mid-span is now discussed. Similar to the bending behaviour of the simply-supported

Simply-supported microbeam subjected to concentrated load at its mid-span

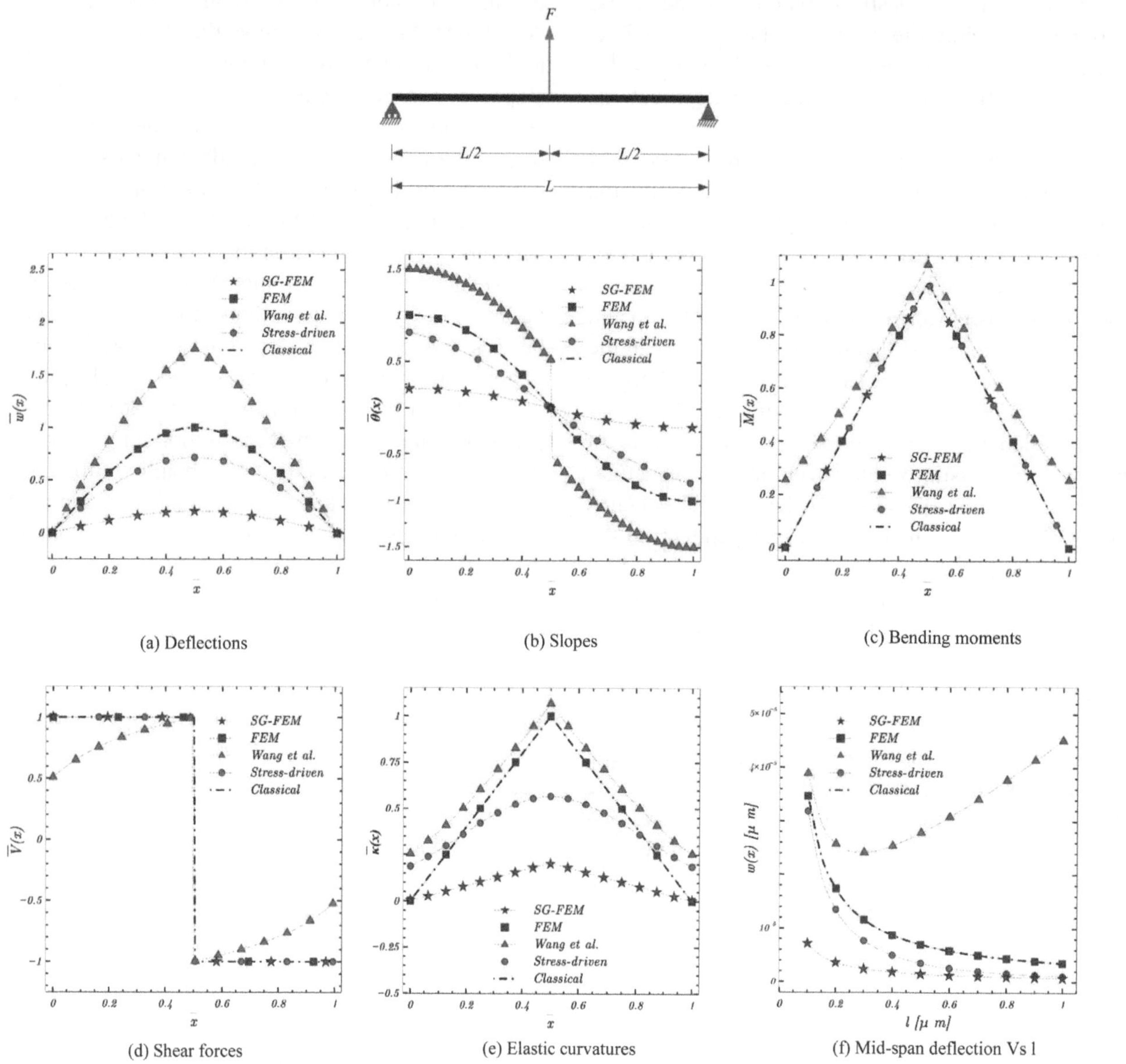

Figure 6.5 Comparison of non-dimensionalised quantities for a simply-supported microbeam subjected to concentrated load at its mid-span
Source: Authors

microbeam discussed in the previous case study, the deflections as shown in Figure 6.6a suggests that Wang et al.'s solutions produce deflections which depends substantially on the length scale parameter "l". FEM solutions however coincide with the results corresponding to classical theory. While the beam deflection curves arising out of the SG-FEM and stress-driven calculations comply with each other and display a significantly stiffened response. When the slopes are evaluated as depicted in Figure 6.6b, it can be seen that a sudden jump discontinuity occurs in the slopes computed according to Wang et al.'s solutions at $x = \frac{L}{2}$. The remaining methods do not display any such discontinuities. Bending moments are shown in Figure 6.6c. Here, the bending moments as per Wang et al.'s solutions do not satisfy the moment equilibrium equations due its convolution integral based definition.

Moment equilibrium is however satisfied by SG-FEM and FEM based solutions. Stress-driven

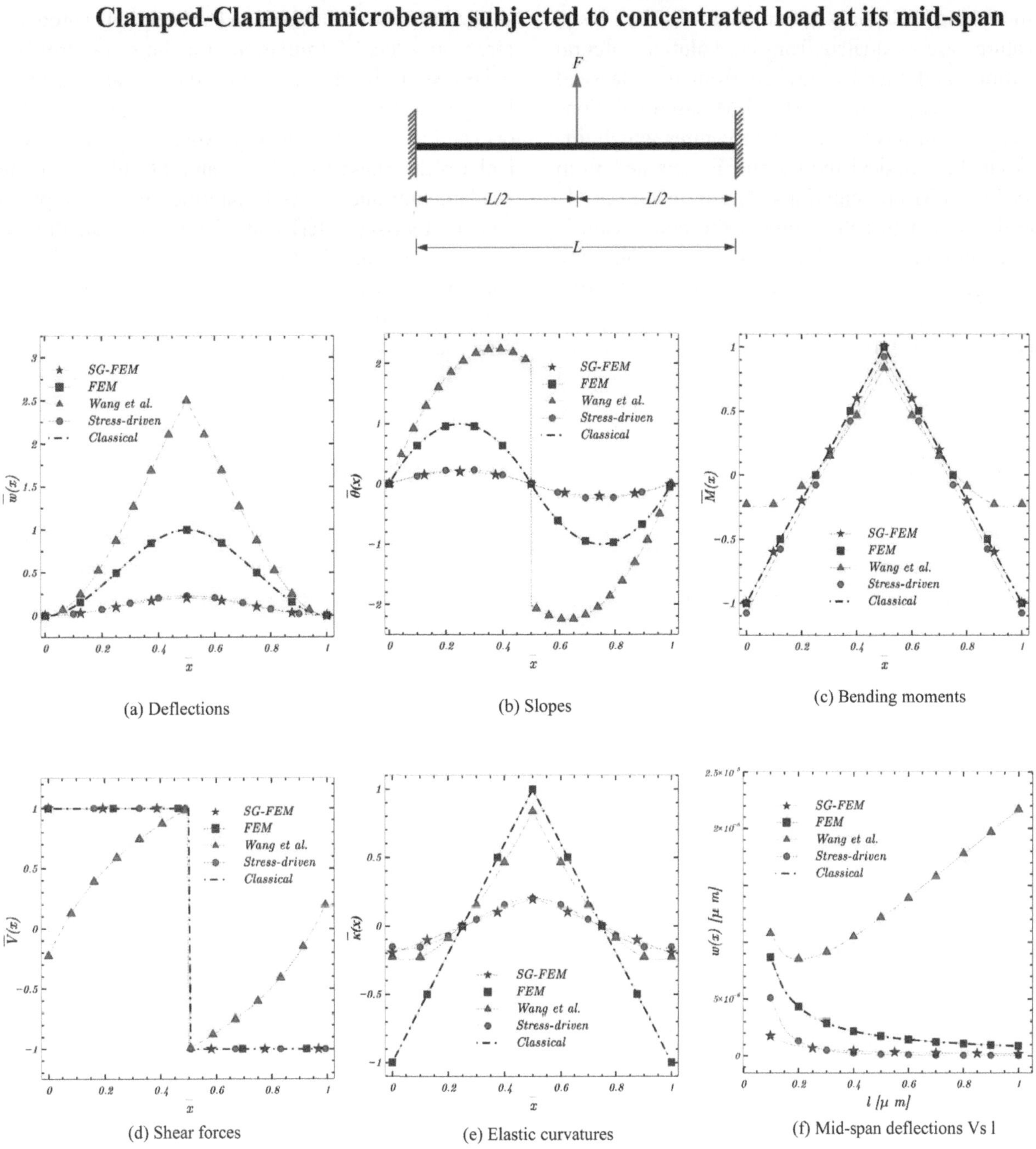

Figure 6.6 Comparison of non-dimensionalised quantities for a clamped-clamped microbeam subjected to concentrated load at its mid-span
Source: Authors

solutions for bending moments is slightly offset from the moment equilibrium equations as it now depends on the length scale parameter owing to the fact that the bending field is defined by considering the equilibrium equations, compatibility equations and the constitutive relations which render the bending moment to be a dependant on the length scale parameter "l". Shear forces are shown in Figure 6.6d. Except for Wang et al.'s solutions, the remaining methods result in shear forces estimates which comply with the force equilibrium equations. Solutions for the elastic curvatures are indicated in Figure 6.6e. The FEM solution for elastic curvature matches with the classical solutions. Wang et al.'s solutions and the stress-driven solutions for elastic curvatures are evaluated from convolution integral definitions and thereby deviate from the classical elasticity-based solutions. SG-FEM based solutions also yield solutions for elastic curvatures which differ from the classical theory and is obtained from direct SG-FEM computations. The influence of "l" upon the mid-span deflections of the microbeam is depicted in Figure 6.6f. Similar to the previous case of the simply-supported microbeam already discussed in the previous section, Wang et al.'s solutions for this case too displays a stiffening trend for $l \in [0.1–0.2]$ µm and thereafter displays a gradual softening trend which again brings to light, the inconsistent nature of Wang et al.'s solution to capture the size effects.

Conclusion

In the present work, the static responses of microbeams in three different configurations and when subjected to concentrated load are studied. Three different non-local formulations for the same problem have been utilised. These formulations are based on Eringen's strain-driven theory, strain gradient theory, and stress-driven theory. Analytical and numerical results (based on FEM and SG-FEM) of the fundamental and derived quantities were compared and summarised. Merits and demerits of each of the methods are highlighted. Except for the analytical solutions of Wang et al. [12], the remaining methods displayed a consistent stiffening effect in the microbeam. Hence, innovative solution techniques must be developed to resolve the issues associated with each of the methods.

Acknowledgement

The authors would like to thank the Department of Aerospace Engineering at IIST for providing the computational facilities and support needed to conduct this research.

Appendix

Shape functions for standard FEM

$$N_1(x) = \frac{1}{{L_e}^3}(2x^3 - 3x^2L_e + {L_e}^3)$$

$$N_2(x) = \frac{1}{{L_e}^3}(x^3L_e - 2x^2{L_e}^2 + x{L_e}^3)$$

$$N_3(x) = \frac{1}{{L_e}^3}(-2x^3 + 3x^2L_e)$$

$$N_4(x) = \frac{1}{{L_e}^3}(x^3L_e - x^2{L_e}^2)$$

Shape functions for SG-FEM

$$N_1(s) = 1 + \frac{1}{d}\{q(q\sinh q - 2\cosh q + 2)[6(\cosh qs - 1) + q^2(2s^3 - 3s^2)] + 6q(q\cosh q - 2\sinh q + 1)(qs - \sinh qs)\}$$

$$N_2(s) = L_e s + \frac{L_e}{d}\{4(q^2 sinhq - 3qcoshq + 3sinhq)(coshq\, s - 1) + (4q^2coshq - 12qsinhq + 2q^2 + 12coshq - 12)(qs - sinhqs) + q^4 sinhq(s^3 - 2s^2) + 2q^3coshq\,(3s^2 - s^3) + 2s^3 - (6q^2 sinhq)s^2\}$$

$$N_3(s) = \frac{{L_e}^2}{d}\{(q^2\sinh q - 4q\cosh q + 6\sinh q - 2q)(coshq\, s - 1) + (q^2coshq - 4qsinhq + 6coshq - 6)(qs - sinhqs) + q^3(s^3 - s^2) + q^3coshq(s^3 - 2s^2) + q^2 sinhq(qs^2 - 4s^3) + q(Coshq - 1)(4s^3 - 12s^2)\}$$

$N_4(s) = \frac{1}{d}\{q\ (q\ sinhq - 2\ coshq + 2)[6\ (1 - \cosh qs) + q^2(3s^2 - 2s^3)] + 6q(-q\ coshq + 2\ sinhq - q)(qs - \sinh qs)\}$

$N_5(s) = \frac{L_e}{d}\{(2q^2 \sinh q - 12 \sinh q + 12q)(\cosh q - 1) + 2(q^2 \cosh q - 6 \cosh q + 2q^2 + 6)(qs - \sinh qs) + (6q^2 \sinh q)s^2 - (2q^3 \cosh q)s^3 + q^3(2s^3 - 6s^2)\}$

$$N_6(s) = \frac{{L_e}^2}{d}\{(2q^2 \cosh\ q - 6 \sinh q + 4q)(1 - \cosh qs) + (2q \sinh q - 6 \cosh q + q^2 + 6)(\sinh qs - qs) + q^3 \cosh q(s^2 - s^3) + q^2 \sinh q(4s^3 - 3s^2) + q^3(2s^2 - s^3) - 4q(\cosh q - 1)s^3\}$$

$$d = \mathrm{q}[(q^3 + 24q) \sinh q - (8q^2 + 24) \cosh q - 4q^2]$$

References

[1] Cosserat, E., Cosserat, P. Theorie des corps deformables, Paris, 1909.

[2] Edelen, D., Laws, N. On the thermodynamics of systems with nonlocality. *Arch. Ration. Mech. Anal.*, 1971;43:24–35.

[3] Kroner, E. On the physical reality of torque stresses in continuum mechanics. *Internat. J. Engg. Sci.*, 1963;1(2):261–278.

[4] Mindlin, R., Tiersten, H. Effects of couple-stresses in linear elasticity. *Arch. Ration. Mech. Anal.*, 1962;11:415–448.

[5] Toupin, R. Elastic materials with couple-stresses. *Arch. Ration. Mech. Anal.*, 1962;11(1):385–414.

[6] Eringen, C. Linear theory of nonlocal elasticity and dispersion of plane waves. *Internat. J. Engg. Sci.*, 1972;10(5):425–435.

[7] Eringen, Edelen, D. On nonlocal elasticity. *Internat. J. Engg. Sci.*, 1972;10(3):233–248.

[8] Eringen. Screw dislocation in non-local elasticity. *J. Phy. D Appl. Phy.*, 1977;10(5):671–678.

[9] Romano, G., Barretta, R. Nonlocal elasticity in nanobeams: the stress-driven integral model. *Internat. J. Engg. Sci.*, 2017;115:14–27.

[10] Romano, G., Barretta, R. Stress-driven versus strain-driven nonlocal integral model for elastic nano-beams. *Comp. Part B Engg.*, 2017;114:184–188.

[11] Romano, G., Luciano, L., Barretta, R., Diaco, M. Nonlocal integral elasticity in nanostructures, mixtures, boundary effects and limit behaviours. *Contin. Mech. Thermodyn.*, 2018;30:641–655.

[12] Wang, Q., Liew, K. Application of nonlocal continuum mechanics to static analysis of micro-and nano-structures. *Phy. Lett. A*, 2007;363(3):236–242.

[13] Wang, Q., Shindo, Y. Nonlocal continuum models for carbon nanotubes subjected to static loading. *J. Mech. Mater. Struct.*, 2006;1(4):663–680.

[14] Khanra, G. K., I., Krishna, P., Raveendranath, P. A generalized functions based approach for stress-driven nanobeam bending problems subjected to point loads. *Mech. Adv. Mater. Struct.*, 2023:1–25.

[15] Kong, S., Zhou, S., Nie, Z., Wang, K. Static and dynamic analysis of micro beams based on strain gradient elasticity theory. *Internat. J. Engg. Sci.*, 2009;47(4):487–498.

[16] Phadikar, J., Pradhan, S. Variational formulation and finite element analysis for nonlocal elastic nanobeams and nanoplates. *Comput. Mater. Sci.*, 2010;49(3):492–499.

[17] Peddieson, J., Buchanan, G. R., McNitt, R. P. Application of nonlocal continuum models to nanotechnology. *Internat. J. Engg. Sci.*, 2003;41(3–5): 305–312.

[18] Fernandez-Saez, J., Zaera, R., Loya, J., Reddy, J. Bending of Euler--Bernoulli beams using Eringen's integral formulation: a paradox resolved. *Internat. J. Engg. Sci.*, 2016;99:107–116.

[19] Tuna, M., Kirca, M. Exact solution of Eringen's nonlocal integral model for bending of Euler--Bernoulli and Timoshenko beams. *Internat. J. Engg. Sci.*, 2016;105:80–92.

[20] Wang, Y., Zhu, X., Dai, H. Exact solutions for the static bending of Euler-Bernoulli beams using Eringen's two-phase local/nonlocal model. *Aip Adv.*, 2016;6(8):85–114.

[21] Khodabakhshi, P., Reddy, J. A unified integro-differential nonlocal model. *Internat. J. Engg. Sci.*, 2015;95:60–75.

[22] Eringen, C. On differential equations of nonlocal elasticity and solutions of screw dislocation and surface waves. *J. Appl. Phy.*, 1983;54(9):4703–4710.

[23] Lam, D., Yang, F., Chong, A., Wang, J., Tong, P. Experiments and theory in strain gradient elasticity. *J. Mech. Phy. Solids*, 2003;51(8):1477–1508.

[24] Fleck, N., Hutchinson, J. Strain gradient plasticity. *Adv. Appl. Mech.*, 1997;33:295–361.

[25] Fleck, N., Hutchinson, J. A reformulation of strain gradient plasticity. *J. Mech. Phy. Solids*, 2001;49(10):2245–2271.

[26] Yang, F., Chong, A., Lam, D., Tong, P. Couple stress based strain gradient theory for elasticity. *Internat. J. Solids Struct.*, 2002;39(10):2731–2743.

[27] Wazwaz. Linear and nonlinear integral equations. Berlin: Springer, 2011;639. 35–36.

[28] Karlicic, D., Murmu, T., Adhikari, S., McCarthy, M. Non-local structural mechanics. Hoboken: John Wiley & Sons, 2015. 518.

[29] Providas, E. Closed-form solution of the bending two-phase integral model of Euler-Bernoulli nanobeams. *Algorithms*, 2022;15(5):151.

[30] Polyanin, Manzhirov, A. Handbook of integral equations. New York: Chapman and Hall/CRC, 2008. 1144.

[31] Romano, G., Barretta, R., Diaco, M., de Sciarra, F. Constitutive boundary conditions and paradoxes in nonlocal elastic nanobeams. *Internat. J. Mech. Sci.*, 2017;121:151–156.

[32] H. K. M., A. M., T. A. M. Strain gradient beam element. *Finite Elem. Anal. Design*, 2013;68:63–75.

7 A comparative analysis on the thin-walled anisotropic blades with symmetric and unsymmetrical aero foil cross-sections

G. Deepak Kumar[a] *and Brajesh Panigrahi*[b]

Birla Institute of Technology and Science, Pilani, Hyderabad Campus, India

Abstract

It is paramount to understand the behaviour of thin-walled beams with different airfoil sections as they find their applications in the critical areas such as space, wind turbine blades and aeronautics. In the present work, we have adopted our previously developed coupled mathematical model for comparing the static and dynamic attributes of a thin-walled anisotropic beam with various airfoil cross-sections. In this regard, four different sections namely box (rectangular), biconvex, symmetric and unsymmetric airfoil profiles are considered. The impact of structural tailoring technique on these beams with different sections is presented for circumferential asymmetric stiffness (CAS) configuration. All the non-classical effects such as primary and secondary warping and transverse shear were considered. Proper care is taken to maintain the accuracy of the present model by validating thoroughly with the results available in the literature. A parametric comparison is presented on the static and dynamic fronts among these sections. The static responses of the beams with various sections are presented at different ply angles. The variation of stiffness parameters for various sections considered, is also displayed. The free vibration behaviour of these beams for the first four modes is presented at different ply angles. The impact of bending-torsion coupling for both static and dynamics is also highlighted and presented for all the sections.

Keywords: Thin-walled, airfoil, CAS, biconvex, anisotropic

Introduction

From the past one to two decades the usage of composite thin-walled (TW) structures is rapidly increasing in various fields of engineering implying for most rigorous design of these structures. These structures have an advantage of high strength to weight ratio and additionally, allows the design engineer for tailoring structurally in order to improve the structural integrity of these structures. From last decade and half there is a significant improvement in the numerical modelling of these structures considering the complexities involved in it. Further the more understanding of these structures enables to model efficient condition monitoring system. The following paragraph addresses the works associated with the modelling of these structures available in the literature.

Vlasov [1] has established the basic TW structure theory which eventually has become the basis for many works. Jung et al. [2] has presented the static analysis on TW box beam with single and double cells. Experimental and numerical analysis of TW box beam subjected to static loads is done by Chandra et al. [3]. Few works [4, 5] have given a basic understanding of the free vibration behaviour of TW box beams. Carrera et al. [6] have presented a unified formulation (CUF) for analysing the rotating composite blades. A FEM-based on Jacobi functions is presented by Carrera and Scano [7] for understanding the free vibration behaviour of plates and shells. Xu et al. [8] has established an efficient CUF-based formulation using Lagrange polynomial expansion for TW structures. Deepak and Panigrahi [9, 10] have presented an energy-based formulation for analysing the free vibration characteristics of a rotating anisotropic box and biconvex TW beams. Pagani et al. [11] has analysed the free vibration behaviour of a TW beam using higher order theories and radial basis functions. Dynamic analysis of a rotating cracked FGM beam is presented in the work by Kumar and Panigrahi [12]. In-plane and out of plane vibrations of a TW beam is presented by Tan et al. [13] using 1-D beam theory. Panigrahi and Pohit [14, 15] presented an energy-based formulation for finding the non-linear free vibration behaviour of a

[a]p20200462@hyderabad.bits-pilani.ac.in, [b]brajesh@hyderabad.bits-pilani.ac.in

DOI: 10.1201/9781003770442-7

rotating cracked FGM beam. Free vibrations of a TW biconvex beam is done by Eken and Kaya [16] for flexural-torsional behaviour. Eken [17] has presented a formulation on free vibration of TW airfoil structures. A coupled non-linear dynamic analysis on TW box beam for both CUS and CAS configuration is done in the work by Kumar and Panigrahi [18]. Non-linear free vibration behaviour of a cracked cantilever beam under the influence of gravitational loading is given by Deepak and Panigrahi [19]. Flexural analysis of a laminated composite beam with open cross sections is presented in the study did by Vukasović and Pavazza [20]. Free vibration behaviour of TW single celled beams by including various non-classical effects is analysed in work by Ovesy and Khaneh Masjedi [21]. Fillipi and Carrera [22] has presented the aerodynamic analysis of composite wings using CUF. Murugan et al. [23] has executed free vibration behaviour of a woven fabric glass laminated beam experimentally and numerically. Zappino et al. [24] has presented the free vibration of a TW beam with longitudinal stiffeners. A bionic bend-twist coupled TW beam analysis inspiring from rice stem is done in thw work did by Huang et al. [25]. Dynamic analysis of a cracked FGM beam using harmonic incrimination method is seen in the work did by Panigrahi and Pohit [26]. Non-linear free vibration behaviour of a rotating isotropic beam is presented in the work did by Panigrahi [27].

From literature it is for sure that over the years the modelling capabilities has been increased but the analysis has been restricted to box beams. In general, several applications in the areas like aerospace, wind turbine blades, etc., demands the structures other than open sections and box beams to enable the advantage of aerodynamics. So, there is need to understand these structures with airfoil sections both static and dynamic regimes. With this regard, the present work proposes a comparative analysis on different sections of TW beams. Three closed sections namely box (rectangular), biconvex and airfoil (NACA021, NACA064) are considered for this work. The static behaviour of all these TW sections is presented in flapping, lagging and extensional directions. In addition, a free vibration analysis is also presented and compared for the first three frequencies. It is observed that the NACA sections are having less stiffness for flapping load while biconvex is experiencing larger deformations for lagging load. Over all box section has higher stiffness compared to all sections. When it comes to free vibration responses, the curved sections like biconvex and airfoil sections are experiencing mode changes at around 90–95° ply angle. While box sections it is uniform at all ply angles.

Beam model

Energy equations

Let us consider a TW beam as shown in Figure 7.1. The corresponding displacement field and strain displacement equations are considered from the work did by Kumar and Panigrahi [9]. Classical laminate theory is adopted as the material is considered to be anisotropic in nature. The constitutive equations were also considered as mentioned by Kumar and Panigrahi [9]. As shown in Figure 7.1 three closed sections are considered for the present work.

The corresponding strain energy can be derived as

$$S.E = \int_0^L \oint \int_0^h \sigma^T \varepsilon \, dndsdz \tag{1}$$

where n, s and z are the coordinates from the generalised coordinate system considered on the mid surface of the contour as shown in the work did by Kumar and Panigrahi [9]. The coordinate n represents thickness, s is contour while z is axial coordinate which is common for both global and generalised coordinate system. By using constitutive equations the strain energy equation mentioned in Equation 1 can be further derived as

$$\begin{aligned} S.E = \int_0^L \Big[& \left(\oint N_{zz} w ds\right) + \left(\oint \left[y N_{zz} - M_{zz}\frac{dx}{ds}\right]\theta_x \, ds\right) \\ & + \left(\oint \left[x N_{zz} + M_{zz}\frac{dy}{ds}\right]\theta_y \, ds\right) \\ & + \left(\oint \left[N_{sz}\frac{dy}{ds} + N_{nz}\frac{dx}{ds}\right](u + \theta_y)\, ds\right) \\ & + \left(\oint \left[N_{sz}\frac{dy}{ds} + N_{nz}\frac{dx}{ds}\right](v + \theta_x)\, ds\right) \\ & + \left(\oint 2N_{sz}\psi\phi ds\right) \\ & - \left(\oint [\bar{F} N_{zz} + \bar{\bar{F}} M_{zz}]\phi'' ds\right)\Big] ds \end{aligned} \tag{2}$$

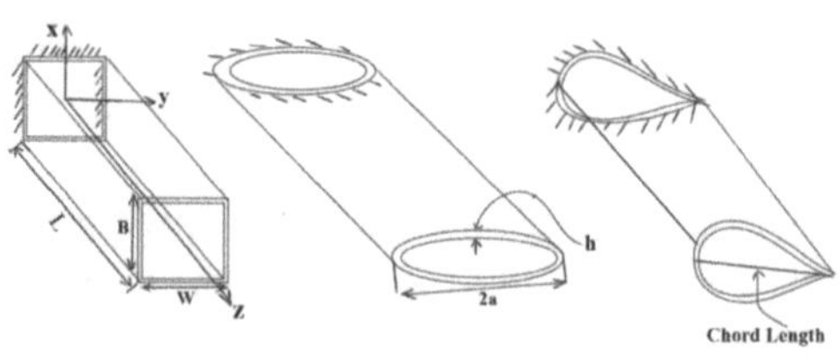

Figure 7.1 Geometry of TW beam closed (a) box beam (b) biconvex (c) airfoil sections
Source: Authors

By substituting the in-plane force and moment constitutive equations mentioned in the work did by Kumar and Panigrahi [9] in Equation 2 can be further simplified in terms of partial differential equation written with respect to longitudinal coordinate as given in the works [9, 10].

Similarly, the kinetic energy equation for the present system in the generalised coordinate system is given as

$$K.E = \frac{\rho}{2}\int_0^L \oint \int_0^h \left[\left(\frac{\partial U}{\partial t}\right)^2 + \left(\frac{\partial V}{\partial t}\right)^2 + \left(\frac{\partial W}{\partial t}\right)^2\right] dndsdz \quad (3)$$

where U, V and W represents the displacements in the global coordinate system written in terms of generalised coordinate system. This equation can further be simplified by considering the integral in thickness direction (n-coordinate) as

$$(m_0, m_1, m_2) = \rho \int_0^h (1, n, n^2) dn \quad (4)$$

Similarly, the integral corresponding to generalised thickness coordinate in Equation 1 will be taken care by writing as product of reduced stiffness parameters and ply thickness as displayed in Equation 5.

$$A_{ij} = \sum_{i=1}^{p} \int_{n(i-1)}^{n(i)} \overline{Q_{ij}}\, dn,\ B_{ij} = \sum_{i=1}^{p} \int_{n(i-1)}^{n(i)} \overline{Q_{ij}}\, ndn,\ D_{ij} = \sum_{i=1}^{p} \int_{n(i-1)}^{n(i)} \overline{Q_{ij}}\, n^2 dn \quad (5)$$

It is assumed to be that the beam is experiencing the harmonic motion with frequency ω rad/s.

Now, the derived strain and kinetic energy equations are non-dimensionalised using the scheme provided below,

$$\xi = \frac{z}{L}, u_0 = \frac{u}{h}, v_0 = \frac{v}{h}, w_0 = \frac{w}{h},$$

$$\theta_x = \theta_x, \theta_y = \theta_y,$$

$$\phi = \phi, \bar{\omega}^2 = \frac{b_1 L^2 \omega^2}{a_{11}} \quad (6)$$

By using Equation 6 the strain and kinetic energy equations are non-dimensionalised as mentioned in the work did by Kumar and Panigrahi [9] The terms a_{11} and b_1 are provided in the appendix of refs [9, 10].

Solving procedure

The solution procedure followed for solving the non-dimensional strain and kinetic energy equations is explained in this section.

Static analysis

The static analysis in the present work is executed by considering three different types of static loading. Figure 7.2 showcases the same. The non-dimensionalised strain energy equation is solved with the aid of trail functions as given in Equation 7.

$$u_0 = \sum_{j=1}^{N} A_j \xi^{j+1};\ v_0 = \sum_{j=1}^{N} B_j \xi^{j+1};\ w_0 = \sum_{j=1}^{N} C_j \xi^{j+1};\ \theta_x = \sum_{j=1}^{N} D_j (j+1)\xi^j;\ \theta_y = \sum_{j=1}^{N} E_j (j+1)\xi^j;\ \phi = \sum_{j=1}^{N} F_j \xi^{j+1} \quad (7)$$

By substituting these trail functions in the non-dimensionalised strain energy equation results in

$$[K]\{X\} = [F] \quad (8)$$

where K is stiffness matrix, X is displacement vector while F is force vector. The size of these matrices depends on the value of N. The value of N in Equation 7 is decided by performing a convergence study which is provided in the work by Kumar and Panigrahi [18].

Dynamic analysis

Along with the static analysis, results pertaining to the free vibration are also presented in this work. As mentioned in Section 3.1 along with non-dimensionalised strain energy equation, kinetic energy equation will also be considered. The trail functions as given in Equation 7 are also substituted in the kinetic energy equation and then an energy potential is established as shown in Equation 9,

$$\Pi = S.E - K.E \quad (9)$$

Then the energy potential is optimised with respect to the displacement coefficients as shown in Equation 7.

$$\frac{\partial \Pi}{\partial A_j} = 0;\ \frac{\partial \Pi}{\partial B_j} = 0;\ \frac{\partial \Pi}{\partial C_j} = 0;\ \frac{\partial \Pi}{\partial D_j} = 0;\ \frac{\partial \Pi}{\partial E_j} = 0;\ \frac{\partial \Pi}{\partial F_j} = 0 \quad (10)$$

This results in an expression leading to an Eigen value problem as given in Equation 11

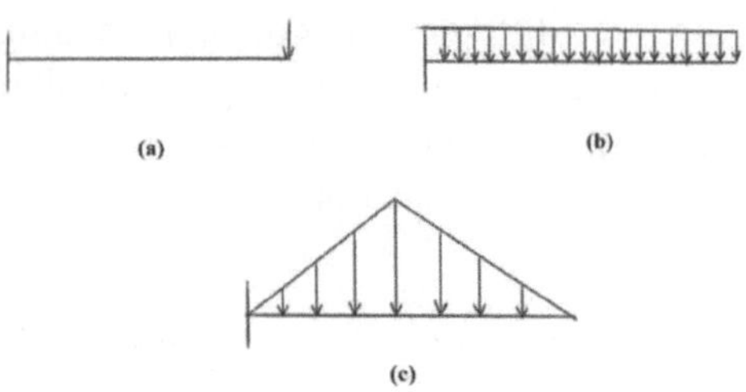

Figure 7.2 Types of static loading (a) Point load (b) UDL (c) UVL
Source: Authors

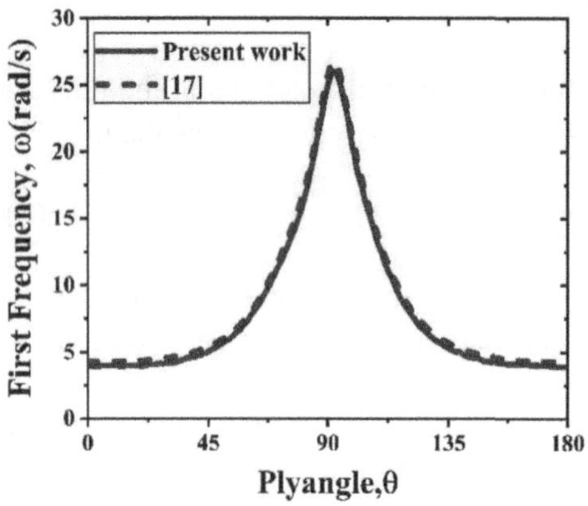

Figure 7.3 Validation of first frequency with [17]
Source: Authors

$$[K]\{X\} = \varpi^2[M]\{X\} \tag{11}$$

where, K and M are stiffness and mass matrices, respectively, where X is a displacement vector and is non-dimensionalised natural frequency as mentioned in Equation 6.

Validation

To understand the efficacy of the present model several validations have been reported in the works of refs [9, 10, 18]. Those validations were provided for dynamic analysis of box beam [9], biconvex section beam [10] and the static analysis of box beam [18]. As the present work involves employing the airfoil section, a validation is provided under this section compared to the results provided in the work by Eken [17].

From the Figures 7.3 and 7.4, it is evident that the results obtained using the present model is matching with the ones in the work by Eken [17] for the first two frequencies. With these validations it is evident that the present model is capable.

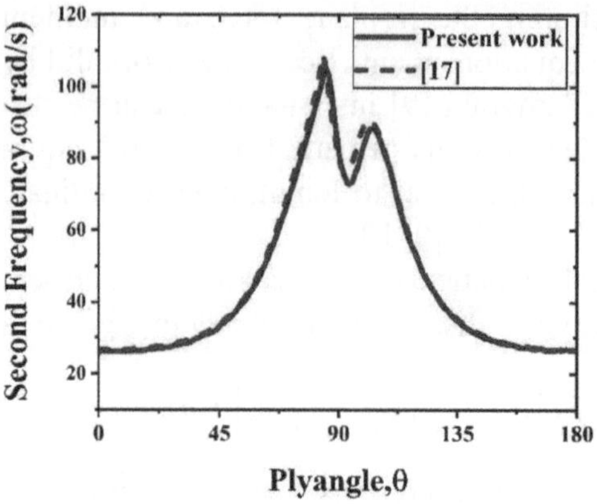

Figure 7.4 Validation of second frequency with [17]
Source: Authors

Results and discussions

Under this section, the results obtained using the present model are presented and discussed briefly. The material properties considered are same as mentioned in the study by Eken [17]. The geometrical properties for the airfoil sections are also same [17] but when it comes to biconvex section, the major diameter is same of the chord length of airfoil section which is 0.8 m. While the rectangular box section has width equal to the chord length of the airfoil. The ratio of the cross-section dimensions is maintained as 4.5 as it is the average ratio found for both the NACA profiles. The ply angle is measured with respect to s-coordinate which his perpendicular to longitudinal coordinate. Hence 90° ply angle in this work means the fibres are oriented in longitudinal direction.

Static responses

Figures 7.5–7.11 show cases the response of an anisotropic TW beam with different sections subjected to various types of static loads. The X-axis is ply angle while the Y-axis is displacement in a given direction in m. Figures 7.5–7.8 displays the response in flapping direction. NACA 64 is experiencing more displacement while box beam is having less displacement with larger stiffness of all the sections. Also, the displacements are more in the case of uniformly distributed load compared to other two loads. One can understand the reason, that the load being constant throughout the span of the beam.

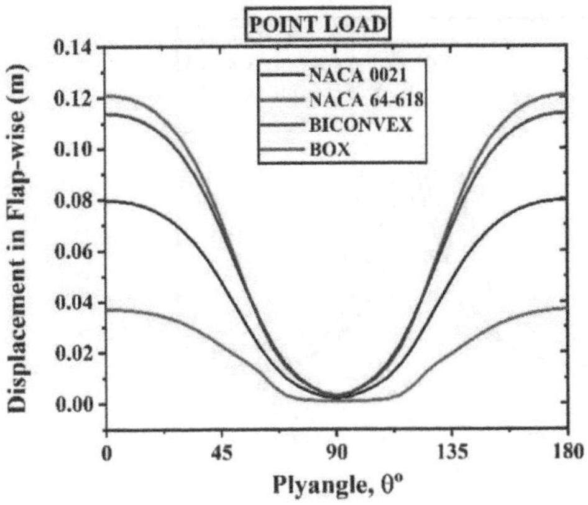

Figure 7.5 Static response in flapping direction under point load
Source: Authors

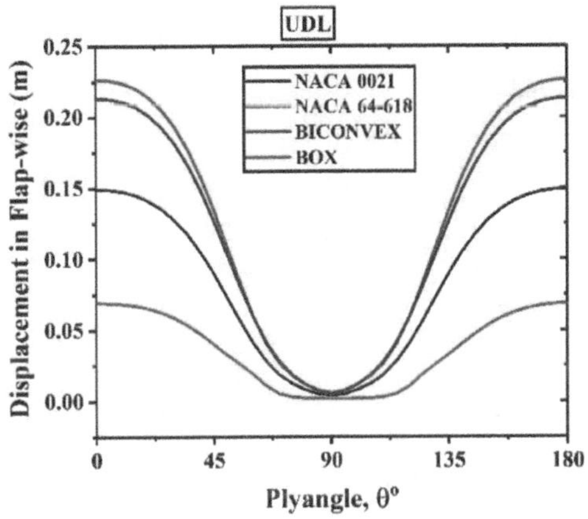

Figure 7.6 Static response in flapping direction under uniformly distributed load (UDL)
Source: Authors

Same type of pattern in the response can be observed from the Figures 7.6 and 7.7 as seen in Figure 7.5. The next set of figures from Figures 7.8–7.10 show cases the response to the load acting in lagging direction. When the load is applied in lagging, the beam with biconvex section is experiencing more displacement, while NACA series aero foils are exhibiting same amount of displacement with box section remaining the stiffer one.

The dip at 90° ply angle is due to the higher stiffness of the beam as the fibres are oriented longitudinally. As the fibres move away from longitudinal axis the stiffness gets reduced and hence experiences more displacement. The response under uniformly varying load in lagging direction also follows the same pattern as shown for other loads with biconvex

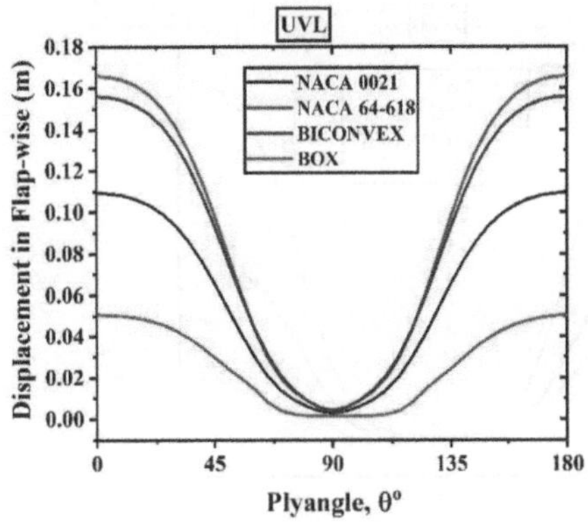

Figure 7.7 Static response in flapping direction under uniformly varying load (UVL)
Source: Authors

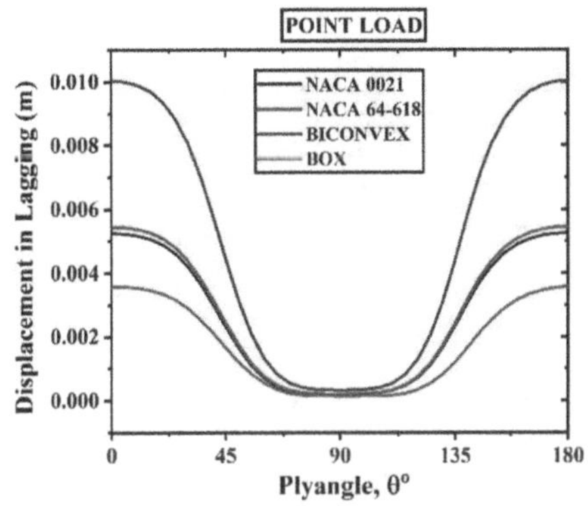

Figure 7.8 Static response in lag direction under point load
Source: Authors

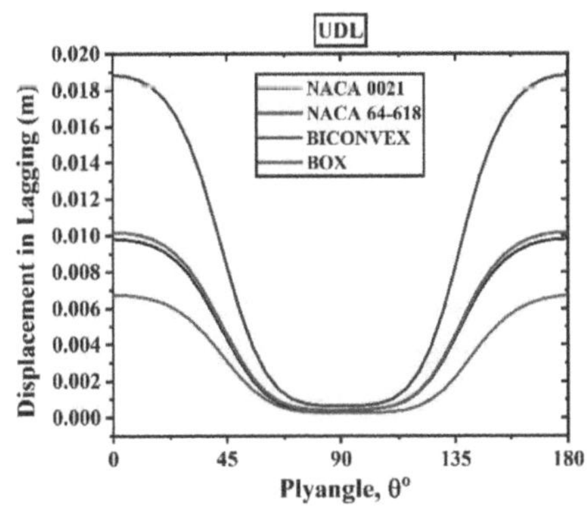

Figure 7.9 Static response in lag direction under uniformly distributed load
Source: Authors

experiencing more displacement. It is obvious that the displacement in flapping will be more compared

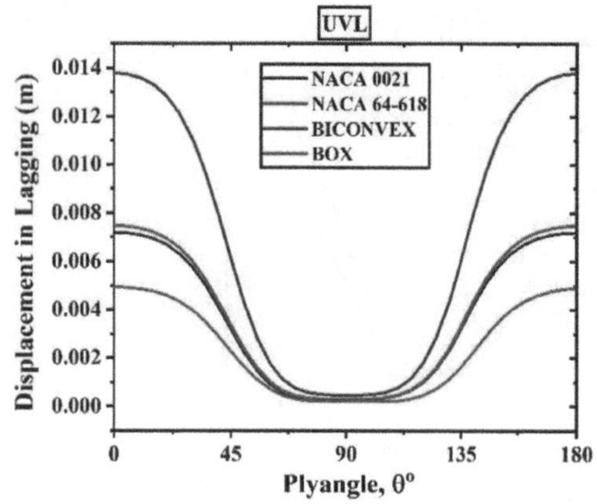

Figure 7.10 Static response in lag direction under uniformly distributed load
Source: Authors

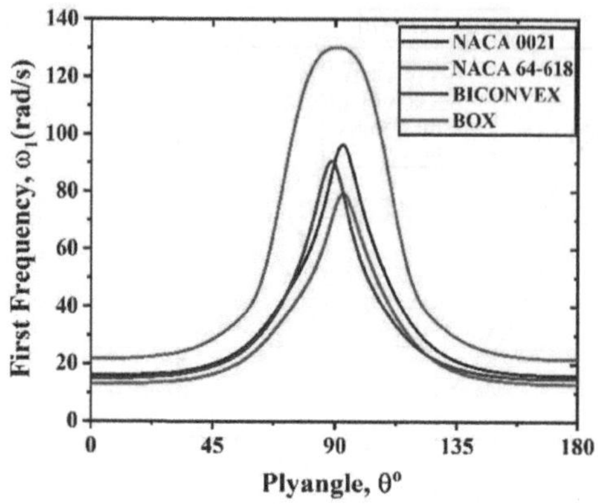

Figure 7.12 Variation in the first frequency at different ply angles
Source: Authors

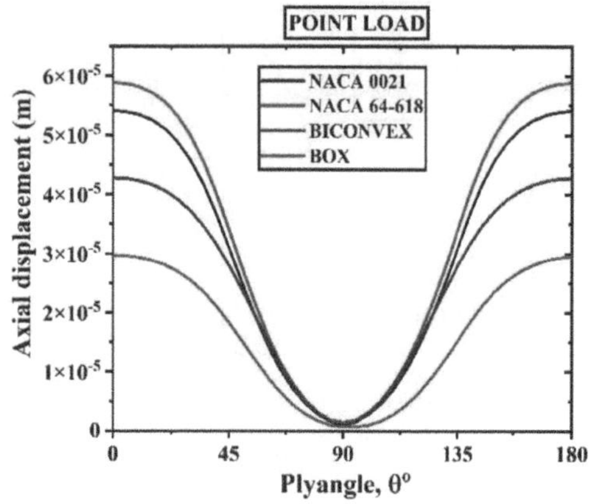

Figure 7.11 Static response in axial direction under point load
Source: Authors

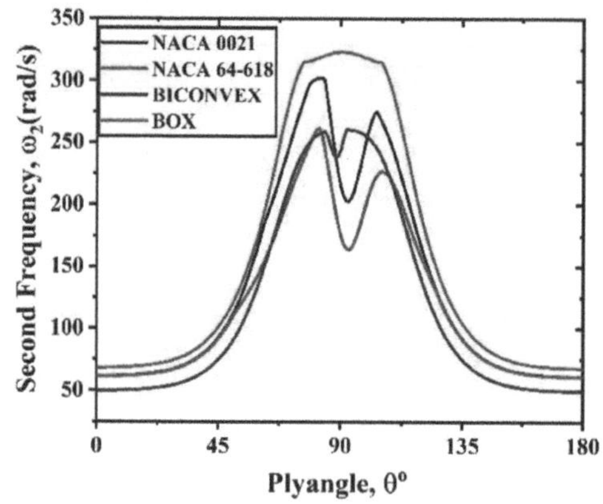

Figure 7.13 Variation in the second frequency at different ply angles
Source: Authors

to displacements in lagging and axial directions. Figure 7.11 shows the response of the beam towards the axial loading. The displacement is very less as the stiffness in the axial direction is more compared to the other two transverse directions. All the results pertaining to the static analysis are in line with the basic physics of these structures. One key observation is that when the load is in lagging direction, the biconvex section is experiencing more displacement when compared to other two NACA profiles even though biconvex section is having same chord length as the NACA profiles.

Dynamic response

Under this section, the results obtained for the free vibration analysis of the anisotropic TW beam with different sections are show cased. Figures 7.12 and 7.13 displays the first two frequencies at different ply angles. The abscissa in Figures 7.12 and 7.13 represents the ply angles while the ordinate represents the natural frequencies in rad/s.

The first mode being the flapping mode for all the sections the variation is almost same. Except that the NACA profiles due to their unsymmetric nature the peak shifts towards right. The biconvex and box beam will have peak frequencies at 90° ply angle. While the box section is dominating the other sections due to high stiffness, the unsymmetric NACA 64-618 profile is having lesser magnitudes of frequencies. Figure 7.13 displays the variation of second frequencies over various ply angles.

The box beam experiences second flapping mode at all the ply angles, while the other sections like NACA profiles and biconvex experiences first torsional mode at around 90–95° ply angle. Hence one can observe a dip in the frequency for the red, blue and black curves. Here also, the unsymmetric nature of the NACA sections can be observed quite clearly when compared to the variation with the biconvex section. The interesting thing about the second frequency is that for the curved profiles, at maximum stiffness condition there will be change in mode from second flapping to first torsional unlike the box beam which has a uniform mode throughout all the ply angles.

Conclusion

A coupled mathematical model for analysing the static and dynamic behaviour of the anisotropic TW beams with different sections namely box, biconvex and NACA series airfoil sections (NACA 0021 – Symmetric about x-axis, NACA 64-618 – Unsymmetric about both axes). The static and free vibration responses of these sections are compared and presented logically. The following are the conclusions drawn from the present work:

i. Under flapping load, the unsymmetric NACA profile NACA 64-618 is exhibiting more deformation.
ii. Under lagging load, the biconvex profile is exhibiting more deformation compared to NACA profiles.
iii. Of all the profiles, box (rectangular) is stiffer and hence has lesser deflections in all directions.
iv. For free vibration responses, due to high stiffness, the box section has higher frequencies compared to all other sections.
v. The first mode being a flapping first mode, for all the sections at all ply angles the variation remains same for all the sections.
vi. For the second mode, the curved profiles experience a change in mode at 90–95° ply angle from second flapping to first torsional. Unlike the box section which experiences a uniform mode at all ply angles.

Future scope

The present work can be extended by including the stiffeners for the curve profiles which makes the problem more complex and interesting. Further, by including the aero elastic forces and non-linearity the problem can be extended.

References

[1] Vlasov, V. Z. Thin-walled elastic beams. *PST Catalog.*, 428, 1–493, 1959.
[2] Jung, S. N., Park, I. J., Shin, E. S. Theory of thin-walled composite beams with single and double-cell sections. *Comp. Part B Engg.*, 2007;38(2):182–192.
[3] Chandra, R., Stemple, A. D., Chopra, I. Thin-walled composite beams under bending, torsional, and extensional loads. *J. Aircraft*, 1990;27(7):619–626.
[4] Song, O., Librescu, L. Free vibration of anisotropic composite thin-walled beams of closed cross-section contour. *J. Sound Vibr.*, 1993;167(1):129–147.
[5] Vo, T. P., Lee, J. Free vibration of thin-walled composite box beams. *Comp. Struct.*, 2008;84(1):11–20.
[6] Carrera, E., Filippi, M., Zappino, E. Free vibration analysis of rotating composite blades via Carrera Unified Formulation. *Comp. Struct.*, 2013;106:317–325.
[7] Carrera, E., Scano, D. Finite elements based on Jacobi shape functions for the free vibration analysis of beams, plates, and shells. *Mech. Adv. Mater. Struct.*, 2024;31(1):4–12.
[8] Xu, X., Fallahi, N., Yang, H. Efficient CUF-based FEM analysis of thin-wall structures with Lagrange polynomial expansion. *Mech. Adv. Mater. Struct.*, 2022;29(9):1316–1337.
[9] Kumar, G. D., Panigrahi, B. A coupled model to analyse dynamic behaviour of non-prismatic rotating anisotropic thin-walled beams. *Mech. Adv. Mater. Struct.*, 2024:1–14.
[10] Kumar, G. D., Panigrahi, B. Influence of flexural-torsional couplings and rotational effects on vibration behaviour of tapered thin-walled anisotropic biconvex beam. *J. Sound Vibr.*, 2024;570:118100.
[11] Pagani, A., Carrera, E., Ferreira, A. J. M. Higher-order theories and radial basis functions applied to free vibration analysis of thin-walled beams. *Mech. Adv. Mater. Struct.*, 2016;23(9):1080–1091.
[12] Kumar, G. D., Panigrahi, B. Nonlinear dynamic analysis on centrifugal stiffening of rotating cracked tapered functionally graded beam for flap-wise vibrations. *J. Vibr. Control*, 2024;30(9–10):2154–2170.
[13] Tan, M., Guo, D., Yang, Q., Yang, L., Luo, D. In-plane and out-of-plane free vibration analysis of thin-walled box beams based on one-dimensional higher-order beam theory. *Mech. Adv. Mater. Struct.*, 2024;31(22):5638–5652.
[14] Panigrahi, B., Pohit, G. Nonlinear modelling and dynamic analysis of cracked Timoshenko functionally

graded beams based on neutral surface approach. *Proc. Instit. Mech. Engrs., Part C J. Mech. Engg. Sci.*, 2016;230(9):1486–1497.

[15] Panigrahi, B., Pohit, G. Effect of cracks on nonlinear flexural vibration of rotating Timoshenko functionally graded material beam having large amplitude motion. *Proc. Instit. Mech. Engrs., Part C J. Mech. Engg. Sci.*, 2018;232(6):930–940.

[16] Eken, S., Kaya, M. O. Flexural-torsional coupled vibration of anisotropic thin-walled beams with biconvex cross-section. *Thin-Walled Struct.*, 2015;94:372–383.

[17] Eken, S. Free vibration analysis of composite aircraft wings modeled as thin-walled beams with NACA airfoil sections. *Thin-Walled Struct.*, 2019;139:362–371.

[18] Kumar, G. D., Panigrahi, B. Nonlinear free vibrations of a structurally tailored anisotropic pre-twisted thin-walled beam subjected to large deformations. *J. Sound Vibr.*, 2024:118766.

[19] Gudlavalleti, D. K., Singh, K., Panigrahi, B. The influence of gravitational stiffening and de-stiffening on nonlinear dynamic behaviour of vertical beams with cracks. *Nondestruc. Test. Eval.*, 2024;39(2):221–237.

[20] Vukasović, M., Pavazza, R. Flexural analysis of thin-walled laminated composite beams with arbitrary open sections and shear influence. *Mech. Adv. Mater. Struct.*, 2022;29(26):5041–5058.

[21] Ovesy, H. R., Khaneh Masjedi, P. Investigation of the effects of constitutive equations on the free vibration behaviour of single-celled thin-walled composite beams. *Mech. Adv. Mater. Struct.*, 2014;21(10):836–852.

[22] Filippi, M., Carrera, E. Aerodynamic and mechanical hierarchical aeroelastic analysis of composite wings. *Mech. Adv. Mater. Struct.*, 2016;23(9):997–1004.

[23] Murugan, R., Ramesh, R., Padmanabhan, K. Investigation of the mechanical behaviour and vibration characteristics of thin walled glass/carbon hybrid composite beams under a fixed-free boundary condition. *Mech. Adv. Mater. Struct.*, 2016;23(8):909–916.

[24] Zappino, E., Cavallo, T., Carrera, E. Free vibration analysis of reinforced thin-walled plates and shells through various finite element models. *Mech. Adv. Mater. Struct.*, 2016;23(9):1005–1018.

[25] Huang, J., Deng, X., Liu, W. Bionic design of bend-twist coupled thin-walled beam based on the structure of rice stem. *Mech. Adv. Mater. Struct.*, 2022;29(26):5177–5190.

[26] Panigrahi, B., Pohit, G. Study of non-linear dynamic behaviour of open cracked functionally graded Timoshenko beam under forced excitation using harmonic balance method in conjunction with an iterative technique. *Appl. Mathem. Model.*, 2018;57:248–267.

[27] Panigrahi, B. A parametric study on influence of slenderness ratio on the nonlinear dynamic behaviour of rotating tapered beams. *Mater. Today Proc.*, 108 (2024): 1–8.

8 Analysis of surface wave propagation in composite structures with corrugated interfaces

N. Bhengra[a] *and M. Rodge*[b]

Department of Mathematics, Visvesvaraya National Institute of Technology(VNIT), Nagpur, Maharashtra, India

Abstract

A theoretical framework is constructed to examine the transmission of shear-horizontal (SH) waves in a stacked composite configuration. The system consists of a functionally graded piezoelectric material (FGPM) layer positioned between a porous piezoelectric medium (PPM) and a pre-stressed orthotropic half-space, with the interfaces joined through a corrugated boundary. The displacement components in all three medium are obtained using the separation of variables method. For both electrically open and short-circuit surface conditions, frequency relation is obtained. Several limiting cases are also derived. To validate the theoretical model, numerical simulations and graphical illustrations are provided. The study further explores the influence of various factors, including the gradient coefficient of the FGPM layer, its width, piezoelectric and dielectric constants, dielectric coupling between two phases of porous aggregate, and the corrugation parameter. The findings reveal that the phase velocity of SH-waves shows strong dependence on these parameters. In general, the suggested model provides a theoretical foundation as well as possible practical uses, especially in the advancement of surface acoustic wave (SAW) devices.

Keywords: Orthotropic substrate, functionally graded piezoelectric layer, porous piezoelectric material, initial stress, surface acoustic wave (SAW) device

Introduction

SAW devices are manufactured mainly by smart composite materials such as shape memory materials, piezoelectric substrate, photoactive materials, chromoactive materials, magnetorheological materials, etc. Depending on factors such as frequency range, sensitivity and environmental conditions, different type of acoustic sensors helps to sense mass and mechanical properties that working on phenomenon of transducing mechanical signals to electrical and vice versa which examine how the elasticity amplitude of an acoustic wave changes with depth within a material, usually decaying exponentially. Application of surface acoustic wave (SAW) with the use of electronic system, including decay lines, filters, oscillators, transformers, resonator, 2D material, microfluidics and flow measurement, corrolator, AC to DC convertor which provides a potential field in radar system, communication system widely used in the sectors of health monitoring, frequency control, automobiles, medicine, the military, and aviation. Some of the key benefits of SAW devices are their capability to operate at high frequencies with low insertion loss and high selectivity which works in cruel environments without using batteries.

Recently, working with smart material has been in trend, due to its abilities like easy assimilation, swift action and customisability of piezoelectric material attracting a large number of researches in smart structures. In the vigorous environments with abrupt gradients of the pressure and temperature, these materials can be used for many advantageous properties. Traditional laminated composite structure is replaced by the new kind of innovative functionally graded piezoelectric material. To achieve the required resultant properties for particular application having regularly varying directions with structure and properties of material, a different kind of material functionally graded piezoelectric material (FGPM) a novel material developed. New kind of innovative functionally graded piezoelectric materials are more efficient and effective shows the diverse nature. Mondal and Sahu [1] examined an FGPM layer with surface corrugations over a piezomagnetic half-space for shear-horizontal (SH)-wave propagation, while Mondal [2] investigated layered system consisting of FGMs and

[a]neelimabhengra@mth.vnit.ac.in, [b]mayurirodge18@gmail.com

DOI: 10.1201/9781003770442-8

porous piezoelectric media for Love-type waves. Hemalatha et al. [3] analysed Rayleigh waves at imperfectly corrugated interfaces, and Hemalatha and Kumar [4] studied transverse wave reflection and transmission at corrugated FGPM interfaces. Sahu et al. [5] examined polarised SH-wave in FGPM layers bounded by corrugated piezomagnetic and elastic media, and Sahu and Baroi [6] considered surface waves in corrugated piezomagnetic layers over inhomogeneous half-spaces. Li et al. [7] studied advance Laguerre polynomial method for Rayleigh waves in FGPM half-spaces, while Jiangong et al. [8] studied wave propagation in spherically curved FGPM plates. Biswas and Sahu [9] investigated Love-type acoustic waves in FG piezomagnetic plates between elastic layers and Kundu et al. [10] examined SH-waves in heterogeneous viscoelastic layers with porous anisotropic coatings under initial stress. Abd-Alla and Ahmed [11] focused on non-homogeneous orthotropic layers under stress for Love wave transmission and Manna et al. [12] analysed Love waves in piezoelectric layers on inhomogeneous substrates. Chaki et al. [13] studied SH-wave behaviour in FGPM layered structures with abrupt thickness variations and imperfect bonding, while Verma et al. [14] analysed Rayleigh-type waves in viscoelastic layers with rigid and yielding foundations. Singh et al. [15] presented Love waves in piezo-poroelastic FG layers with electrode boundaries and abrupt interfaces, whereas Salah et al. [16] examined theoretically modelled Rayleigh waves in FGPMs. Cao et al. [17] derived Love wave dispersion in FGPM layered composites and Singhal [18] provided approximate frequency relations for piezo-composite surface waves. Chen et al. [19] addressed non-local effects in magneto-electro-elastic multi-layered plates, and Qian et al. [20] examined piezoelectric Love waves in FGPM layered structures. Chaudhary et al. [21] studied imperfect interface effects on Love waves in FGPM layers, while Tani and Liu [22] examined SH-wave transmission in FGPM plates. Finally, Manukyan et al. [23] explored FGPM layered systems, emphasising the role of graded properties in acoustic applications using Love wave propagation.

The combine effect of electrical and magnetic field is known as piezoelectric effect. Porous piezoelectric material is completely made up of piezoelectric material with porosity showing great piezoelectrical effect. It has wide variety of application in engineering which is the porous material with high open porosity. Mercadelli et al. [24] discussed porous piezoelectric ceramics. By using analytical and numerical method, Martínez-Ayuso et al. [25] discovered the homogenisation of porous electric material. Energy harvesting is one of the applications of porous piezoelectric material, to take information and knowledge about its application researchers are working on it since decades. A detailed non-linear analysis of energy harvesters of the functionally graded piezoelectric was discovered by Derayatifar et al. [26]. In a related contribution, Heshmati and Amini [27] studied an extensive investigation of energy harvesting in functionally graded piezoelectric beams subjected to vibrations induced by traveling multi-oscillators. Mostly all the researchers are concentrated towards the anisotrophic material, but nowadays due to advancement and highly use of composite material it authorised us to examine the characteristic of the waves in the material, which gives the valid information for the future work.

It is crucial to manage and use the surface wave since it travels a great distance with high frequency and high energy along the semi-infinite elastic media's free surface. Its amplitude also falls down rapidly inside the medium. Corrugation is related to shaping surface into parallel ridges and grooves this can be achieved by the folding, moulding or any other manufacturing methods. Research has been done on the corrugated FGPM layer with various condition for propagation of SH-wave, a detailed analysis is required to achieve more knowledge about abilities and characteristics of FGPM for propagation of surface waves. It requires thoroughly knowledge of all the characteristics to get the perfect solution for the waves in terms of displacement and velocity which gives the numerous advantages in the practical problems. Recently, surface acoustic waves in biosensing applications studied by Huang et al. [28]. SAW devices and their wireless communications applications demonstrated in the work did by Ruppel et al., and Liu et al. [29, 30]. Qian et al. [31] discussed surface acoustic wave devices for sensor applications.

In this paper, the detailed study of displacement component of FGPM medium, porous piezoelectric medium and orthotropic half-space is calculated by variable separable approach. By applying the boundary condition frequency equations are formed for electrically closed and open cases. Additionally, the graphical behaviour of phase velocity with different parameter were evaluated, giving excellent result in numerical and graphical analysis which is helpful for the designing and manufacturing the SAW devices with functionally graded piezoelectric material in the composite corrugated configuration.

Mathematical framework of the problem

A three-layered structural configuration is considered, where a FGPM layer of uniform thickness is placed among two different media. The lower medium is modelled as a pre-stressed orthotropic substrate, whereas upper medium corresponds to a corrugated porous piezoelectric layer. The Cartesian system is defined with *z*-axis aligned to the polarisation direction of the FGPM layer and perpendicular to the *xy*-plane (see Figure 8.1). The displacement field is then described by *u*, *v*, and *w* along the *x*, *y*, and *z*-directions, respectively.

Fundamental equations and solutions for the FGPM medium

In the non-existence of free charges and body forces, the equilibrium equations for elasticity together with Gauss's law of electrostatics are represented as:

$$\tau_{mn,n} = \rho \ddot{u}_m, \quad D_{m,m} = 0, \quad (m, n = 1 \ to \ 3) \tag{1}$$

where $\tau_{mn,n}$, D_m, ρ is the stress tensor, electric displacement and density of mass for the piezoelectric medium, respectively.

Assume SH-waves travel in the *y*–direction, the electric potential and the entire displacement out of the plane is shown as,

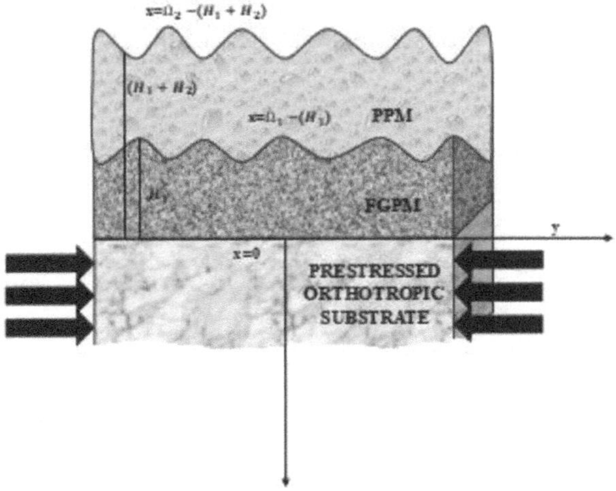

Figure 8.1 Schematic representation of layers
Source: Authors

$$w = w, \ u = 0, \ \phi = \phi, \ v = 0 \tag{2}$$

From Equation (2) in Equation (1), we get

$$\frac{\partial \tau_{zx}}{\partial x} + \frac{\partial \tau_{yz}}{\partial y} = \rho \frac{\partial^2 w}{\partial t^2}, \quad \frac{\partial D_x}{\partial x} + \frac{\partial D_y}{\partial y} = 0 \tag{3}$$

The functionally graded piezoelectric material's constitutive equations are expressed as follows:

$$\tau_{yz} = \bar{C}_{44}\frac{\partial w}{\partial y} + \bar{e}_{15}\frac{\partial \phi}{\partial y}, \ \tau_{zx} = \bar{C}_{44}\frac{\partial w}{\partial x} + \bar{e}_{15}\frac{\partial \phi}{\partial x} \tag{4}$$

and

$$D_x = \bar{e}_{15}\frac{\partial w}{\partial x} - \bar{\Phi}_{11}\frac{\partial \phi}{\partial x}, \ D_y = \bar{e}_{15}\frac{\partial w}{\partial y} - \bar{\Phi}_{11}\frac{\partial \phi}{\partial y}. \tag{5}$$

For the formulation of governing equations of the piezoelectric layer substitute Equations (4) and (5) into Equation (3), as follows:

$$\frac{\partial}{\partial x}\left[\bar{C}_{44}\frac{\partial w}{\partial x} + \bar{e}_{15}\frac{\partial \phi}{\partial x}\right] + \frac{\partial}{\partial y}\left[\bar{C}_{44}\frac{\partial w}{\partial y} + \bar{e}_{15}\frac{\partial \phi}{\partial y}\right] = \rho\frac{\partial^2 w}{\partial t^2} \tag{6}$$

and

$$\frac{\partial}{\partial x}\left[\bar{e}_{15}\frac{\partial w}{\partial x} - \bar{\Phi}_{11}\frac{\partial \phi}{\partial x}\right] + \frac{\partial}{\partial y}\left[\bar{e}_{15}\frac{\partial w}{\partial y} - \bar{\Phi}_{11}\frac{\partial \phi}{\partial y}\right] = 0 \tag{7}$$

The properties of material are described as:

$$\bar{C}_{44} = \bar{C}_{44}^0 e^{\beta x}, \ \ \bar{e}_{15} = \bar{e}_{15}^0 e^{\beta x}, \ \ \bar{\Phi}_{11} = \bar{\Phi}_{11}^0 e^{\beta x}, \ \ \rho = \rho^0 e^{\beta x} \tag{8}$$

In the material coefficients, the superscript 0 indicates the *x* direction factors, whereas β is the exponential factor that characterises the degree of the material gradient in the *x*–direction. $\beta = 0$ clearly represents the scenario of homogenous material.

The governing equations are written as,

$$\bar{C}_{44}^0\left(\frac{\partial^2 w}{\partial x^2} + \frac{\partial^2 w}{\partial y^2}\right) + \bar{e}_{15}^0\left(\frac{\partial^2 \phi}{\partial x^2} + \frac{\partial^2 \phi}{\partial y^2}\right) + \bar{C}_{44}^0\beta\frac{\partial w}{\partial x} + \bar{e}_{15}^0\beta\frac{\partial \phi}{\partial x} = \rho^0\frac{\partial^2 w}{\partial t^2} \tag{9}$$

and

$$\bar{e}_{15}^0\left(\frac{\partial^2 w}{\partial x^2} + \frac{\partial^2 w}{\partial y^2}\right) - \bar{\Phi}_{11}^0\left(\frac{\partial^2 \phi}{\partial x^2} + \frac{\partial^2 \phi}{\partial y^2}\right) + \bar{e}_{15}^0\beta\frac{\partial w}{\partial x} - \bar{\Phi}_{11}^0\beta\frac{\partial \phi}{\partial x} = 0. \tag{10}$$

Now we assume,

$$\psi = \phi - \frac{\bar{e}_{15}}{\bar{\Phi}_{11}} w \tag{11}$$

Equation (10) can be rewritten as follows,

$$\bar{e}_{15}^0\left(\beta\frac{\partial \phi}{\partial x} + \nabla^2\phi\right) = \frac{\bar{e}_{15}^0}{\bar{\Phi}_{11}^0}\left[\beta\bar{e}_{15}^0\frac{\partial w}{\partial x} + \bar{e}_{15}^0\nabla^2 w\right] \tag{12}$$

Now from Equations (9–12), we get

$$\left(\beta\frac{\partial w}{\partial x}+\nabla^2 w\right)=\frac{1}{C_F^2}\frac{\partial^2 w}{\partial t^2} \tag{13}$$

and

$$\left(\beta\frac{\partial \psi}{\partial x}+\nabla^2 \psi\right)=0 \tag{14}$$

where, $\overline{C_{44}}=\left(\bar{C}^0_{44}+\frac{(\bar{e}^0_{15})^2}{\bar{\Phi}^0_{11}}\right)$ and $c_F^2=\frac{\overline{C_{44}}}{\rho^0}$.

Assuming the solution of Equations (13) and (14), we get

$$\psi=\psi(x)e^{i\kappa(y-ct)} \tag{15}$$

$$w=w(x)e^{i\kappa(y-ct)} \tag{16}$$

Now from Equations (13–16), we obtain

$$\frac{d^2w}{dx^2}+\beta\frac{dw}{dx}+\kappa^2\left(\frac{c^2}{c_F^2}-1\right)w(x)=0 \tag{17}$$

and

$$\frac{d^2\psi}{dx^2}+\beta\frac{d\psi}{dx}-\kappa^2\psi(x)=0. \tag{18}$$

Now the solution of Equations (17) and (18), we have

$$\psi(x)=(A_1e^{r_1x}+A_2e^{r_2x}) \tag{19}$$

and

$$w(x)=e^{sx}(M_7\cos\lambda x+M_8\sin\lambda x) \tag{20}$$

where, $s=\frac{-\beta}{2}$, $\lambda=\frac{1}{2}\sqrt{4\kappa^2\left(\frac{c^2}{c_F^2}-1\right)-\beta^2}$ and $r_{1,2}=\frac{-\beta\pm\sqrt{\beta^2+4\kappa^2}}{2}$.

Substituting Equations (19) and (20) into Equation (11), we get

$$\phi=\mathrm{M}_1e^{r_1x}+\mathrm{M}_2e^{r_2x}+\frac{\bar{e}^0_{15}}{\bar{\Phi}^0_{11}}[e^{sx}(M_7\cos\lambda x+M_8\sin\lambda x)]. \tag{21}$$

Thus, it is possible to determine the FGPM layer's electric potential and mechanical displacement as,

$$w(x)=e^{sx}(M_7\cos\lambda x+M_8\sin\lambda x)e^{i\kappa(y-ct)} \tag{22}$$

$$\phi(x,y,t)=\mathrm{M}_1e^{r_1x}+\mathrm{M}_2e^{r_2x}+\frac{\bar{e}^0_{15}}{\bar{\Phi}^0_{11}}[e^{sx}(M_7\cos\lambda x+M_8\sin\lambda x)]e^{i\kappa(y-ct)} \tag{23}$$

Fundamental equation and solution for the porous piezoelectric medium (PPM)

The stress-strain relation for PPM:

$$\begin{aligned}\sigma_{mn}&=F_{mnkl}\varepsilon_{kl}+m_{mn}\varepsilon^*-e^p_{kmn}E_k-\zeta_{kmn}E^*_k,\\ \sigma^*&=\overline{m}_{mn}\varepsilon_{mn}+R\varepsilon^*-\zeta_kE_k-e^*_kE^*_k,\\ D_i&=e^p_{mkl}\varepsilon_{kl}+\zeta_m\varepsilon^*+T_{ml}E_l+A^p_{ml}E^*_l,\\ D^*_m&=\zeta_{mkl}\varepsilon_{kl}+e^*_m\varepsilon^*+A^p_{ml}E_l+T^*_{ml}E^*_l\end{aligned} \tag{24}$$

where, σ_{mn} (σ^*) and ε_{mn} (ε^*) are the stress and strain tensor components for the solid (fluid) phase in PPM. D_i (D^*_m) and E_m (E^*_m) are the electric displacement and electric field vector for the solid (fluid) phase in PPM. Also F_{mnkl}, $\overline{m}_{mn}$, R are elastic constants e_{mnk}, e^*_m, ζ_{mnk}, ζ_m are the piezoelectric constants and T_{mn}, T^*_{mn}, A^p_{mn} are dielectric constants.

The motion equations corresponding to PPM, neglecting surface charge density and body forces, are expressed as:

$$\begin{aligned}\sigma_{mn}&=(\rho_{11})_{mn}\ddot{u}_n+(\rho_{12})_{mn}\ddot{U}_n\\ \sigma^*_{,m}&=(\rho_{12})_{mn}\ddot{u}_n+(\rho_{22})_{mn}\ddot{U}_n\\ D_{m,m}&=0,\quad D^*_{m,m}=0.\end{aligned} \tag{25}$$

Further, we have

$$\varepsilon_{mn}=\frac{1}{2}(u_{m,n}+u_{n,m}),\ \varepsilon^*=u^*_{m,m},\ E_m=-\phi_{,m},\ E^*_m=-\phi^*_{,m} \tag{26}$$

The stress-strain relation used for the material is expressed as,

$$\begin{aligned}\sigma_{11}&=F_{11}\varepsilon_{11}+F_{12}\varepsilon_{22}+F_{13}\varepsilon_{33}+\overline{m}_{11}\varepsilon^*-e^p_{31}E_3\\ \sigma_{22}&=F_{12}\varepsilon_{11}+F_{11}\varepsilon_{22}+F_{13}\varepsilon_{33}+\overline{m}_{11}\varepsilon^*-e^p_{31}E_3\\ \sigma_{33}&=F_{13}\varepsilon_{11}+F_{13}\varepsilon_{22}+F_{33}\varepsilon_{33}+\overline{m}_{33}\varepsilon^*-e^p_{33}E_3\\ \sigma_{32}&=2F_{44}\varepsilon_{32}-e^p_{15}E_2\\ \sigma_{31}&=2F_{44}\varepsilon_{31}-e^p_{15}E_1\\ \sigma_{12}&=2F_{66}\varepsilon_{12}\\ \sigma^*&=\overline{m}_{11}\varepsilon_{11}+\overline{m}_{11}\varepsilon_{22}+\overline{m}_{33}\varepsilon_{33}+R\varepsilon^*\\ D_1&=2e^p_{15}\varepsilon_{13}+T_{11}E_1+A^p_{11}E^*_1\\ D_2&=2e^p_{15}\varepsilon_{23}+T_{11}E_2+A^p_{11}E^*_2\\ D_3&=e^p_{31}\varepsilon_{11}+e^p_{31}\varepsilon_{22}+e^p_{33}\varepsilon_{33}+\zeta_3\varepsilon^*+T_{33}E_3+A^p_{33}E^*_3\\ D^*_1&=A^p_{11}E_1+T^*_{11}E^*_1\\ D^*_2&=A^p_{11}E_2+T^*_{11}E^*_2\\ D^*_3&=e^*_3\varepsilon^*+A^p_{33}E_3+T^*_{33}E^*_3\end{aligned} \tag{27}$$

where, $F_{66}=\frac{(F_{11}-F_{12})}{2}$.

The electric potential and mechanical displacements for propagating SH wave in *xy*-plane is expressed as:

$$\phi^*_1=\phi^*_1, U^*_3=W^*_2, \phi_1=\phi_1, u_1=0, u_2=0, u_3=w_2. \tag{28}$$

Using Equations (24–28), we get

$$F_{44}\nabla^2w_2+e^p_{15}\nabla^2\phi_1=\rho^p\frac{\partial^2w_2}{\partial t^2} \tag{29}$$

$$e^p_{15}\nabla^2w_2-T^p_{11}\nabla^2\phi_1=0 \tag{30}$$

where, $\rho^p=\rho_{11}-\frac{\rho^2_{12}}{\rho_{22}}$ and $T^p_{11}=T_{11}-\frac{(A^p_{11})^2}{T^*_{11}}$ and $\nabla^2\phi^*_1=-\frac{A_{11}}{T^*_{11}}\nabla^2\phi_1$.

We assumed the formal solution of Equations (29) and (30) as follows,

$$w_2 = W_2(x)e^{[i\kappa(y-ct)]}\ ,\ \phi_1 = \phi_1(x)e^{[i\kappa(y-ct)]} \quad (31)$$

here, the wave number is defined as $\kappa(=\frac{\omega}{c})$ where, c represents phase velocity and ω is angular frequency of wave propagation.

Using Equation (31) in Equation (29) yields,

$$W_2 = \left(M_3e^{\delta x} + M_4e^{-\delta x}\right) \quad (32)$$

where, $\delta = \sqrt{\kappa^2\left(1 - \frac{c^2}{c_p^2}\right)}$, $\overline{F_{44}} = \left(F_{44} + \frac{(e_{15}^p)^2}{T_{11}^p}\right)$ and $c_p^2 = \frac{\overline{F_{44}}}{\rho^p}$.

Therefore, the mechanical displacement for the porous piezoelectric medium is given by,

$$w_2 = \left(M_3e^{\delta x} + M_4e^{-\delta x}\right)e^{i\kappa(y-ct)} \quad (33)$$

Now to solve Equation (30), we assume

$$\psi_1 = \phi_1 - \frac{e_{15}^p}{T_{11}^p}w_2. \quad (34)$$

Now, from Equations (30), (31) and (34), we have

$$\nabla^2\psi_1 = 0. \quad (35)$$

Now to solve Equation (35), we assume

$$\psi_1 = \psi_1(x)e^{[i\kappa(y-ct)]}. \quad (36)$$

Now from Equations (35) and (36), we have

$$\psi_1(x) = \left(M_5e^{\kappa x} + M_6e^{-\kappa x}\right)e^{[i\kappa(y-ct)]} \quad (37)$$

With the aid of Equations (34) and (37), we get the electric potential functions which is given by

$$\phi_1(x,y,t) = \left[M_5e^{\kappa x} + M_6e^{-\kappa x} + \frac{e_{15}^p}{T_{11}^p}\left(M_3e^{\delta x} + M_4e^{-\delta x}\right)\right]e^{[i\kappa(y-ct)]} \quad (38)$$

Fundamental equation and solution for the orthotropic half-space

For the orthotropic elastic solid medium, the equation of motion assuming negligible body forces are given by:

$$\begin{aligned}
&\tau_{11,1} + \tau_{12,2} + \tau_{13,3} - P\left(\omega_{3,2} - \omega_{2,3}\right) = \rho_2\ddot{u}_3,\\
&\tau_{21,1} + \tau_{22,2} + \tau_{23,3} - P\omega_{3,1} = \rho_2\ddot{v}_3,\\
&\tau_{31,1} + \tau_{32,2} + \tau_{33,3} - P\omega_{2,1} = \rho_2\ddot{w}_3
\end{aligned} \quad (39)$$

where, ρ_2 are considered as density of the medium and $\ddot{u}3$, $\ddot{v}3$ and $\ddot{w}3$ are used as the double time derivative components in the respective direction of x, y and z–axis.

where,

$$\omega_1 = \frac{1}{2}\left(w_{3,y} - v_{3,z}\right),\ \omega_2 = \frac{1}{2}\left(u_{3,z} - w_{3,x}\right),\ \omega_3 = \frac{1}{2}\left(v_{3,x} - u_{3,y}\right) \quad (40)$$

The relationship governing the incremental stress and strain components is formulated as,

$$\begin{aligned}
&\tau_{11} = B_{11}e_{xx} + B_{12}e_{yy} + B_{13}e_{zz},\\
&\tau_{12} = 2Q_3e_{xy},\\
&\tau_{22} = B_{21}e_{xx} + B_{22}e_{yy} + B_{23}e_{zz},\\
&\tau_{23} = 2Q_1e_{yz},\\
&\tau_{33} = B_{31}e_{xx} + B_{32}e_{yy} + B_{33}e_{zz},\\
&\tau_{13} = 2Q_2e_{zx}.
\end{aligned} \quad (41)$$

For the propagation of SH-type waves, the resulting equation of motion is obtained as:

$$\left(Q_2 + \frac{P}{2}\right)\frac{\partial^2 w_3}{\partial x^2} + Q_1\frac{\partial^2 w_3}{\partial y^2} = \rho_2\frac{\partial^2 w_3}{\partial t^2}. \quad (42)$$

To solve Equation (42), we assume

$$w_3 = W_3(x)e^{[i\kappa(y-ct)]} \quad (43)$$

Now with the help of Equations (42) and (43), we get the solution

$$w_3 = \left(M_9e^{mx} + E_2e^{-mx}\right)e^{i\kappa(y-ct)} \quad (44)$$

where, $m^2 = \kappa^2\left(\frac{1}{\left(\frac{Q_2}{Q_1} + \frac{P}{2Q_1}\right)} - \frac{c^2}{c_0^2} \times \frac{1}{\left(\frac{Q_2}{Q_1} + \frac{P}{2Q_1}\right)}\right)$

and $c_0^2 = \frac{Q_1}{\rho_2}$.

Boundary conditions (BC's)

Furthermore, the subsequent boundary requirements have to be fulfilled when the SH-type wave travels through the stacked structure, as represented in Figure 8.1.

1. The free surface is affected by both mechanical and electrically open conditions i.e., at $x = \Omega_2(y) - (H_1 + H_2)$,

$$\sigma_{xz} - \Omega'_2\sigma_{yz} = 0,\ D_1 + D_1^* = 0. \quad (45)$$

2. The free surface is subjected to both mechanical and electrically short conditions i.e., at $x = \Omega_2(y) - (H_1 + H_2)$ are expressed as follows,

$$\sigma_{xz} - \Omega'_2\sigma_{yz} = 0,\ \phi_1 = 0. \quad (46)$$

3. At the interface the condition of continuity i.e., at $x = \Omega_1(y) - (H_1)$ between the PPM and FGPM are written as,

$$w_2 = w, \sigma_{xz} - \Omega'_1\sigma_{yz} = \tau_{xz} - \Omega'_1\tau_{yz}, \phi_1 = \phi, D_1 + D_1^* = D_x. \quad (47)$$

4. The condition of continuity at the interface i.e., at $x = 0$ between the FGPM and prestressed orthotropic half-space are written as,

$$w = w_3, \tau_{xz} = \sigma_{yz}^1, \phi = 0. \quad (48)$$

Governing frequency relation for the electrically open condition

Using the BC's Equations (45–48) with the help of the entire displacement component, electric potential, we will get the following:

$$\varkappa_{11}M_3 - \varkappa_{12}M_4 + \varkappa_{13}M_5 - \varkappa_{14}M_6 = 0 \quad (49)$$

$$-\varkappa_{23}M_5 + \varkappa_{24}M_6 = 0 \quad (50)$$

$$\varkappa_{31}M_3 + \varkappa_{32}M_4 - \varkappa_{35}M_7 - \varkappa_{36}M_8 = 0 \quad (51)$$

$$\varkappa_{41}M_3 - \varkappa_{42}M_4 + \varkappa_{43}M_5 - \varkappa_{44}M_6 - \varkappa_{45}M_7 - \varkappa_{46}M_8 - \varkappa_{47}M_1 - \varkappa_{48}M_2 = 0 \quad (52)$$

$$\varkappa_{51}M_3 + \varkappa_{52}M_4 + \varkappa_{53}M_5 + \varkappa_{54}M_6 - \varkappa_{55}M_7 - \varkappa_{56}M_8 - \varkappa_{57}M_1 - \varkappa_{58}M_2 = 0, \quad (53)$$

$$-\varkappa_{63}M_5 + \varkappa_{64}M_6 + \varkappa_{67}M_1 + \varkappa_{68}M_2 = 0 \quad (54)$$

$$\varkappa_{75}M_7 + \varkappa_{76}M_8 - \varkappa_{79}M_9 = 0 \quad (55)$$

$$\varkappa_{85}M_1 + \varkappa_{86}M_2 + \varkappa_{87}M_7 + \varkappa_{88}M_8 - \varkappa_{89}M_9 = 0 \quad (56)$$

$$\varkappa_{95}M_7 + \varkappa_{96}M_8 + \varkappa_{97}M_1 + \varkappa_{96}M_2 = 0. \quad (57)$$

Equations (49–57) collectively satisfy the necessary boundary conditions under electrically open conditions.

By eliminating M_i's, *where* $i = 1 - 9$, we have

$$|\varkappa_{mn}| = 0, \qquad m, n = 1 \text{ to } 9. \quad (58)$$

Appendix I contains the only non-zero components of the aforementioned determinant.

Governing frequency relation for the electrically short condition

Using the BC's Equations (45–48) with the help of the entire displacement component, electric potential, we will get the following:

$$\gamma_{11}M_3 - \gamma_{12}M_4 + \gamma_{13}M_5 - \gamma_{14}M_6 = 0 \quad (59)$$

$$\gamma_{21}M_3 + \gamma_{22}M_4 + \gamma_{23}M_5 + \gamma_{24}M_6 = 0 \quad (60)$$

$$\gamma_{31}M_3 + \gamma_{32}M_4 - \gamma_{35}M_7 - \gamma_{36}M_8 = 0 \quad (61)$$

$$\gamma_{41}M_3 - \gamma_{42}M_4 + \gamma_{43}M_5 - \gamma_{44}M_6 - \gamma_{45}M_7 - \gamma_{46}M_8 - \gamma_{47}M_1 - \gamma_{48}M_2 = 0 \quad (62)$$

$$\gamma_{51}M_3 + \gamma_{52}M_4 + \gamma_{53}M_5 + \gamma_{54}M_6 - \gamma_{55}M_7 - \gamma_{56}M_8 - \gamma_{57}M_1 - \gamma_{58}M_2 = 0, \quad (63)$$

$$-\gamma_{63}M_5 + \gamma_{64}M_6 + \gamma_{67}M_1 + \gamma_{68}M_2 = 0 \quad (64)$$

$$\gamma_{75}M_7 + \gamma_{76}M_8 - \gamma_{79}M_{10} = 0 \quad (65)$$

$$\gamma_{85}M_1 + \gamma_{86}M_2 + \gamma_{87}M_7 + \gamma_{88}M_8 - \gamma_{89}M_{10} = 0 \quad (66)$$

$$\gamma_{95}M_7 + \gamma_{96}M_8 + \gamma_{97}M_1 + \gamma_{96}M_2 = 0. \quad (67)$$

Equations (59–67) collectively satisfy the necessary boundary conditions under electrically short conditions.

By eliminating M_i's,, *where* $i = 1$ *to* 10, we have

$$|\gamma_{mn}| = 0, \qquad m, n = 1 \text{ to } 9. \quad (68)$$

Appendix I contains the only non-zero components of the aforementioned determinant.

Special cases

Case I

For the electrically open scenario, we get the dispersion relation when the model simplifies to a structure made up of a FGPM layer covering an elastic substrate

$$\begin{aligned} &e^{-H_1(r_1+r_2+s+2\beta)}\,\overline{C_{44}}[r_2\, e^{r_1H_1}\{\kappa\lambda(-\bar{e}_{15}^0\Phi_{11}^s\overline{C_{44}}r_1 + C_{44}^s d_2\bar{\Phi}_{11}^0(\kappa\Phi_{11}^s + r_1\bar{\Phi}_{11}^0))\cos\lambda H_1 \\ &\quad -(-\bar{e}_{15}^0\Phi_{11}^s\overline{C_{44}}\kappa r_1 s + \kappa\Phi_{11}^s\bar{\Phi}_{11}^0(\kappa s C_{44}^s d_2 + \overline{C_{44}}(\lambda^2 + s^2)) + \\ &(\bar{\Phi}_{11}^0)^2 r_1(C_{44}^s d_2\kappa s + \overline{C_{44}}(\lambda^2 + s^2)))\sin\lambda H_1\} \\ &+e^{r_2H_1}\{\kappa\lambda(-\bar{e}_{15}^0\Phi_{11}^s\overline{C_{44}}r_1 + C_{44}^s d_2\bar{\Phi}_{11}^0(\kappa\Phi_{11}^s + r_1\bar{\Phi}_{11}^0))\cos\lambda H_1 \\ &\quad -(-\bar{e}_{15}^0\Phi_{11}^s\overline{C_{44}}\kappa r_1 s + \kappa\Phi_{11}^s\bar{\Phi}_{11}^0(\kappa s C_{44}^s d_2 + \overline{C_{44}}(\lambda^2 + s^2)) + \\ &(\bar{\Phi}_{11}^0)^2 r_1(C_{44}^s d_2\kappa s + \overline{C_{44}}(\lambda^2 + s^2)))\sin\lambda H_1\}] \end{aligned} \quad (69)$$

which is the dispersion relation for a SH-wave propagating in a FGPM layer lying over an elastic substrate.

Case II

For the electrically short situation, we get the dispersion relation when the model simplifies to a structure made up of a FGPM layer covering an elastic substrate.

$$\begin{aligned}
&\frac{1}{(\bar{\Phi}_{11}^0)^2} e^{-H_1(r_1+r_2+s+2\beta)}[-\bar{\Phi}_{11}^0 \bar{e}_{15}^0 \Phi_{11}^s \overline{C_{44}}(e^{2sH_1}\bar{e}_{15}^0 + e^{r_1H_1+r_2H_1}\overline{C_{44}})\kappa\lambda(r_1 - r_2) \\
&+e^{sH_1}\bar{\Phi}_{11}^0 \lambda \overline{C_{44}}\{-e^{r_1H_1}[C_{44}^s d_2 \bar{\Phi}_{11}^0 \kappa(\kappa\Phi_{11}^s + r_1\bar{\Phi}_{11}^0) + \bar{e}_{15}^0(-\Phi_{11}^s \overline{C_{44}}\kappa r_1 + \bar{e}_{15}^0 \Phi_{11}^s \kappa r_2 + \Phi_{11}^s \bar{\Phi}_{11}^0 r_1 r_2)] \\
&\quad +e^{r_2H_1}[C_{44}^s d_2 \bar{\Phi}_{11}^0 \kappa(\kappa\Phi_{11}^s + r_2\bar{\Phi}_{11}^0) + \bar{e}_{15}^0(\bar{e}_{15}^0 \Phi_{11}^s \kappa r_1 - \overline{C_{44}}\Phi_{11}^s \kappa r_2 + \bar{e}_{15}^0 \bar{\Phi}_{11}^0 r_1 r_2)]\}\cos\lambda H_1 \\
&+\{e^{H_1(s+r_1)}[(\bar{e}_{15}^0)^3 \Phi_{11}^s \overline{C_{44}}\kappa r_1 r_2 - \bar{e}_{15}^0 \bar{\Phi}_{11}^0 \Phi_{11}^s \kappa(C_{44}^s d_2 \bar{e}_{15}^0 \kappa r_2 + \overline{C_{44}} s(\overline{C_{44}} r_1 + \bar{e}_{15}^0 r_2)) \\
&\quad +(\bar{\Phi}_{11}^0)^3 \overline{C_{44}} r_1(\kappa s d_2 C_{44}^s + \overline{C_{44}}(\lambda^2 + s^2)) + (\bar{\Phi}_{11}^0)^2(-(\bar{e}_{15}^0)^2 r_1 r_2(C_{44}^s d_2\kappa + \overline{C_{44}} s) + \Phi_{11}^s \overline{C_{44}}\kappa(\kappa s d_2 C_{44}^s + \overline{C_{44}}(\lambda^2 + s^2)))] \\
&\quad -e^{H_1(s+r_2)}[(\bar{e}_{15}^0)^3 \Phi_{11}^s \overline{C_{44}}\kappa r_1 r_2 - \bar{e}_{15}^0 \bar{\Phi}_{11}^0 \Phi_{11}^s \kappa(C_{44}^s d_2 \bar{e}_{15}^0 \kappa r_1 + \overline{C_{44}} s(\overline{C_{44}} r_2 + \bar{e}_{15}^0 r_1)) + \\
&+(\bar{\Phi}_{11}^0)^3 \overline{C_{44}} r_2(\kappa s d_2 C_{44}^s + \overline{C_{44}}(\lambda^2 + s^2)) + (\bar{\Phi}_{11}^0)^2(-(\bar{e}_{15}^0)^2 r_1 r_2(C_{44}^s d_2\kappa + \overline{C_{44}} s) + \\
&\Phi_{11}^s \overline{C_{44}}\kappa(\kappa s d_2 C_{44}^s + \overline{C_{44}}(\lambda^2 + s^2)))]\}\sin\lambda H_1]
\end{aligned} \tag{70}$$

The above expression gives the dispersion relation of an SH-wave transmission in a FGPM layer placed over in elastic substrate.

Numerical and graphical analysis

The composite system under consideration consists of an FGPM layer positioned in the middle of the porous piezoelectric medium and an orthotropic half-space subjected to initial stress. For this configuration, the frequency relation connecting the wave number (κH_1) and phase velocity $(\frac{c}{c_F})$ is derived for both boundary conditions i.e., electrically short and open, which serve as the basis for the numerical analysis. The results are presented graphically, where the electrically short case represented by solid curves, while electrically opens case indicates the dashed curves. These plots are generated using MATHEMATICA and taken data of the specified constants of material.

Dimensionless phase velocity represented by (c/c_F) and wave number represented by (κH_1) for the graphical representation.

In this model the material constants used are summarised in Table 8.1.

Effect of dielectric constant associated with porous piezoelectric material

The influence of dielectric constants T_{11} and T_{11}^* with respect to the connection between $(\frac{c}{c_F})$ and (κH_1) is shown in Figures 8.2 and 8.3. From Figure 8.2, it follows that, $(\frac{c}{c_F})$ undergoes minimal change at lower wave numbers, but exhibits considerable variation with increasing wave number. The results further reveal that a rise in T_{11} lowers the phase velocity. Figure 8.3 demonstrates that the effect of T_{11} is less significant. Its influence becomes observable at higher frequencies but reduces as T_{11} increases.

Effect of piezoelectric constant associated with FGPM layer and porous piezoelectric material of the medium

Figures 8.4 and 8.5 illustrate the graph of $(\frac{c}{c_F})$ with respect to (κH_1) for considering the distinct values of the piezoelectric constant, taken as the key influencing parameter for the FGPM layer and the porous piezoelectric medium, respectively. In Figure 8.4, the effect of the e_{15}^F on $(\frac{c}{c_F})$ is found to be minimal. The results indicate that phase velocity decreases as e_{15}^F increases. Conversely, Figure 8.5 shows the influence of the piezoelectric constant e_{15}^p. In this case, the curves are more widely separated and clearly distinguishable, and the $(\frac{c}{c_F})$ increases with higher values of e_{15}^p.

Effect of width of the FGPM layer and porous piezoelectric material of the medium

Figures 8.6 and 8.7 illustrate regarding the relation between the $(\frac{c}{c_F})$ and (κH_1) during SH-wave propagation through the FGPM layer. Figure 8.6 considers the width of the FGPM layer H_1, as the governing parameter. It is observed that all curves maintain a consistent trend across the range and the angular velocity enhances the phase velocity. Figure 8.7, on the other hand, focuses on the width of the porous

Table 8.1 Dielectric, elastic and piezoelectric constants of FGPM layer, porous piezoelectric and initially stressed orthotropic half-space.

FGPM Layer	Porous Piezoelectric Material ($BaTiO_3$)	Orthotropic Half-Space
$\bar{C}_{44}^0 = 2.30 \times 10^9$ N/m^2 $\rho^0 = 7500$ kg/m^3	$F_{11} = 150.4$ GPa, $e_{15}^p = 11.4$ C/m^2, $T_{11} = 10.8$ nC/Vm $F_{12} = 65.63$ GPa, $e_{31}^p = -4.32$ C/m^2, $T_{33} = 13.1$ nC/Vm $F_{13} = 65.94$ GPa, $e_{33}^p = 17.4$ C/m^2, $T^*_{11} = 11.8$ nC/Vm $F_{33} = 145.5$ GPa, $\zeta_{15} = 4.56$ C/m^2, $T^*_{33} = 13.9$ nC/Vm $F_{44} = 43.86$ GPa, $\zeta_{31} = -1.728$ C/m^2, $A_{11}^p = 12.8$ nC/Vm $\bar{m}_{11} = 8.8$ GPa, $\zeta_{33} = 6.96$ C/m^2, $A_{33}^p = 15.1$ nC/Vm $\bar{m}_{33} = 5.2$ GPa, $e^*_3 = -3.6$ C/m^2 $R = 20$, $\zeta_3 = -7.5$ C/m^2	$Q_1 = 3.99 \times 10^{10}$ N/m^2 $Q_2 = 5.82 \times 10^{10}$ N/m^2 $\rho_3 = 4500$ kg/m^3

Source: Authors

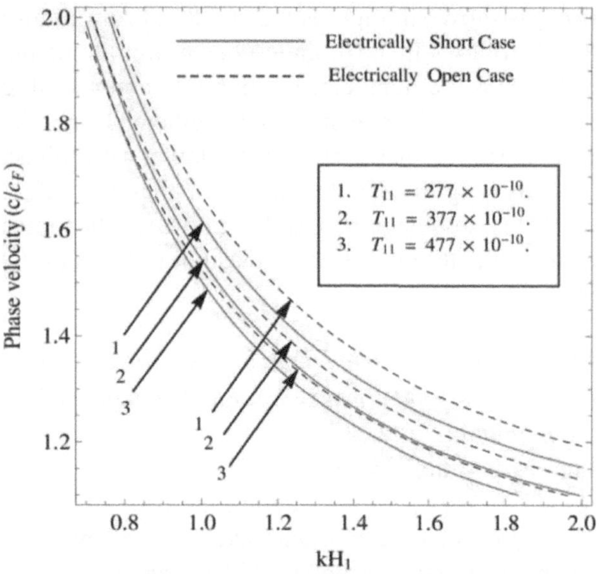

Figure 8.2 Plot of $(\frac{c}{c_F})$ with respect to (κH_1) for different dielectric constant values T_{11}

Source: Authors

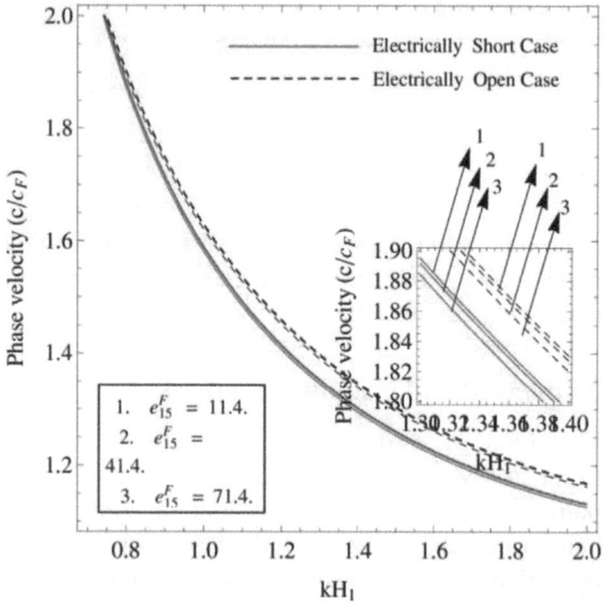

Figure 8.4 Plot of $(\frac{c}{c_F})$ with respect to (κH_1) for different magnitude of piezoelectric connected with FGPM layer

Source: Authors

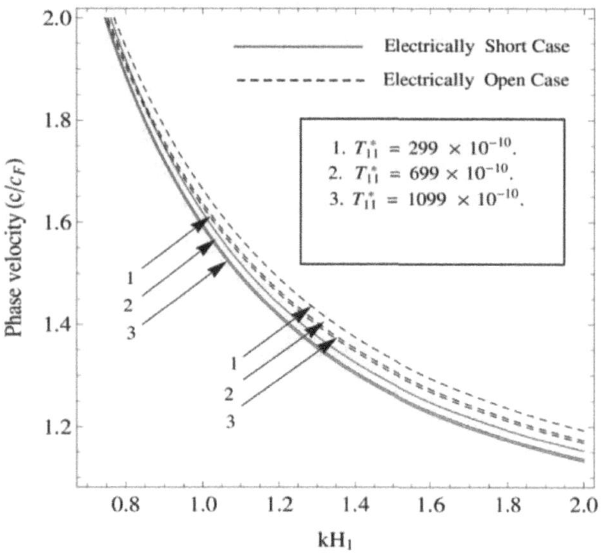

Figure 8.3 Plot of $(\frac{c}{c_F})$ with respect to (κH_1) for different values of dielectric constant T^*_{11}

Source: Authors

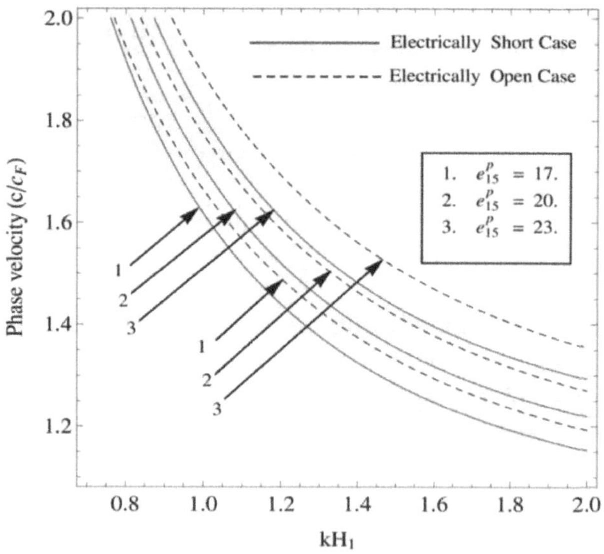

Figure 8.5 Plot of $(\frac{c}{c_F})$ against (κH_1) for different magnitude of piezoelectric constant constant connected with porous piezoelectric material medium

Source: Authors

piezoelectric layer, H_2. While the overall behaviour of the curves mirrors that of Figure 8.6, increasing H_2 leads to a reduction in $(\frac{c}{c_F})$. It is also evident that the electrically open condition produces consistently higher phase velocity than the electrically short condition.

Effect of position parameter and corrugation parameter associated with lower and upper corrugated interface

Considering distinct values of position and the initial surface flatness coefficient pertaining to the upper

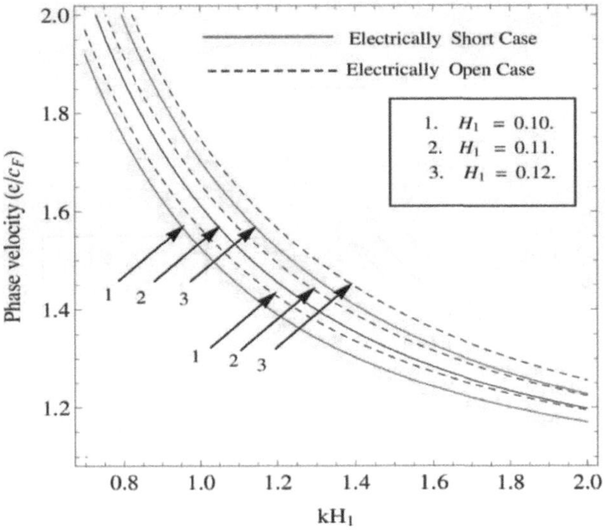

Figure 8.6 Plot of $(\frac{c}{c_F})$ with respect to (κH_1) for different values of H_1
Source: Authors

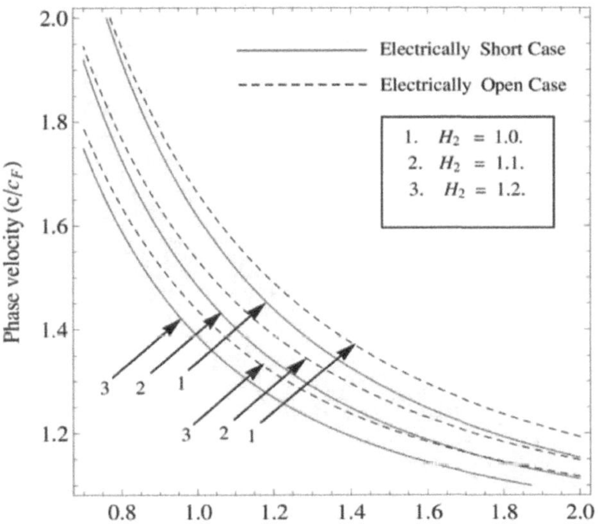

Figure 8.7 Plot of $(\frac{c}{c_F})$ with respect to (κH_1) for different values of H_2
Source: Authors

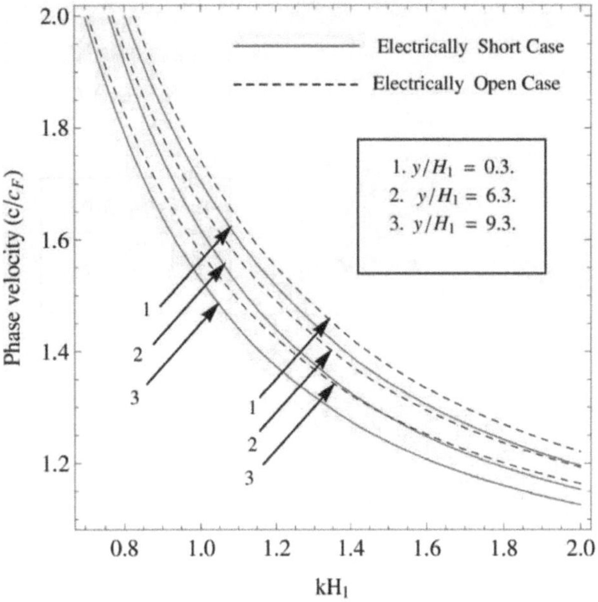

Figure 8.8 Plot of $(\frac{c}{c_F})$ with respect to (κH_1) for different values of $(\frac{y}{H_1})$
Source: Authors

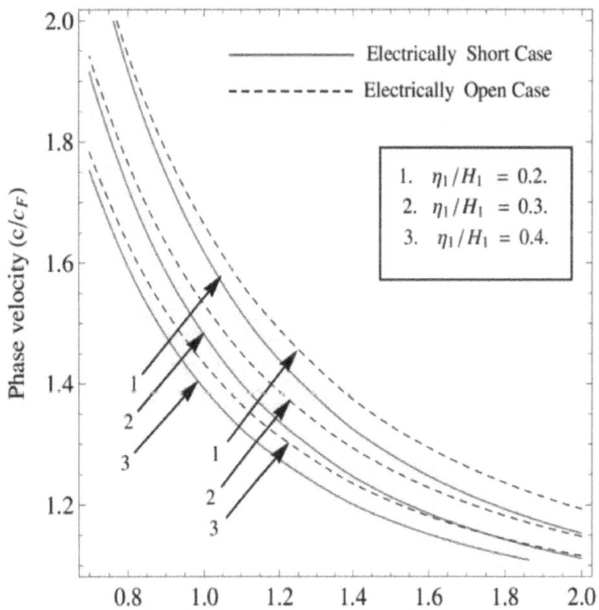

Figure 8.9 Plot of $(\frac{c}{c_F})$ with respect to (κH_1) for different values of $(\frac{\eta_1}{H_1})$
Source: Authors

and lower corrugated interfaces for $(\frac{c}{c_F})$, with respect to the (κH_1) represented in the Figures 8.8–8.10. Across all cases, the electrically open configuration consistently exhibits higher phase velocities compared to the electrically short configuration. Figure 8.8 highlights the effect of the position parameter $(\frac{y}{H_1})$. For the increasing values of $(\frac{y}{H_1})$, $(\frac{c}{c_F})$ decreases. Figure 8.9 depicts the influence of the initial flatness parameter $(\frac{\eta_1}{H_1})$, corresponding to the lower corrugated interface. The curves demonstrate a consistent trend across the frequency range, with $(\frac{c}{c_F})$ decreasing as $(\frac{\eta_1}{H_1})$ increases. The influence of the initial flatness parameter $(\frac{\eta_2}{H_1})$, for the upper corrugated interface is shown

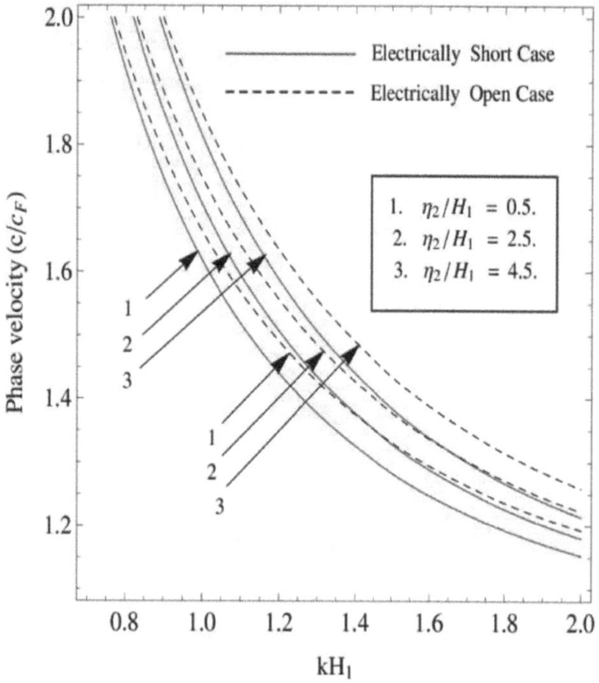

Figure 8.10 Plot of $(\frac{c}{c_F})$ with respect to (κH_1) for different values of $(\frac{\eta_2}{H_1})$
Source: Authors

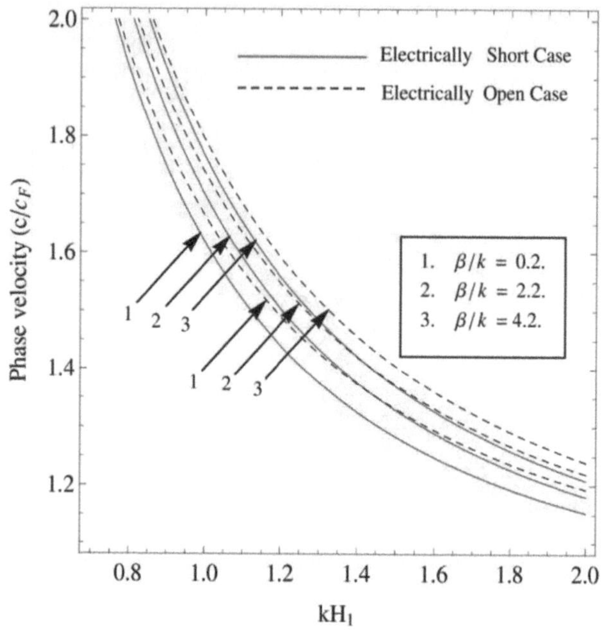

Figure 8.11 Plot of $(\frac{c}{c_F})$ with respect to (κH_1) for different values of $(\frac{\beta}{\kappa})$
Source: Authors

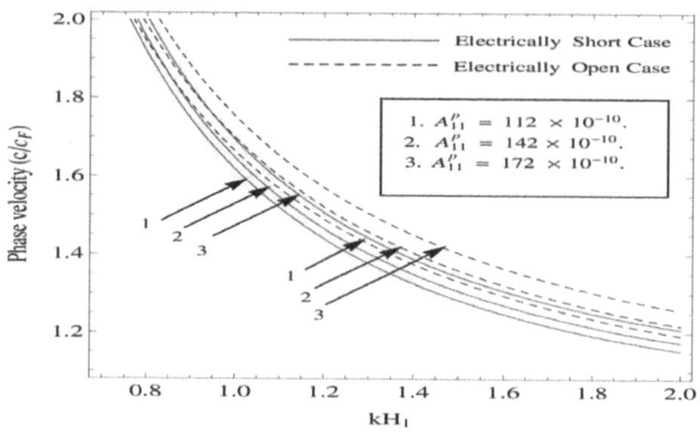

Figure 8.12 Plot of $(\frac{c}{c_F})$ against (κH_1) for distinct values of A_{11}^{p}
Source: Authors

in Figure 8.10. In contrast to the lower interface, $(\frac{c}{c_F})$ increases with increasing $(\frac{\eta_2}{H_1})$, indicating a positive impact of this parameter on SH-wave propagation.

Effect of gradient coefficients and dielectric coupling on phase velocity

The gradient coefficients of the FGPM layer on the $(\frac{c}{c_F})$ against (κH_1) on SH-wave given in the Figure 8.11. It is observed that for both electrically short and electrically open conditions increasing the gradient coefficient results in higher $(\frac{c}{c_F})$.

For both electrical conditions impact of the dielectric coupling constant between the two phases of the porous aggregate, denoted by A_{11}^{p}, on the dimensionless phase velocity $(\frac{c}{c_F})$ of SH-waves shown in the Figure 8.12. The findings reveal that as the dielectric coupling constant increases, the phase velocity of SH-waves also rises.

Concluding remarks

The developed theoretical model has been employed to examine the impact of gradient coefficients, piezoelectric constants and corrugation parameters on the propagation of SH-waves in a FGPM layer embedded between a porous piezoelectric medium and an initially stressed orthotropic substrate. The analysis reveals that the selected parameters show great impact on the phase velocity of the SH-waves. Several limiting cases have also been examined to validate and enrich the study. The major findings can be summarised as follows:

1. The phase velocity of SH-wave is strongly affected by dielectric constants, piezoelectric constants, layer thickness and corrugation parameters in both short-circuit and electrically open conditions.

2. The dielectric constants (T_{11}, T^*_{11}), piezoelectric constant (e^F_{15}), width of the PPM layer (H_2) and the lower interface parameters (η_1/H_1) tend to reduce the phase velocity under both boundary conditions.
3. In contrast, on increasing the piezoelectric constant (e^P_{15}), width of FGPM layer (H_1), upper interface parameter (η_2/H_1), gradient coefficient (β/κ) and dielectric coupling constant (A^ρ_{11}), SH-wave velocity increases.
4. The present investigation offers useful insights for the construction of composite structures in SAW devices, which play an important role in filters, oscillators, transformers, sensors and other acoustic wave-based systems.
5. The proposed model has practical applications in the development of SAW-based technologies, including communication devices, SAW filters for GPS, television broadcasting and potentially in mobile phone systems.

Appendix

$$\varkappa_{11} = \gamma_{11} = \left(\delta\overline{F_{44}} - \Omega'_2 i\kappa\overline{F_{44}}\right)e^{\delta(\Omega_2-(H_1+H_2))}, \varkappa_{12} = \gamma_{12} = -\left(\delta\overline{F_{44}} + \Omega'_2 i\kappa\overline{F_{44}}\right)e^{-\delta(\Omega_2-(H_1+H_2))},$$
$$\varkappa_{13} = \gamma_{13} = \left(ke^p_{15} - e^p_{15}\Omega'_2 i\kappa\right)e^{\kappa(\Omega_2-(H_1+H_2))}, \varkappa_{14} = \gamma_{14} = \left(\kappa e^p_{15} - e^p_{15}\Omega'_2 i\kappa\right)e^{-\kappa(\Omega_2-(H_1+H_2))},$$

$$\gamma_{21} = \frac{e^p_{15}}{T^p_{11}}e^{\delta(\Omega_2-(H_1+H_2))}, \gamma_{22} = \frac{e^p_{15}}{T^p_{11}}e^{-\delta(\Omega_2-(H_1+H_2))}, \gamma_{23} = e^{\kappa(\Omega_2-(H_1+H_2))}, \gamma_{24} = e^{-\kappa(\Omega_2-(H_1+H_2))},$$

$$\varkappa_{23} = -T^p_{11}\kappa e^{\kappa(\Omega_2-(H_1+H_2))}, \varkappa_{24} = T^p_{11}\kappa e^{-\kappa(\Omega_2-(H_1+H_2))},$$

$$\varkappa_{31} = \gamma_{31} = e^{\gamma(\Omega_1-(H_1))}, \varkappa_{32} = \gamma_{32} = e^{-\gamma(\Omega_1-(H_1))},$$
$$\varkappa_{35} = \gamma_{35} = e^{s(\Omega_1-(H_1))}\cos\lambda(\Omega_1 - (H_1)), \varkappa_{36} = \gamma_{36} = e^{s(\Omega_1-(H_1))}\sin\lambda(\Omega_2 - (H_1 + H_2))$$

$$\varkappa_{41} = \gamma_{41} = \overline{F_{44}}(\delta - \Omega'_1 i\kappa)e^{\delta(\Omega_1-(H_1))}, \varkappa_{42} = \gamma_{42} = \overline{F_{44}}(\delta - \Omega'_1 i\kappa)e^{-\delta(\Omega_1-(H_1))},$$
$$\varkappa_{43} = \gamma_{43} = e^p_{15}(\kappa - \Omega'_1 i\kappa)e^{\kappa(\Omega_1-(H_1))}, \varkappa_{44} = \gamma_{44} = e^p_{15}(\kappa - \Omega'_1 i\kappa)e^{-\kappa(\Omega_1-(H_1))},$$
$$\varkappa_{45} = \gamma_{45} = \overline{C_{44}}e^{\beta(\Omega_1-(H_1))}e^{s(\Omega_1-(H_1))}\big[s\cos\lambda(\Omega_1 - (H_1)) - \lambda\sin\lambda(\Omega_1 - (H_1)) - i\kappa\Omega'_1\cos\lambda(\Omega_1 - (H_1))\big],$$
$$\varkappa_{46} = \gamma_{46} = \overline{C_{44}}e^{\beta(\Omega_1-(H_1))}e^{s(\Omega_1-(H_1))}\big[s\sin\lambda(\Omega_1 - (H_1)) + \lambda\cos\lambda(\Omega_1 - (H_1)) - i\kappa\Omega'_1\sin\lambda(\Omega_1 - (H_1))\big],$$
$$\varkappa_{47} = \gamma_{47} = \bar{e}^0_{15}e^{\beta(\Omega_1-(H_1))}e^{r_1(\Omega_1-(H_1))}(r_1 - i\kappa\Omega'_1),$$
$$\varkappa_{48} = \gamma_{48} = \bar{e}^0_{15}e^{\beta(\Omega_1-(H_1))}e^{r_2(\Omega_1-(H_1))}(r_2 - i\kappa\Omega'_1),$$

$$\varkappa_{51} = \gamma_{51} = \frac{e^p_{15}}{T^p_{11}}e^{\delta(\Omega_1-(H_1))}, \varkappa_{52} = \gamma_{52} = \frac{e^p_{15}}{T^p_{11}}e^{-\delta(\Omega_1-(H_1))}, \varkappa_{53} = \gamma_{53} = e^{\kappa(\Omega_1-(H_1))}, \varkappa_{54} = \gamma_{54} = e^{-\kappa(\Omega_1-(H_1))},$$
$$\varkappa_{55} = \gamma_{55} = \frac{\bar{e}^0_{15}}{\Phi^0_{11}}e^{s(\Omega_1-(H_1))}\cos\lambda(\Omega_1 - (H_1)), \varkappa_{56} = \gamma_{56} = \frac{\bar{e}^0_{15}}{\Phi^0_{11}}e^{s(\Omega_1-(H_1))}\sin\lambda(\Omega_1 - (H_1)),$$
$$\varkappa_{57} = \gamma_{57} = e^{r_1(\Omega_1-(H_1))}, \varkappa_{58} = \gamma_{58} = e^{r_2(\Omega_1-(H_1))},$$

$$\varkappa_{63} = \gamma_{63} = -T^p_{11}\kappa e^{\kappa(\Omega_1-(H_1))}, \varkappa_{64} = \gamma_{64} = T^p_{11}\kappa e^{-\kappa(\Omega_1-(H_1))}, \varkappa_{67} = \gamma_{67} = \bar{\Phi}^0_{11}r_1 e^{(\beta+r_1)(\Omega_1-(H_1))},$$
$$\varkappa_{68} = \gamma_{68} = \bar{\Phi}^0_{11}r_2 e^{(\beta+r_2)(\Omega_1-(H_1))},$$

$$\varkappa_{75} = \gamma_{75} = 1, \varkappa_{76} = \gamma_{76} = 0, \varkappa_{79} = \gamma_{79} = 1, \varkappa_{85} = \gamma_{85} = \overline{C_{44}}s, \varkappa_{86} = \gamma_{86} = \overline{C_{44}}\lambda,$$
$$\varkappa_{87} = \gamma_{87} = r_1 e^0_{15}, \varkappa_{88} = \gamma_{88} = r_2\bar{e}^0_{15},$$

$$\varkappa_{95} = \gamma_{95} = \frac{\bar{e}^0_{15}}{\Phi^0_{11}}, \varkappa_{96} = \gamma_{96} = 0, \varkappa_{97} = \gamma_{97} = 1, \varkappa_{98} = \gamma_{98} = 1.$$

References

[1] Mondal, S., Sahu, S. A. Propagation of SH waves in corrugated FGPM layer lying over a piezomagnetic half-space. *Mech. Adv. Mater. Struct.*, 2019;26(1):29–34.

[2] Mondal, S., Sahu, S. A., K. K. Pankaj. Transference of Love-type waves in a bedded structure containing a functionally graded material and a porous piezoelectric medium. *Appl. Mathem. Mech.*, 2019;40:1083–1096.

[3] Hemalatha, K., Kumar, S., Akshaya, A. Rayleigh wave at imperfectly corrugated interface in FGPM structure. *Coupl. Sys. Mech.*, 2023;12(4):337.

[4] Hemalatha, K., Kumar, S . Reflection and transmission of transverse waves at a corrugated interface of two functionally graded piezoelectric half-spaces. 2023. https://doi.org/10.21203/rs.3.rs-3706183/v1.

[5] Sahu, S. A., Mondal, S., Dewangan, N. Polarized shear waves in functionally graded piezoelectric material layer sandwiched between corrugated piezomagnetic layer and elastic substrate. *J. Sandwich Struct. Mater.*, 2019;21(8):2921–2948.

[6] Sahu, S. A., Baroi, J. Analysis of surface wave behaviour in corrugated piezomagnetic layer resting on inhomogeneous half-space. *Mech. Adv. Mater. Struct.*, 2019;26(7):639–650.

[7] Li, K., et al. The complex rayleigh waves in a functionally graded piezoelectric half-space: An improvement of the laguerre polynomial approach. *Materials*, 2020;13(10):2320.

[8] Jiangong, Y., et al. Wave propagation in functionally graded piezoelectric spherically curved plates. *Physica Status Solidi*, 2007;244(9):3377–3389.

[9] Biswas, M., Sahu, S. A. Analysis of Love-type acoustic wave in a functionally graded piezomagnetic plate sandwiched between elastic layers. *Acta Mechanica*, 2022;233(11):4295–4310.

[10] Kundu, S., et al. Impacts on the propagation of SH-waves in a heterogeneous viscoelastic layer sandwiched between an anisotropic porous layer and an initially stressed isotropic half space. *J. Mech.*, 2017;33(1):13–22.

[11] Abd-Alla, A. M., Ahmed, S. M. Propagation of Love waves in a non-homogeneous orthotropic elastic layer under initial stress overlying semi-infinite medium. *Appl. Mathem. Comput.*, 1999;106(2–3):265–275.

[12] Manna, S., Kundu, S., Gupta, S. Love wave propagation in a piezoelectric layer overlying in an inhomogeneous elastic half-space. *J. Vibr. Control*, 2015;21(13):2553–2568.

[13] Chaki, M. S., Singh, A. K. A study on SH wave propagation in FGPM layered structure due to abrupt thickening and imperfect bonding at the interface. *Adv. Struct. Engg. Mech.*, 2019.

[14] Verma, A. K., et al. Rayleigh-type wave propagation on a transversely isotropic viscoelastic layer with yielding and rigid foundations. *Mech. Adv. Mater. Struct.*, 2019;26(2):107–118.

[15] Singh, A. K., Rajput, P., Chaki, M. S. Analytical study of Love wave propagation in functionally graded piezo-poroelastic media with electroded boundary and abruptly thickened imperfect interface. *Waves Random Comp. Media*, 2022;32(1):463–487.

[16] Salah, I. B., Njeh, A., Ghozlen, M. H. B. A theoretical study of the propagation of Rayleigh waves in a functionally graded piezoelectric material (FGPM). *Ultrasonics*, 2012;52(2):306–314.

[17] Cao, X., et al. Propagation of Love waves in a functionally graded piezoelectric material (FGPM) layered composite system. *Internat. J. Solids Struct.*, 2009;46(22–23):4123–4132.

[18] Singhal. Approximation of surface wave frequency in piezo-composite structure. *Comp. Part B Engg.*, 2018;19:19–28.

[19] Chen, J., Guo, J., Pan, E. Wave propagation in magneto-electro-elastic multilayered plates with nonlocal effect. *J. Sound Vibr.*, 2017;400:550–563.

[20] Qian, Z.-H., Jin, F., Hirose, S. Piezoelectric Love waves in an FGPM layered structure. *Mech. Adv. Mater. Struct.*, 2011;18(1):77–84.

[21] Chaudhary, S., Singhal, A., Sahu, S. A. Influence of the imperfect interface on Love-type mechanical wave in a FGPM layer. *J. Solid Mech.*, 2019;11(1):201–209.

[22] Tani, J., Liu, G.-R. SH surface waves in functionally gradient piezoelectric plates. *JSME Internat. J. Ser. A Mech. Mater. Engg.*, 1993;36(2):152–155.

[23] Manukyan, G. A., Manukyan, N. K., Manukyan, Z. K. Propagation of love waves in a functionally graded piezoelectric material (FGPM) layer system. *Internat. J. Solids Struct.*, 2009;46(22).

[24] Mercadelli, E., Sanson, A., Galassi, C . Porous piezoelectric ceramics. INTECH Open Access Publisher, 2010. 111–128.

[25] Martínez-Ayuso, G., et al. Homogenization of porous piezoelectric materials. *Internat. J. Solids Struct.*, 2017;113:218–229.

[26] Derayatifar, M., Tahani, M., Moeenfard, H. Nonlinear analysis of functionally graded piezoelectric energy harvesters. *Comp. Struct.*, 2017;182:199–208.

[27] Heshmati, M., Amini, Y. J. A. M. M. A comprehensive study on the functionally graded piezoelectric energy harvesting from vibrations of a graded beam under travelling multi-oscillators. *Appl. Mathem. Model.*, 2019;66:344–361.

[28] Huang, Y., Das, P. K., Bhethanabotla, V. R. Surface acoustic waves in biosensing applications. *Sensors Actuat. Rep.*, 2021;3:100041.
[29] Ruppel, Clemens, C. W., Reindl, L., Weigel, R. SAW devices and their wireless communications applications. *IEEE Microw. Magaz.*, 2002;3(2):65–71.
[30] Liu, B., et al. Surface acoustic wave devices for sensor applications. *J. Semicond.*, 2016;37(2):021001.
[31] Qian, Z. H., Jin, F., Lu, T., Kishimoto, K. Transverse surface waves in an FGM layered structure. *Acta Mechanica*, 2009;207(3):183–193.

9 Experimental and numerical investigation of face-core debonding in honeycomb sandwich panels

Kumar[1,a], R. Ganesh Narayanan[1,b] and N. Muthu[1,2,c]

[1]Department of Mechanical Engineering, IIT Guwahati, Guwahati, Assam, India

[2]Department of Mechanical Engineering, IIT Palakkad, Palakkad, Kerala, India

Abstract

This work presents an experimental and numerical study to characterise the fracture from debonding between the core and facesheet of honeycomb sandwich double cantilever beam (DCB).The experimental setup was carefully designed to facilitate accurate testing. To simulate the experimental tests, numerical simulations were conducted using a cohesive zone model to replicate crack propagation. An inverse method was used to find out the cohesive properties by correlate the numerical load-displacement curves to the experimental data. Establishing the cohesive strength and fracture energy values allowed for the effective characterisation of the cohesive law for mode I interface fracture. A strong correlation was shown between the numerical predictions and the experimental data for the load-displacement curves. The finite element method (FEM) analysis provided deeper understanding into the stress distribution and delamination behaviour of the joints. The stress analysis at the crack tip indicated that the DCB test involves a mix-mode (I/II) debonding process, with mode I fracture being the dominant mechanism. Furthermore, the study thoroughly explored the crack creation and propagation mechanism at the interface of face and core, providing detailed information on the physical phenomena governing the debonding mechanism in sandwich structures.

Keywords: Honeycomb, sandwich panel, cohesive zone modelling, mode I

Introduction

Sandwich structures are integral to several sectors due to their high strength and stiffness. Their design typically consists of two thin, stiff outer panels that bear the primary in-plane loads (e.g., bending and tensile stresses) and a thick inner core. Honeycomb cores, in particular, are favoured for their ultralight weight and superior mechanical properties [1, 2]. This combination provides exceptional sound insulation, energy absorption and heat resistance, making them indispensable for ensuring the efficiency and safety of high-speed trains, ships, and airplanes. Meanwhile, the honeycomb core plays a crucial role in resisting shear forces, allowing the structure to handle complex and varied loads.

While offering significant advantages, the combination of two distinct materials in sandwich structures does not always result in synergistic performance. One of the key challenges is structural instability, which can lead to catastrophic failure [3]. Sandwich structures are particularly prone to various types of instability, such as local and global buckling, depending on their core thickness. Ji and Waas [4] demonstrated that core thickness plays a critical role in determining failure modes during impact loading. In thick-core sandwich structures, the two skins tend to respond independently to impact, leading to localised buckling at the impacted end. This localised buckling often initiates debonding at core and skin, a failure that can progressively propagate through the structure. Conversely, in thinner sandwich structures, global buckling typically occurs first, causing an overall collapse. As the mechanical properties of the facesheets and core differ significantly, facesheet-core contact is often the weakest point in a sandwich construction. This interface is highly susceptible to failure under load, making it a critical area of concern in design and performance assessments.

When sandwich structures experience debonding failure, the load transmission between the facesheets and the core is disrupted, resulting in a major loss of load carrying capacity and stiffness. Numerous testing methodologies have been suggested to characterise opening, shearing, and mixed-mode skin-core debonding behaviours. To examine

[a]avneesh.kumar@iitg.ac.in, [b]ganu@iitg.ac.in, [c]nelsonm@iitpkd.ac.in

DOI: 10.1201/9781003770442-9

mode I debonding, Prasad and Carlsson [5] adjusted the DCB test for composite plates and modified it for foam core sandwich beams. According to their research, greater fracture lengths lead to higher stress intensity factors, while core thickness has less impact on energy release rate and stress intensity factors [6]. To prevent crack kinking into core, Li and Carlsson [7] introduced the debonding test, ensuring that cracks propagate near the face-core interface. Furthermore, Cantwell and Davies [8] introduced the SCB test, which was developed from the DCB test, to assess the strength of the face-core interfacial. Quispitupa et al. [9] extended mixed-mode bending (MMB) test, which was initially created for unidirectional composites, to foam-core sandwich structures because sandwich panels are frequently exposed to mixed-mode loading in practical applications. This extension provides a framework for calculating compliance and strain energy release rates.

CZM offers an approach to the analysis of face-core debonding in sandwich sheets. Initially introduced by Dugdale [10], CZM has since been applied to numerous fracture problems. CZM is often used to model debonding, in contrast to standard fracture mechanics, which enables the prediction of layer separation and the generation and progression of fractures in bi-material structures. It captures the damage mechanisms at crack tip by defining a constitutive connection between traction and displacement. Various CZM methods have been used in the literature, including polynomial [11], bi-linear [12], and exponential [13]. For instance, Mohammed and Liechti [14] used CZM in numerical and experimental studies to predict crack formation at aluminium-epoxy corners. CZM has shown accurate predictions of failure loads. Sridharan and Li [15] used cohesive layers in ABAQUS to simulate face-core debonding in foam core sandwich sheet under mode-I and compression load. Hower et al. [16] developed a traction-separation law that successfully simulates mode I behaviour in aluminium honeycomb structures under SCB testing, while Gopalakrishnan et al. [17] investigated debonding in metallic honeycomb sandwich structures by combining CZM with experiments.

This research aims to examine the debonding fracture in honeycomb sandwich sheets, focusing on the facesheet-core interface during DCB tests. The study assesses the effects of key parameters—such as fracture toughness, facesheet and core thickness, on load-displacement behaviour and crack propagation. Numerical simulations using the CZM determine the cohesive strength and the fracture energy by fitting numerical curves to the experimental data. The research characterises the cohesive law for mode I interfacial fracture, providing insights into stress distribution and delamination behaviour to optimise mechanical performance and reliability.

Experimental

Material

The sandwich structures under consideration consist of a 6 mm honeycomb core (AA3003) and the 0.95 mm thick facesheet (AA5052-H32). Both the skin and core materials are commercially available, and the assembly of the sandwich structures was achieved using epoxy adhesive. The material characteristics of the core and facesheet are detailed as follows.

- The core and facesheet exhibit Young's modulus of 70 GPa.
- Both materials have a Poisson ratio of approximately 0.3.
- Yield strength of facesheet is 189 MPa, the honeycomb core has a yield strength of 185 MPa [18].

Fabrication process

Figure 9.1 illustrates the joining process of the sandwich structure. The bonding was achieved using an epoxy adhesive to ensure a robust bond between the honeycomb core and the facesheets. Before adhesive application, both the skin and core were wiped with acetone to eliminate any contaminants. The adhesive was applied uniformly on the skin, and the core

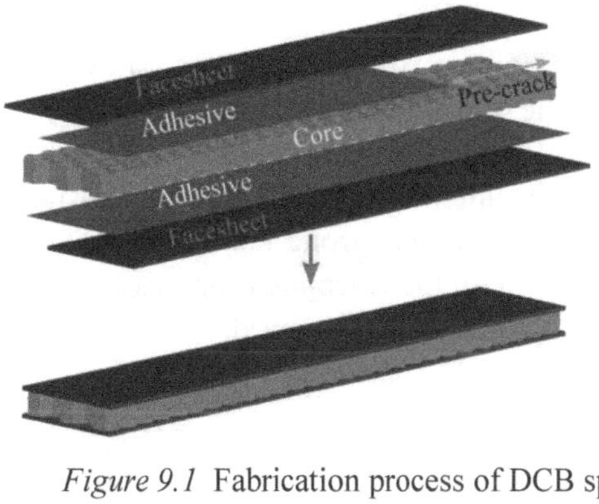

Figure 9.1 Fabrication process of DCB specimen
Source: Authors

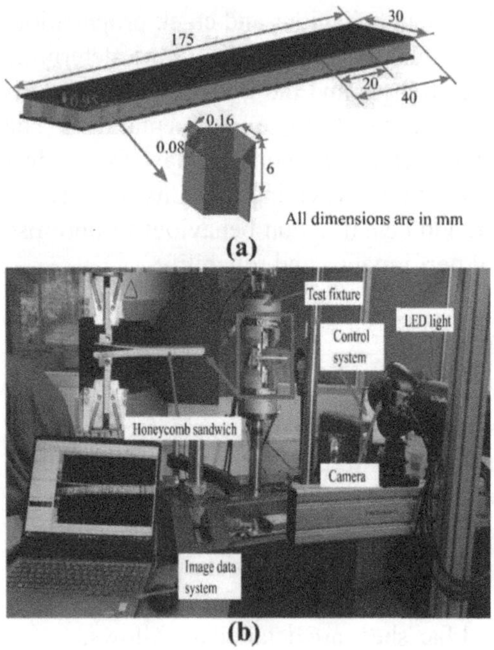

Figure 9.2 DCB test (a) specimen dimension and (b) test setup
Source: Authors

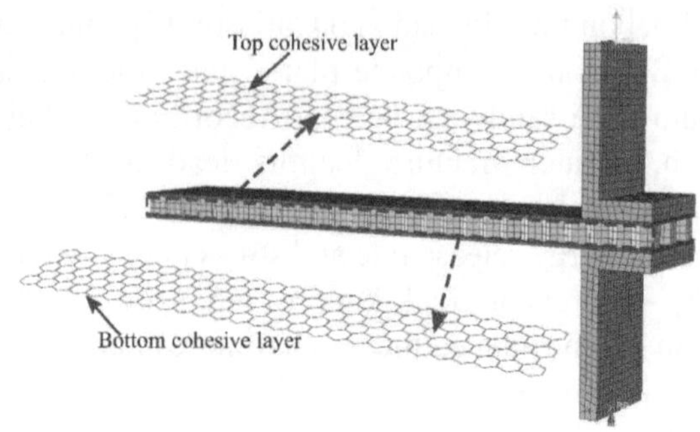

Figure 9.3 FE model of DCB test of sandwich structure
Source: Authors

was properly aligned and positioned over the adhesive-applied area. To ensure a solid bond, even pressure was applied through clamping during the curing process. The fabrication procedure was conducted in two stages. First, the lower skin was bonded with honeycomb and left to cure for 1 day at ambient temperature. After the lower sheet securely attached, the similar process was followed for the upper facesheet. This two-step method was used to prevent adhesive flow through the honeycomb core and ensure a clean, precise bond between all components. A pre-crack was formed by inserting a non-adhesive nylon sheet at the interface of the two surfaces.

DCB test

The DCB specimen dimensions are shown in Figure 9.2a. The preparation process involved attaching end blocks with Araldite adhesive, followed by clamping the specimens in a bench vice for a 24-hour curing period to ensure a strong bond. As shown in Figure 9.2b, where the specimens subjected to a loading rate of 1 mm/min ram speed.

Numerical analysis

ABAQUS software was used to model DCB sandwich test specimens, incorporating CZM to simulate the propagation of cracks. Cohesive layer finite elements, which were inserted between the upper facesheet and core to represent the interfacial behaviour, were used to construct the cohesive damage model. The main work of these simulations was to apply an inverse method, in which the cohesive parameters were precisely determined by fitting numerical findings to experimental load-displacement curves.

DCB test FE model

FE simulations were performed using ABAQUS® 2017 to model DCB tests and compare the results with experimental data. R3D4 elements were used to simulate rigid crossheads applying the load, while C3D8 elements were applied to mesh the skin and core. To simulate damage progression in the adhesive layer, COH3D8 cohesive elements were utilised. A small viscosity value of 0.00001 was introduced to prevent convergence problems during the simulation, ensuring stable damage evolution. The cohesive layer was connected to the lower and upper skins and the honeycomb core using a "tie" constraint, maintaining proper bonding between components. Figure 9.3 presents the FE simulation setup, including the boundary conditions that were used to closely replicate the experimental conditions in DCB tests.

CZM

The bonding at interfaces, particularly between the skin and the core, was characterised using CZM. The CZM principles assert that cohesive forces act to resist surface separation at these interfaces. The behaviour of this separation and the corresponding resistive forces are governed by the traction-separation law, which defines the connection between the separation distance (δ) and the cohesive stress

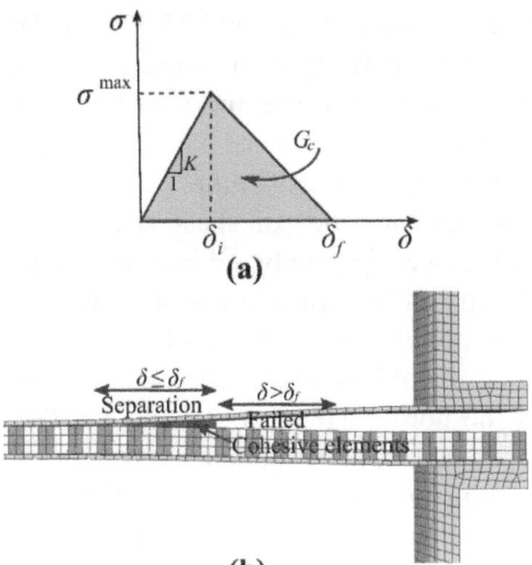

Figure 9.4 (a) Traction separation curve and (b) propagation of crack in DCB specimen
Source: Authors

(σ) that opposes it. A commonly used and straightforward type of traction-separation law is shown in Figure 9.4a. In this model, during the initial separation phase, the cohesive stress (σ) increases as the separation (δ) grows. The maximum cohesive strength (σ^{max}) occurs at the displacement , marking the point where the interface starts to degrade. The σ gradually reduces with further increase in separation, finally reaching zero at the displacement , which signifies full separation. This entire crack propagation is depicted in Figure 9.4b.

The interface toughness is shown by the area contained by the traction-separation curve, G_c, which is a critical parameter in assessing the durability and performance of bonded interfaces in sandwich structures. In the context of a triangular traction-separation curve, the interface toughness can be expressed mathematically.

$$G_c = \frac{1}{2}\sigma^{max}\delta_f \tag{1}$$

$$\sigma = \begin{cases} k\delta, & ,\delta < \delta_i \\ k\delta(1-D) & ,\delta_i \le \delta \le \delta_f \\ 0 & ,\delta_f < \delta \end{cases} \tag{2}$$

Where k and D is the penalty stiffness and damage variable [19–23].

D can be illustrated as

$$D = \frac{\delta_f(\delta-\delta_i)}{\delta(\delta_f-\delta_i)} \tag{3}$$

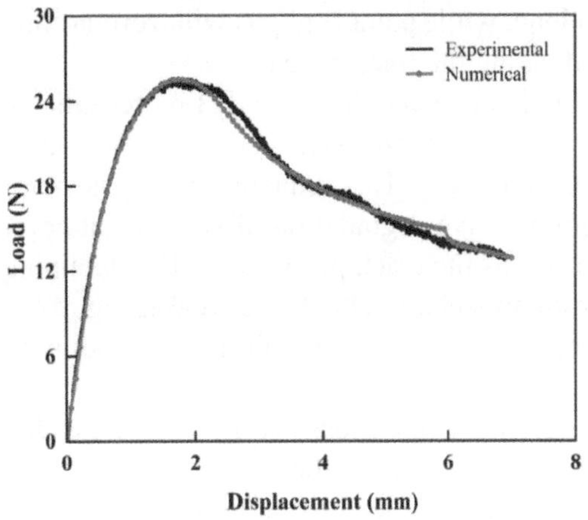

Figure 9.5 Load-displacement behaviour of DCB test of the sandwich
Source: Authors

Results and discussion

DCB results

Figure 9.5 presents the load evolution curves recorded during the DCB test, showcasing experimental data and FE simulation results for adhesive-bonded sandwich structure. The results demonstrate a strong correlation in the experimental and numerical results, validating the accuracy of FE model used in this analysis. At the beginning of the test, the load response exhibits a linear trend, indicating that the adhesive joint behaves elastically under the applied load. As the loading continues, the behaviour transitions into a non-linear phase, reflecting the complex interactions within the adhesive layer with core and skins. This non-linear behaviour continues until the maximum load is attained, showcasing the peak strength of the adhesive bond. After reaching this maximum load, the curve indicates a decrease in load as the crack propagates through the adhesive. This decrease in load is expressive of damage occurring within the adhesive, leading to a reduction in the structural integrity of the joint. Overall, the observed load evolution highlights the effectiveness of the adhesive bond in resisting initial loads and the critical nature of crack propagation in determining the overall performance of the sandwich.

Figure 9.6 shows the crack tip shear and normal stress, along with the related load-displacement behaviour. A and B are two key load points highlighted on load-displacement curve: point A relates to the maximum normal stress, indicating the damage initiation load, while point B aligns with zero normal stress, indicating the fracture load.

The region between points A and B represents the phase in which the normal stress gradually decreases from its peak to zero. This reduction in stress signifies the progressive degradation of stiffness at core-face interface as the crack propagates. The declining stress during this phase reflects the weakening of the joint in the facesheet and core, ultimately leading to failure.

Figure 9.7 shows the shear and normal stress distribution at the middle of the cohesive layers (for load points A and B), demonstrating that both modelling approaches accurately capture the joint's critical stress concentrations. Understanding these stress distributions is essential for evaluating joint performance and reliability, as they directly influence the likelihood of damage and failure. Damage analysis was also conducted, focusing on the stiffness degradation (SDEG) of cohesive elements. The SDEG within the CZM quantifies the reduction in stiffness of the adhesive during DCB test.

The distributions of shear and normal stress across joint's top interface are plotted in Figure 9.8 for load points A and B. At these load positions, the highest normal stress is recorded at zero and 25.97 mm. The normal stress at point B starts to propagate cracks when it drops to zero at 0 mm and peaks at 25.97 mm. The difference between these points represents the damage zone length of 25.97 mm.

Furthermore, the presence of shear stress along the length of the overlap, combined with its smaller magnitude compared to normal stress, indicates that the debonding process observed in the DCB test involves a coupling of both mode (I/II) fracture [24]. However, as the normal stress dominates, the fracture is primarily governed by mode I.

Figure 9.9 illustrates the progression of damage in the sandwich sheets during the DCB test at load points A and B. At load point B, the damage value

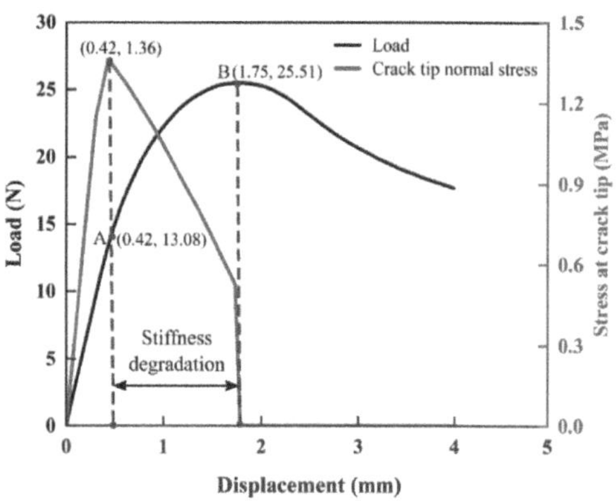

Figure 9.6 Crack tip stress
Source: Authors

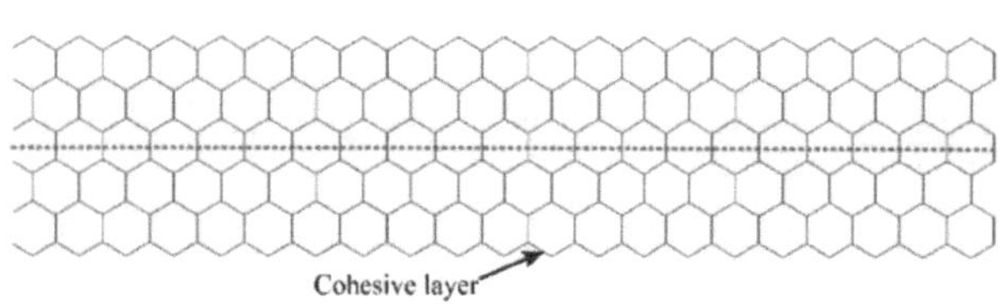

Figure 9.7 Stress and damage response at cohesive mid-layer
Source: Authors

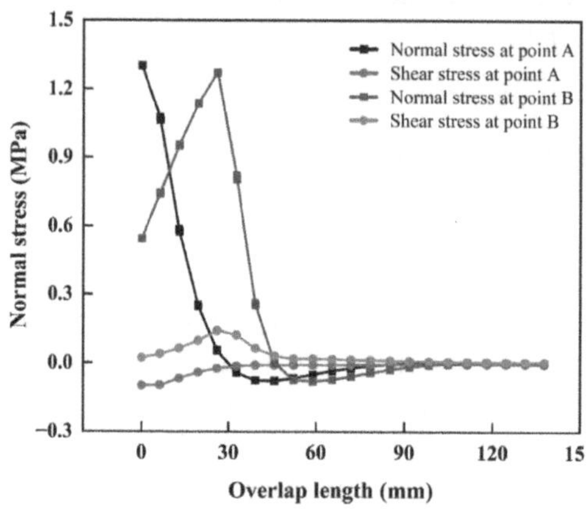

Figure 9.8 Shear and normal stress distribution of the sandwich structure of DCB test at points A and B of Figure 9.6
Source: Authors

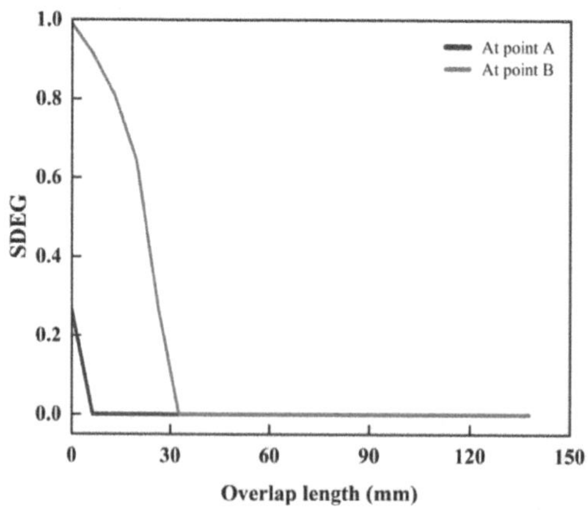

Figure 9.9 Damage behaviour of sandwiches at points A and B of Figure 9.6
Source: Author

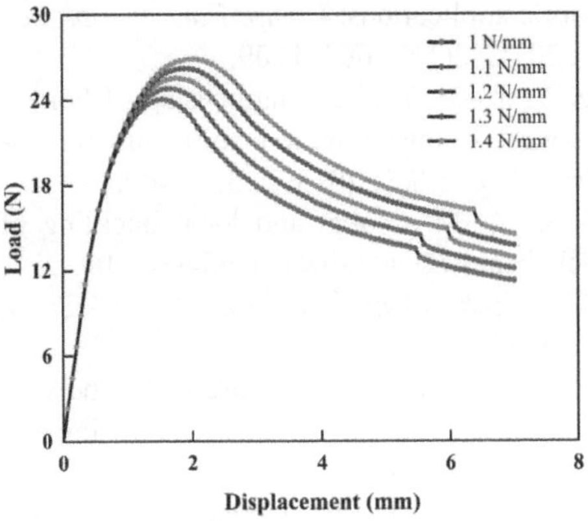

Figure 9.10 Effect of varying fracture toughness on load-displacement curve
Source: Authors

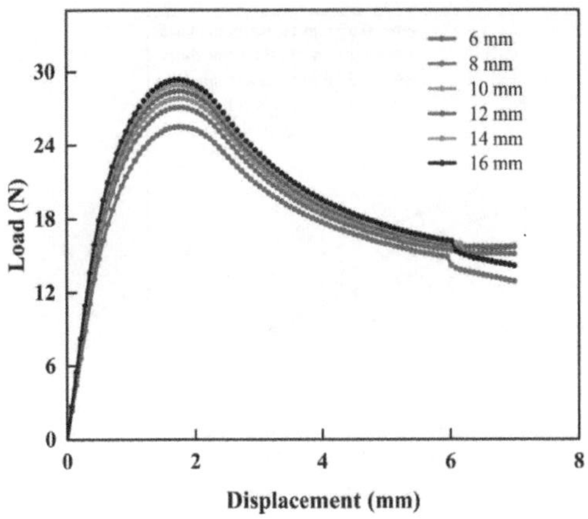

Figure 9.12 Effect of the thickness of honeycomb core on load-displacement curve
Source: Authors

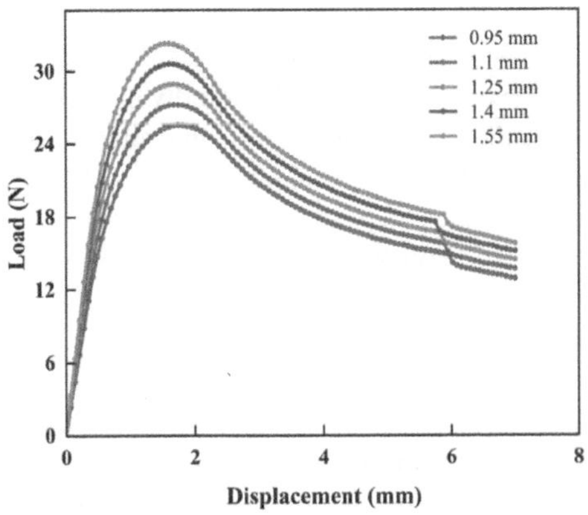

Figure 9.11 Effect of facesheet thickness on load-displacement
Source: Authors

approaches unity, signifying the start of crack propagation. This value indicates that the adhesive at the interface has reached its critical failure point, marking the transition from damage initiation to active crack growth within the sandwich structure.

Effect of key parameters on sandwich structures
The numerical study was performed to find the effect of fracture toughness, facesheet thickness, and core thickness on the mechanical response of the sandwich. In this study, one parameter was varied at a time while keeping two other two parameters constant. The values chosen for the study were 1.2 N/mm for fracture toughness, 0.95 mm for facesheet thickness, and 6 mm for honeycomb core thickness.

As depicted in Figures 9.10–9.12, fracture toughness, facesheet, and core thickness all have an important impact on load-displacement response of the sandwich. Figure 9.10 illustrates that increased fracture toughness leads to higher fracture loads and displacements. This trend indicates that materials with higher toughness exhibit enhanced resistance to crack initiation and propagation [24]. In Figure 9.11, the fracture load increases as facesheet thickness increases, highlighting the importance of thicker facesheets in enhancing the load-bearing capacity. However, the fracture displacement remains constant for all facesheet thicknesses. The load-displacement curve slope is notably influenced by the thickness of the facesheet, since an increase in thickness results in greater stiffness, leading to a more pronounced slope variation. In Figure 9.12, as the thickness of core increases, both fracture load and stiffness rise, indicating that thicker cores contribute to a stronger structure. However, similar to facesheet thickness, the fracture displacement remains nearly the same across different core thicknesses. It was also observed that as core thickness continues to increase, the fracture load approaches a stable value, suggesting a limit to the load-carrying capacity achievable through core thickness alone.

Furthermore, Figure 9.13 shows, the impact of core and facesheet thicknesses on interfacial strength are examined. The findings demonstrate that as thickness of the facesheet increases, so does strength. On the other hand, the strength is inversely proportional to core thickness. However, the strength fluctuation is rather little; for instance, the interfacial strength only drops by around 1.88% when the core thickness changes from 6 to 16 mm while facesheet thickness

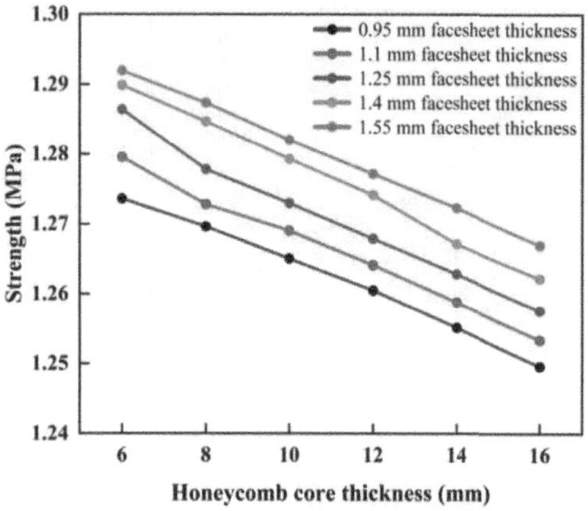

Figure 9.13 Strength analysis at different core and facesheet thicknesses
Source: Authors

remains at 0.95 mm. The DCB test shows features of both mode (I/II) debonding processes, as was previously mentioned. Consequently, the simultaneous effects of shear and normal stresses occurring on the interface affect the interfacial strength.

Conclusion

This work investigates the debonding fracture mechanisms in honeycomb core sandwich structures using experimental and numerical processes using the DCB test. To simulate the experimental results, numerical simulations were conducted employing the CZM to replicate crack propagation, utilising the inverse method to estimate the cohesive parameters by correlating the numerical load-displacement curves to the experimental curve. The numerical model allowed for the extraction of critical parameters such as crack tip stress and damage initiation load. Furthermore, the investigation revealed that the normal and shear stresses indicated a coupled mode (I/II) debonding, the normal stress consistently higher than shear stress, confirming that the mode I fracture primarily governs the DCB test. Following this, a comprehensive study was conducted to find out the effects of fracture toughness, core, and facesheet thickness on the load-displacement behaviour and strength of the sandwich structures.

References

[1] Witik, R. A., Payet, J., Michaud, V., Ludwig, C., Månson, J. A. E. Assessing the life cycle costs and environmental performance of lightweight materials in automobile applications. *Comp. Part A Appl. Sci. Manufac.*, 2011;42(11):1694–1709.

[2] Zhang, Q., et al. Bioinspired engineering of honeycomb structure - Using nature to inspire human innovation. *Prog. Mater. Sci.*, 2015;74:332–400.

[3] JI, W., Waas, A. M. Global and local buckling of a sandwich beam: Generalized analysis. *Am. Soc. Comp. 21st Tech. Conf. Am. Soc. Comp. 2006*, 2006;3:1505–1519.

[4] Ji, W., Waas, A. M. Dynamic failure of a sandwich structure subjected to an in-plane axial impact. *Comp. Struct.*, 2017;180:751–759.

[5] Prasad, S., Carlsson, L. A. Debonding and crack kinking in foam core sandwich beams-I. Analysis of fracture specimens. *Engg. Frac. Mech.*, 1994;47(6):813–824.

[6] Muthu, N., Maiti, S. K., Falzon, B. G., Guiamatsia, I. Computation of stress intensity factors in functionally graded materials using partition-of-unity meshfree method. *Aeronaut. J.*, 2012;116(1186):1263–1287.

[7] Li, X., Carlsson, L. A. The tilted sandwich debond (TSD) specimen for face/core interface fracture characterization. *J. Sandwich Struct. Mater.*, 1999;1(1):60–75.

[8] Cantwell, W. J., Davies, P. A test technique for assessing core-skin adhesion in composite sandwich structures. *J. Mater. Sci. Lett.*, 1994;13(3):203–205.

[9] Quispitupa, A., Berggreen, C., Carlsson, L. A. Design analysis of the mixed mode bending sandwich specimen. *J. Sandwich Struct. Mater.*, 2010;12(2):253–272.

[10] Chen, C., Fleck, N. A., Lu, T. J. Mode I crack growth resistance of metallic foams. *J. Mech. Phy. Solids*, 2001;49(2):231–259.

[11] Tvergaard, V. Effect of fibre debonding in a whisker-reinforced metal. *Mater. Sci. Engg. A*, 1990;125(2):203–213.

[12] Tvergaard, V., Hutchinson, J. W. The relation between crack growth resistance and fracture process parameters in elastic-plastic solids. *J. Mech. Phy. Solids*, 1992;40(6):1377–1397.

[13] Needleman, A. An analysis of tensile decohesion along an interface. *J. Mech. Phy. Solids*, 1990;38(3):289–324.

[14] Mohammed, I., Liechti, K. M. Cohesive zone modeling of crack nucleation at bimaterial corners. *J. Mech. Phy. Solids*, 2000;48(4):735–764.

[15] Sridharan, S., Li, Y. Static and dynamic delamination of foam core sandwich members. *AIAA J.*, 2006;44(12):2937–2948.

[16] Höwer, D., Lerch, B. A., Bednarcyk, B. A., Pineda, E. J., Reese, S., Simon, J. W. Cohesive zone modeling for mode I facesheet to core delamination

of sandwich panels accounting for fiber bridging. *Comp. Struct.*, 2018;183(1):568–581.

[17] Gopalakrishnan, K. C., Kumar, R. R., Lal, S. A. Cohesive zone modeling of coupled buckling - Debond growth in metallic honeycomb sandwich structure. *J. Sandwich Struct. Mater.*, 2012;14(6):679–693.

[18] Zhang, Y., Yan, L., Zhang, W., Su, P., Han, B., Guo, S. Metallic tube-reinforced aluminum honeycombs: Compressive and bending performances. *Comp. Part B Engg.*, 2019;171:192–203.

[19] Camanho, P. P., Davila, C. G., de Moura, M. F. Numerical simulation of mixed-mode progressive delamination in composite materials. *J. Comp. Mater.*, 2003;37(16):1415–1438.

[20] Alfano, G., Crisfield, M. A. Finite element interface models for the delamination analysis of laminated composites: Mechanical and computational issues. *Internat. J. Num. Methods Engg.*, 2001;50(7):1701–1736.

[21] Rana, P. K., Narayanan, R. G. Numerical and experimental response of FSSW of AA5052-H32/epoxy/AA5052-H32 sandwich sheets with varying core properties. *Internat. J. Mater. Form.*, 2021;14(6):1297–1317.

[22] Kumar, A., Muthu, N., Ganesh Narayanan, R. Prediction of peel strength of sandwich sheets made of aluminium alloys fabricated by friction stir spot welding based hybrid process using cohesive zone modeling and finite element simulations. *Engg. Failure Anal.*, 2024;162:1–24.

[23] Kumar, A., Narayanan, R. G., Muthu, N. Enhancing the performance of honeycomb core sandwich panels through friction stir spot welding strategies and fracture of lap shear specimen. *Mech. Adv. Mater. Struct.*, 2024:3123–3143.

[24] Xue, P., Wei, X., Li, Z., Xiong, J. Face-core interfacial debonding characterization model of an all-composite sandwich beam with a hexagonal honeycomb core. *Engg. Frac. Mech.*, 2022;269:108554.

10 Damage evaluation to containment of nuclear power plant subjected to impact loading of different aircraft

Alim Rawsan, Sher Bahadur Singh[a] *and Pabitra Ranjan Maiti*

Department of Civil Engineering, Indian Institute of Technology (BHU), Varanasi, India

Abstract

For the safety evaluation of a nuclear power plant (NPP) containment, actions under various operations and extreme conditions, shall be assessed. This study uses a finite element (FE) approach to present the structural behaviour of nuclear containment under different aircraft. A simplified model of an actual BWR Mark-III nuclear containment of 67 m height and 42 m inner diameter is fully modelled. To evaluate the damages due to aircraft crashes, four types of aircraft have been considered (Boeing 707-320, Phantom F4, Airbus A320 and Boeing 767-400). The behaviour of structure such as stress, plastic strain, deformation, damage and scabbing area, are investigated. The collision angle is normal to the target surface. The response of concrete is characterised using the concrete damaged plasticity (CDP) model, while the response of the reinforcement is predicted by the Johnson-Cook model to account for their respective material properties. The finite element analysis of the model has been done using ABAQUS (explicit) to describe the actions of the NPP. The results showed the most damaging aircraft is Boeing 767-400 and Phantom F4 is responsible for maximum deformation while minimum deformation was found for Airbus A320. Finally, it is concluded local damage was observed for all the aircrafts.

Keywords: Nuclear containment structure, tension damages, Phantom F4, Boeing 767-400, CPD, Johnson-Cook model

Introduction

In the dawn of the 21st century, the calculated collisions of substantial commercial aircraft into nuclear power plant (NPP) edifices and other critical structures incited intense scrutiny worldwide, especially following the cataclysmic events of September 11, 2001. Preceding this incident, the structural integrity of nuclear installations was not engineered to endure the formidable onslaught of external forces, whether from diminutive or massive commercial airliners. A standard view of Phantom F4 aircraft crash on RCC plate has been displayed in Figure 10.1.

Presently, there are 64 nuclear reactors either expanding or under construction, predominantly situated in Asia. A robustly constructed reinforced concrete (RC) containment vessel is highly effective in shielding nuclear reactors from external threats, including seismic events. In the past, the seismic resilience of these containment vessels has been investigated through both quantitative assessments and experimental analysis. Riera [1, 2] started the work on this topic at first time to get the response of NPP structure. Abbas et al. [3] investigated on the same topic. In conclusion, significant studies have now been undertaken to address the implications of accidental aircraft impacts on critical structures, such as nuclear power plant (NPP) facilities, including those conducted by Kukreja [4], Lin and Tang [5], Jiang and Chorzepa [6], Sadique et al. [7]. Nuclear containment walls generally fall into two categories: internal and external. The internal wall serves to regulate the radioactive energy, on the other hand, the external wall prevents the structure from outer encroachments. Due to the high cost of experimental testing on NPP wall responses, some researchers have opted to numerically simulate containment wall models to analyse their behaviour under the impact of an airplane crash. Sugano et al. [8] and Abbas [9] researched external wall of NPP against the impact of Boeing 707-320 aircraft using load history curve. Sugano et al. [8] also done an experiment on RCC plate for Phantom F4 fighter jet to get the actual behaviour of aircraft damage and induced effects which is shown in Figure 10.1. Few researches on this type of problem were performed before. But damage calculation was not considered in any type of analysis. In certain instances, a force history curve was employed, while in others, aircraft models were taken into account to assess the response. Several literature suggested that the stress-strain behaviour

[a]sherbahadursingh.rs.civ21@iitbhu.ac.in

DOI: 10.1201/9781003770442-10

Figure 10.1 Phantom F4 aircraft crash on RCC plate
Source: Authors

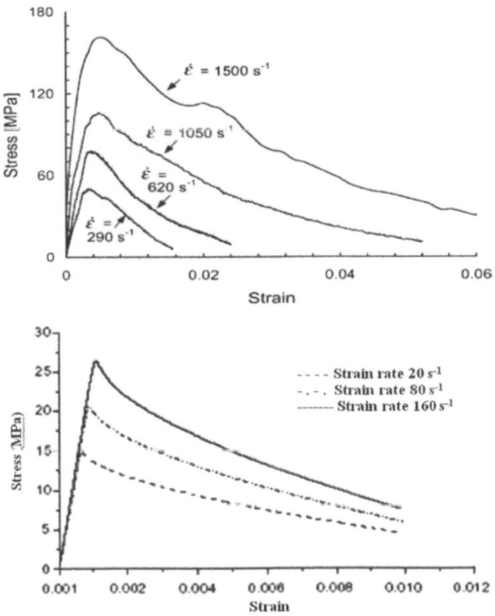

Figure 10.2 Concrete behaviour with various strain rates under compression (top) and tension (bottom)
Source: Authors

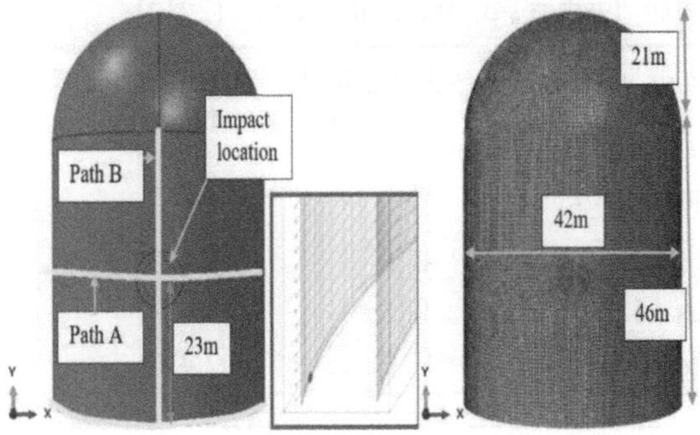

Figure 10.3 Dimension and impact location, reinforcement, and meshing details
Source: Authors

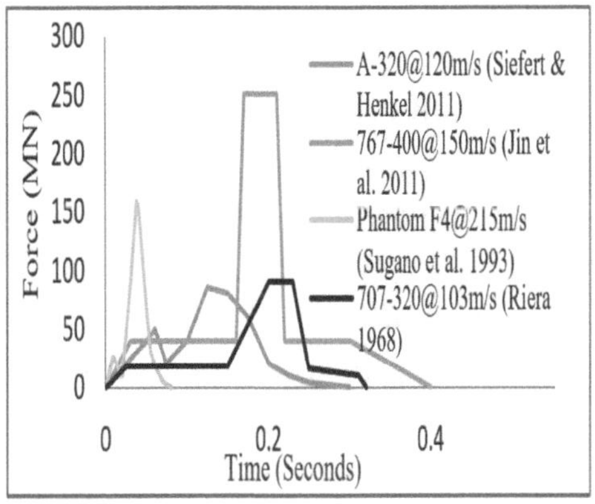

Figure 10.4 Force-time history curves of different aircraft
Source: Authors

depends largely on the nature of force, and rate of loading etc. which is presented in Figure 10.2 [10, 11]. This study utilises damage plasticity model to incorporate the actual behaviour of concrete and Johnson-Cook model for reinforcing bars, available in ABAQUS/Explicit [12].

Mathematical model

Containment and loading details

The boiling water reactor Mark-III (BWR Mark-III) structure is employed in this study. The NPP is designed using 3D solid deformable elements, while the reinforcement utilises 3D deformable wire elements. The total height of structure was 67 m, with 46 m of cylindrical section and 21 m of dome section, and has an inner diameter of 42 m, as illustrated in Figure 10.3. The other details are adopted from the literature of Saberi et al. [13].

For aircraft, the force-time history curves are utilised for imposing the load upon the structure [14]. Figure 10.4 represents the force history curve of Boeing 707-320/767-400, Phantom F4 and Airbus-A320. To predict the most damaging aircraft on NPP containment structure, various aircraft are considered and two paths (Path A and Path B) are selected to show the results which are shown in Figure 10.3. The features of all aircraft are given in Table 10.1.

To address the unnecessary distortion of the elements, the model has meshed with great attention

Table 10.1 Features of different aircrafts [2, 14].

Name of aircrafts	Length of aircraft (m)	Radius of fuselage (m)	Velocity of impact (m/s)	Peak impact force (MN)	Total impact time (s)	Wingspan (m)
Boeing 707-320	46.6	3	103	90	0.3472	44.42
Boeing 767-400	48.5	5.03	150	250	0.4	47.5
Airbus A320	37.57	3.95	120	85	0.3	34.10
Phantom F4	17.74	2.42	215	158	0.08	11.77

Source: Authors

for the simulation. The structural meshing details are shown in Figure 10.3. The meshing details and interaction have been adopted from Rawsan and Maiti [14].

Materials model and properties

The damage models considered in this study are based on the works of Rawsan and Maiti [14].

Results and discussion

The response of NPP containment to an aircraft collision is contingent on various factors such as the aircraft's velocity, mass, length, impact surface area, and prevailing weather conditions. The location of impact for all the aircraft is considered at 23 m height from the base of containment structure in the present study. This height has been taken because maximum deformation was observed in this location that means it is the most critical location of NPP structure. The explicit impact analysis is performed for simulation of model and the deformation and damage is measured for different aircrafts. The maximum deformations 0.168 m for Boeing 767-300, 251 m for Phantom F4, 0.09 m for Boeing 707-320 and 0.039 meters for Airbus A320 are observed which is illustrated in Figures 10.5–10.8. The aircraft Boeing 767-400 has the maximum damaging potential and Airbus A320 has the minimum damaging potential, as displayed in Figures 10.5 and 10.8, respectively. The deformation is high, but the damage is not maximum for the Phantom F4 fighter jet because it has maximum impact velocity but minimum in wing's span (area of impact).

Due to the larger wing span in Boeing 767-400, this aircraft is responsible for more damage in concrete. When the graph of deformation of an element in impact location has been drawn for every aircraft

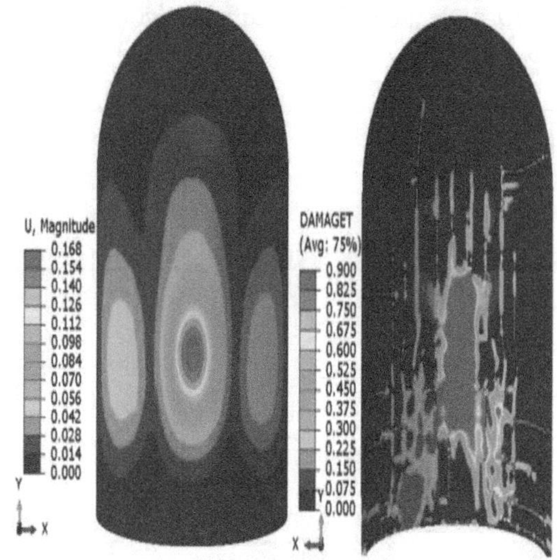

Figure 10.5 Maximum deformation and tension damages in containment for Boeing 767-400
Source: Authors

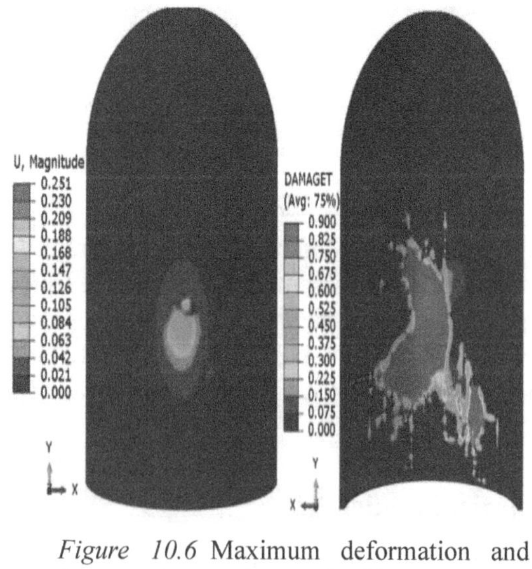

Figure 10.6 Maximum deformation and tension damages in containment for Phantom F4
Source: Authors

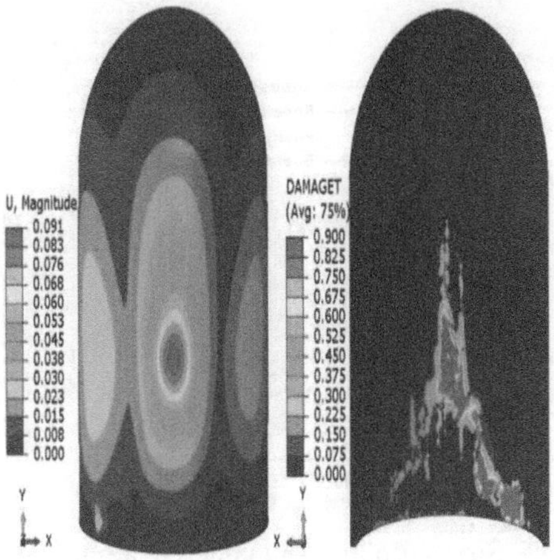

Figure 10.7 Maximum deformation and tension damages in containment for Boeing 707-320
Source: Authors

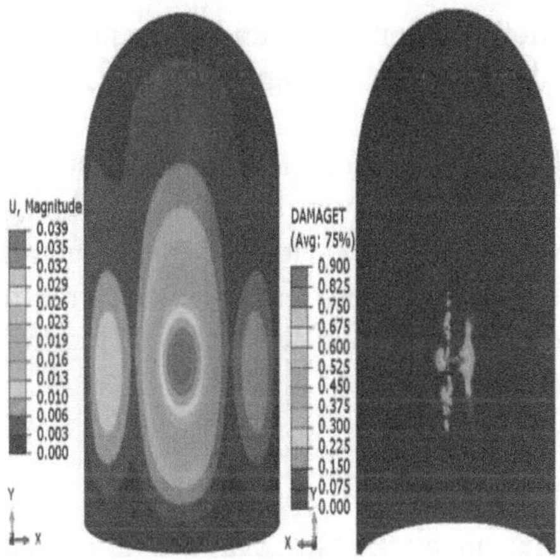

Figure 10.8 Maximum deformation and tension damages in containment for Airbus A320
Source: Authors

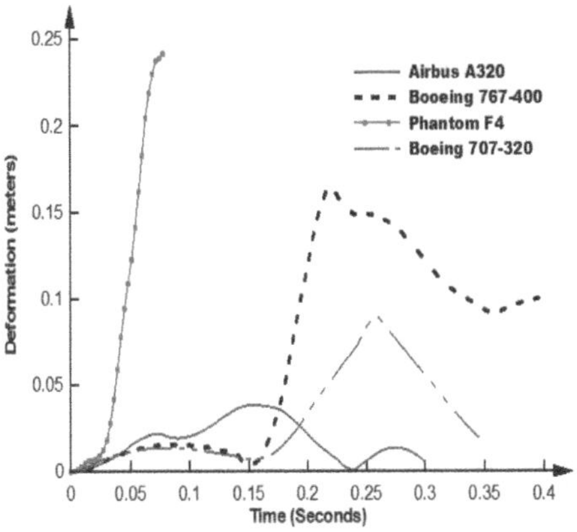

Figure 10.9 Deformation of an element at impact location for different aircrafts
Source: Authors

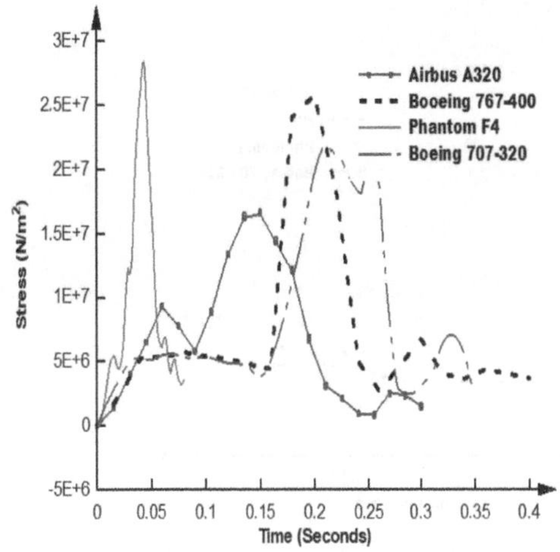

Figure 10.10 Stress of an element at impact location for different aircraft
Source: Authors

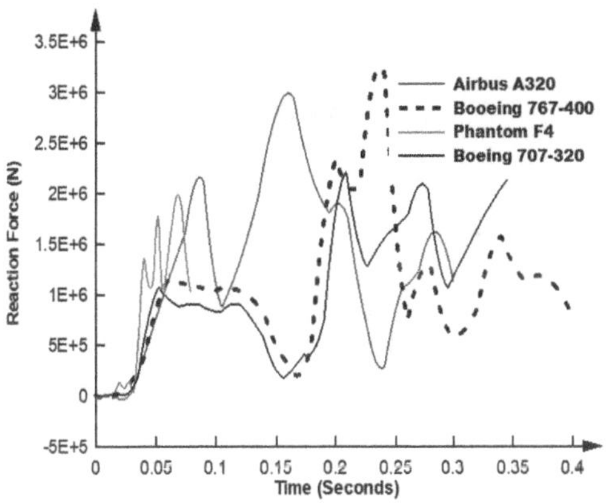

Figure 10.11 Reaction force at the base of containment produced by different aircraft
Source: Authors

throughout the total time of impact, it is observed that Phantom F4 provided large deformation (Figure 10.9) and it is a local phenomenon. When the stress profile has been plotted for an element in impact location, the maximum stress is found for Phantom F4 and when the reaction profile for an element (located at the containment's base) has been plotted, the maximum reaction is found for Boeing 767-400 (Figures 10.10 and 10.11). From Figure 10.12, it may be concluded when damage starts in structural elements, the plastic strains are also observed. The deformation and stress behaviour on containment along path A and path B is maximum for Boeing 767-400 and it is a global phenomenon (Figures 10.13–10.15).

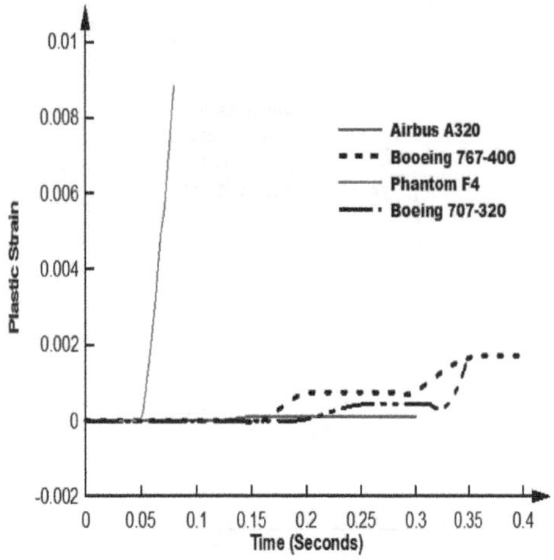

Figure 10.12 Plastic strain of an element at impact location for different aircraft
Source: Authors

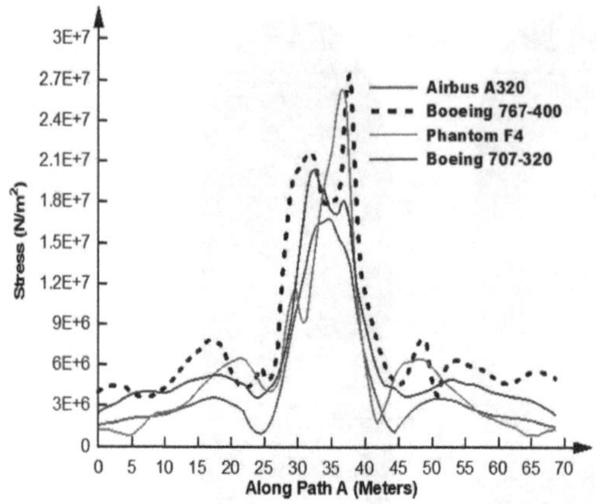

Figure 10.15 Stress produced in concrete along path A
Source: Authors

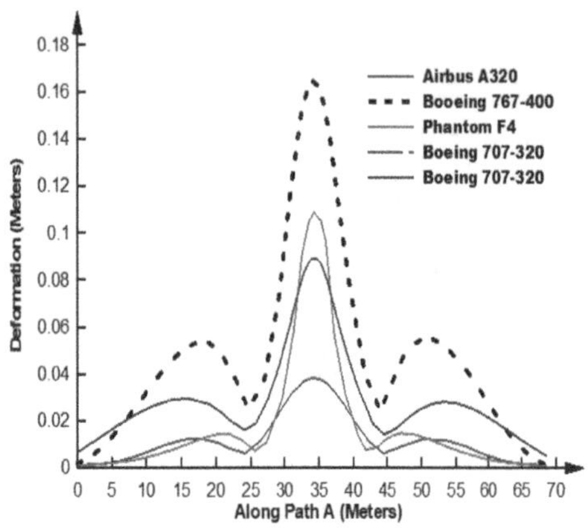

Figure 10.13 Global deformation in concrete along path A
Source: Authors

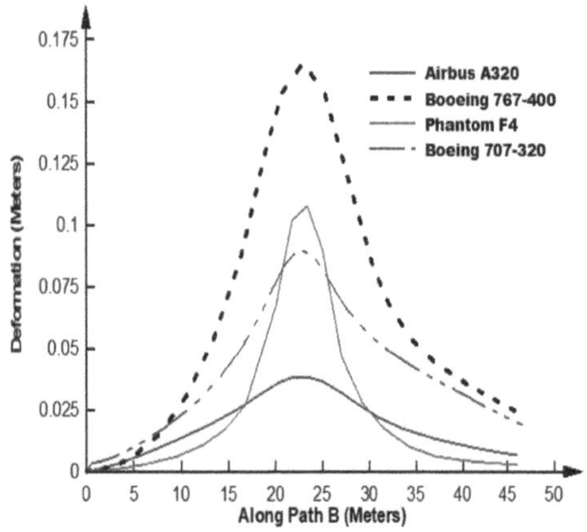

Figure 10.14 Global deformation in concrete along path B
Source: Authors

As the base of containment is fixed, deformation is zero but some deformation was found in the junction between cylinder and dome (Figure 10.14). When a single element has been taken in impact location for deformations and stresses plotting with time, it is also clear that the maximum deformation occurs when wings come in contact with the object and all the plots of deformation and stress for every aircraft is similar to the force-time history curves (Figures 10.9 and 10.10).

Conclusion

Current study is an attempt to evaluate the most damaging aircraft on NPP structure (BWR Mark-III) from different aircraft crash (Boeing 707-320, 767-400, Phantom F4 and Airbus A320) using finite element analysis. The subsequent conclusions are drawn from the analysis:

a) The maximum and minimum deformation is observed 0.251 m for Phantom F4 and 0.039 m for Airbus A320, respectively when the impact load is applied at 23 m height from the base of containment.
b) The maximum tension damages are observed at impact location for Boeing 767-400 due to more impact area.
c) The minimum tension damages are observed at impact location for Airbus A320 due to less impact force.
d) The contact of the aircraft's wings with the NPP containment induces peak deformation and stress levels.

e) After some time of impact, the elements experience plastic strains and get damaged.
f) The damage due to aircraft over the containment structure is not a global failure.
g) Finally, it may be concluded that the Boeing 767-400 aircraft has more damaging potential.

References

[1] Riera, J. D. On the stress analysis of structures subjected to aircraft impact forces. *Nucl. Eng. Des.*, 1968;8:415–426.

[2] Riera, J. D. A critical reappraisal of nuclear power plant safety against accidental aircraft impact. *Nucl. Eng. Des.*, 1980;57(1):193–206.

[3] Abbas, H., Paul, D. K., Godbole, P. N., Nayak, G. C. Reaction-time response of aircraft crash. *Comput. Struct.*, 1995;55(5):809–817.

[4] Kukreja, M. Damage evaluation of 500 MWe Indian pressurized heavy water reactor nuclear containment for aircraft impact. *Nucl. Engg. Des.*, 2005;235(17–19):1807–1817.

[5] Lin, F., Tang, H. Nuclear containment structure subjected to commercial aircraft crash and subsequent vibrations and fire. *Nucl. Engg. Des.*, 2017;322:68–80.

[6] Jiang, H., Chorzepa, M. G. Aircraft impact analysis of nuclear safety-related concrete structures: A review. *Eng. Fail. Anal.*, 2014;46:118–133.

[7] Sadique, M. R., Iqbal, M. A., Rawsan, A., Gupta, N. K. Response of outer containment of an NPP against aircraft crash and induced fire. *Thin-Walled Struct.*, 2018;126:182–192.

[8] Sugano, T., et al. Full-scale aircraft impact test for evaluation of impact force. *Nucl. Engg. Des.*, 1993;140(3):373–385.

[9] Abbas, H., Paul, D. K., Godbole, P. N., Nayak, G. C. Aircraft crash upon outer containment of nuclear power plant. *Nucl. Engg. Des.*, 1996;160(1–2):13–50.

[10] Grote, D. L., Park, S. W., Zhou, M. Dynamic behavior of concrete at high strain rates and pressures: I. experimental characterization PII: S 0 7 3 4-7 4 3 X (0 1) 0 0 0 2 0-3. *Int. J. Impact Engg.*, 2001;25:869–886.

[11] Lu, Y., Xu, K. Modelling of dynamic behaviour of concrete materials under blast loading. *Int. J. Solids Struct.*, 2004;41(1):131–143.

[12] Dassault Systèmes Simulia. Getting Started With ABAQUS Inactive Edition Version 6.8. 2008:1–621.

[13] Saberi, R., Alinejad, M., Mahdavi, M. O., Sepanloo, K. Numerical analysis of nuclear power plant structure subjected to aircraft crash. *Int. J. Adv. Struct. Engg.*, 2017;9(4):341–352.

[14] Rawsan, A., Maiti, P. R. Behavior of nuclear power plant containment under aircraft crash. *Iran. J. Sci. Technol. Trans. Civ. Engg.*, 2021;45(1):207–218.

[15] Rawsan, A., Maiti, P. R. Nuclear containment wall under aircraft crash. *Int. J. Recent Technol. Engg.*, 2019;8(3):3249–3256.

11 Sand inclusion composite structures for enhanced ballistic impact resistance

Manas Thakur[a], Nakka Nishika[b], Bommiditha Jyothsnavi[c], Surkanti Sai Sahasra[d] and Srikant Sekhar Padhee[e]

Department of Mechanical Engineering, Indian Institute of Technology Ropar, Rupnagar, India

Abstract

With the ever-increasing threat of ballistic impact, it is essential to provide a solution that is not only effective but also economical. In recent years polymer matrix sand composites (PMSCs) have emerged as a potential cost-effective solution. In this on-going effort, current investigation aims at offering more robust protection against varied ballistic impacts against potential ballistic threats. This research investigates the enhancement of ballistic impact resistance in polymer matrix composites through the inclusion of sand in a graded fashion. Variable material properties are obtained along the thickness by varying the sand inclusions size and weight fraction in the polymer matrix. The gradation offers a stepwise structure that has a prime base impact zone containing abrasive sand inclusions densely graded with a range of typical inclusion sizes which is brittle and hard enough to erode the projectile. Further gradation involves a less dense region providing tensile strength which is able to reflect the stress wave and reduce the impact energy with providing a backing to the frontal brittle region. The barely dense or neat matrix regions prevents backlash and offers cushioning effect. The addition of sand particles increases the surface area of the composite and improves adherence to the inclusions and matrix. This enhanced adhesion ensures efficient load transfer across the surface, which raises the composite's overall hardness. PMSCs were fabricated for observing the effects on two matrix compositions improvising manufacturing procedures and subjected to a battery of tests, including tensile testing, Izod impact testing, Shore-D hardness testing, etc. in order to evaluate their mechanical properties comprehensively. The experimental results reveal that varying the weight fraction of sand content markedly influences the mechanical properties up to a certain weight fraction for a typical size range of inclusions. Experimental and simulation studies were involved in order to extract the mechanical properties and to observe the material behaviour against ballistic threats. The study reveals underscoring the potential of PMSCs as cost-effective and environmentally sustainable composites for advanced ballistic protection applications.

Keywords: Ballistic impacts, polymer matrix sand composites (PMSCs), impact resistance, inclusion, sand

Introduction

Effective ballistic resistant structures often exhibit an amalgamation of several layers with either different material configurations or distinguished structural configurations which ultimately provides high hardness, toughness, brittle, lightweight, and high energy absorbent characteristics in the structures. Monolithic structures, which often consist of a single material like iron, steel, or aluminium alloys, are commonly used for ballistic-resistant structures. However, these structures tend to have high areal densities, resulting in excessive weight, making them inefficient in terms of weight management. Additionally, they usually exhibit low multi-hit capabilities, which reduces their effectiveness when subjected to repeated impacts. Moreover, the overall cost of such monolithic structures is often high, further limiting their suitability for ballistic impact applications [1–3]. Studies have been carried out in order to reduce the weight of the monolithic structure by incorporation of multi-layered structures such as sandwich structures comprising of regions with significant characteristics which combinedly work as the bullet pierces through the thickness of the structure. In order to restrict penetration, ballistic sandwich designs are usually constructed with a hard frontal layer, usually made of ceramics or Rolled Homogeneous Armor (RHA), to absorb initial impact energy. This causes bullet deformation and spreads contact stresses [4]. Energy-absorbent middle layers, such as composite metal foams (CMFs) or corrugated structures, are used to

[a]manasthakur159@gmail.com, [b]nishika1028@gmail.com, [c]jyothsnavi987@gmail.com, [d]saisahasrasurkanti@gmail.com, [e]sspadhee@iitrpr.ac.in

DOI: 10.1201/9781003770442-11

produce light weight structures and provide effective backing to frontal regions. Such arrangement helps mitigate catastrophic failure of the structure [5]. Further to distribute the impact forces and reduce the back face signature enabling impact damage mitigation, the last layer is usually made up of a thin ductile layer [6]. However, each material layer presents unique limitations. Brittle frontal layers, especially ceramics, are costly, heavy, and have limited multi-hit capacity, as they require substantial support to withstand fragmentation. Energy-absorbent cores, such as corrugated sheets and CMFs, face manufacturing challenges and limited multi-hit durability due to cell deformation. Optimising core density and managing delamination issues also remain areas of concern.

In order to achieve effective strength-to-weight ratio, high-strength fibres such as aramid, carbon, and UHMWPE are often used in fibre-reinforced composites to provide multi-hit resistance and efficient energy dissipation [7]. Also, natural fibres such as flax and jute and synthetic fibres combine to create hybrid composites, which offer economical and environmentally beneficial solutions. The natural fibres help absorb energy, while the synthetic fibres increase load resistance [8]. Further improvements in ballistic performance can be achieved by (a) selection of tuned resin, (b) tailoring the fibre orientation and (c) tuning the fibre-matrix bonding through surface treatment. Despite its effectiveness, polymer matrix fibre composites' structural integrity is compromised by matrix cracking, delamination, and fibre pull-out. Fibre composites and sandwich constructions frequently have significant manufacturing costs, and fibres like aramid are moisture-sensitive and degrade in harsh environments [20, 21].

To improve structural integrity, functionally graded materials (FGMs) enable smooth property transitions between layers, allowing for direction-specific mechanical responses under high-velocity impacts. By maximising material qualities where they are most required, this gradient design allows the structure to endure a variety of impact circumstances [9]. When combined, these techniques produce lightweight, multipurpose, ballistic-resistant structures that are appropriate for high-impact uses.

Inclusion composites, using fillers like sand, ceramics, or metallic particles within a polymer matrix, are promising for ballistic protection [10]. These inclusions boost hardness and energy dissipation, enhancing resistance to high-velocity impacts. Sand-filled composites, in particular, increase stiffness and hardness at a low cost, though optimising particle type, size, and distribution remains one of the daunting tasks. The densification of sand within the matrix limits voids, aiding impact resistance. Key factors like sand preparation, segregation, and curing processes significantly influence these composites' effectiveness [11, 12].

In this research work, the influence of sand inclusions in two types of polymer matrix compositions forming a composite has been studied. Refinement of the composite fabrication process has been carried out to obtain the necessary mechanical properties with variations in sand inclusion size and weight fraction. To extract these properties, tests such as uniaxial tensile, I-Zod impact, and Shore-D hardness were conducted, with samples prepared by optimising the fabrication process in each condition. Properties thus extracted are incorporated in decision parameters for an effective graded stacking sequence for ballistic threats. The stacking sequence is arranged to achieve variable mechanical properties in a single structure by leveraging the tailor-ability of PMSCs through sand inclusions and polymer matrix. The layered structure provides an initial, densely graded sand region for impact absorption and abrasive tip deterioration, mitigating localised stresses through increased contact area. A less dense backing region provides additional residual energy absorption and stress transfer, while the final, minimally dense region acts as a cushion by absorbing bullet momentum. As it is difficult to conduct experimental studies for all the particle sizes and volume fractions, a simulation assailed studies have been carried out. DIGIMAT has been used to evaluate the effective properties of representative volume elements (RVE) by varying inclusion size and volume fraction. The thus formed composites exhibit superior interfacial bonding between the matrix and the fillers, preventing delamination under high-velocity impacts. This structural integrity ensures that the layers remain intact, even under extreme stress, thereby enhancing the material's overall ballistic resistance and energy absorption capabilities.

Experimental

Materials

Epoxy has been chosen as the matrix for its non-toxicity, ease of handling, and room-temperature curing

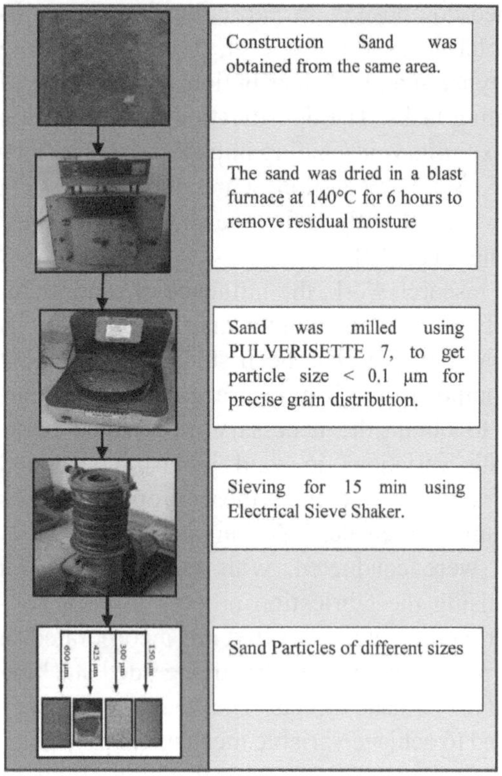

Figure 11.1 Sand pre-processing
Source: Authors

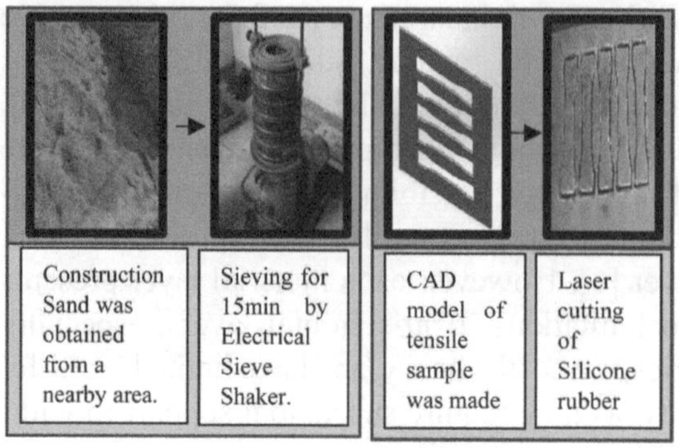

Figure 11.2 Sand pre-processing and mould preparation
Source: Authors

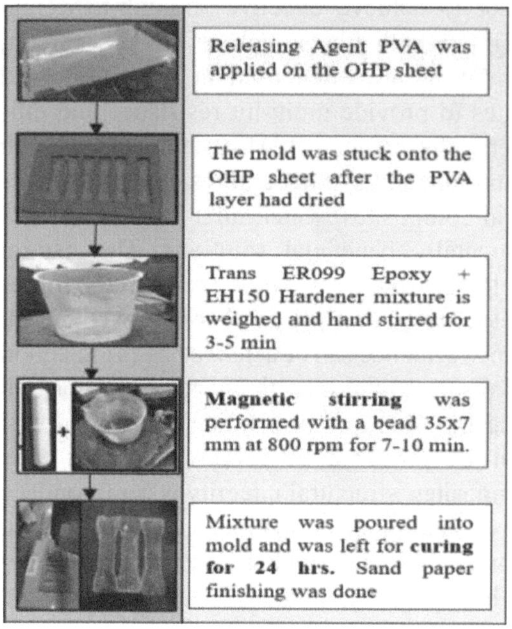

Figure 11.3 Testing specimen preparation process
Source: Authors

capability [14]. The reactive epoxide rings form a strong network, offering excellent mechanical strength and rigidity. The primary binding agent for initial sand inclusions was ER099 Epoxy Resin with EH150 hardener, valued for its strength, chemical resistance, low shrinkage, non-crystallising properties, availability, and cost-effectiveness. This combination tolerates temperatures up to 60°C with a long pot life. In the second experiment, LY556 Bisphenol-A-grade epoxy with HY951 hardener was used for its mechanical stability, thermal resistance up to 80°C, and adjustable reactivity [13]. Construction sand, primarily silica, has been selected for its irregular morphology, enhancing adhesion with minimal voids for better and homogenous structural integrity [14]. Its chemical inertness ensures compatibility with the epoxy matrix, and its abrasive nature improves tapped density, aiding in energy dissipation during projectile impact and reducing fragmentation [14, 15].

Sand processing

Two distinct sand processing procedures were developed for composite samples with ER099 and LY556 epoxy resins. For ER099-based samples, sand was air-dried at 30°C for 30–35 hours, effectively removing moisture and organic impurities. It was then sieved using an electric sieve shaker for 15 minutes to obtain specific particle sizes of 0.15 mm, 0.3 mm, 0.425 mm, and 0.6 mm, enhancing packing density and mechanical strength. Finer particles filled voids, while larger particles improved flowability. The fractions were stored in airtight containers to maintain quality (Figures 11.1–11.3). The improved procedure for LY556 samples involved additional steps to enhance sand purity and performance. Sun-drying on a polyethylene layer reduced moisture, while remaining organic impurities (e.g., leaves, roots) were manually removed. An isopropyl alcohol wash further purified the sand, followed by a high-temperature treatment at 140°C in a blast furnace for 6 hours to ensure thorough moisture removal. Milling with the PULVERISETTE 7 achieved uniform particle sizes under 0.1 μm, enhancing packing density and interlocking. Sieving of sand was performed similar to the previous procedure.

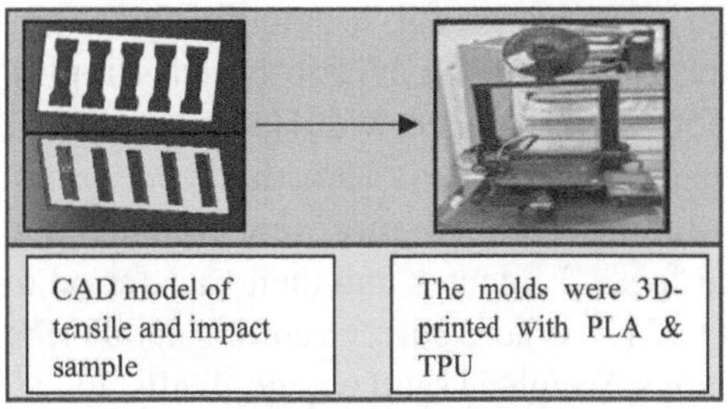

Figure 11.4 Mould preparation
Source: Authors

Mould preparation

Two mould preparation methods were applied for composite testing: laser cutting and additive manufacturing. For tensile moulds, silicone rubber was selected due to its flexibility, durability, and resistance to epoxy resins, avoiding chemical reactions. A 5 mm thick silicone rubber sheet was cut using the BSM laser cutting and engraving machine, known for its high-speed DSP control and precision, producing reliable tensile moulds (Figure 11.4).

The improved procedure for LY556 samples involved moulds which were created with a 3D printer from Shenzhen Creality 3D technology Co., Ltd. (Figure 11.5). For thermoplastic polyurethane (TPU) moulds, a 0.2 mm nozzle, bed temperature of 230°C, nozzle temperature of 70°C, and speed of 65 mm/s were set. TPU's elasticity minimises shrinkage and simplifies demoulding, ideal for tensile, Izod impact, and Shore-D hardness tests. The moulds are reusable, enhancing cost efficiency. For ballistic impact samples, PLA with 90% infill density was used, along with a 0.4 mm nozzle, 200°C nozzle temperature, 60°C bed temperature, and 50 mm/s speed for improved rigidity and thermal stability. PLA moulds include clamping features, ensuring secure and accurate testing. These optimised 3D printing parameters and materials improve mould durability, efficiency, and composite testing precision.

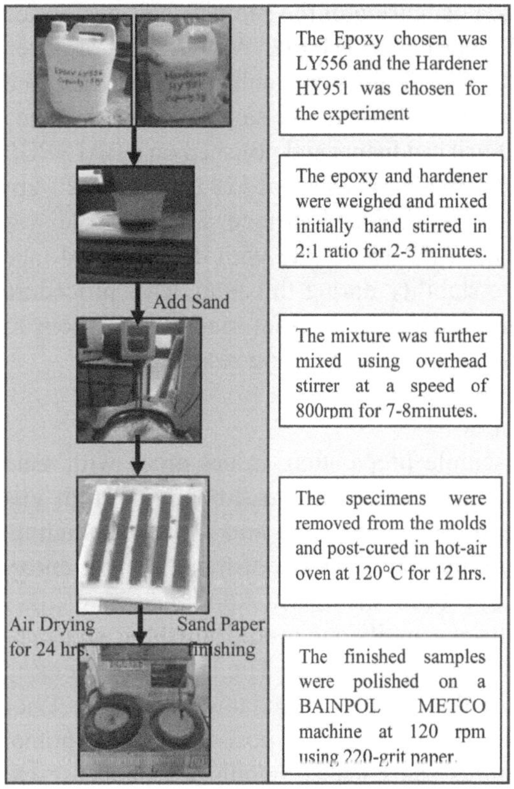

Figure 11.5 Specimen preparation procedure
Source: Authors

Sample preparation

Two specimen preparation methods were implemented using ER099 and LY556 epoxies to create durable composite samples with sand filler. For the ER099-based samples, Epoxy trans ER099 spl (Bisphenol-A grade) was weighed using an iScale machine with precise minimum (20/40/100 g) and maximum (10/20/30 kg) weight capacities. Hardener trans EH150 was added at 10% of the epoxy's weight, and sand was weighed to match the target weight fraction in the mixture. For example, in a 100 g epoxy-hardener blend with 20% sand content, 20 g of sand was added. This epoxy-hardener mix was hand-stirred for 3–5 minutes, followed by mixing on a Springer Magnetic Stirrer with hot plate at 70% speed for 7–10 minutes to achieve thorough sand dispersion. Moulds were lined with an OHP sheet coated in polyvinyl alcohol as a releasing agent, ensuring easy removal after curing. The sand-epoxy mixture was poured into the moulds and left to air-dry for at least 24 hours. Post-curing, the samples were carefully removed and finished with medium and finer grit sandpaper for a smooth, even surface (Figure 11.5). For the LY556-based samples, Epoxy LY556 was combined with Hardener HY951 (Triethylenetetramine) at a 10% hardener ratio. The epoxy and hardener mixture were weighed and hand-stirred for 2–3 minutes and then transferred to a REMI RQ-121 overhead stirrer, running at 800 rpm for 5–6 minutes. Weighted sand was gradually added, extending the mixing time to 20–22 minutes for consistent sand distribution. The moulds were prepared similarly, lined with OHP sheets as per section 3.1.3, and the sand-epoxy mix was poured in, air-drying for 24 hours. Post curing of the samples was performed in a GAYATRI SCIENTIFIC WORKS hot-air oven at 100°C for 12 hours to enhance durability and adhesion. After curing, the samples were finished with 120-grit sandpaper and polished on a BAINPOL METCO polishing machine at 120 rpm with 220-grit sandpaper for a refined surface. Each sample was secured on a flat base plate with double-sided tape to ensure stability during finishing. This procedure was replicated across different sand compositions to maintain consistency in testing results.

Drawbacks

During sample preparation, issues arose with sand particle settling [15] and weak bonding. High viscosity leads to reduced voids and improved bonding. Therefore, the low viscosity of trans ER-099 epoxy might have been the issue. Additional challenges included uneven silicone mould finishing and bubbles from magnetic stirring. The lack of post-curing weakened the samples' mechanical strength. Prior studies [14] suggest that post-curing at optimal temperatures can enhance bonding, toughness, and strength, highlighting a key area for improvement

Tensile test

Tensile strength is a characteristic that defines the ability of the material to resist breaking under tension. This property is influenced by the molecular structure of the epoxy resin and the nature of the reinforcing materials, such as sand inclusions. In epoxy-based composites, tensile strength directly correlates with the crosslink density of the cured resin, the quality of the bonding between the epoxy matrix and the reinforcement [14]. The tensile tests were performed on both neat epoxy samples with different sizes and weight fractions of sand inclusions in accordance with the ASTM Standard D638. The testing was performed using Zwick/Roell Z010. A load cell of 10 kN was used and the testing was performed corresponding to the speed 5 mm min^{-1}. The specimens measured 19 mm in width, 50 mm in gage length and the distance between the grips was 115 mm.

Impact test

Impact strength is a characteristic that defines the ability of the material to resist sudden forces or shock by determining the energy absorbed during the fracture. Izod impact test was performed on both neat epoxy samples with different sizes and weight fractions of sand inclusions in accordance with the ASTM Standard D256-23. The specimens measured 63.5 mm in length, 12.7 mm in width, and 5 mm in thickness. A V-notch of angle 67.5° was given at the centre of the specimen. Each test specimen was placed atop a vertical cantilever beam and fractured by a single pendulum striking the centre of the sample.

Shore-D hardness test

Hardness is the measure of the material's resistance to plastic deformation by indentation. The hardness of both neat epoxy samples with different sizes and weight fractions of sand inclusions was measured using Shore D durometer in accordance with ASTM standard D2240-15. The specimens are of dimensions 35 mm in length, 25 in width and 6 mm in thickness.

Effective properties evaluation

In order to analyse the effect of inclusion size and volume fraction on the macro-scale effective properties were extracted from the RVE which was done using commercial FEA tool DIGIMAT FEA. Spherical inclusions with a random but uniform distribution over the matrix were considered. The variation has been carried out with respect to inclusion size and volume fraction. Displacement boundary condition has been imposed as negligible variance in effective properties has been imposed by other boundary conditions. Effective properties are extracted by volume averaging method where the stresses and strains in the respective directions as per the boundary conditions imposed are averaged out over the volume which ultimately gave effective properties in the particular direction and plane. Some typical configurations and distributions are as shown in Figure 11.6.

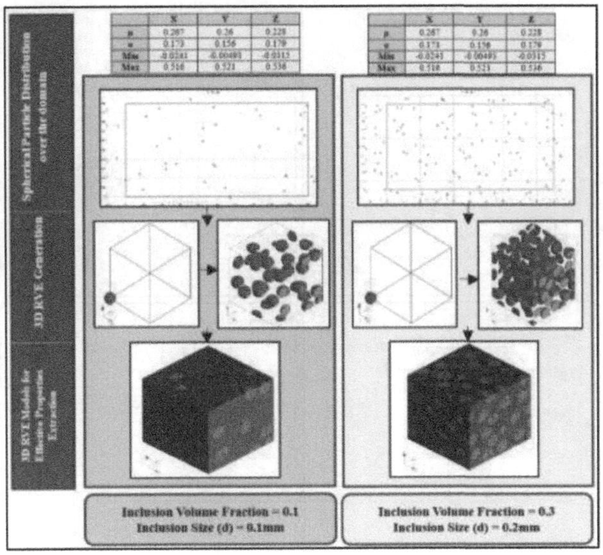

Figure 11.6 Modelling of RVE for varying inclusion size
Source: Authors

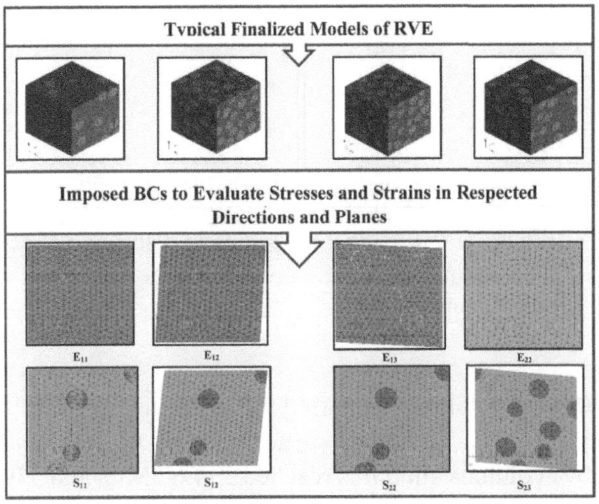

Figure 11.7 BCs applied to extract effective properties by varying inclusion size and volume fraction
Source: Authors

The isotropic properties assigned to the constituents i.e., epoxy and sand were as in refs [13, 16].

Effective properties were extracted by imposing appropriate boundary conditions on varied RVEs where the homogenous isotropic properties for spherical sand inclusions are considered as in study did by Das [16]. A schematic of the different BCs applied on the RVE in order to extract effective properties in respective directions and planes is shown in Figure 11.7. The simulation study reveals that

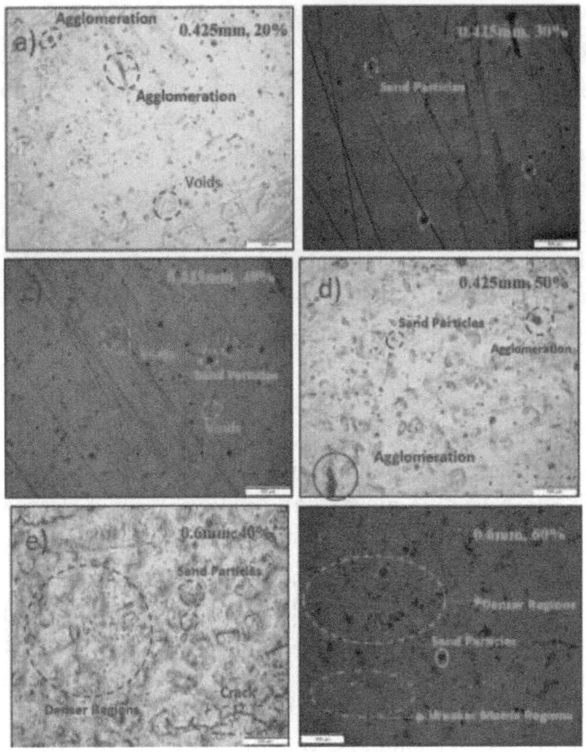

Figure 11.8 Images captured through optical microscope of PMSCs
Source: Authors

through this approach, a comprehensive understanding of how these parameters influence critical material properties was developed.

Results and discussion

Microstructural analysis

Optical images generated using an Olympus base microscope was classified based on sand inclusion size and volume fraction. The refined manufacturing process enabled a more uniform sand distribution up to a certain limit, with fewer agglomerations in samples with smaller grain sizes and lower volume fractions, as observed in Figure 11.8a. Although larger particle sizes did not significantly increase agglomerations, higher volume fractions led to unevenly dense regions due to particle clustering, as shown in Figure 11.8d. Voids were generally observed across samples as shown in Figure 11.8c, though they were minimal and likely resulted from minor inconsistencies in filling or dispersion during manufacturing. Under load, cracks formed in the less dense, neat epoxy areas, which were structurally weaker than the sand-reinforced regions as shown in Figures 11.8e, f.

Effective properties outputs

The effective properties for varied RVEs were extracted where variation in the inclusion size from 0.1 to 1 mm in diameter and the volume fraction from 0.2 to 0.8 in most of the simulations were carried out.

The simulation results for RVE clearly demonstrate a significant increase in the elastic moduli in the three principal directions of the PMSCs E_1, E_2, and E_3, as both the inclusion diameter and volume fraction were increased as depicted in Figures 11.9a–c. The properties showed minimal variation in the principal directions demonstrating a quasi-isotropic Nature of PMSCs. An increase in moduli results from larger inclusions contributing greater stiffness, effectively restricting matrix deformation. With increased size or volume, inclusions act as rigid supports, enhancing the composite's overall stiffness. On the contrary, Poisson's ratios (ν_{21}, ν_{13}, etc.) decrease with larger inclusion size and volume fraction, as stiffer inclusions limits lateral deformation, aligning with increased PMSCs rigidity. Additionally, shear moduli G_{12}, G_{23}, G_{13} increase due to inclusions effectively resisting shear by evenly transferring loads Figures 11.9d–f.

In conclusion, the study effectively captures the influence of inclusion size and volume fraction on the mechanical properties of the RVE. The results underline the critical role of these parameters in enhancing the stiffness and shear moduli while reducing lateral deformation, contributing to the material's overall robustness. This approach of varying inclusion size and volume fraction can serve as a powerful design tool in optimising composite materials for high-performance applications.

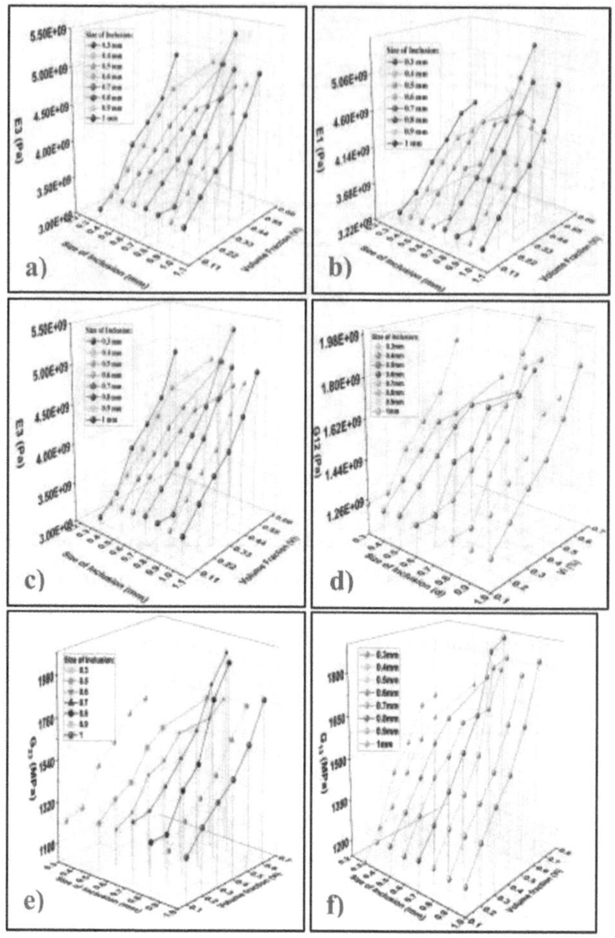

Figure 11.9 Effective properties in respective directions and planes
Source: Authors

Experimental material properties

Experimental testing including uni-axial tensile test, Izod impact test, and Shore-D hardness test was performed adopting a refined manufacturing process as discussed in Section 2 with appropriate standard specimen sizes and dimensions specified in Sections 2.6, 2.7, 2.8. The major comparison was made between two grades of polymer matrixes and refinement has been carry out each iteration in order to obtain the consistency of obtained results and thereby select the best suitable grade with the optimised manufacturing process.

Tensile testing of ER 099 (epoxy) with EH 150 (hardener) and 0.3 mm sand particles, varying sand volume fractions from 0.30 to 0.50, revealed that a 30% sand volume increased the modulus by 50.20% to 366 MPa compared to neat epoxy. At 40% and 50% volumes, modulus reached 534.67 MPa and 539 MPa, marking 119% and 121.20% increases, respectively as shown in Figure 11.10a. Impact strength for the same configuration rose to 110 J/m and 135 J/m at 40% and 50% sand volume, showing 7% and 32% improvements over the neat mixture represented in Figure 11.10b. While higher sand volume fractions present challenges due to manufacturing irregularities discussed in Section 2.5, these insights highlighted areas for improvement, particularly in mitigating agglomeration to reduce localised stress in weaker matrix regions.

A more viscous, harder epoxy grade Section 2.1 was used, offering advantages over the previous grade. Tensile tests with 0.3 mm and 0.425 mm sand particles at increasing volume fractions showed

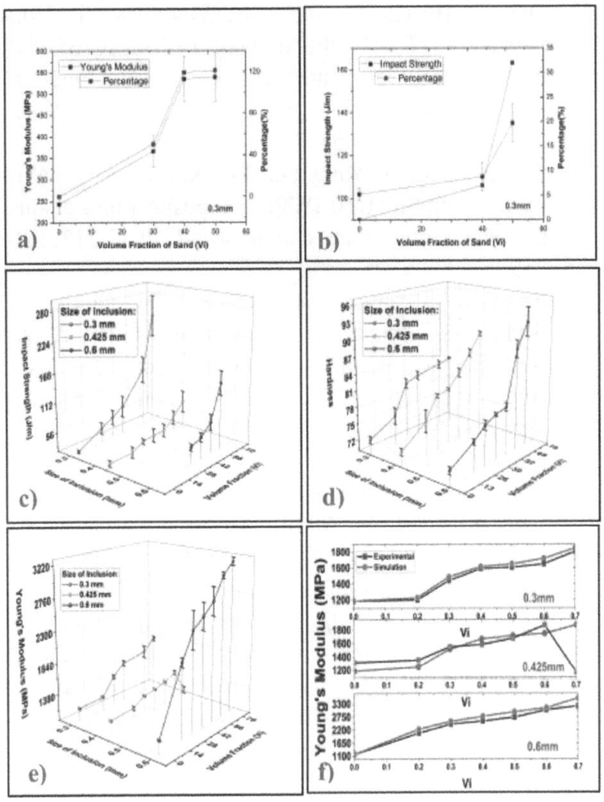

Figure 11.10 Experimental properties
Source: Author

property enhancement. For 0.3 mm sand, Young's Modulus increased by 51% (1810 MPa) at 70% volume, while for 0.425 mm sand, it rose to 1445 MPa i.e., 25% increase at 60% volume, as shown in Figure 11.10e. Higher sand volumes up to 70% for 0.425 mm sand size degraded modulus, suggesting manufacturing improvements. Impact strength reached 210 J/m for 0.3 mm sand at 70% volume and showed gradual increments for 0.45 mm and 0.6 mm sand 100 J/m and 150 J/m at 60% and 70% volumes Figure 11.10c. Shore-D hardness tests indicated values of 83.5, 89, and 92.5 for 0.3 mm, 0.425 mm, and 0.6 mm sand at 70% volume, Figure 11.10d with a 13.7% increase over the pure matrix. The comparison between the simulation and experimental properties extracted are as in Figure 11.10f.

The use of the second grade of epoxy hardener mixture significantly improved the performance of the PMSCs, with results aligning closely with findings in relevant literature [11, 12] and the homogenised properties extracted in Section 3.2. This improvement in the manufacturing procedure demonstrates a viable solution for ballistic applications, offering

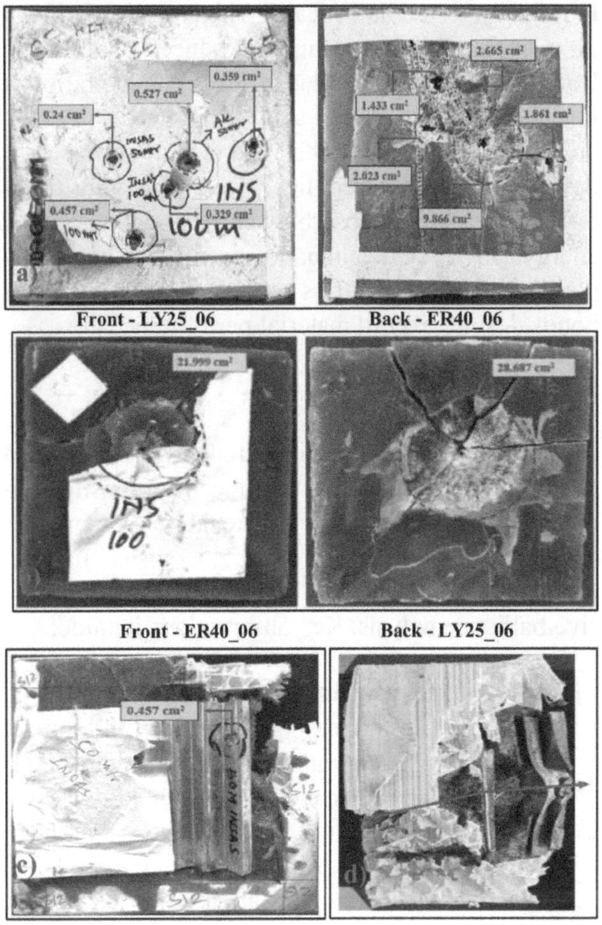

Figure 11.11 Ballistic testing on configured samples
Source: Authors

lightweight, cost-effective, and readily available materials for ballistic structures.

Ballistic samples assertions

The testing setup apparatus was with different firing distances along with the guns used to perform testing, with sample configurations detailed in ref [17]. Couurogated sheets and mesh were used to act as a stiff structure along with potent of mesh to further distribute stresses over larger area. Both samples i.e., Figures 11.11a, b underwent ballistic testing with an INSAS rifle from 100 m. Comparative analysis shows a larger bullet impact area in sample LY25_06 than in ER40_06, indicating greater bullet deformation upon impact with LY25_06, likely due to the higher stiffness of the LY556 epoxy. The stronger polymer-inclusion bonding in ER40_06 contributed to enhanced material integrity, as evidenced by more visible cracks in LY25_06, suggesting it absorbed a larger share of impact energy. Figure 11.11c configures as in ref [17], tested ballistically with an INSAS rifle at 50 m increased layer counts introduced complexity, leading to delamination due to stress concentration at interfaces. Figure 11.11d shows sample LY80_04_1 [17], tested with a SIG rifle at 50 m. Bullet trajectory deviation was observed, likely due to varying properties across layers, absorbing and redirecting impact energy. However, delamination occurred in both samples due to interface stress concentration, weakening bonds despite overall material performance [18].

Conclusion

This work presents a detailed study on developing a novel sand-based functionally graded composite optimised for ballistic resistance. The fabrication process and experimentally verified homogenised material properties are discussed, followed by a comprehensive ballistic analysis. Key observations include:

1) Mechanical properties improved significantly, with modulus increasing up to 125% and impact strength by 32% with increase in volume fraction of sand as observed in Section 4.3
2) With an increase in inclusion size it has been observed that the stiffness had increased, restricting matrix deformation and decreasing Poisson's ratio, contributing mainly to lateral deformation. Both elastic and shear moduli also increased as reported in Section 4.2.
3) LY-556 ballistic samples showed greater impact areas due to higher stiffness, improving impact absorption compared to ER-099. However, increased layering led to delamination from stress concentration at interfaces as observed in Section 4.4.
4) One of the critical observations is that gradation proved superior to uniform distribution, reducing bullet's kinetic energy over shorter intervals and decreasing penetration depth as observed in Section 4.4.

Further analyses are being carried out to improve and optimise the mechanical properties which will be communicated in the future articles.

References

[1] Børvik, T., Langseth, M., Hopperstad, O. S., Malo, K. A. Ballistic penetration of steel plates. *Internat. J. Impact Engg.*, 1999;22(9–10):855–886.

[2] Dey, S., Børvik, T., Teng, X., Wierzbicki, T., Hopperstad, O. S. On the ballistic resistance of double-layered steel plates: an experimental and numerical investigation. *Internat. J. Solids Struct.*, 2007;44(20):6701–6723.

[3] Shen, Z., Hu, D., Yang, G., Han, X. Ballistic reliability study on SiC/UHMWPE composite armor against armor-piercing bullet. *Comp. Struct.*, 2019;213:209–219.

[4] Pawar, M. J., Patnaik, A., Biswas, S. K., Pandel, U., Bhat, I. K., Chatterjee, S., Mukhopadhyay, A. K., Banerjee, R., Babu, B. P. Comparison of ballistic performances of Al2O3 and AlN ceramics. *Int. J. Impact Engg.*, 2016;98:42–51.

[5] Marx, J., Portanova, M., Rabiei, A. Performance of composite metal foam armors against various threat sizes. *J. Comp. Sci.*, 2020;4(4):176.

[6] Cui, T., Zhang, J., Li, K., Peng, J., Chen, H., Qin, Q., Poh, L. H. Ballistic limit of sandwich plates with a metal foam core. *J. Appl. Mech.*, 2022;89(2):021006.

[7] Gellert, E. P., Cimpoeru, S. J., Woodward, R. L. A study of the effect of target thickness on the ballistic perforation of glass-fibre-reinforced plastic composites. *Internat. J. Impact Engg.*, 2000;24(5):445–456.

[8] Safri, S. N. A., Sultan, M. T. H., Jawaid, M., Jayakrishna, K. Impact behaviour of hybrid composites for structural applications: A review. *Comp. Part B Engg.*, 2018;133:112–121.

[9] Gunes, R., Aydin, M., Apalak, M. K., Reddy, J. N. The elasto-plastic impact analysis of functionally graded circular plates under low-velocities. *Comp. Struct.*, 2011;93(2):860–869.

[10] Benveniste, Y. A new approach to the application of Mori-Tanaka's theory in composite materials. *Mech. Mater.*, 1987;6(2):147–157.

[11] Chian, S. C., Tan, B. C. V., Sarma, A. Reprint of – Projectile penetration into sand: Relative density of sand and projectile nose shape and mass. *Internat. J. Impact Engg.*, 2017;105:80–88.

[12] Kumar, T. S., Joladarashi, S., Kulkarni, S. M., Doddamani, S. Influence of sea sand reinforcement on the static and dynamic properties of functionally graded epoxy composites. *Iran. Poly. J.*, 2024:1–18.

[13] https://www.farix.hu/pdf/1360669728.pdf.

[14] https://idrlib.iitbhu.ac.in/xmlui/bitstream/handle/123456789/2511/06_Chapter%2002.pdf?sequence=6&isAllowed=y.

[15] Ghadr, S. Effect of grain size on undrained anisotropic behaviour of sand–fibre composite. *Trans. Geotech.*, 2020;22:100323.

[16] Molina, O., Vilarrasa, V. and Zeidouni, M., 2017. Geologic carbon storage for shale gas recovery. Energy Procedia.

[17] [17 https://github.com/Manas15959/Samples_Configurations.

[18] Zemani, K., May, A., Khatir, S., Gilson, L., Cuong-Le, T., Wahab, M. A., Slamani, H. Numerical analysis of an experimental ballistic test of Al/SiC functionally graded materials. *Comp. Struct.*, 2024;333:117909.

[19] Nunes, L. M., Paciornik, S., d'Almeida, J. R. M. Evaluation of the damaged area of glass-fiber-reinforced epoxy-matrix composite materials submitted to ballistic impacts. *Comp. Sci. Technol.*, 2004;64(7–8):945–954.

[20] Wambua, P., Vangrimde, B., Lomov, S., Verpoest, I. The response of natural fibre composites to ballistic impact by fragment simulating projectiles. *Comp. Struct.*, 2007;77(2):232–240.

[21] Lane, R. A. Under the microscope: understanding, detecting, and preventing microbiologically influenced corrosion. *J. Failure Anal. Prev.*, 2005;5:10–12.

12 Failure analysis of adhesive bonded joints with defects using non-local continuum damage models in FEM

S. Sai Kumar[1,a], *M. Venkatesh*[1,b] *and N. Muthu*[1,2,c]

[1]Department of Mechanical Engineering, IIT Guwahati, North Guwahati, Assam, India

[2]Department of Mechanical Engineering, IIT Palakkad, Palakkad, Kerala, India

Abstract

The failure behaviour of adhesive bonded joints, predicted using continuum damage models through the finite element method (FEM), is one of several effective approaches. However, despite its widespread use for modelling various failure mechanisms, conventional FEM often produces mesh-dependent results and sometimes fail to converge, particularly in complex scenarios such as joints with adhesive defects, including sharp cracks. Non-local approaches, which account for the influence of surrounding regions, effectively address these challenges and produce numerically converged structural responses. In this work, integral-based non-local strain theories are employed to analyse the failure behaviour of problems across various modes where FEM encounters difficulties. A mixed-mode continuum damage model, utilising a quadratic stress and linear fracture energy criterion, is adopted to capture the linear damage evolution in adhesively bonded joints. To illustrate the capability of the proposed methodology in achieving mesh independence and stable iterative convergence for relatively medium sized meshes, numerical examples of adhesive joints with an interfacial sharp crack are analysed under both independent and mixed-mode loading conditions.

Keywords: Adhesive bonded joints, non-local integral approach, mixed-mode continuum damage model

Introduction

Over the past few years, considerable attention has been given to the development of various non-homogeneous materials, including functionally graded materials [1], porous and lattice structures [2, 3], composites [4], and others, as well as exploring their capabilities. Traditional materials such as concrete, metals and their alloys are typically assumed to be homogeneous and isotropic, where their behaviour is effectively described by classical continuum theories in different states of deformation and material response [5, 6]. However, the aforementioned modern materials physically exhibit non-local effects that cannot be neglected due to their consistent long-range interactions at cross-scales (nano, micro, meso), which classical theories fail to capture.

As a result, specialised (non-local) continuum theories have been formulated to account for these long-range interactions. These models are generally classified into two main types: gradient formulations and integral theories. In gradient theory [7], non-local effects are modelled using higher order gradient terms, which enhance accuracy in capturing mechanical disturbances. However, these terms often lack a clear physical link to the material's microstructure, making the approach more indirect and phenomenological. While integral theories capture non-local effects by applying weighted averaging that physically accounts for interactions within a specific horizon, reflecting the collective behaviour at finer scales and incorporated them into the continuum models [8].

Eringen [9, 10] introduced the non-local elasticity in linear elastic fracture mechanics, providing valuable information on the behaviour of materials in regions with sharp cracks, where stress concentrations can lead to singularities. Later, Polizzotto [11] developed variational principles that incorporate non-local theories, demonstrating the thermodynamic consistency of the constitutive law on a global scale. Building on those non-local models, Pisano et

[a]sai176103019@iitg.ac.in, [b]venkateshmarpina@gmail.com, [c]nelsonm@iitpkd.ac.in

DOI: 10.1201/9781003770442-12

al. [12] extended them to finite element implementations, developing algorithms for numerical analysis and solving simple 1D and 2D examples. Research in non-local integral theories is primarily divided into two main directions. The first emphasises the numerical development of non-local theories to enhance their performance. Zhang et al. [13] developed a method for assembling the non-local stiffness matrix that minimises computation time by using weakly singular kernels instead of conventional non-singular ones. Later, Patnaik et al. [14] utilised fractional-order kernels to introduce fractional-order finite element methods (f-FEM), which they applied to solve benchmark examples in both 1D and 2D non-local problems.

The second one focuses on the application of non-local theories in various engineering contexts. Friswell et al. [15] introduced a novel approach to beam analysis that incorporates non-local theories and capitalises on the advantages of FEM. Ansari et al. [16] investigated the bending of embedded nanoplates, using Eringen's non-local integral approach for the analysis. One of the major applications of non-local theories, where significant research advancements have been made, is in addressing strain localisation or strain softening issues through damage mechanics models. Initial research on non-local damage mechanics began by studying the failure behaviour of concrete [17]. Later, Jirasek [18] and Borino et al. [19] developed procedures for finite element implementation, integrating regularised damage models. Andrade et al. [20] applied non-local integral theories by regularising the damage variable to model ductile deformations while ensuring adherence to thermodynamic principles. Kumar and Muthu [21] recently extended the application of the strain difference approach to damage mechanics, taking advantage of a symmetric non-local stiffness matrix, which significantly reduces computational time during the assembly of the stiffness matrix in each iteration.

However, because of the numerous advantages of adhesive-bonded joints, extensive research has been dedicated to predicting their strength. Various approaches have been explored, including fracture mechanics, cohesive zone modelling (CZM), continuum mechanics, continuum damage mechanics (CDM), and the extended finite element method (XFEM) [22]. García et al. [23] examined the onset and progression of cracks in wind turbine joints using a CDM model. Their approach utilised the Drucker-Prager exponential criterion to characterise the elasto-plastic behaviour of the adhesive, incorporating linear softening. Chousal and de Moura [24] developed a mixed-mode damage model using continuum elements with a triangular traction-separation law. They applied quadratic stress and fracture energy criteria to determine damage onset and mixed-mode displacements.

Regarding defects in adhesive-bonded joints and their failure analysis, considerable research has been conducted on the impact of bond line defects or cavities [25, 26]. However, to the best of the author's knowledge, numerical studies addressing flaws caused by air entrapment between the adhesive and the adherend, leading to weak contact in the form of interfacial cracks [27], remain less explored. Moreover, it is observed that analysing the strength of joints with these initial cracks using CDM concepts within conventional FEM poses challenges in achieving mesh-independent results, especially for mixed-mode problems. Therefore, in this work, non-local integral theories are incorporated into the finite element framework to tackle these challenges, with Eringen's non-local elasticity model being adapted to account for the behaviour of damaged materials. The concepts of CDM have been utilised to predict damage evolution in the adhesive region. The proposed method is validated against established results from mode I double cantilever beam (DCB) tests and is further applied to mixed-mode bending (MMB) problems.

Non-local strain approach

As discussed in Section 1, the non-local integral approach employs kernel functions to define the intensity of interactions between a material point and its neighbours within a specific range, known as the interaction radius. In this study, strain is selected as the non-local variable because of its importance in representing material behaviour throughout both the linear elastic and the softening phases.

In 2D Euclidean space ($V \in \mathbb{R}^2$) with Cartesian co-ordinates $x = \{x,y\}$, Eringen proposed the non-local stress-strain relationship as [28]:

$$\boldsymbol{\sigma}(\mathbf{x}) = \mathbb{C} : \bar{\boldsymbol{\varepsilon}}(\mathbf{x}) \tag{1}$$

where, $$\bar{\boldsymbol{\varepsilon}}(\mathbf{x}) = \int_V \alpha(\mathbf{x},\mathbf{x}')\,\boldsymbol{\varepsilon}(\mathbf{x}')\mathrm{d}V(\mathbf{x}') \tag{2}$$

Here, $\mathbb{C}$ represents the fourth-order elasticity tensor, while $\boldsymbol{\sigma}$ and $\bar{\boldsymbol{\varepsilon}}$ stand for the stress and non-local strain at the point x. The extent of the interaction with neighbours ($\mathbf{x'}$) is governed by the influence radius L_R, where the weighting function gradually decreases as the distance between $\mathbf{x}$ and $\mathbf{x'}$ grows, with the interaction being strongest when $\mathbf{x'}$ coincides with $\mathbf{x}$.

It is well established that for a uniform strain field, Equation (1) should reduce to its local form (i.e., Hooke's law). This can be verified by demonstrating that $\int_V \alpha(x,x')dV(x') = 1$. However, this condition is valid only in regions far from boundaries. Near the boundaries, $\int_V \alpha(x,x')\, dV(x') < 1$, leading to increase in the stress levels compared to those in uniformly stressed areas. To circumvent these boundary effects, the weight functions employed in this research are formulated as follows:

$$\bar{\boldsymbol{\varepsilon}}(\mathbf{x}) = \int_V W(\mathbf{x},\mathbf{x}')\,\boldsymbol{\varepsilon}(\mathbf{x}')dV(\mathbf{x}')$$
$$W(\mathbf{x},\mathbf{x}') = \frac{\alpha(\mathbf{x},\mathbf{x}')}{\int_V \alpha(\mathbf{x},\mathbf{x}')\,dV(\mathbf{x}')} \quad (3)$$

In the present work, the constitutive relation in Equation (1) for a damaged material is expressed as:

$$\boldsymbol{\sigma}(\mathbf{x}) = \mathbb{C}(1-D) : \bar{\boldsymbol{\varepsilon}}(\mathbf{x}) \quad (4)$$

Continuum damage model

This study models the damage evolution in adhesive-bonded joints based on the approach presented by [24]. In this framework, both the adhesive layer and the adherend are treated as continuous materials, and a mixed-mode bilinear traction-separation law (TSL) is applied to the adhesive. The initiation of damage in the adhesive material occurs when the non-local stresses in the relevant modes reach their critical limits, which are determined using a quadratic stress criterion expressed as follows:

$$\left(\frac{\sigma_I}{\sigma_u}\right)^2 + \left(\frac{\tau}{\tau_u}\right)^2 = 1 \quad (5)$$

Here, σ_I and τ indicate the non-local stresses associated with the normal and shear directions, respectively. The corresponding ultimate stresses are denoted as σ_u and τ_u. Based on Equation (5), the components of stress that trigger damage onset under mixed-mode loading conditions (denoted as σ_{eI} and σ_{eII}) can be defined as follows:

$$\text{If } \sigma_I > 0, \quad \sigma_{eI} = \frac{\sigma_u \tau_u}{\sqrt{\tau_u^2 + \beta_s^2 \sigma_u^2}} \quad (6)$$

$$\text{If } \sigma_I < 0, \quad \sigma_{eII} = \frac{\sigma_u \tau_u}{\tau_u \sqrt{\left(\frac{1}{\beta_s^2} + \frac{\alpha_s^2}{\beta_s^2}\right) + \sigma_u^2}} \quad (7)$$

Referring to Hooke's law, the relative displacements in the corresponding modes are established as follows:

$$\text{If } \sigma_I > 0, \quad \delta_{o,Im} = \frac{\sigma_{eI}}{E}(1-\nu\alpha_s)l_c, \quad \delta_{o,IIm} = \frac{2\sigma_{eI}(1+\nu)}{E}\beta_s l_c \quad (8)$$

$$\text{If } \sigma_I < 0, \quad \delta_{o,Im} = \frac{\sigma_{eII}}{E\,\beta_s}(1-\nu\alpha_s)l_c, \quad \delta_{o,IIm} = \frac{2\sigma_{eII}(1+\nu)}{E} l_c \quad (9)$$

Equivalent relative displacement:

$$(\delta_{o,m}) = \sqrt{\delta_{o,Im}^{\ 2} + \delta_{o,IIm}^{\ 2}}$$

β_s is expressed as the absolute value of shear stress divided by the normal stress, represented as $\frac{|\tau|}{\sigma_I}$. Additionally, α_s denotes the ratio of stresses along the x and y directions, with the mode I component appearing in the denominator.

This study employs a bilinear traction separation law (see Figure 12.1), allowing for the calculation of damage evolution based on the linear fracture energy

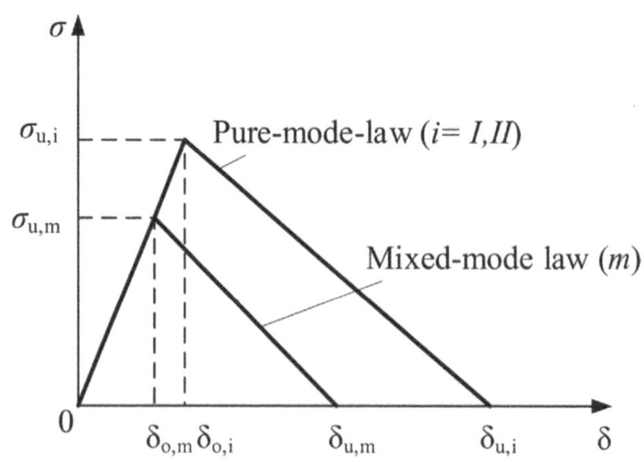

Figure 12.1 Mixed mode traction-separation law
Source: Authors

Table 12.1 Material properties used for DCB test.

	E (GPa)	*ν*	σ_u (MPa)	τ_u (MPa)	G_{Ic} (N/mm)	G_{IIc} (N/mm)
Adhesive	4	0.3	20	20	0.3	0.6
Adherend	210	0.3	-	-	-	-

Source: Authors

Table 12.2 Mesh sizes used for DCB test.

	Adherend (mm)	Adhesive (mm)	No. of elements
Adhesive	0.25 × 0.75	0.25 × 0.2	1880
Adherend	0.1 × 0.75	0.1 × 0.2	4700

Source: Authors

criterion, $\frac{G_I}{G_{Ic}} + \frac{G_{II}}{G_{IIc}} = 1$, where the energy components G_I and G_{II} are given by:

$$\text{If } \sigma_I > 0, \quad G_I = \frac{\sigma_{eI}\delta_{u,Im}}{2}, \quad G_{II} = \frac{|\tau|\,\delta_{u,IIm}}{2} \tag{10}$$

$$\text{If } \sigma_I < 0, \quad G_I = \frac{\sigma_I\delta_{u,Im}}{2}, \quad G_{II} = \frac{\sigma_{eII}\delta_{u,IIm}}{2} \tag{11}$$

Equivalent relative displacement:

$$(\delta_{u,m}) = \sqrt{\delta_{u,Im}^{\;2} + \delta_{u,IIm}^{\;2}}$$

Using the above Equations (6–11), the damage evolution for mixed mode cases in the adhesive region is expressed as:

$$D = D_m = \frac{\delta_{um}(\delta_m - \delta_{om})}{\delta_m(\delta_{um} - \delta_{om})} \tag{12}$$

Numerical examples

In this section, mode I problem is first solved to verify the formulations and methodologies of the non-local approach presented in this study. Next, an example is provided to highlight the benefits of non-local integral methods compared to traditional finite element analysis.

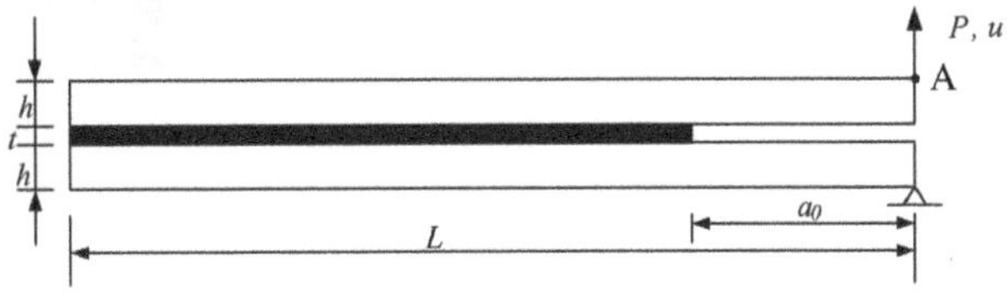

Figure 12.2 Adhesively-bonded DCB specimen: geometry, dimensions, loading and boundary conditions

Source: Authors

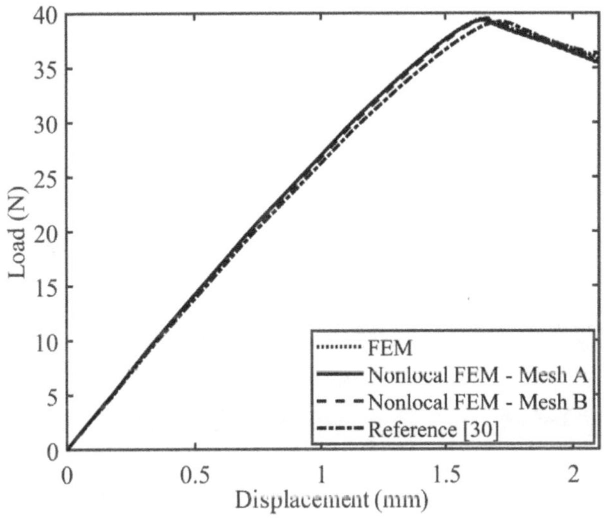

Figure 12.3 Load-displacement response: comparison of FEM, non-local FEM with reference results

Source: Authors

Double cantilever beam (DCB) test

The continuum damage model under mixed-mode conditions is applied to a DCB under pure mode I loading. The geometry of the DCB test, as shown in Figure 12.2 [24], has the following dimensions: L = 100 mm, a_0 = 30 mm, h = 1.5 mm, t = 0.2 mm, and width B = 10 mm. The material properties of both the adherend and adhesive are presented in Table 12.1.

Numerical analysis is performed using four-node quadrilateral elements for both the adherend and

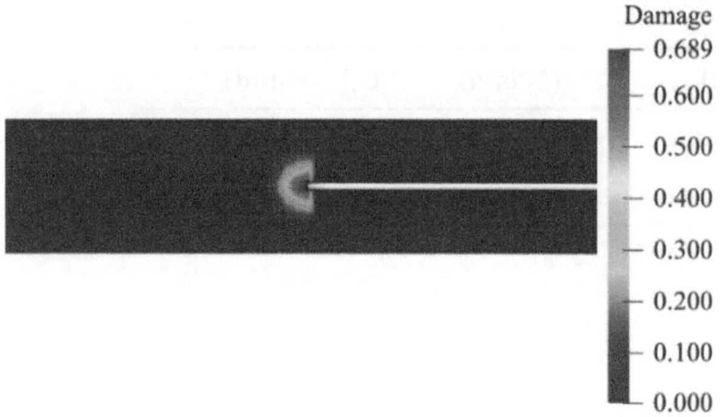

(a)

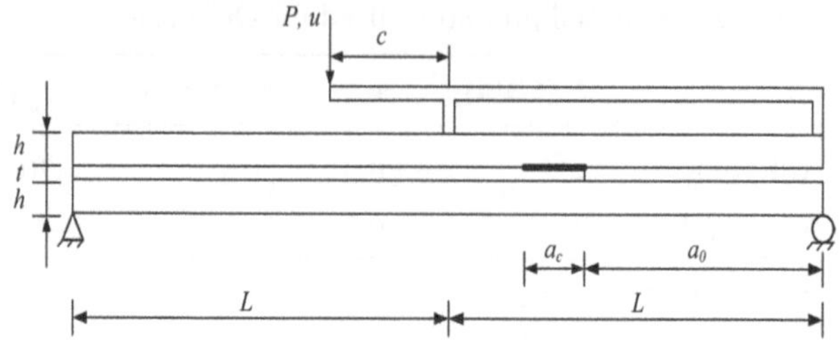

Figure 12.5 Adhesively bonded DCB specimen subjected to mixed-mode loading conditions
Source: Authors

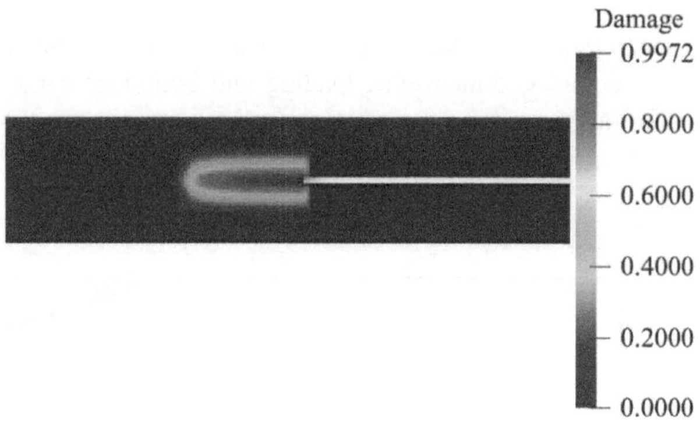

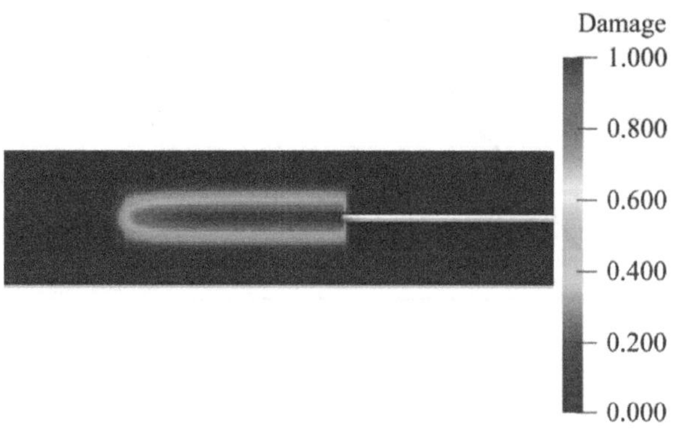

(c)

Figure 12.4 Damage evolution at a displacement of (a) 0.55 mm, (b) 1.65 mm, and (c) 2.00 mm
Source: Authors

adhesive, under the assumption of plane stress conditions. Mesh details are outlined in Table 12.2. An influence radius (L_R) of 0.1 is adopted, encompassing neighbouring elements to account for stress computation and quartic weight functions are employed in the case studies presented in this work.

Figure 12.3 compares the load-displacement responses of both non-local and local FEM analyses with reference data. The close match between the non-local results, conventional FEM, and reference values confirms the accuracy of the proposed numerical approach. These validated methods are reliable for use in further case studies. To demonstrate the mesh independence of the non-local FEM, two different mesh configurations, labelled A and B, are employed.

Figure 12.4 presents the damage evolution in the refined mesh, illustrated by contour plots. During the analysis, damage to the adhesive elements begins at a displacement of 0.27 mm and progresses from there. Figure 12.4a shows the damage evolution at a displacement of 0.55 mm, with a maximum damage of 0.689. In Figure 12.4b, at the peak load and a displacement of 1.65 mm, the maximum damage reaches 0.9972, after which the stiffness begins to decrease. Figure 12.4c illustrates the damage evolution at a displacement of 2.00 mm, where some adhesive elements have reached a damage level of 1, indicating that their stiffness has become zero.

Mixed-mode bending (MMB) problem

In this study, the same geometry is used, but the loading conditions are modified as shown in Figure 12.5 to enable mixed-mode conditions in the specimen. Additionally, a defect in the form of a pre-crack (a_c) of 5 mm length is introduced at the interface between the top adherend and the adhesive. To optimise the computational analysis of joints while ensuring the accuracy of the results, bias effects in the mesh size classification are implemented. The material properties of the adherend and adhesive are specified in Table 12.3.

To investigate the impact of mesh sensitivity on numerical accuracy, a comprehensive mesh study is conducted using conventional FEM (see Figure

Table 12.3 Material properties used for mixed-mode problem.

	E (GPa)	ν	**σ_u (MPa)**	**τ_u (MPa)**	**G_{Ic} (N/mm)**	**G_{IIc} (N/mm)**
AV138	4.89	0.35	39.45	30.2	0.2	0.38
Adherend	210	0.3	-	-	-	-

Source: Authors

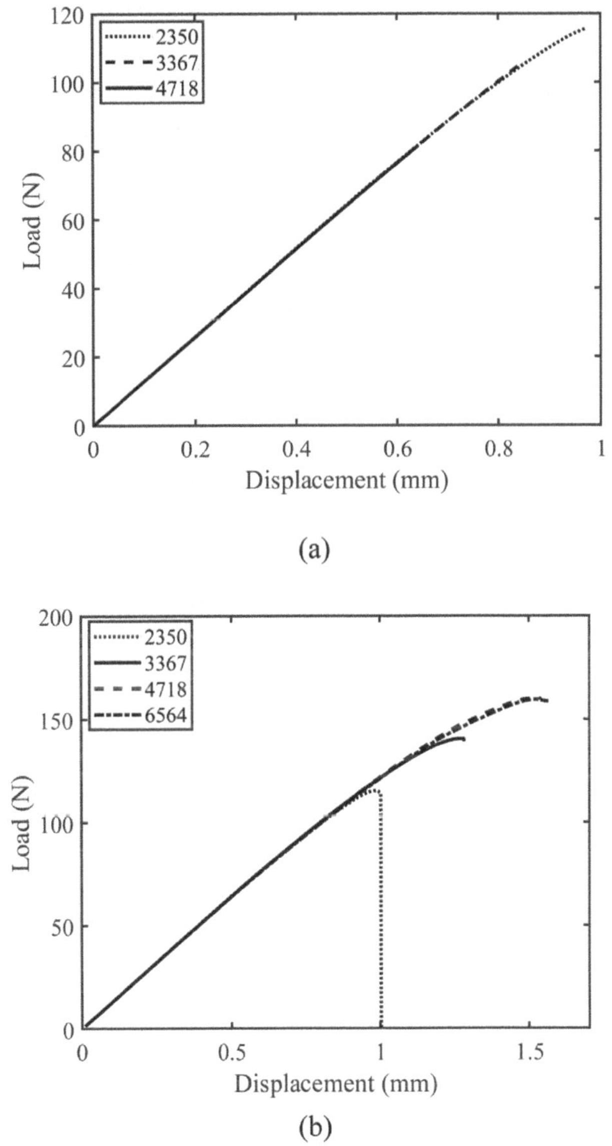

Figure 12.6 Load-displacement responses: (a) FEM and (b) non-local FEM

Source: Authors

12.6a). The initial coarse mesh, consisting of 2,350 elements, achieves a peak load of 115.494 N at a displacement of 0.972 mm, but instability arises at this point. Subsequent refinements, using meshes of 3,367 and 4,718 elements, result in instability at reduced displacements of 0.84 mm and 0.64 mm, respectively, along with a decrease in peak load that is physically unrealistic. These findings highlight the reliability of conventional FEM results, emphasising the challenges of instability and mesh dependency that lead to non-convergent outcomes.

However, the non-local strain-based approach is utilised to address these limitations. First, the choice of influence radius (L_R) in the formulations is based on two key factors. First, it must be selected to ensure that the stiffness of the load-displacement response aligns with the conventional FEM results, maintaining the integrity of the analysis. Second, it should optimise computational efficiency while delivering accurate and reliable results. Considering these factors, the influence radius is set at 0.025 mm.

Figure 12.6b presents the mesh sensitivity analysis for the proposed non-local FE methodology. The coarse mesh with 2,350 elements yielded results comparable to conventional FEM, indicating joint failure at the same displacement. The second mesh, with 3,367 elements, produced a peak load of 140.29 N at a displacement of 1.283 mm, reflecting expected trends in mesh refinement. Subsequent analyses of finer meshes showed convergence at a displacement of 1.53 mm, with a peak load of 159.446 N, demonstrating consistent softening and failure behaviour across all configurations.

Conclusion

In this work, a methodology integrating non-local theories with the finite element framework is implemented to accurately capture failure behaviour in adhesive-bonded joints. A continuum damage model under mixed-mode conditions is employed, based on a quadratic stress criterion and a linear fracture energy approach, to describe the damage evolution in these joints. The non-local FEM results closely match the reference data and conventional FEM, particularly in the load-displacement responses, demonstrating the reliability of the proposed model for future case studies and practical applications. By addressing challenges in FEM such as mesh sensitivity and instability, especially in joints with pre-existing defects or cracks at the adhesive-adherend interface, the non-local FEM delivers mesh-independent results in mixed-mode damage problems, proving its robustness over conventional methods.

Acknowledgement

The work is supported by the DST-SERB and VSSC, ISRO of the project titled "Functionality Enhancement through Design and Development of Advanced Finite Element Algorithms for STR tools" under IMPRINT.IIC (IMP/2019/000276) scheme.

References

[1] Parameswaran, V., Shukla, A. Processing and characterization of a model functionally gradient material. *J. Mater. Sci.*, 2000;35(1):21–29.

[2] Patnaik, S., Jokar, M., Ding, W., Semperlotti, F. Distillation of non-locality in porous solids. *Proc. Royal Soc. A Mathemat. Phy. Engg. Sci.*, 2023;479(2275):20220770.

[3] Russillo, A. F., Failla, G. Wave propagation in stress-driven nonlocal Rayleigh beam lattices. *Int. J. Mech. Sci.*, 2022;215:106901.

[4] Kanouté, P., Boso, D. P., Chaboche, J. L., Schrefler, B. A. Multiscale methods for composites: A review. *Arch. Comput. Methods Engg.*, 2009;16(1):31–75.

[5] Mazars, J., Lemaitre, J. Application of continuous damage mechanics to strain and fracture behavior of concrete. Springer, 1985. 507:520.

[6] Malcher, L., Mamiya, E. N. An improved damage evolution law based on continuum damage mechanics and its dependence on both stress triaxiality and the third invariant. *Int. J. Plast.*, 2014;56:232–261.

[7] Lim, C. W., Zhang, G., Reddy, J. N. A higher-order nonlocal elasticity and strain gradient theory and its applications in wave propagation. *J. Mech. Phys. Solids*, 2015;78:298–313.

[8] Jirásek, M. Nonlocal damage mechanics. *Revue Européenne de Génie Civil*, 2007;11(7–8):993–1021.

[9] Eringen, A. C., Kim, B. S. Stress concentration at the tip of crack. *Mech. Res. Commun.*, 1974;1(4):233–237.

[10] Eringen, A. C. Line crack subject to shear. *Int. J. Fract.*, 1978;14(4), 367:379.

[11] Polizzotto, C. Nonlocal elasticity and related variational principles. *Int. J. Solids Struct.*, 2001;38(42–43):7359–7380.

[12] Pisano, A. A., Sofi, A., Fuschi, P. Nonlocal integral elasticity: 2D finite element based solutions. *Int. J. Solids Struct.*, 2009;46(21):3836–3849.

[13] Zhang, X., Gunzburger, M., Ju, L. Quadrature rules for finite element approximations of 1D nonlocal problems. *J. Comput. Phys.*, 2016;310:213–236.

[14] Patnaik, S., Sidhardh, S., Semperlotti, F. Fractional-Order models for the static and dynamic analysis of nonlocal plates. *Commun. Nonlinear. Sci. Numer. Simul.*, 2021;95:105601.

[15] Friswell, M. I., Adhikari, S., Lei, Y. Non-local finite element analysis of damped beams. *Int. J. Solids Struct.*, 2007;44(22–23):7564–7576.

[16] Ansari, R., Torabi, J., Norouzzadeh, A. Bending analysis of embedded nanoplates based on the integral formulation of Eringen's nonlocal theory using the finite element method. *Physica B Condens Matter*, 2018;534:90–97.

[17] Pijaudier-Cabot, G., Asce, S. M., Basant, Z. P., Asce, F. Nonlocal damage theory. *J. Engg. Mech.*, 1987;113(10):1512–1533.

[18] Jirásek, M. Computational aspects of nonlocal models. *Eur. Conf. Comput. Mech.*, 1999. 1:10.

[19] Borino, G., Failla, B., Parrinello, F. A symmetric nonlocal damage theory. *Int. J. Solids Struct.*, 2003;40(13–14):3621–3645.

[20] Andrade, F. X. C., César De Sá, J. M. A., Andrade Pires, F. M. A ductile damage nonlocal model of integral-type at finite strains: Formulation and numerical issues. *International Journal of Damage Mechanics*. 2011;20(4):515–557.

[21] Sai Kumar, S., Muthu, N. A thermodynamically consistent strain difference-based nonlocal damage mechanics approach for failure analysis of quasi-brittle materials. *Theoret. Appl. Frac. Mech.*, 2024;131:104351.

[22] Ramalho, L. D. C., Campilho, R. D. S. G., Belinha, J., da Silva, L. F. M. Static strength prediction of adhesive joints: A review. *Int. J. Adhes. Adhes.*, 2020;96:102451.

[23] García, J. A., Chiminelli, A., García, B., Lizaranzu, M., Jiménez, M. A. Characterization and material model definition of toughened adhesives for finite element analysis. *Int. J. Adhes. Adhes.*, 2011;31(4):182–192.

[24] Chousal, J. A. G., De Moura, M. F. S. F. Mixed-mode I+II continuum damage model applied to fracture

characterization of bonded joints. *Int. J. Adhes. Adhes.*, 2013;41:92–97.

[25] Ribeiro, F. M. F., Campilho, R. D. S. G., Carbas, R. J. C., da Silva, L. F. M. Strength and damage growth in composite bonded joints with defects. *Compos. B Engg.*, 2016;100:91–100.

[26] Feraren, P., Jensen, H. M. Cohesive zone modelling of interface fracture near flaws in adhesive joints. *Engg. Fract. Mech.*, 2004;71(15):2125–2142.

[27] Chadegani, A., Batra, R. C. Analysis of adhesive-bonded single-lap joint with an interfacial crack and a void. *Int. J. Adhes. Adhes.*, 2011;31(6):455–465.

[28] Eringen, A. C., Edelen, D. G. B. On nonlocal elasticity. *Int. J. Engg. Sci.*, 1972;10(3):233–248.

13 Modelling of hollow cylinders with auxetic lattice structures

Pranjali Nema[a], Soumya Ranjan Sahoo[b], Suman Deb[c] and Rithika Reddy Rondla[d]
School of Mechanical Sciences, Indian Institute of Technology Bhubaneswar, India

Abstract

This study investigates the axial compression behaviour of hollow auxetic cylinder structures (HACS) with different auxetic lattice patterns such as arrowhead, star, and chiral auxetic cells. Cylindrical models of the same dimensions are prepared and their deformation patterns are examined. Poisson's ratio is calculated through quasi-static finite element models. The influences of the structural parameters such as wall thickness on the compression characteristics, and minimum negative Poisson's ratio (NPR) value of HACS are also investigated. The finite element model has been validated by replicating previous research works. The influence of two polymeric materials, ABS (acrylonitrile butadiene styrene) and PLA (polylactic acid), on Poisson's ratio was investigated. The results indicate that Poisson's ratio remains largely unaffected by the choice of material.

Keywords: Auxetic, 3D printing, finite element models, quasi-static compression

Introduction

The accelerated progress in science and technology over recent decades has made the development of advanced, application-specific materials with tailored properties quite necessary. Materials that can exhibit extraordinary properties not generally found naturally can often allow for a wider range of applications than previously thought possible. Lightweight and durable materials have pushed the aerospace and automotive industries to achieve safer and faster travel. With the advancements in computational modelling and additive manufacturing, newer possibilities have opened for more complex structures and geometries that can provide better mechanical resilience.

Metamaterials are special materials artificially engineered to have exotic properties beyond those of naturally occurring materials. They achieve their unique characteristics through carefully engineered geometric patterns and arrangements. A wide variety of metamaterials can be engineered depending on their intended applications.

Auxetic structures are a distinct class of metamaterials that have attracted considerable attention due to their unconventional mechanical properties and an expanding spectrum of applications. These structures expand in the perpendicular axis under a tensile load and compress laterally under compressive load, thereby exhibiting a negative Poisson's ratio (NPR) [1].

First reported in 1987 [2], these special types of materials were originally named auxetic, derived from the Greek "auxetikos", meaning "to tend to increase" [3]. Studies have shown that auxetic structures, in comparison to conventional materials, have improved energy absorption capabilities [4] and bending performance [5].

Tubular geometries are among the most prevalent shapes in mechanical components and are widely utilised across various applications due to their versatility and functional significance. While it is known that tubular structures are excellent for energy absorption [6] and crashworthiness [7]; integrating auxetic geometries into them offers the potential to improve their performance significantly.

Through this paper, we present a comprehensive study of the influence of some geometrical parameters and materials on the compression characteristics of a hollow auxetic cylinder structure (HACS).

[a]21me01034@iitbbs.ac.in, [b]soumyasahoo@iitbbs.ac.in, [c]sumandeb@iitbbs.ac.in, [d]21me01029@iitbbs.ac.in

DOI: 10.1201/9781003770442-13

Design and modelling

Design of auxetic geometries

This section introduces the design process and specific parameters used for the HACS. Initially, a 2D sketch was created in SOLIDWORKS 2015 software. Regular repeating cells for all structures were developed to ensure equal length overall. All the 2D sketches are shown in Figure 13.1. This single layer of sketch was then wrapped around a cylinder with an equal circumference, and the cells were cut out. First, all structures were created outer diameter of 70 mm and a wall thickness of 2 mm. A gap of 5 mm was provided both at the top and bottom of the HACS to ensure proper force distribution. The total length of the structures is 100 mm. Figure 13.2 presents all the HACS and provides simpler notations used further.

Material parameters

The models were studied for two materials: ABS-Acrylonitrile Butadiene Styrene and PLA-Polylactic Acid. These were selected because they are the most commonly used materials for prototype testing using 3D printing. They also allow for a consistent baseline for comparing structural designs with pre-existing research studies. Table 13.1 lists the material properties used in the analysis.

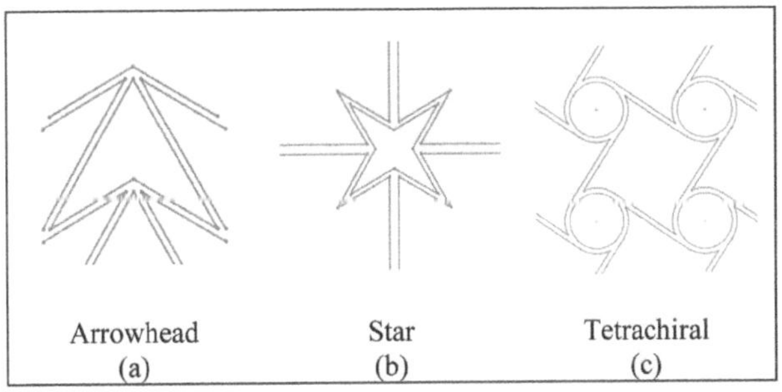

Figure 13.1 Sketches of the auxetic geometries
Source: Authors

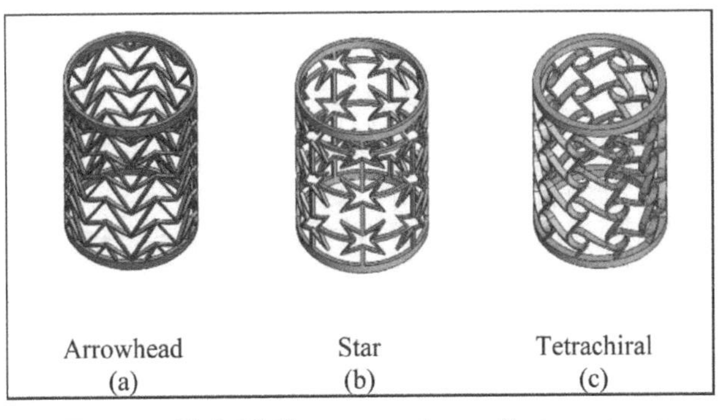

Figure 13.2 Hollow auxetic cylinder structures (HACS)
Source: Authors

Numerical simulation

1. **Finite element modelling in ANSYS**

   A quasi-static numerical analysis was performed on the HACS using the "2020 R2" version of the finite element analysis software ANSYS. The materials were simulated using an elastic constitutive model. The Parasolid geometry was imported from SOLIDWORKS 2015, and material properties were assigned. A uniform ramped force was applied across one face of the cylinder while constraining the opposite face in all degrees of freedom. The deformations were captured in the axial and lateral directions. Poisson's ratio (v) was calculated through the given equation by using the deformation values obtained previously:

$$v = -\frac{\epsilon_{lateral}}{\epsilon_{axial}}$$

   Here, ϵ is the strain in lateral and axial directions.

Table 13.1 Properties of materials used in this study.

Material	Density (kg/m^3)	Young's modulus (Pa)	Poisson's ratio
PLA	1250	3.5E+09	0.36
ABS	1100	2.2E+09	0.37

Source: Authors

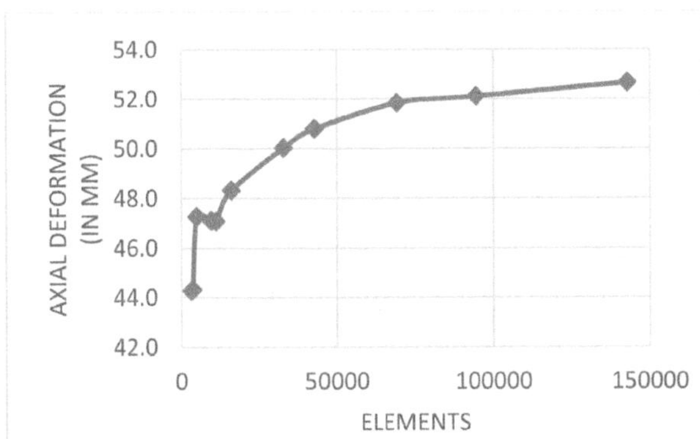

Figure 13.3 Convergence study for validation model
Source: Authors

2. **Validation study**
   To validate the model, an analysis was performed using the results obtained by Yang et al. [8] through an investigation of the material behaviour of auxetic pads. The convergence study for the model is shown in Figure 13.3. Using an appropriate mesh technique, the study was successfully replicated with results obtained to be within a tolerance of 5%. Figure 13.4 compares the results of present model with the validation results obtained previously.
3. **Convergence study**
   In the finite element modelling of HACS, a tetrahedral mesh was employed to effectively capture the complex geometry of the structure. SOLID187, a 10-noded tetrahedral element was used in the analysis due to its suitability for deformation studies of complex 3D geometries. To better incorporate the fine-scale features, a custom-sized mesh was necessary. Therefore, a convergence study was performed on structure M1. The total axial deformation was monitored for varying element sizes until stable results could be obtained. As observed in Figure 13.5 consistent deformation results could be obtained for about 15000 elements. Corresponding to this, an element size of 2 mm was used for meshing in all further analyses (Figure 13.6).

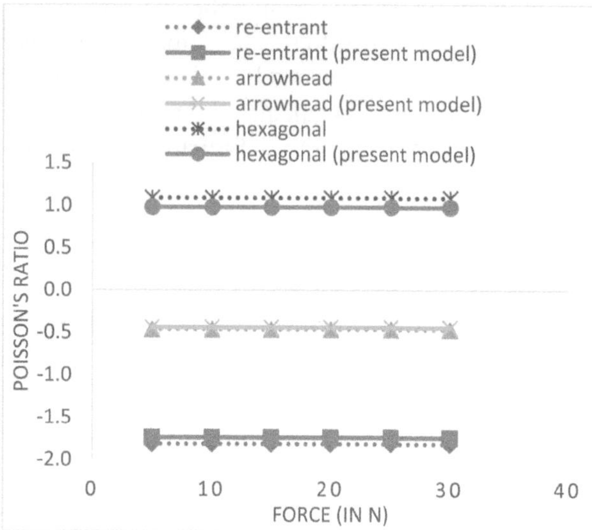

Figure 13.4 Validation of Poisson's curve [9]
Source: Authors

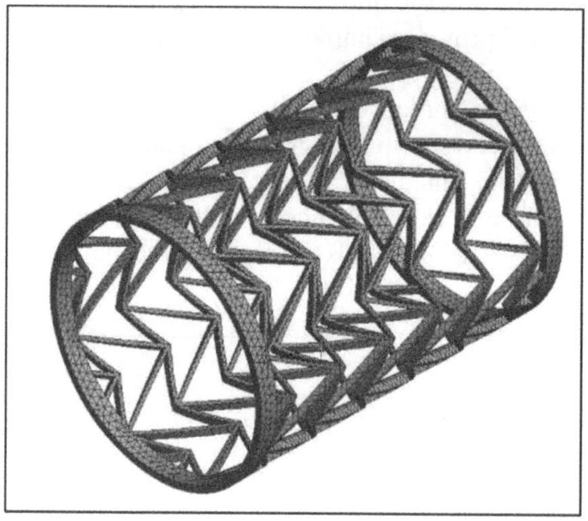

Figure 13.6 Meshing of arrowhead structure with element size 2 mm
Source: Authors

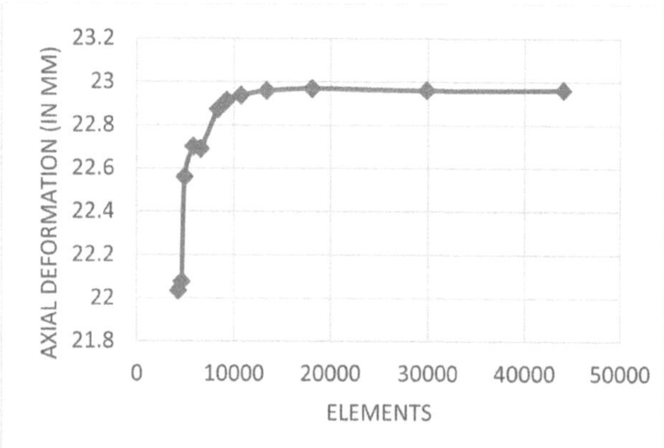

Figure 13.5 Convergence study for arrowhead under a compressive load of 10 N
Source: Authors

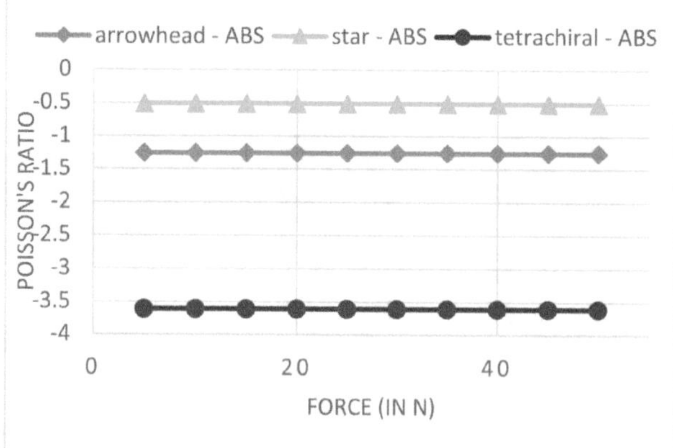

Figure 13.7 Poisson's ratio for HACS with material ABS under compressive load
Source: Authors

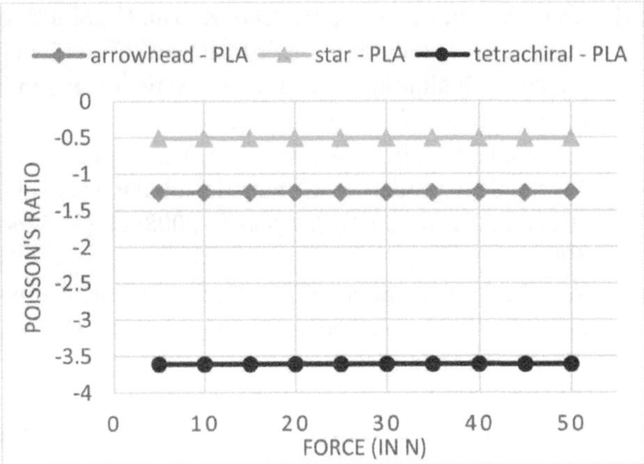

Figure 13.8 Poisson's ratio for HACS with material PLA under compressive load
Source: Authors

Results and discussion

Poisson's ratio for different materials

Figures 13.7 and 13.8 depict the Poisson's ratio of the three materials under different compressive loads. By keeping the length-to-diameter ratio constant for the models, we can easily compare the results across different geometries. We observed all three structures, as expected, showed a negative Poisson's ratio. The tetrachiral geometry showcased the most negative value of all three, while the star-shaped geometry had the least negative value. This can be attributed to the large degree of rotation possible for a cell in tetrachiral geometry while the cells in a star or arrowhead geometry can only move in translation.

The structures were examined for two different materials – ABS and PLA. The materials, however, had an almost negligible effect on Poisson's ratio.

Varying geometric parameters

To determine the effects of changing geometric parameters, multiple models were made with different wall thickness. For each HACS the thickness was varied from 1 mm to 10 mm while keeping the outer diameter constant. A load of 50 N was then applied to one end of the cylinder.

The HACS with arrowhead and star geometries showed a slight increase in the negative Poisson's ratio; the drop was more significant in case of the arrowhead geometry. The Poisson's ratio for arrowhead geometry begins to converge after a thickness of 6 mm and remains constant at a value of -1.42.

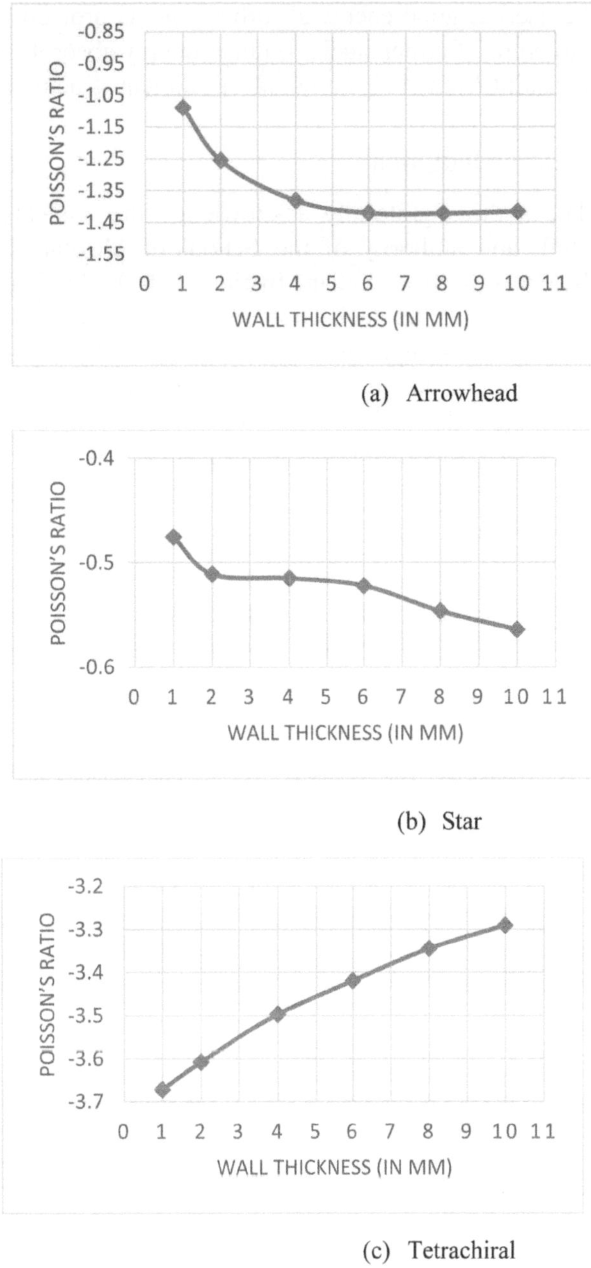

(a) Arrowhead

(b) Star

(c) Tetrachiral

Figure 13.9 Variation in Poisson's ratio for HACS with varying wall thickness
Source: Authors

The tetrachiral geometry on the other hand displayed a drop in the negative Poisson's ratio value (Figure 13.9).

Conclusion

The present work is a comparative study of the behaviour of various types of HACS structures concerning their Poisson's ratio. These structures can be used as good energy absorbers and as protective structures. Further studies on their energy absorption capabilities can help design more efficient structures.

Acknowledgment

The authors gratefully acknowledge the students, staff, and authority of the School of Mechanical Sciences at the Indian Institute of Technology Bhubaneswar for their cooperation in the research. This work is supported by SRIC IIT Bhubaneswar (project no. SP135 and SP137). We also deeply appreciate the facilities and resources provided by the CADCAM Laboratory at IIT Bhubaneswar.

References

[1] Wu, W., Liu, P., Kang, Z. A novel mechanical metamaterial with simultaneous stretching-and compression-expanding property. *Mater. Design*, 2021;208:109930.

[2] Lakes, R. Foam structures with a negative Poisson's ratio. *Science*, 1987;235(4792):1038–1040.

[3] Evans, K. E., Nkansah, M. A., Hutchinson, I. J., Rogers, S. C. Molecular network design. *Nature*, 1991;353(6340):124–124.

[4] Yang, S., Qi, C., Wang, D., Gao, R., Hu, H., Shu, J. A comparative study of ballistic resistance of sandwich panels with aluminum foam and auxetic honeycomb cores. *Adv. Mech. Engg.*, 2013;5:589216.

[5] Scarpa, F., Smith, C. W., Ruzzene, M., Wadee, M. K. Mechanical properties of auxetic tubular truss-like structures. *Physica Status Solidi*, 2008;245(3):584–590.

[6] Wei, Y., Fan, Y., Shen, X., Gong, J., Hao, W. A novel cylindrical mechanical metamaterial: Design, fabrication, and compressive properties. *Structures*, 2024;63:106336.

[7] Gomes, R. A., de Oliveira, L. A., Francisco, M. B., Gomes, G. F. Tubular auxetic structures: A review. *Thin-Walled Struct.*, 2023;188:110850.

[8] Yang, C., Vora, H. D., Chang, Y. Behavior of auxetic structures under compression and impact forces. *Smart Mater. Struct.*, 2018;27(2):025012.

[9] Qi, C., Jiang, F., Yang, S., Remennikov, A. Multiscale characterization of novel re-entrant circular auxetic honeycombs under quasi-static crushing. *Thin-Walled Struct.*, 2021;169:108314.

14 Modelling of piezoelectric sandwich energy harvester with auxetic cores

Rithika Reddy Rondla[a], *Soumya Ranjan Sahoo*[b] *and Pranjali Nema*[c]
School of Mechanical Sciences, Indian Institute of Technology Bhubaneswar, India

Abstract

This paper presents a comprehensive study on the modelling and optimisation of a piezoelectric sandwich energy harvester featuring auxetic cores compared to isotropic cores, aimed at enhancing vibrational energy harvesting efficiency. The proposed auxetic piezoelectric energy harvester with a tapered thickness is engineered to achieve a more uniform stress distribution, leading to significant improvements in power output and energy density compared to conventional uniform-thickness designs. Through finite element analysis (FEA) and experimental validation, the optimised TAEH demonstrates enhanced electromechanical performance. Additionally, a numerical model is developed to simulate the dynamic behaviour of the energy harvester with auxetic substrates. Simulations reveal that the auxetic core induces biaxial strain in the piezoelectric material, generating higher electrical output without increasing mechanical stress. The model is refined through iterative simulations to accurately predict the harvester's structural and electrical performance, showcasing the potential of auxetic designs to advance the field of sustainable vibrational energy harvesting in mechanical systems.

Keywords: Auxetic, vibrations, piezoelectric, tapered thickness

Introduction

Recently, harvesting energy from ambient vibrations has gained attention as an effective method to generate electrical power, especially for operating low-power devices like embedded systems and wireless sensor networks. Amongst the various energy harvesting technologies, considerable interest has been directed toward piezoelectric materials, as they enable the direct conversion of mechanical stress into electrical output. However, enhancing the efficiency of piezoelectric energy harvesters remains a challenging task, especially when dealing with the inherent limitations of conventional isotropic core structures and uniform-thickness designs, which often lead to uneven stress distributions and suboptimal power output. In response to these challenges, novel structural designs such as auxetic materials, known for their unique negative Poisson ratio, have been introduced to improve the mechanical properties and overall energy conversion efficiency of piezoelectric harvesters. Auxetic materials exhibit the counterintuitive property of expanding perpendicular to applied tension, which can be leveraged to achieve a more uniform stress distribution in piezoelectric energy harvesters. This characteristic makes auxetic cores an intriguing alternative to conventional isotropic cores for improving the performance of energy harvesting devices. This paper presents a comprehensive study on the modelling, optimisation, and performance comparison of piezoelectric sandwich energy harvesters with auxetic cores and conventional isotropic cores. The proposed design incorporates a piezoelectric energy harvester with a tapered thickness profile, aimed at further enhancing the stress distribution and increasing power output. By employing finite element analysis (FEA) and conducting simulation validation, this work investigates the electromechanical performance of the auxetic-core energy harvester, emphasising its potential to generate higher electrical output without introducing additional mechanical stress. The iterative refinement of the model enables accurate predictions of the energy harvester's structural and electrical performance, demonstrating the viability of auxetic designs in advancing the efficiency and sustainability of vibrational energy

[a]21me01029@iitbbs.ac.in, [b]soumyasahoo@iitbbs.ac.in, [c]21me01034@iitbbs.ac.in

DOI: 10.1201/9781003770442-14

harvesting in mechanical systems. In the following sections, we will detail the design methodology, simulation results, and validation, highlighting the significant improvements offered by the auxetic-core piezoelectric energy harvester over conventional designs.

Theory

In this section, we present the theoretical foundation for the modelling of the auxetic piezoelectric energy harvester (AEH) with re-entrant hexagonal auxetic cores. According to Erturk and Inman [1], the piezoelectric energy harvesting process relies on the electromechanical coupling of mechanical strain and electrical potential, governed by piezoelectric constitutive equations. For this work, we investigate the performance of the harvester using both auxetic and conventional isotropic cores, analysing their influence on stress distribution and electrical output. The auxetic structure's unique property of negative Poisson's ratio induces a biaxial strain in the piezoelectric layers, enhancing the electromechanical coupling without significantly increasing mechanical stress. We derive the governing equations for the dynamic behaviour of the energy harvester, incorporating both mechanical and electrical domains, and perform FEA to validate the system's performance. Additionally, the voltage generated by the piezoelectric layers under dynamic loading is given as follows by Erturk and Inman [2]:

$$V(t) = \frac{-e_{31}\, bd}{h_p C_p} \int_0^L \frac{\partial^2 w(x,t)}{\partial x^2}\, dx \tag{1}$$

where $w(x, t)$ is the transverse displacement of the beam, e_{31} is the piezoelectric coefficient, b is the width of the beam, h_p is the thickness of the piezoelectric layer, and Cp is the capacitance.

The model shown in Figure 14.1 indicates the model used in the following simulations. Piezo patch is shown in blue, and auxetic core in grey in the Figure 14.1.

Design and modelling

Design of auxetic substructures for energy harvesters

The auxetic structure in the energy harvester is designed using a re-entrant hexagonal, honeycomb, and oval geometry, known for its negative Poisson's ratio, which enhances the uniformity of strain distribution across the piezoelectric layers. This geometry enables biaxial expansion under tension, leading to more efficient mechanical-to-electrical energy conversion. The core is tapered in further designs to further optimise stress distribution, reducing stress concentrations that typically occur in uniform-thickness designs. Finite element modelling is used to refine the auxetic core's dimensions, ensuring optimal performance under dynamic loading conditions.

Material parameters

The designs of the substructure for the cantilevered energy harvesters were chosen from existing popular auxetic shapes used in research. Polylactic acid (PLA) of Ingeo Grade 4060D was utilised as the material of the beam. This material was chosen due to its popularity and availability for 3D Printing applications. The structure is complete with a layer

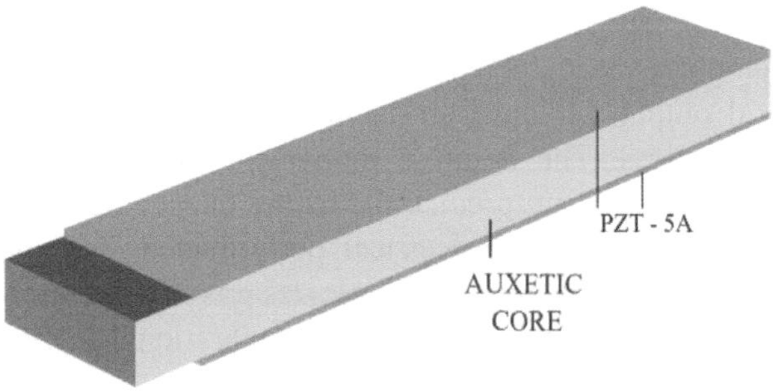

Figure 14.1 Schematic of cantilever energy harvester using auxetic core
Source: Authors

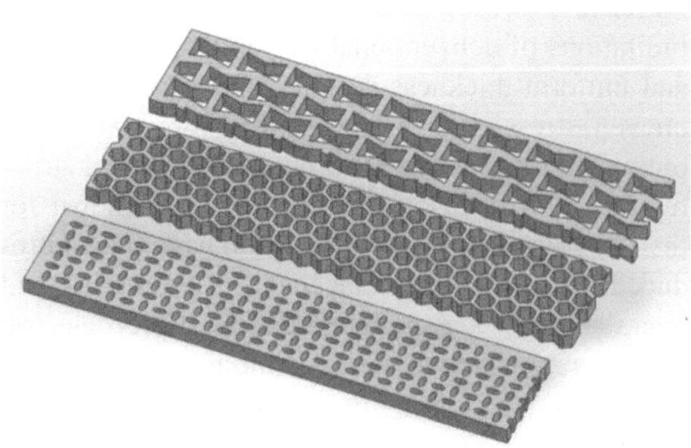

Figure 14.2 Schematic diagram of auxetic core structure
Source: Authors

Table 14.1 Geometric parameters of cantilever structure.

Symbol	Parameter	Value
L	Bimorph length	60 mm
B	Bimorph width	12 mm
H	Thickness of substrate	2 mm
T	Thickness of piezo patch	0.3 mm

Source: Authors

Table 14.2 Material parameters of PZT-5A.

Symbol	Parameters	Value
P	Density	7750 kg/m^3
Y_p	Young's Modulus	63 GPa
e_{31}	Piezo stress coefficient	-6.5 C/m^2
$\varepsilon_{33}/\varepsilon_0$	Relativity permittivity	2664.8
η_s	Isotropic loss factor	0.01748

Source: Authors

Table 14.3 Material parameters of PLA.

Symbol	Parameter	Value
ρ	Density	1240 kg/m^3
V	Poisson's Ratio	0.36

Source: Authors

of Ciba-Geigy Epoxy Resin joining the PZT-5A patch with the PLA Substructure.

Figure 14.2 gives the auxetic structures analysed in the further study. The auxetic cores were fixed at one end to act as a cantilever and then subjected to dynamic loading of 1 g force. The study was done in the frequency domain, varying frequency across the Eigen frequency of the cantilever harvester.

The piezoelectric bimorph cantilever is composed of the upper and lower piezoelectric layers of PZT-5A and the intermediate auxetic substrate layer of PLA. The substrate layer and piezoelectric patch are joined by an epoxy layer of Ciba-Geigy as mentioned earlier. The parameters in Tables 14.1 and 14.2 are suitable for 3D finite element modelling and distributed parameter modelling. For the definition of PZT-5A we are using 3D Material parameters given in material library of COMSOL. In general, piezoelectric

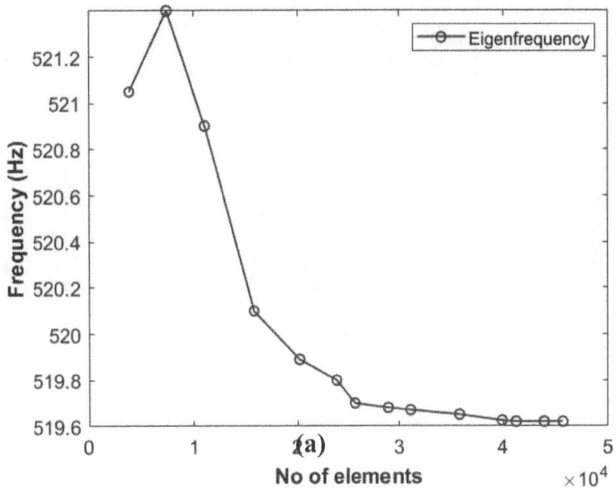

Figure 14.3 Convergence graph obtained for validation model of Erturk and Inman [1] bimorph cantilever

Source: Authors

material parameters (Table 14.3), the parameters of distributed parameter model are simplified in the form of 1D by Euler-Bernoulli beam theory. Using the stress-charge constitutive relation, the elastic matrix of PZT-5A is

$$\begin{bmatrix} 120.3 & 75.18 & 75.09 & 0 & 0 & 0 \\ 75.18 & 120.3 & 75.09 & 0 & 0 & 0 \\ 75.09 & 75.09 & 110.88 & 0 & 0 & 0 \\ 0 & 0 & 0 & 21.05 & 0 & 0 \\ 0 & 0 & 0 & 0 & 21.05 & 0 \\ 0 & 0 & 0 & 0 & 0 & 22.57 \end{bmatrix} GPa$$

The electromechanical coupling matrix of PZT-5A is

$$\begin{bmatrix} 0 & 0 & 0 & 0 & 12.295 & 0 \\ 0 & 0 & 0 & 12.295 & 0 & 0 \\ -5.35 & -5.35 & 15.7835 & 0 & 0 & 0 \end{bmatrix} C/m^2$$

The relative permittivity matrix of PZT-5A is

$$\begin{bmatrix} 919.1 & 0 & 0 \\ 0 & 919.1 & 0 \\ 0 & 0 & 826.6 \end{bmatrix}$$

Finite element modelling in COMSOL

In the finite element modelling of a cantilever beam with an auxetic core and PZT-5A piezoelectric patches using COMSOL 3D, the auxetic core's mechanical properties such as negative Poisson's ratio are combined with the piezoelectric material properties of the PZT-5A patches to simulate the complex interactions. The model incorporates both mechanical and

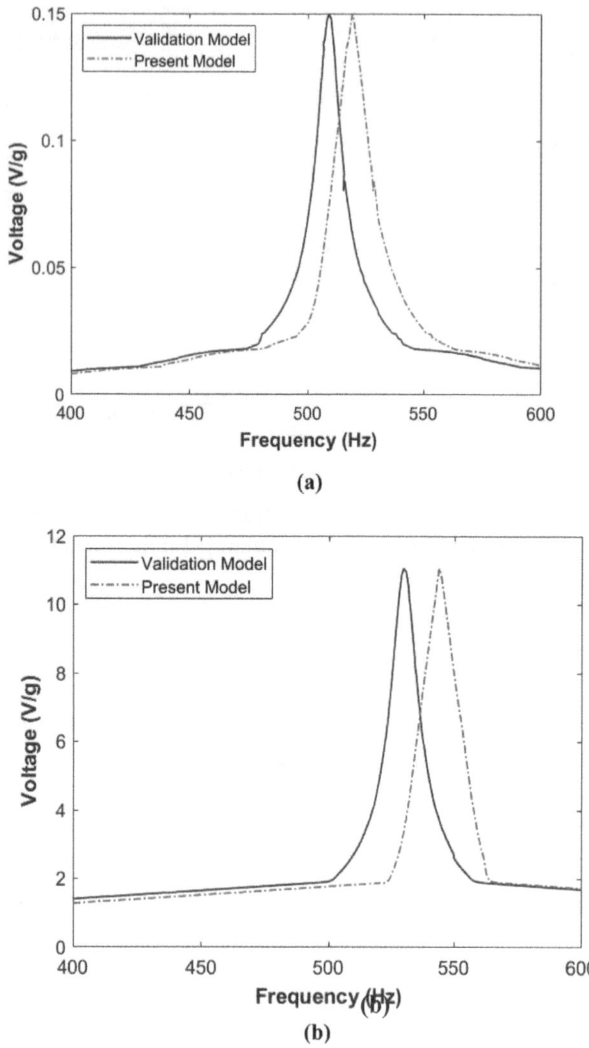

Figure 14.4 Comparison of voltage–frequency response curves of piezoelectric bimorph in series type with validation model of Erturk and Inman [1]. (a) Load resistance of 470 Ω. (b) Load resistance of 1 MΩ

Source: Authors

To validate the model, an additional study was performed using Erturk and Inman [4] bimorph beam as directed in COMSOL by Wang et al. [7]. A bimorph cantilever beam of brass core with PZT-5H was modelled as directed and following convergence study and voltage parameters were obtained.

The convergence study was performed against value of Eigen frequency of the model, which was observed to be 513.55 + 2.2% Hz in study conducted by Wang et al. [7]. The values obtained in Figure 14.3 indicate the validation of model. The stability in results was achieved gradually, as mesh was made finer and stable values were found around 45000 elements.

The voltage frequency response values depicted in Figure 14.4 were also validated according to Wang et al. [7] COMSOL 3D simulation for short circuit and open circuit in series condition (Table 14.4).

In the finite element modelling of the cantilever beam with piezoelectric patches and auxetic substrate, a tetrahedral mesh is utilised to accommodate the complex geometry of the structure. Tetrahedral elements offer greater flexibility in meshing irregular domains and ensure a good fit around curved or detailed regions, such as the auxetic core's intricate geometry. Although requiring more elements for similar accuracy compared to hexahedral meshes, tetrahedral meshing simplifies the automatic discretisation process and adapts well to complex boundary conditions. Auxetic structures are prone to complications in mesh due to their sharp edges on miniature level, hence a custom sized tetrahedral elements mesh was utilised.

Figure 14.5 shows the meshing of re-entrant hexagonal auxetic core energy harvester. Frequency domain analysis was conducted to evaluate the

Table 14.4 Comparing values of Wang et al. [7] model with the present COMSOL 3D model as depicted in Figure 14.4.

	Resistance	Validation model COMSOL 3D	Present model COMSOL 3D	Relative error
Resonant Frequency/Hz	470 Ω	508.6	518.6	+1.96%
	1M Ω	530.2	544	+2.6%
Peak voltage/V	470 Ω	0.1502	0.1497	-0.33%
	1M Ω	11.12	11.052	+0.612%

Source: Authors

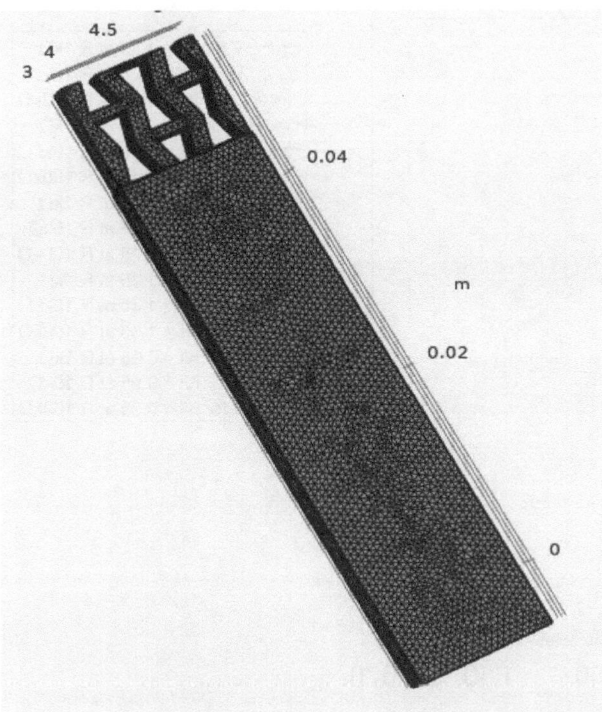

Figure 14.5 Mesh for finite element analysis
Source: Authors

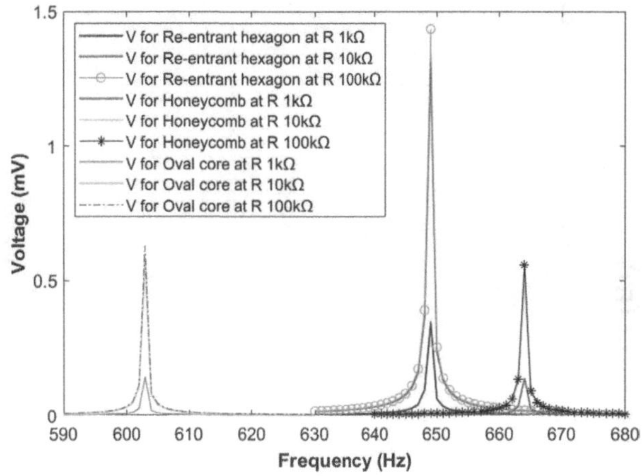

Figure 14.6 Voltage vs. frequency plot for three different auxetic structures cores
Source: Authors

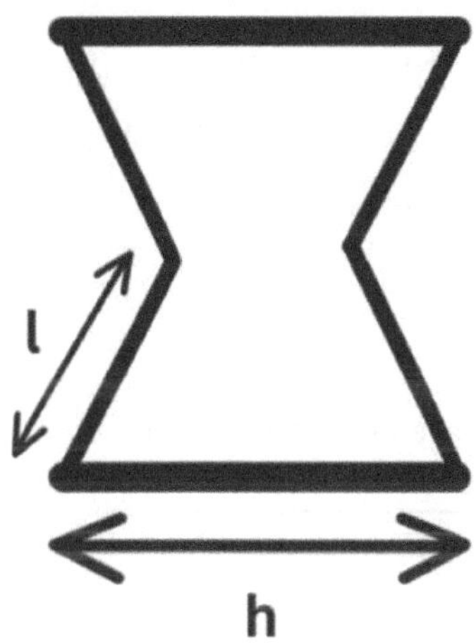

Figure 14.7 Re-entrant hexagonal structure
Source: Authors

dynamic response of the cantilever beam with an auxetic substrate and piezoelectric patches under varying load resistances. By applying harmonic loads at different frequencies, the analysis examines how the structural vibrations influence the output voltage and energy harvesting efficiency of the piezoelectric patches. This study helps identify resonant frequencies and the corresponding impact of load resistance on the system's performance, allowing for optimisation of energy harvesting capabilities. The results provide insight into the optimal load conditions for maximising energy conversion while ensuring structural integrity.

Electromechanical coupling simulation

The electromechanical coupling analysis was performed in COMSOL 3D. An inertial load was applied to the whole structure, with the acceleration of 1 g (9.8 m/s^2) as mentioned in Wang et al. [7]. The analysis was performed at three different load resistances, 1 kΩ, 10 kΩ and 100 kΩ while varying their frequency. In Figure 14.4, the mentioned cores in legend are labelled with re-entrant hexagon for core 1, oval structure for core 2 and honeycomb structure for core 3.

The graph in Figure 14.6 shows results for bimorph energy harvester, with the piezo patches connected in parallel. The graph displays that re-entrant hexagon is the better structure of the three chosen ones for the analysis. Hence further analysis was done by varying the h/l ratio for the re-entrant hexagonal structure.

The re-entrant hexagonal auxetic structure is characterised by inwardly-angled cell walls, which result in a negative Poisson's ratio, allowing it to expand laterally when stretched. By varying the ratio of wall thickness (h) to cell length (l), the mechanical properties, such as stiffness and strain distribution, can be finely tuned to optimise the energy harvesting

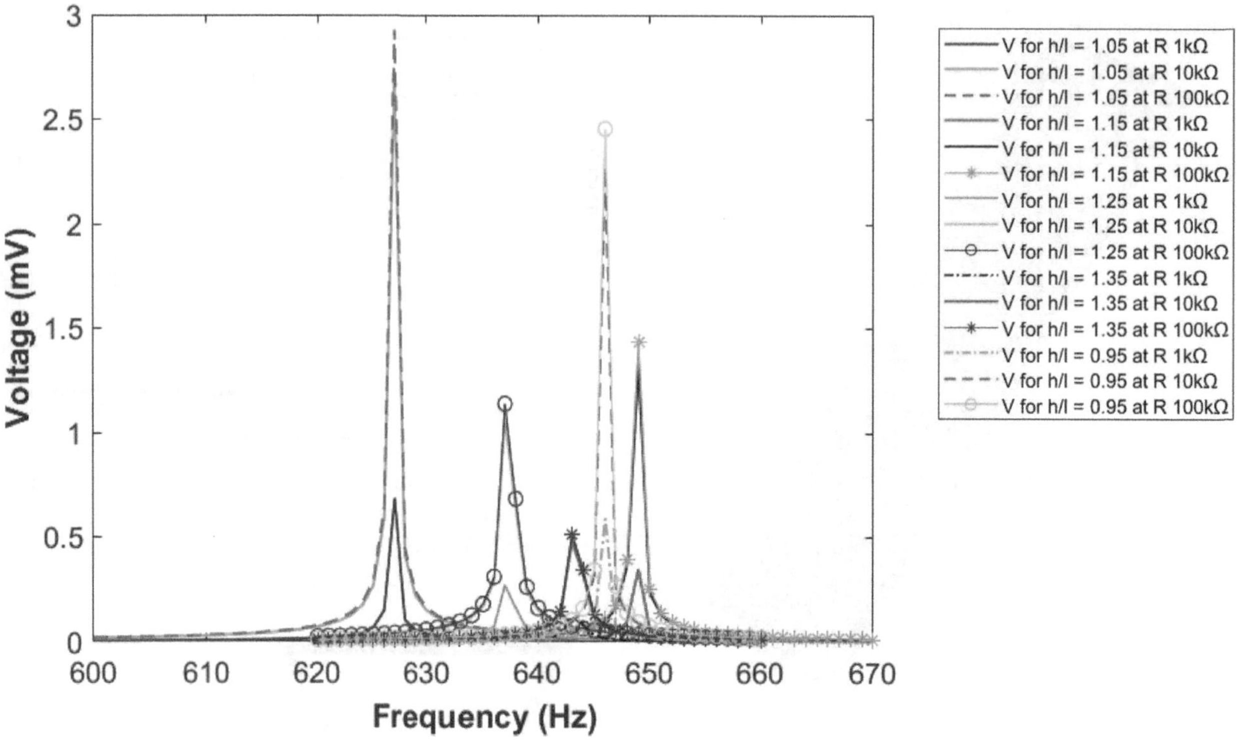

Figure 14.8 Voltage vs. frequency plot for re-entrant hexagonal structures with different h/l ratios
Source: Authors

Table 14.5 Maximum voltage for each re-entrant hexagonal structure.

h/l ratio	Maximum voltage (mV)
0.95	2.45
1.05	2.925
1.15	1.4325
1.25	1.1385
1.35	0.51725

Source: Authors

efficiency of the piezoelectric core. This variation directly influences the stress concentration and biaxial strain experienced by the piezoelectric material, enhancing electrical output.

In Figure 14.7, the parameters h and l are defined. To optimise the design of the auxetic core, the h/l ratio was varied to perform analysis. This is to find the optimal auxetic core for efficient energy harvesting. The h/l values were varied from 0.95 to 1.35 and analysis was performed. All the geometric and material parameters remain same as the previous analysis. The graph shown in Figure 14.8 shows that the highest voltage was recorded for structure with h/l ratio of 1.05.

Hence the optimal design for maximum electrical output can be achieved using the auxetic core of re-entrant hexagonal structure of h/l ratio equal to 1.05 (Table 14.5).

With this re-entrant hexagon structure, we went forward with tapered analysis with auxetic core thickness varying linearly from 1.5 mm to 3 mm. Figure 14.9 depicts the structure schematically.

In the Figure 14.9 tapered structure, due to the linearly varied thickness in the beam, the stress away from the fixed end can be significantly increased, contributing to a more uniform stress distribution compared to auxetic energy harvester of uniform thickness. This has been discussed by Chen et al. [3] for a unimorph energy harvester structure.

In Figure 14.10, we can observe that voltage generated using energy harvester with auxetic core of tapered thickness is four times the voltage generated using energy harvester with auxetic core of uniform thickness.

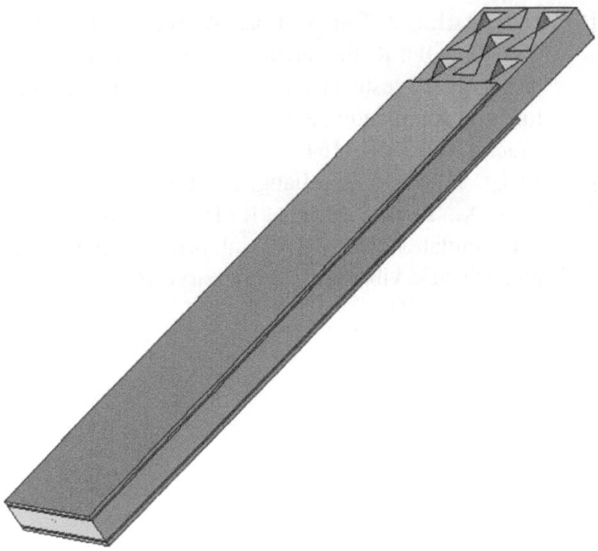

Figure 14.9 Model of tapered auxetic cantilever beam
Source: Authors

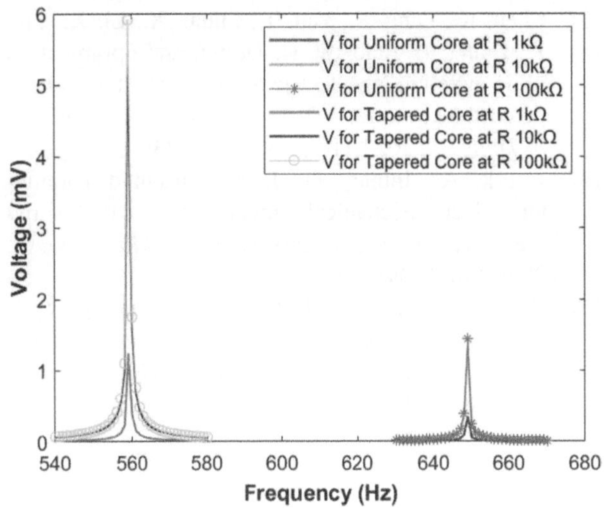

Figure 14.10 Voltage vs. frequency plot for re-entrant hexagonal structures with tapered core and uniform core
Source: Authors

Table 14.6 Maximum voltage under load resistance of 100 kΩ for different structures.

Structure	Maximum voltage (mV)
Uniform beam	
Auxetic honeycomb	0.5603
Auxetic oval	0.6302
Auxetic re-entrant hexagon	1.4325
Tapered beam	
Auxetic re-entrant hexagon	5.9145

Source: Authors

Results

The following is the summary of results for different auxetic structures compared and maximum voltage generated for each structure under 1 g loading.

Table 14.6 results show the importance of using auxetic structures and especially tapered structures for efficient electric voltage difference generation. The auxetic cores also carry advantage of weight reduction compared to metallic cores as the materials we use are types of PLA and hence can be used as harvesters for even minute vibrational disturbance.

Conclusion

The vibrational energy harvesters have been widely popular to power microsensors sustainably. The utilisation of auxetic cored harvesters improves efficiency for same design parameters as an isotropic beam substructure. The design optimisation provided is shown to improve the efficiency as voltage observed was in range of microvolts compared to millivolts in case of auxetic core. The tapered auxetic structure was observed to be the most efficient in energy harvesting.

Acknowledgement

The authors gratefully acknowledge the students, staff, and authority of School of Mechanical Sciences for their cooperation in the research. This work was supported by the SRIC, IIT Bhubaneswar (SP 135).

References

[1] Erturk, A., Inman, D. J. An experimentally validated bimorph cantilever model for piezoelectric energy harvesting from base excitations. *Smart Mater. Struct.*, 2009;18(2):025009.

[2] Erturk, A., Inman, D. J. Issues in mathematical modelling of piezoelectric energy harvesters. *Smart Mater. Struct.*, 2008;17(6):065016.

[3] Chen, K., Fang, S., Yan, T., Chuai, X., Li, X., Lai, Z., Liang, J., Liao, W. H. Design and optimization of an auxetic piezoelectric energy harvester with tapered thickness for IoT applications. *IEEE Internet of Things J.*, 2024;11(23):37831–37839.

[4] Erturk, A., Inman, D. J. A distributed parameter electromechanical model for cantilevered piezoelectric energy harvesters. *J. Vibr. Acoust.*, 2008;130(4):041002.

[5] Zheng, B., Chang, C. J., Gea, H. C. Topology optimization of energy harvesting devices using piezoelectric materials. *Struct. Multidiscip. Optimiz.*, 2009;38:17–23.

[6] Gao, Q., Lu, Y., Shi, Y., Liao, W. H., Yin, G., Li, J., Xiao, F., Qiu, R. Enhancing the output performance of energy harvesters using hierarchical auxetic structure and optimization techniques. *IEEE Trans. Indus. Elec.*, 2023;71(9):11641–11649.

[7] Wang, L., Zhao, L., Jiang, Z., Luo, G., Yang, P., Han, X., Li, X., Maeda, R. High accuracy comsol simulation method of bimorph cantilever for piezoelectric vibration energy harvesting. *AIP Adv.*, 2019;9(9):095067.

15 Multistable shape-adaptive metamaterial with tailored material composition and geometry

Abhijeet Kumar[a] *and Ayan Haldar*[b]

Civil Engineering Department, Indian Institute of Technology (BHU) Varanasi, Varanasi, India

Abstract

Metamaterials exhibit distinctive properties that do not exist in natural materials. These properties result from the material's structure rather than its composition. Shape-adaptive metamaterials are engineered with structural components that can change shape when triggered by external inputs, such as temperature, pressure, or electric fields. These materials have applications in adaptive structures in robotics, aerospace, automotive engineering, and deployable structures. However, in most applications, a continuous energy source is required to maintain a desired shape.

To address this challenge, a new self-locking metamaterial employing the concept of multistability has been developed, which exhibits both translational and bending motion. These metamaterials consist of an arrangement of two different types of bistable unit cells: straight and inclined. The straight unit cells are responsible for translational behaviour, while the inclined unit cells contribute to the bending behaviour of the lattice. Each unit cell is made of four bistable anisotropic composite laminates. Composite laminate exhibits bistability due to induced residual stress generated from cool down from elevated curing temperature to room temperature.

To analyse the metamaterial, a finite element model has been developed for both unit cells and lattices. Upon the application of a load, the straight stable configuration snaps into the curved stable configuration. An extensive parametric analysis has been conducted to observe the effect of different inclination angles of the unit cells on the bending behaviour of the lattice. It is observed that the maximum upward deflection of 0.43 m (along the Y-axis) for a 45° inclination angle of the laminate in the inclined unit cell. This bending mechanism can be utilised in applications such as aircraft flaps, grippers, and bascule bridges.

Keywords: Metamaterial, multistable, fibre-reinforced composite, snap-through

Introduction

The demand for advanced materials that can meet the evolving challenges of modern technology has increased rapidly in recent years. Industries ranging from aerospace and telecommunications to healthcare and renewable energy are increasingly seeking innovative solutions to enhance performance, efficiency, and adaptability. This quest for new materials has led to significant interest in engineered materials or metamaterials that possess properties not typically found in naturally occurring substances. By tailoring the design of the structure at lower levels, metamaterials enable unique functionalities such as negative refraction [1], modified Poisson's ratio [2, 3], and tailored mechanical properties or stiffness [4], offering exceptional capabilities for various applications in fields such as aerospace [5], energy harvesting [6], and biomedical devices [7]. A particularly promising area within metamaterials is shape-adaptive metamaterials, which can change its shapes in response to external stimuli such as temperature, pressure, or electric fields. Sakovsky et al. [8] developed a shape-adaptable metamaterial that exhibits selective deformability (about 60%), allowing them to adapt their shape while maintaining structural integrity. This capability enables high strain without permanent deformation, making them suitable for applications that require significant shape changes. Rafsanjani et al. [9] developed a shape-adaptive metamaterial consisting of snapping unit cells that undergo large deformations under tensile stress. The material transitions from an undeformed wavy configuration to a deformed diamond-like structure. It is particularly designed for applications in programmable, non-linear deformation solids for energy absorption

[a]abhijeetkumar.rs.civ21@itbhu.ac.in, [b]ayan.civ@itbhu.ac.in

DOI: 10.1201/9781003770442-15

and impact mitigation, allowing for precise control of its mechanical behaviour under different loading conditions.

Most concepts of shape-adaptive metamaterials developed so far require a continuous external force to maintain different configurations. Such concepts are energetically demanding and require additional mechanisms to hold a particular configuration. To overcome this limitation, novel designs of multistable metamaterials have been developed, which offer multiple stable configurations that can be maintained without any constant external input. These materials can transition from one stable shape to another through a snap-through mechanism, eliminating the need for holding forces. Multistability in isotropic materials can be achieved through geometric modifications such as introducing initial curvature [10], grid patterns [11], and dimpled or buckled surfaces [12], and perforations or holes. In anisotropic materials, multistability can be induced through prestressing during fabrication [13] or thermal cooling [14, 15], particularly in unsymmetric composite laminates, where cooling from curing temperatures to room temperature causes the generation of residual stress, resulting in multiple stable configurations. Yang et al. [16] developed a multistable 2D metamaterial with unit cells made of U-shaped curved beams. These segments exhibit snapping behaviour, which allows them to switch between different stable configurations. They also noted that due to differences in stiffness between supporting and snapping segments, there were significant morphological changes in shape, resulting in a zero Poisson ratio. By stacking these 2D metamaterials, they created a 3D version with enhanced energy absorption properties, particularly useful in applications like shock absorbers. Shan et al. [17] developed a shape-adaptive metamaterial composed of bistable tilted beams designed for the controlled trapping of elastic strain energy. Through parametric analysis of the tilt angle (θ) and the thickness-to-length ratio (t/L), they observed that snap-through instability occurs at $\theta = 20°$ and $t/L = 0.11$, but without trapping energy at these specific values. This concept shows promise for applications in reconfigurable safety gear and crash mitigation technologies in aircraft and automobiles. Tao et al. [18] developed an intelligent multistable metamaterial that is shape-reconfigurable by utilising shape memory polymers (SMPs). This metamaterial comprises of two distinct types of elements: flexible elements made of RGD8530 and rigid elements made of VircWhite, connected in a wavy configuration. The researchers observed that the metamaterial undergoes significant changes in response to temperature variations; specifically, when the temperature increases, the material expands. Notably, complete expansion of the structure is achieved at 70°C, demonstrating its effective shape reconfiguration capabilities. The performance characteristics of the metamaterials, such as their mechanical properties and shape configurations, are often fixed after manufacturing. Consequently, if the requirements for their use change or there's a need for different functional behaviours such as variable stiffness, damping or load-bearing capacity, existing designs may not adapt effectively without undergoing extensive re-engineering or remanufacturing. To address these limitations, Mao et al. [19] developed a reprogrammable metamaterial that consists of a demountable bar with a fixed frame containing two bistable curved beams. The multistable behaviour can be changed by inserting the bar in a frame with different lengths. The major limitation of existing metamaterials is the challenge of meeting the conflicting requirements of shape adaptability, lightweight construction, and load-carrying capacity simultaneously. Kumar et al. [20] addressed this limitation by developing self-locking multistable metamaterials as shown in Figure 15.1a. Their design features a shape-adaptive metamaterial that exhibits translational motion (shape-adaptivity) in the longitudinal direction while supporting loads in the transverse direction. The unit cell consists of two bistable elements (multi-sectioned laminates) and two monostable elements (symmetric laminates). Bistability in the multi-sectioned (unsymmetric sectioned in the middle and symmetric sections at the ends (Figure 15.1b) composite laminates was achieved by inducing residual stress through thermal cooling from curing temperature to room temperature. Parametric analysis showed that the induced strain and load-carrying capacity are highly dependent on the geometries of both the bistable and monostable composite laminates.

Extending our earlier work [20], we propose a new metamaterial with tailored material composition and geometry that exhibits both bending and translational behaviours during snap-through, as illustrated in Figure 15.1c. The proposed metamaterial consists of alternating straight and inclined unit cells made from carbon fibre composite laminates (Figure 15.1c). The

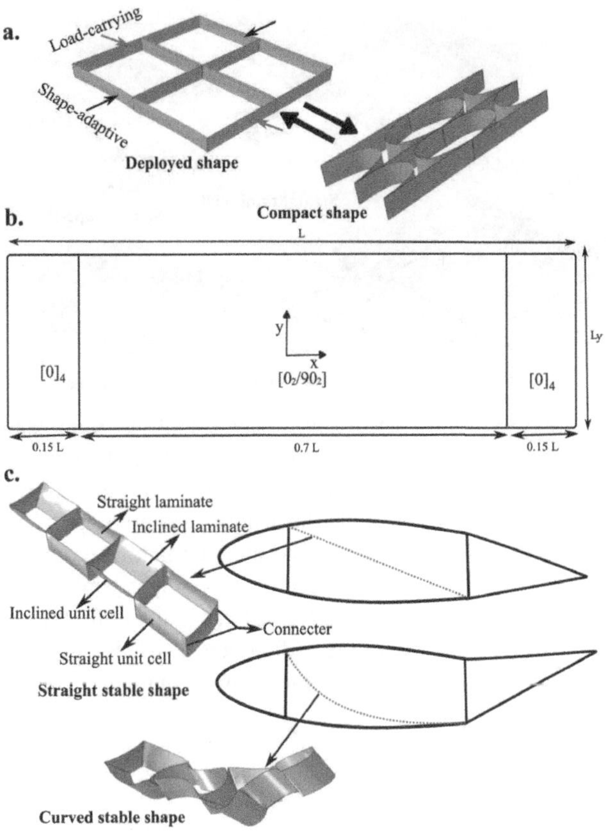

Figure 15.1 (a) The shape-adaptive metamaterial that demonstrates translational motion. It has two stable configurations – Deployed and compact [20]. (b) The three-sectioned laminate consists of an unsymmetrical region in the middle and a symmetric section at the end [20]. (c) The proposed metamaterial can be integrated into the larger structure (demonstrate translational and bending motion). It has two stable configurations – Straight and curve
Source: Authors

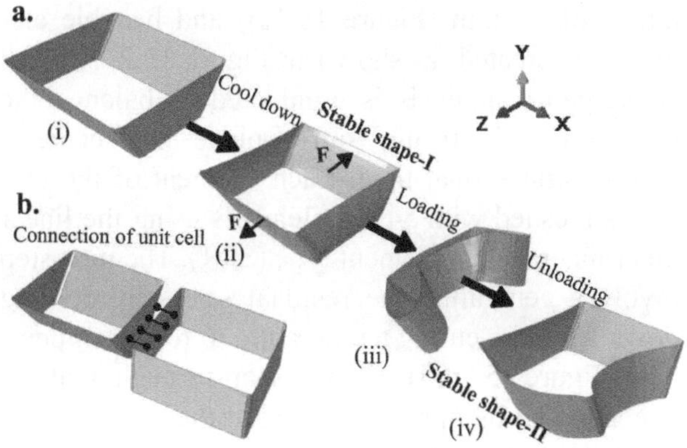

Figure 15.2 (a) Stable configurations of the unit cells: (i) Initial configuration of the monolithic unit cell, (ii) Stable shape-I with residual stress after cooling down, (iii) Deformed configuration after applying a load, and (iv) Stable shape-II after removing the load. (b) Connection between two-unit cells (centerline nodes)
Source: Authors

straight unit cells are responsible for translational motion, while the inclined unit cells provide bending motion to the lattice. Each straight unit cell consists of four straight members connected by a connecter (Figure 15.1c), while each inclined unit cell consists of two inclined and two straight members, also connected by a connecter. All composite elements are multi-sectioned laminates with piece-wise variations in fibre orientation, designed to remain bistable under simply supported boundary (SS) conditions at the ends of the laminate [20]. This proposed metamaterial can be integrated into a larger structure, as shown in Figure 15.1c.

The manuscript is organised as follows: Section 1 consists background and objectives. Section 2 details the numerical modelling, including how the unit cell and lattice are designed in the finite element (FE) software ABAQUS. Section 3 details the analysis of numerical modelling, highlighting the behaviour of the unit cell and the lattice, with a focus on their bending behaviour and deformation. Parametric analysis is also conducted to examine bending characteristics. Section 4 concludes the study by summarising the key findings and outlining future research directions.

Numerical model

A series of finite element modelling studies have been conducted to analyse the behaviour of the designed metamaterials. The finite element modelling is performed using the commercially available software ABAQUS, considering geometrical nonlinearities. The numerical modelling is conducted in a hierarchical fashion, starting with the analysis of the unit cell and then progressing to the analysis of the lattice. The different steps involved are described in detail as follows:

Analysis of unit cell

As a first step, the geometry of the unit cell, which consists of a connector (created by using a fillet radius of 10 mm (Figure 15.1c)) and bistable elements, is created, as shown in Figure 15.2. A mesh convergence analysis is conducted to balance the precision of the results (out-of-plane displacement) with computational time. Each segment of the unit cell is meshed with 54×18 elements using the linear quadrilateral shell element type (S4R). The next step involves generating the residual stress by cooling down from the curing temperature to room temperature (Figure 15.2a, i). Then, a snap-through analysis is conducted by applying a load at the centre of the laminate (Figure 15.2(a, ii–iv)).

Cool down

To simulate the cooling process of the unit cell, a static finite element method with geometric non-linearity has been employed. An artificial damping (damping factor set to 0.0002) has been used for numerical stabilisation, ensuring convergence of the simulation while without any changes in the final results. The model assumes a consistent surface temperature across the unit cell, applying a uniform temperature field ($\Delta T = 180°C$) across the whole domain to mimic the curing phase. As the unit cell cools down from curing temperature to room temperature, residual stress is generated, as shown in Figure 15.2a, i, which is further used in the snap-through analysis. During the analysis, the centre node of each laminate of the unit cell has been fixed to prevent rigid body motion.

Snap-through

Predicting the snap-through behaviour of the unit cell is important for designing a shape-adaptive structure. In this study, a concentrated load is applied to the centre of the laminate to initiate snap-through, as illustrated in Figure 15.2a, ii–iv. The snap-through analysis includes the following steps:

1. The initial boundary conditions are applied, where the centre node of the laminate is fixed; this is removed first, and then the mid-edge of the unit cell is clamped (Figure 15.3).
2. The snap-through behaviour is examined by applying a load at the centre of a laminate of the unit cell in stable shape-I, as depicted in Figure 15.2a, ii, and observing the resulting deformed shape (Figure 15.2(a, iii)). This load must exceed the snap-through threshold to cross the limiting point.

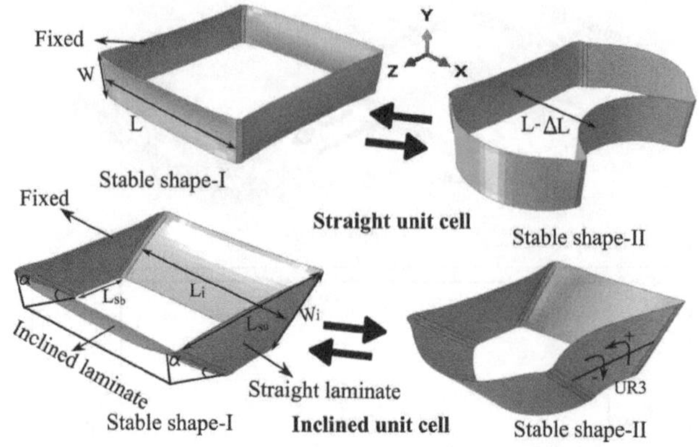

Figure 15.3 The stable shapes of straight and inclined unit cells are illustrated. The dimensions of the straight unit cell and the inclined unit cell are provided, where L is the length of the straight unit cell, L_i is the length of the inclined laminate, and L_s (L_{sb} and L_{su}) is the length of the straight laminate in the inclined unit cell. ΔL represents the deformation of the unit cell in the x-direction, and UR3 denotes the rotation of the straight laminate in the unit cell in z-direction

Source: Authors

3. In the subsequent phase of the analysis, the applied force is removed, allowing the laminate to stabilise and reach another stable shape (stable shape-II), as illustrated in Figure 15.2a, iv.

Analysis of lattice

Once the unit cell is analysed, a lattice structure consisting of a 4×1 unit cell arrangement is modelled. The first step in modelling the lattice involves arranging the unit cells together, followed by generating residual stress by cooling down from the curing temperature to room temperature in each unit cell. Further, the unit cells are connected together, as shown in Figure 15.2b. In the next step, an equilibrium step has been applied to the redistribution of stress between the unit cells. Furthermore, similar to the analysis of unit cells, a snap-through analysis has been conducted, and two stable shapes are observed, as shown in Figure 15.1c.

Results

In this work, a monolithic unit cell has been constructed using a connector, with its material properties detailed in Table 15.1, where T_c is the thickness

Table 15.1 Material properties for connecter NTPT Thin ply TP402/T700S.

E_1 (GPa)	E_2 (GPa)	G_{12} (GPa)	ν_{12}	α_{11} (/°C)	α_{22} (/°C)	T_c (mm)
157	9.4	5.2	0.3	-9×10^{-8}	2.6×10^{-8}	0.04

Source: Authors

Table 15.2 Material properties for carbon/epoxy.

E_1 (GPa)	E_2 (GPa)	G_{12} (GPa)	ν_{12}	α_{11} (/°C)	α_{22} (/°C)	T_p (mm)
161	11.8	5.17	0.3	-1.8×10^{-8}	3×10^{-8}	0.131

Source: Authors

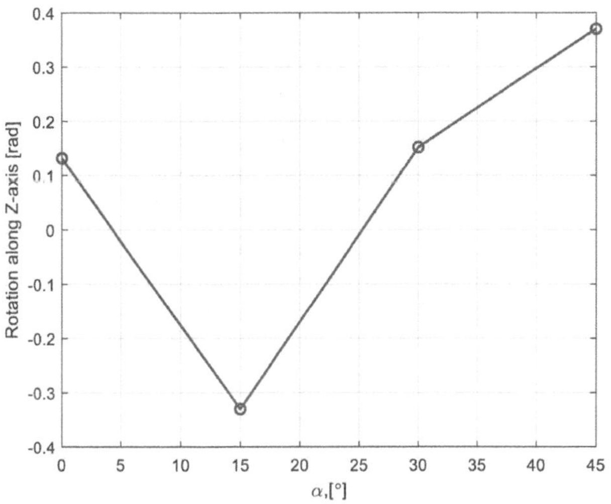

Figure 15.4 The effect of the rotation angle (α in degree) of the inclined laminate in the inclined unit cell on the rotation of the straight laminate (UR3 in rad.) has been observed
Source: Authors

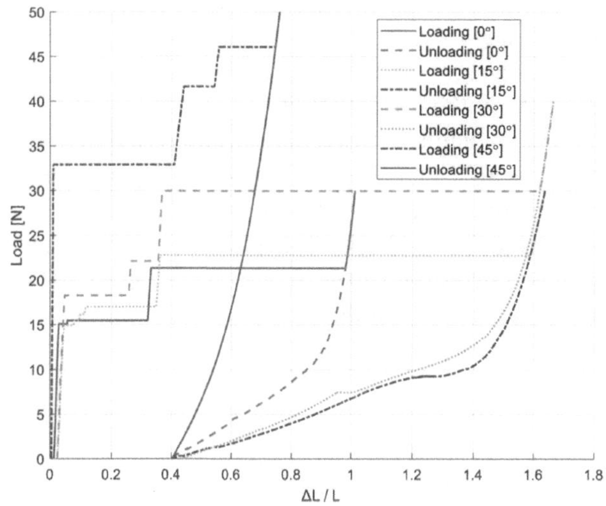

Figure 15.5 The snap-through load is calculated for varying curvature at different inclination angles (α) of the inclined laminate in the inclined unit cell. Additionally, strain is calculated for different values of α
Source: Authors

of the connector ply. Two types of unit cells, straight and inclined, have been used to develop the bending metamaterial, as shown in Figure 15.1c. The straight unit cell consists of four straight bistable elements, while the inclined unit cell consists of two inclined and two straight bistable elements, as shown in Figure 15.3. Their material properties are provided in Table 15.2 [20], where T_p is the thickness of the ply of the bistable element.

Analysis of unit cell

The straight unit cell consists of straight laminates with dimensions L = 0.3 m and W = 0.1 m. The inclined unit cell consists of both straight and inclined elements. The dimensions of the straight elements are L_{su} = 0.3 m, L_{sb} = 0.3 - 0.2 Tan[α] where α is the inclination angle of an inclined laminate of the inclined unit cell, as shown in Figure 15.3. The dimensions of the inclined elements are L_i = 0.3 m and W_i = 0.3 Sec[α]. The total thickness of bistable elements is 0.524 mm (four layers), and the total thickness of the connector is 0.08 mm (two layers) for both types of unit cells.

The analysis of the unit cell has been conducted by applying a load at the centre of the laminate in stable shape-I (Figure 15.3), the unit cell deforms, and upon removing the load, it returns to stable shape-II. The effect of inclination angle (α) of inclined laminate of unit cell (with α = 0° corresponding to the straight unit cell) on the rotation of straight laminate of unit cell (UR3 of centreline node (Figure 15.3)) is shown in Figure 15.4. It is observed that as α increases, UR3 initially decreases up to α = 15° (rotating in the anti-clockwise direction, Figure 15.3) and then starts to increase (rotating in the clockwise direction). The maximum rotation is observed for α = 45°, which is about 0.38 rad (Figure 15.4).

Further, the analysis has been extended to include snap-through behaviour and strain, conducted by varying the value of α from 0° to 45°, as shown in Figure 15.5. It is observed that there are no significant changes in the strains within the unit cell due to the changes in the inclination angle (α) of the laminate in the inclined unit cell (about 0.4). As the inclination angle increases, the width of the inclined laminate also increases (as α increases W_i increases), but this does not affect the strain of the laminate. It has already been observed that the effect of width on strains is negligible [20]. However, it is noted that as α increases, the snap-through load also increases due to an increase in the width of the inclined laminate [20].

Analysis of lattice

A lattice is constructed using both types of unit cells, inclined and straight, arranged alternately, as shown in Figure 15.6a. When a load is applied to the lattice (in its straight, stable shape), the laminates undergo snap-through (Figure 15.6b), and after the load is removed, the lattice stabilises in a curved, stable shape. To observe the effect of the inclination angle (α) on the bending behaviour of the lattice, a lattice is constructed using inclined unit cells where α varies from 0° to 45.

The transverse displacement (y-direction, L_y) of the centre node of the laminates in the unit cells (U1, U2, U3, U4, U5 in Figure 15.6a) of the lattice is plotted against the varying length of the lattice, as shown in Figure 15.7. It is observed that for intermediate α values (15° and 30°), bending below the axis occurs due to the anti-clockwise rotation (UR3) of the inclined unit cell (initial unit cell, Figure 15.3). For higher and lower α values, bending above the axis is observed (clockwise rotation of inclined unit cell, UR3). A similar pattern is also seen in the analysis of individual unit cells (Section 3.1).

Further, the analysis has been extended, and snap-through analysis has been conducted for lattices by varying the value of α from 0° to 45°. The snap-through load of a laminate of the unit cell has been plotted with the total strains in the lattice, as shown in Figure 15.8. Not much changes in strains for varying inclination angles (α) have been observed (about -0.1, which shows that there is not much effect of α on translational behaviour of lattice), while the snap-through load increases as α increases up to 30° because the width of inclined laminate also increases. It is already observed in a previous analysis by Kumar et al. [20] that as the width of laminate increases, snap-through load increases. For α = 45°,

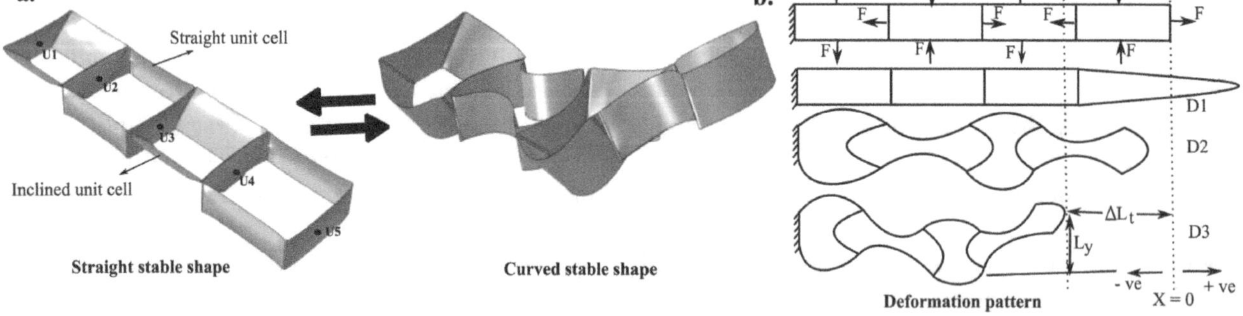

Figure 15.6 (a) The stable shapes of the lattice are shown: straight stable shape and curved stable shapes. (b) The deformed pattern of the lattice is presented, where F is the applied load, and D1, D2, D3, and D4 represent the deformation patterns. ΔL_t and L_y are the deformed length of the lattice in the x-direction and y- direction respectively in curved stable state

Source: Authors

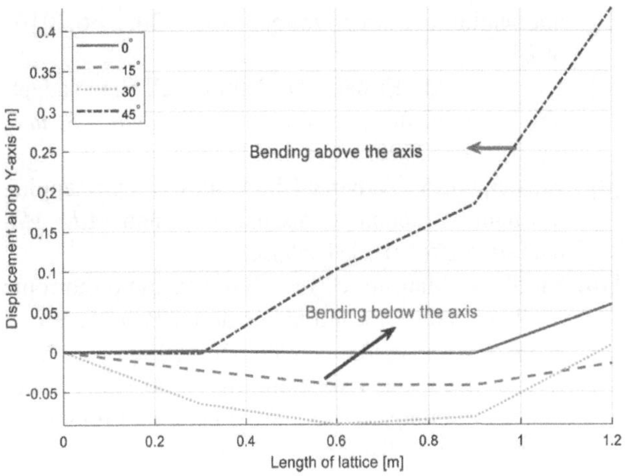

Figure 15.7 The transverse displacement (Y-displacement, L_y) of the center node of the unit cells in the lattice is plotted for varying lattice lengths at different values of α. Lattice bending above the x-axis is referred to as "bending above the axis," while lattice bending below the x-axis is referred to as "bending below the axis"
Source: Authors

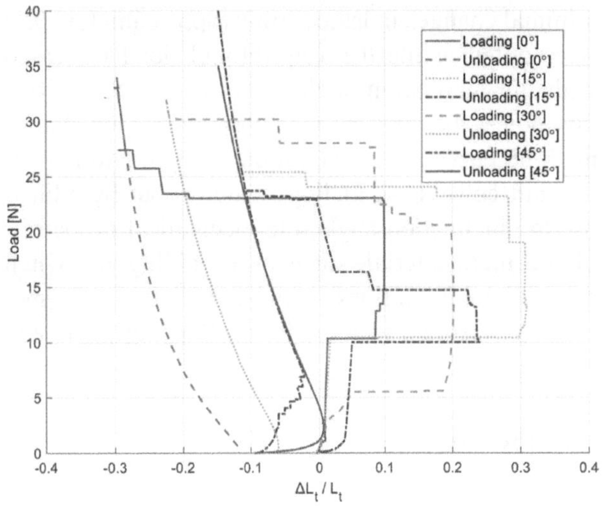

Figure 15.8 Load-strain curves of lattices are plotted for varying inclination angles (α) of the inclined laminate in the unit cell
Source: Authors

the snap-through load suddenly decreases due to the maximum clockwise rotation observed in the straight laminate of the inclined unit cell (Figure 15.3), which helps in the bending of the lattice (reducing snap-through load). According to Figure 15.8, as the load increases, the strains increase (graph in positive x-direction), which shows that the initial deformation happens in the positive x-direction (D1, Figure 15.6 (b)). After that, as the load increases, the following strains decrease (graph in the negative x direction, (D2, Figure 15.6b)), and at the end, a slight curve is observed, which shows the bending of the lattice (D3, Figure 15.6b).

Conclusion

This study successfully demonstrates the development of a shape-adaptive metamaterial that exhibits both translational and bending behaviour. The proposed metamaterial has two stable configurations: a straight shape and a curved shape, enabling it to switch between these configurations in response to the externally applied load.

The analysis was conducted in a hierarchical manner, beginning with the investigation of the unit cell and extending to the lattice structure. For the unit cell, the effect of the inclination angle (α) of the inclined laminate of the unit cell on the rotation of the straight laminate of the unit cell was thoroughly examined, revealing its utility in curving the metamaterial. The maximum rotation was observed for α = 45°. Additionally, the analysis highlighted the effect of inclination angle on strain within the unit cell, as well as its impact on the snap-through load required to transition between stable shapes. Minimal changes in strain were observed across different inclination angles, while the snap-through load increased as the inclination angle (α) increased, with the maximum snap-through load observed at α = 45°, measuring approximately 47 N.

Further analysis was conducted on the lattice, composed of alternating straight and inclined unit cells. The effect of the inclination angle of the inclined laminates on the bending behaviour of the entire lattice was also observed. It was found that the inclination angle significantly influences the lattice's bending behaviour: at lower inclination angles (15° and 30°), the lattice curves inward, whereas at higher inclination angles (45°), it bends outward. The maximum and minimum displacements along the y-axis were 0.43 m and -0.08 m, respectively.

Additionally, the effect of the inclination angle on strain distribution and snap-through load within the lattice was examined. Similar to the unit cell, minimal changes in lattice strain (approximately 0.1) were noted, while the snap-through load increased with the inclination angle up to $\alpha = 30°$. Beyond this point, the snap-through load decreased due to the high curvature arising in the straight laminate of the unit cell. These findings suggest that by adjusting the inclination angle, the mechanical properties of the metamaterial—such as its ability to switch between configurations—can be finely tuned for specific applications such as aircraft flaps, grippers and bascule bridges.

Acknowledgement

I would like to express my sincere gratitude to the Ministry of Human Resource Development (MHRD) for providing the fellowship that made this work possible.

References

[1] Padilla, W. J., Basov, D. N., Smith, D. R. Negative refractive index metamaterials. *Mater. Today*, 2006;9(7–8):28–35.

[2] Kolken, H. M. A., Zadpoor, A. A. Auxetic mechanical metamaterials. *RSC Adv.*, 2017;7(9):5111–5129.

[3] Kumar, A., Haldar, A. Auxetic metamaterial using bistable composite laminates. *Recent Dev. Struct. Engg.*, 2024;1:525–533.

[4] Zhai, Z., Wang, Y., Jiang, H. Origami-inspired, on-demand deployable and collapsible mechanical metamaterials with tunable stiffness. *Proc. Natl. Acad. Sci. U. S. A.*, 2018;115(9):2032–2037.

[5] Daynes, S., Weaver, P. M. Review of shape-morphing automobile structures: Concepts and outlook. *Proc. Instit. Mech. Engr. Part D J. Autom. Engg.*, 2013;227(11):1603–1622.

[6] Alkurt, F. O., et al. Octagonal shaped metamaterial absorber based energy harvester. *Medziagotyra*, 2018;24(3):253–259.

[7] Kumar, R., Kumar, M., Chohan, J. S., Kumar, S. Overview on metamaterial: History, types and applications. *Mater. Today Proc.*, 2022;56:3016–3024.

[8] Sakovsky, M., Ermanni, P. A thin-shell shape adaptable composite metamaterial. *Compos. Struct.*, 2020;246:112390.

[9] Rafsanjani, A., Akbarzadeh, A., Pasini, D. Snapping mechanical metamaterials under tension. *Adv. Mater.*, 2015;27(39):5931–5935.

[10] Vidoli, S., Maurini, C. Tristability of thin orthotropic shells with uniform initial curvature. *Proc. R. Soc. A Math. Phys. Eng. Sci.*, 2008;464(2099):2949–2966.

[11] Loukaides, E. G. Expanding the build plate: Functional morphing 3D printed structures through anisotropy. *Front. Mater.*, 2020;7:159.

[12] Norman, A. D., Seffen, K. A., Guest, S. D. Multistable corrugated shells. *Proc. R. Soc. A Math. Phys. Eng. Sci.*, 2008;464(2095):1653–1672.

[13] Daynes, S., Potter, K. D., Weaver, P. M. Bistable prestressed buckled laminates. *Compos. Sci. Technol.*, 2008;68(15–16):3431–3437.

[14] Hyer, M. W. Some observations on the cured shape of thin unsymmetric laminates. *J. Compos. Mater.*, 1981:175–194.

[15] Kumar, A., Anilkumar, P. M., Haldar, A. Bifurcation behaviour of general unsymmetric laminates under thermal stress. *Thin-Walled Struct.*, 2025;217:113885.

[16] Yang, H., Ma, L. Multi-stable mechanical metamaterials by elastic buckling instability. *J. Mater. Sci.*, 2019;54(4):3509–3526.

[17] [Shan, S., et al. Multistable architected materials for trapping elastic strain energy. *Adv. Mater.*, 2015;27(29):4296–4301.

[18] Tao, R., et al. 4D printed multi-stable metamaterials with mechanically tunable performance. *Compos. Struct.*, 2020;252.

[19] Mao, J. J., Wang, S., Tan, J. J., Liu, M. Modular multistable metamaterials with reprogrammable mechanical properties. *Engg. Struct.*, 2022;272(3):114976.

[20] Kumar, A., Haldar, A., Weaver, P. M. An adaptive metastructure concept using bistable composite laminates. *Compos. Struct.*, 2024;330:117767.

16 Isogeometric shape optimisation of geometrically non-linear beam

Deepak Kumar[a], *Abhijeet Kumar*[b] *and Ayan Haldar*[c]

Civil Engineering Department, Indian Institute of Technology (BHU), Varanasi, India

Abstract

Design optimisation in structural engineering plays a crucial role in improving the performance structures by minimising material usage and reducing costs. However, shape and topology optimisation of structures undergoing large deformations presents significant challenges, due to its inherent non-linearity. This work presents an advanced isogeometric gradient-based shape optimisation framework tailored for geometrically non-linear beams, by carrying out the design sensitivity analysis within a curvilinear coordinate system, derived from non-uniform rational B-splines (NURBS) space. This enables non-parametric shape optimisation of beam structures, directly integrating computer-aided design (CAD), finite element modelling (FEM), and an optimisation strategy that combines non-linear path-following methods with iterative design modifications.

As an initial step, a 1D isogeometric finite element framework accounting for geometrical non-linearities is developed. Using the isogeometric analysis several benchmark problems accounting of beam undergoing large deformations is solved and validated against analytical results. This framework is integrated with a gradient-based optimisation scheme by employing the adjoint sensitivity method, particularly crucial in scenarios where the number of design variables exceeds the number of constraint equations. We consider objective function of the optimisation as "minimising strain energy", with the B-spline mesh control points serving as the design variables. The interior point method was employed due to its efficiency in handling non-linear optimisation problems.

This developed methodology demonstrates the potential of isogeometric shape optimisation as a powerful tool for improving the performance and efficiency of geometrically non-linear beam structures in diverse engineering applications.

Keywords: Geometrically non-linear beam, sensitivity analysis, isogeometric analysis, curvilinear coordinate systems, shape optimisation

Introduction

The advent of advanced computational capabilities has unlocked new possibilities for the development of numerical techniques aimed at design optimisation. Automated design optimisation has become essential for reducing both design time and costs while facilitating improved engineering outcomes. The integration of finite element methods (FEM) into structural optimisation has more efficient and accurate solutions to structural differential equations. A flexible and robust numerical framework is important for effective structural optimisation, as it allows for precise modelling of structural properties and behaviours [1]. This synergy has expanded the capabilities of existing FEM codes by incorporating diverse optimisation strategies, thereby enhancing applications across civil [2], aeronautical [3], and mechanical engineering [4]. As a result, structural optimisation has emerged as a transformative tool for creating designs that minimise material usage, reduce costs, and improve performance.

Conventional finite element models employ polynomial functions for design representation [5]. While polynomial functions offered some degree of boundary smoothness, their high-degree variants introduced oscillations, limiting their applicability to simple curve families and restricting local control over shape modifications. To address this limitation, B-spline curves were adopted, as Braibant and Fleury [6] demonstrated their effectiveness in designing complex shapes. B-splines provided smooth boundaries and enhanced local control by using control point coordinates as design variables. Further advancements introduced non-uniform rational B-splines (NURBS), as explored by Schramm

[a]deepakkumar.rs.civ23@itbhu.ac.in, [b]abhijeetkumar.rs.civ21@itbhu.ac.in, [c]ayan.civ@itbhu.ac.in

DOI: 10.1201/9781003770442-16

and Pilkey [7], to represent structural shapes. These advancements allowed for improved representation of complex geometries, although structural displacement responses were still approximated using the p-version finite element method. Hughes et al. [8] demonstrated the potential of IGA to seamlessly combine CAD and FEM, enabling precise geometric representation and improved solution accuracy. Isogeometric analysis (IGA) bridges CAD and FEM by using the same NURBS basis functions for both, ensuring accuracy and eliminating geometry discretisation.

However, when addressing geometrically non-linear structures that undergo large deformations, the complexity of the problem increases significantly [9]. Traditional linear assumptions become inadequate in such cases, as structural responses are highly sensitive to loading paths and material distributions [10, [11]. Geometrically non-linear systems exhibit intricate behaviours that require advanced numerical tools to capture accurately. While linear structural responses have been extensively studied, research on NURBS with geometrically non-linear models remains comparatively limited.

Efforts to integrate non-linear structural analysis into optimisation frameworks have seen varied achievements. Early research predominantly focused on continuum-based approaches that employed parametric design spaces. For instance, Santos and Choi [12] developed a unified framework for shape design sensitivity analysis of non-linear trusses and beams, addressing both geometric and material non-linearities. This approach utilised continuum equilibrium equations and domain integrals to derive sensitivity expressions using adjoint variable and direct differentiation methods. By avoiding the differentiation of finite element matrices, it facilitated integration into standard FEM codes. Similarly, Cho and Jung [13] extended this continuum-based sensitivity method to topology optimisation of displacement-loaded non-linear structures. Their method utilised the adjoint variable approach and directly incorporated tangent stiffness matrices from equilibrium iterations, thus mitigating convergence challenges associated with force-loaded systems. Wu Xunlei et al. [14] further emphasised the importance of adaptive meshing techniques to improve accuracy in non-linear finite element analysis.

Discrete methods for parameterised shape optimisation of non-linear structures have also been explored. Bletzinger et al. [15] proposed a discrete optimisation approach that approximates non-linear response functions through triangular simplifications, significantly reducing computational costs. However, this method often compromised accuracy due to the limited number of sample points used in the integration of non-linear curves.

Despite these advancements, most existing studies have not fully addressed the complexities of non-linear behaviour in conjunction with shape sensitivity analysis or optimisation strategies that integrate CAD tools. Consequently, there remains a need for novel frameworks that fill the gap between geometrically non-linear analysis and shape optimisation. This study presents a new isogeometric formulation for geometrically non-linear beam structures, utilising the precision of NURBS within a curvilinear coordinate system. This framework facilitates seamless integration of CAD and FEA, optimising beam geometries directly in NURBS space, which enhances the link between design and analysis. Employing a gradient-based optimisation approach with the adjoint sensitivity method allows for efficient handling of complex geometries, especially when design variables exceed constraint equations in large deformation scenarios. Our innovative strategy addresses both linear and non-linear analyses by utilising B-spline shape functions, treating beam formulations similarly to shell structures. We focus on minimising strain energy, using B-spline control points as design variables, and apply the interior point method to effectively manage non-linearities. This research advances isogeometric analysis for non-linear structural applications, providing a robust tool for enhancing structural resilience and efficiency.

Discretisation of a beam

The beam's initial geometry, $\bar{B}$ is described by a centerline, $(\bar{\Omega})$, and its deformed geometry B is described by a new centerline, (Ω), as shown in Figure 16.1. The location of material points in the system is described by position vectors in both the initial ($\bar{\mathbf{r}} \in \bar{B}$) and deformed ($r \in B$) configurations are parameterised by arc-length $\theta^1 \in \omega \in \mathbb{R}$ and two orthogonal coordinates, $\theta^2 \in [-h/2, h/2]$ and $\theta^3 \in [-b/2, b/2]$, which is the beam's height (h) and width (b).

$$\bar{\mathbf{r}}(\theta^1, \theta^2, \theta^3) = \bar{\mathbf{x}}(\theta^1) + \theta^2 \bar{\mathbf{a}}_2(\theta^1) + \theta^3 \bar{\mathbf{a}}_3$$
$$\mathbf{r}(\theta^1, \theta^2, \theta^3) = x(\theta^1) + \theta^2 \mathbf{a}_2(\theta^1) + \theta^3 \mathbf{a}_3 \tag{1}$$

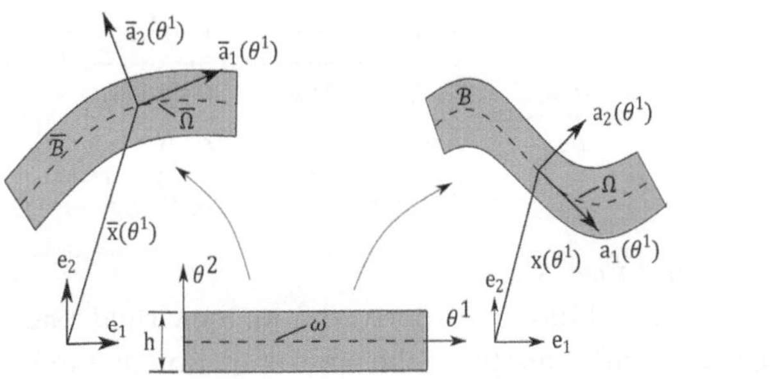

Figure 16.1 Initial (left) and deformed (right) configurations of beam
Source: Authors

The initial and deformed beam centrelines, represented by position vectors $\bar{x}$ and $\mathbf{x}$ and the unit vectors $\bar{a}_2$, a_2 and $\bar{a}_3$, a_3 are orthogonal to the tangent of the center line. These vectors are represented to reflect the rate of change of position along the parameter θ^1. In both reference and deformed configurations, the tangent vector $\bar{a}_1$ and a_1 is defined as the derivatives of the position vector with respect to θ^1, i.e., $\frac{d\bar{x}}{d\theta^1} = \bar{x}'$ and $\frac{dx}{d\theta^1} = x'$.

In the analysis, the beam's centreline is embedded in a 2D space. The binormal vector $\bar{a}_3 = a_3 = e_3$ is taken as a unit vector, i.e., normal to this space. The unit normal vectors $\bar{a}_2$ and a_2 are derived as the cross-product of $\bar{a}_3$, $\bar{a}_1$, and a_3, a_1 respectively and normalised by the magnitude of $\bar{a}_1$ and a_1. Their derivative with respect to θ^1 arc: $\bar{a}_2' = \bar{a}_3 \times \frac{\bar{a}_1'}{|\bar{a}_1|} - \frac{\bar{a}_1 \cdot \bar{a}_1'}{|\bar{a}_1|^2}\bar{a}_2$ and $a_2' = a_3 \times \frac{a_1'}{|a_1|} - \frac{a_1 \cdot a_1'}{|a_1|^2} a_2$.

The covariant base vectors are derived from the position vector with respect to convective coordinates θ^α. In the reference configuration, the covariant base vector $\bar{g}_\alpha$ is obtained by differentiating the reference position vector $\bar{r}$ with respect to θ^1. The co- and contravariant metric tensors, representing the inner products of these base vectors, form diagonal matrices. The contravariant matrix $[\bar{g}^{\alpha\beta}]$, is the inverse of the covariant matrix $[\bar{g}_{\alpha\beta}]$, with the appropriate values placed along the diagonal.

From these matrix components, the Green-Lagrange axial strain E_{11} is derived from the difference between the deformed and reference configuration $E_{11} = \frac{1}{2}(g_{11} - \bar{g}_{11})$, representing axial deformation in the beam. Expanding this expression yields: $E_{11} = \frac{1}{2}(a_1 \cdot a_1 - \bar{a}_1 \cdot \bar{a}_1) - \theta^2(a_2 \cdot a_1' - - \bar{a}_2 \cdot \bar{a}_1')$. Thus, the $E_{11} = \alpha + \theta^2\beta$ where $\alpha = \frac{1}{2}$ $(a_1 \cdot a_1 - \bar{a}_1 \cdot \bar{a}_1)$ and $\beta = -(a_2 \cdot a_1' - \bar{a}_2 \cdot (\bar{a}_1'))$. The strain consists of two parts: membrane straining and beam curvature change. The strain is expanded to show dependence on the curvature $a_2 \cdot a_1'$. This strain formulation helps in capturing the beam's deformation under large displacements.

Weak form of governing equations

The weak form of the momentum balance in the reference configuration is expressed as:

$$\underbrace{\int_{\bar{\mathcal{B}}} \mathrm{S} : \delta \mathrm{E} dV}_{\delta\pi_{\text{int}}} = \underbrace{\int_{\bar{\mathcal{B}}} \hat{\mathrm{b}}_0 \cdot \delta \mathrm{r}\, dV + \int_{\lambda\bar{\mathcal{B}}} \hat{\mathrm{t}}_0 \cdot \delta \mathrm{r}\, dA}_{\delta\pi_{ext}} + \int_{\alpha_N \bar{\mathcal{B}}} \lambda \cdot \delta \mathrm{r}\, dV \quad (2)$$

and

$$\int_{\partial_D \bar{\mathcal{B}}} (\mathrm{r} - \hat{\mathrm{r}}) \cdot \delta\lambda dV = 0$$

This approach uses a generalised weak form to impose Dirichlet boundary conditions with flexibility. A Lagrange multiplier, λ, is introduced to enforce these conditions as a traction field. The internal stress-strain relationship is defined by the 2nd Piola-Kirchhoff stress tensor (S), which is linked to the Green-Lagrange strain tensor (*E*) through a hyperelastic material model. The stress tensor is derived from the partial derivative of the hyperelastic potential energy density (W) with respect to the strain tensor *E*.

The volume integral is performed on the convected parameters (θ^1, θ^2, θ^3) which give:

$$\int_{\bar{\mathcal{B}}} (\cdot) dV = \int_b \int_h \int_\omega (\cdot)\sqrt{\bar{g}}\, d\theta^1\, d\theta^2\, d\theta^3 \quad (3)$$

The Jacobian is determined by the deformation gradient vectors, linking initial and deformed configurations. The beam's curvature and deformation are captured by adjusting the basis vector $\bar{a}_1$ with a θ^2 parameter. The matrix $\bar{g}$ expresses the infinitesimal arc length in the deformed state, scaled by a factor $\bar{\mu}(\theta^1, \theta^2) = 1 - \theta^2\, \bar{\eta}(\theta^1)$, which reflects changes in the beam's surface relative to its original shape. For a straight beam, $\bar{\eta}$ is zero, indicating no initial curvature. Using $\bar{\mu}$, the volume integral is expressed as:

$$\int_{\bar{\mathcal{B}}} (\cdot) dV = \int_b \int_h \int_\omega (\cdot)\bar{\mu} d\bar{\Omega}\, d\theta^2\, d\theta^3 \quad (4)$$

The domain $d\bar{\Omega}$ is discretised by elements, and volume integrals are approximated by summing sub-integrals over the element domains. The centre-line position is discretised using shape functions N^k (θ^1) and nodal position vectors $\bar{x}_k$ and x_k in the initial and deformed configurations.

Internal forces

The internal force at node K (f_{int}^K), is computed by integrating the axial stress component S^{11} over the beam's volume, scaled by the derivative of the axial strain E^{11} with respect to the nodal position x_K. The St. Venant-Kirchhoff model relates the axial stress to the strain using the material stiffness constant, i.e., $S^{11} = C^{1111} E_{11}$ where the material constant C^{1111} is defined based on Young's modulus E and the contravariant metric term $\bar{g}^{11}$, adapting to the beam's configuration. The linear material relationship is introduced to the internal forces, yielding:

$$f_{int}^K = \int_{\bar{\Omega}} \int_h \int_b E^{11} C \frac{\partial E^{11}}{\partial x_K} \bar{\mu}\, d\theta^3\, d\theta^2 d\bar{\Omega} \tag{5}$$

The internal force at the Kth node, f_{int}^K, is calculated by integrating force and moment resultants n and m over the beam's cross-section. Here, $n = CA\alpha$ with $A = bh$ as the cross-sectional area, and $m = CI\beta$, where $I = \frac{1}{12} bh^3$ is the second moment of area for a rectangular cross-section. These expressions account for the material stiffness C and geometry of the beam.

$$f_{int}^K = \int_{\bar{\Omega}} \left(n \frac{\partial \alpha}{\partial x_K} + m \frac{\partial \beta}{\partial x_K} \right) d\bar{\Omega} \tag{6}$$

This formulation integrates the effects of axial and bending stresses across the cross-section, capturing the structural response based on the chosen cross-sectional geometry and material properties. The axial strain component α derivative with respect to the nodal position x_K is expressed as $\frac{\partial \alpha}{\partial x_K} = a_1 \cdot \frac{\partial a_1}{\partial x_K}$.. For the bending strain component β and its derivative is $\frac{\partial \beta}{\partial x_K} = -\frac{\partial a_2}{\partial x_K} \cdot a_1' - a_2 \cdot \frac{\partial a_1'}{\partial x_K}$, where, a_1 and a_2 are the tangent and normal vectors to the beam's centreline, respectively.

The variation of the tangent vector a_1 with respect to the nodal displacements is defined as: $\frac{\partial a_1}{\partial x_K} = N'^J \frac{\partial x_J}{\partial x_K} = N'^J \boldsymbol{I}$.. In index notation, it becomes $\frac{\partial a_1^i}{\partial x_K^j} = N'^J \frac{\partial x_J^i}{\partial x_K^j} = N'^K \delta_j^i$ where N'^J are the shape function derivatives, $\boldsymbol{I}$ is the identity matrix, and δ_j^i is the Kronecker delta. This formulation shows how strain-type variables respond to changes in nodal positions.

The stiffness matrix of internal forces is:

$$k_{int}^{KJ} = \frac{\partial f_{int}^K}{\partial x_L} = \int_{\bar{\Omega}} \left(\frac{\partial n}{\partial x_L} \frac{\partial \alpha}{\partial x_K} + n \frac{\partial^2 \alpha}{\partial x_K \partial x_L} + \frac{\partial m}{\partial x_K} \frac{\partial \beta}{\partial x_K} + m \frac{\partial^2 \beta}{\partial x_K \partial x_L} \right) d\bar{\Omega} \tag{7}$$

External forces

The external forces are derived from the virtual external work after applying the approximation method:

$$\delta \pi_{ext} = f_{ext}^K = \left(\int_{\bar{B}} \hat{b}_0 \cdot \frac{\partial r}{\partial x_K} dV + \int_{\partial_N \bar{B}} \hat{t}_0 \frac{\partial r}{\partial x_K} dA \right) \delta x_K \tag{8}$$

Only linear body loads $\hat{b}_0$, and the end-point loads are considered

$$f_{ext}^K = \int_{\bar{\Omega}} \left(\hat{p} \cdot \frac{\partial x}{\partial x_K} + \hat{m} \frac{a_2}{|a_1|} \cdot \frac{\partial a_1}{\partial x_K} \right) d\bar{\Omega} + \left(\hat{P} \cdot \frac{\partial x}{\partial x_K} + \hat{M} \frac{a_2}{|a_1|} \cdot \frac{\partial a_1}{\partial x_K} \right) \Bigg|_{\theta^1 = \partial\omega} \tag{9}$$

The stiffness matrix of external forces is:

$$k_{ext}^{KJ} = \frac{\partial f_{ext}^K}{\partial x_L} = \int_{\bar{\Omega}} \hat{m} \frac{\partial}{\partial x_L} \left(\frac{a_2}{|a_1|} \cdot \frac{\partial a_1}{\partial x_K} \right) d\bar{\Omega} + \hat{M} \frac{\partial}{\partial x_L} \left(\frac{a_2}{|a_1|} \cdot \frac{\partial a_1}{\partial x_K} \right) \Bigg|_{\theta^1 = \partial\omega} \tag{10}$$

B-spline shape functions

In order to parametrize the curve, the standard B-Spline shape functions are used to parametrize the curve by combining cubic Bezier segments. Each element has a parametric domain ω^e, forming a partition of the beam domain $\omega = U_e\, \omega^e$. The finite element shape function is defined by four control points and is described over the dimensionless parameter $\xi = [0,1]$.

$$N^1(\xi) = \frac{1}{6}(-\xi^3 + 3\xi^2 - 3\xi + 1)$$

$$N^1(\xi) = \frac{1}{6}(3\xi^3 - 6\xi^2 + 4)$$

$$N^1(\xi) = \frac{1}{6}(-3\xi^3 + 3\xi^2 + 3\xi + 1)$$

$$N^1(\xi) = \frac{1}{6}(\xi^3)$$

In applying these shape functions to the beam element, each element requires four nodes for accurate representation of the geometry and kinematics. This strategy enhances continuity across elements while maintaining the structure's overall integrity in terms of displacement and rotation fields.

Boundary condition

Each node represented with two primary degrees of freedom (DOF): axial and transverse displacements, but realistic beams also need rotational DOF. Ghost nodes enable smooth B-spline shape functions but add complexity to boundary condition implementation. A transformation matrix T maps physical constraints onto ghost nodes, ensuring smoothness. Stiffness matrix K is adjusted using $\hat{K} = (T^T)^{-1}.K.T^{-1}$, maintaining accuracy in boundary constraints and global stiffness integrity in isogeometric analysis

Sensitivity analysis

In order to utilise this beam element for shape optimisation, it is necessary to compute its analytical sensitivities. These sensitivities are essential for understanding how changes in the design variables affect the system's behaviour, enabling efficient optimisation of the structure's shape. This involves deriving expressions for derivatives while referencing matrices and vectors used in the calculation of element stiffness matrix. The strain energy serves as the objective function to minimise, with B-spline control points as design variables. This approach systematically links beam element formulations to shape optimisation goals, focusing on reducing strain energy while adhering to the analytical framework of stiffness and deformation.

$$\psi(s,u) = \frac{1}{2}\int_{\bar{B}} (S:E)dV = \frac{1}{2}\int_{\bar{B}} (CA\alpha^2 + CI\beta^2)d\bar{\Omega} \tag{11}$$

Where ψ represent strain energy. The equilibrium equation in non-linearities is $f_{int}(s,u) - f_{ext}(s,u) = 0$. Using Karush-Kuhn-Tucker condition the objective function $\psi(s,u)$ is written as $\psi(s,u) = \psi + \lambda(f_{int} - f_{ext})$ to minimise the strain energy is: $\frac{d\psi(s,u)}{ds_i} = 0 = 0$. By imposing this condition, we get the Lagrange parameter $\lambda = -K_T^{-1}\frac{d\psi}{dx_k}$

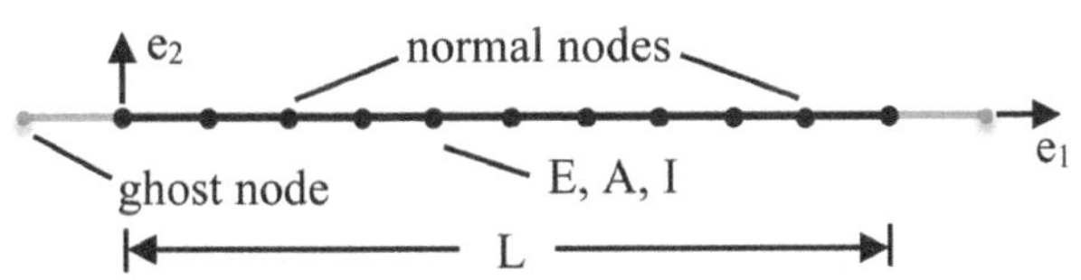

Figure 16.2 1D Beam meshes with ghost nodes
Source: Authors

Finite element models

To verify the new finite element model using B-splines, we analysed two problems: a cantilever beam and the arch undergoing large deformations under point loading, both considering geometric non-linearities in ABAQUS [16]. The beam and arch were meshed with 64 and 150 elements, respectively, using two-node linear beam elements (B21). In the beam model, one end had zero translation in both the x and y directions and zero rotation in the z direction, while the other end was free with no constraints. In the arch model, translation in the x and y directions was restricted at one end, but rotation in the z direction was allowed, while the other end had neither translation nor rotation (fixed).

Results

Geometrically non-linear beam under large deformations has been analysed using isogeometric analysis (IGA) with uniform B-spline basis functions. The code has initially been validated by performing a linear analysis of a beam under cantilever boundary conditions, which has demonstrated accuracy and consistency with established results. Subsequently, non-linear analyses have been carried out on both beam and arch models, addressing large deformation scenarios. Finally, the shape optimisation of the beam has been performed for a general loading case. For the B-spline interpolation, which requires four control points per element, additional "ghost" nodes have been introduced at the boundaries. These nodes have ensured proper continuity of the interpolation across the mesh, as shown in Figure 16.2. The inclusion of ghost nodes has been critical in maintaining the smoothness of the spline functions, particularly at the boundaries.

The analyses were performed using the following geometric properties of beam: E = 2 × 10^5 MPa, b = 100 mm, h = 50 mm, L = 5000 mm.

Linear analysis

To validate the accuracy and reliability of the linear beam formulation using uniform B-splines, numerical analyses were performed on benchmark problems with meshes discretised into 64 elements comparing results with exact solutions.

For a cantilever beam under a small point load (P = 2 N), the B-spline-based solution accurately captured the exact deflection profile, showing excellent agreement with analytical results (Figure 16.3). This

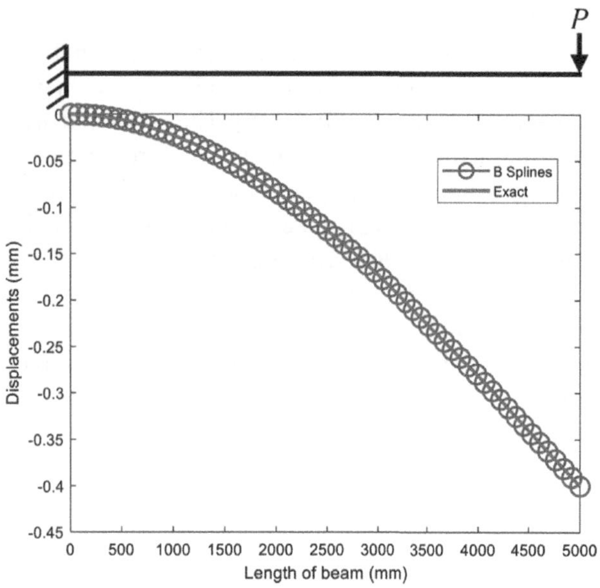

Figure 16.3 Comparison of deflected beam profile obtains using B-spline and exact (formula-based)
Source: Authors

confirms the precision and reliability of the formulation for linear beam analysis.

Non-linear analysis
To further assess the applicability of this approach in non-linear scenarios, higher loads were applied to the beam and arc models to analyse their response under large deformations. The resulting non-linear solutions were compared with FEM results (obtained from ABAQUS), to evaluate the robustness and accuracy in capturing complex non-linear behaviour.

Beam model
For the non-linear analysis, the geometric properties of the beam have been kept consistent with those used in the linear case. However, the load magnitude has been increased to P=20 kN to investigate the effects of geometric non-linearity. The transverse displacement of the beam has been calculated by varying its length, as illustrated in Figure 16.4.

According to the linear analysis, the maximum displacement at the applied load of 20 kN should be 4 m. However, due to the non-linear effects, the maximum displacement has been observed to be 2.7 m, indicating a significant reduction caused by geometric non-linearity.

The graph demonstrates (Figure 16.5) that the B-spline solution aligns precisely with the non-linear FEM results with error of 0.9% (where "*v*" represents displacement in y direction). This close agreement

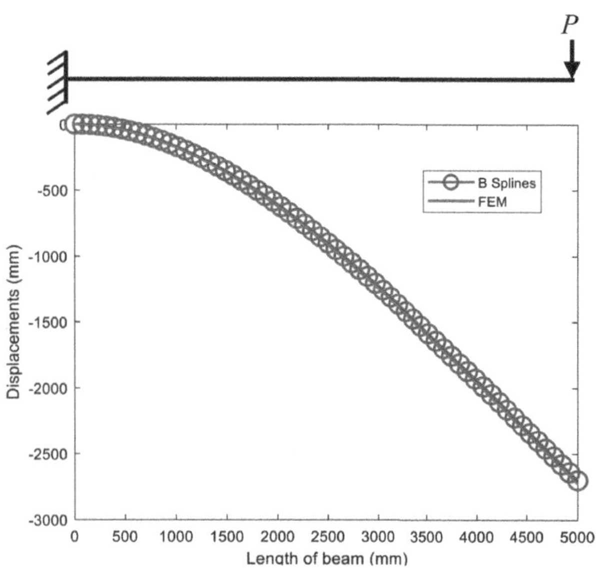

Figure 16.4 Comparison of deflected beam profiles obtained using B-spline and FEM
Source: Authors

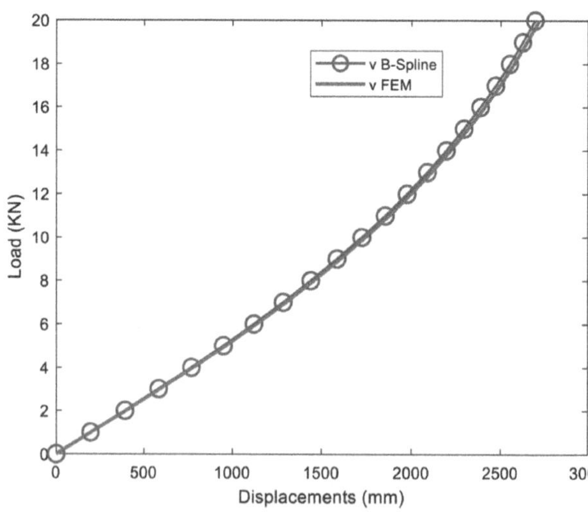

Figure 16.5 Load displacement curve of the geometric non-linear beam
Source: Authors

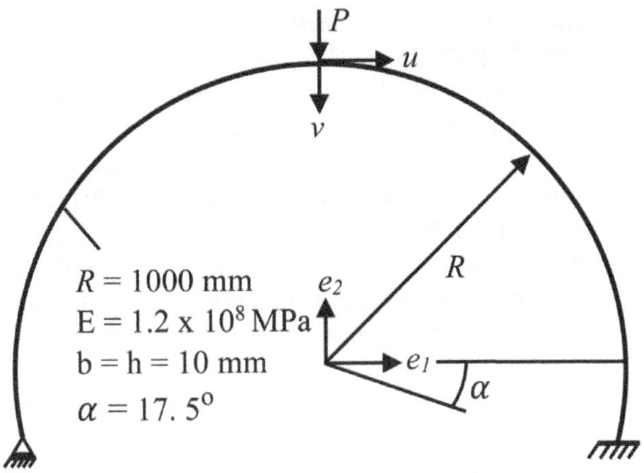

Figure 16.6 Deep arch with non-symmetric supports
Source: Authors

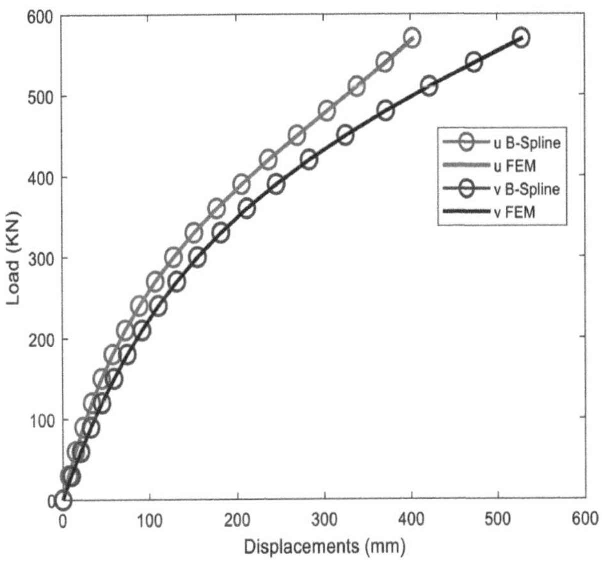

Figure 16.7 Load displacement curve of geometric non-linear arch
Source: Authors

highlights the effectiveness of the B-spline formulation in capturing the non-linear response with high accuracy. The matching curves confirm that the approach reliably handles geometrically non-linear problems, including cases involving large displacements, where traditional linear methods fail to provide accurate predictions [9].

Arch model

The analysis has been further extended to a deep arch buckling problem with non-symmetrical supports, as illustrated in Figure 16.6. Non-linear analysis has

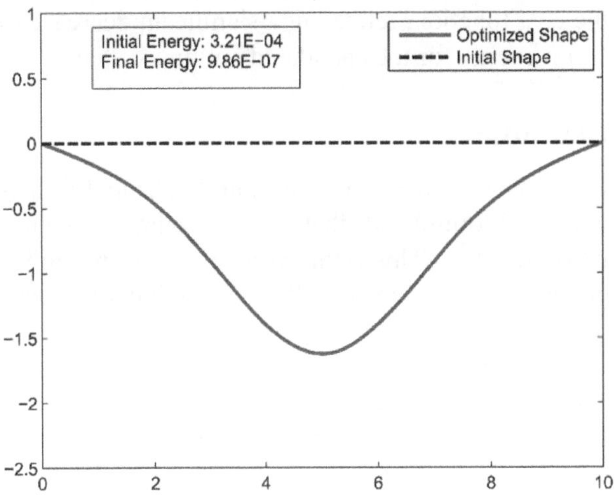

Figure 16.8 Initial and optimised shape for P = –0.02 kN
Source: Authors

been performed using the force control method to study the structural response under increasing loads. Load-deformation curves have been generated to capture both vertical (v) and horizontal (u) displacements, as shown in Figure 16.7.

It has been observed that the displacements in both horizontal (u) and vertical (v) directions increase non-linearly with increasing load. The results show a good agreement between the B-spline-based isogeometric analysis and FEM results obtained using ABAQUS. The error is observed to be 0.5% for horizontal displacement (u) and 0.4% for vertical displacement (v).

Optimisation

Sensitivity analysis has been employed for the non-linear beam under the simply supported boundary condition to determine its optimal shape. The objective is to minimise the strain energy, which has been defined as the objective function. The calculated sensitivities of the objective function were utilised in an optimisation algorithm implemented in MATLAB, applying the interior point method for its capability in solving non-linear optimisation problems. The control points of the B-spline curve serve as the design variables, enabling precise adjustments in the beam's geometry to achieve the optimisation goals.

The optimised shape of a thin, straight beam under loading of (P = 0.02 kN) is shown in Figure 16.8. This optimisation achieves a 99.6% reduction in the cost function, reflecting a significant decrease in strain energy for the calculated optimised shape.

Conclusion

In this work, a new IGA based finite element framework is formulated that can account geometric non-linearities. This framework successfully unifies geometry and analysis, offering significant advantages over standard polynomial-based finite element methods. It enables precise representation of complex shapes with minimal geometric errors and ensures higher-order inter-element continuity.

Initially, isogeometric analysis was validated for a linear beam by comparing the results with analytical solutions. The framework was then extended to non-linear beam and arch models, where it demonstrated excellent agreement with FEM results obtained from ABAQUS, showing a maximum error of 0.9% for the beam and 0.5% for the arch.

Furthermore, the shape optimisation of the non-linear beam was conducted using discrete sensitivity analysis. This approach proved computationally efficient, particularly when the number of design variables exceeded the number of response functions. The calculated sensitivities of the objective function were utilised in an optimisation algorithm implemented in MATLAB, applying the interior point method for its capability in solving non-linear optimisation problems. A significant reduction in strain energy was achieved, with a 99.6% improvement under an applied load of 0.02 kN.

Acknowledgement

The authors would like to acknowledge SERB for receiving funding from SERB SRG project (SRG/2022/001582) to carry out this research. The support for this funding is gratefully acknowledged.

References

[1] Beyer, H. G., Sendhoff, B. Robust optimization - A comprehensive survey. *Comput. Methods Appl. Mech. Engg.*, 2007;196(33–34):3190–3218.

[2] Haldar, A., Jansen, E., Rolfes, R. Analysis of morphing trailing edge flap with embedded multistable variable stiffness laminates. *AIAA Scitech 2019 Forum*, 2019.

[3] Goetzendorf-Grabowski, T. Multi-disciplinary optimization in aeronautical engineering. *Proc. Inst. Mech. Engg. Part G J. Aerosp. Engg.*, 2017;231(12):2305–2313.

[4] Kulkarni, N. K., Patekar, S., Bhoskar, T., Kulkarni, O., Kakandikar, G. M., Nandedkar, V. M. Particle swarm optimization applications to mechanical engineering - A review. *Mater. Today Proc.*, 2015;2(4–5):2631–2639.

[5] Strouboulis, T., Babuška, I., Copps, K. The design and analysis of the generalized finite element method. *Comput. Methods Appl. Mech. Engg.*, 2000;181(1–3):43–69.

[6] Braibant, V., Fleury, C. Shape optimal design using B-splines. *Comput. Methods Appl. Mech. Engg.*, 1984;44(3):247–267.

[7] Schramm, U., Pilkey, W. D. The coupling of geometric descriptions and finite elements using NURBs - A study in shape optimization. *Finite Elem. Anal. Des.*, 1993;15(1):11–34.

[8] Hughes, T. J. R., Cottrell, J. A., Bazilevs, Y. Isogeometric analysis: CAD, finite elements, NURBS, exact geometry and mesh refinement. *Comput. Methods Appl. Mech. Engg.*, 2005;194(39–41):4135–4195.

[9] Romero, I. A comparison of finite elements for non-linear beams: The absolute nodal coordinate and geometrically exact formulations. *Multibody Syst. Dyn.*, 2008;20(1):51–68.

[10] Bagri, A. S., Haldar, A., Chawla, H. Circumventing snap-through instability through alternate loading strategies. *Recent Dev. Struct. Engg.*, Kumar, R., Bakre, S. V., Goel, M. D. (Eds.) Singapore: Springer Nature Singapore, 2025;2:451–459.

[11] Haldar, A., Groh, R. M. J., Jansen, E., Weaver, P. M., Rolfes, R. An efficient semi-analytical framework to tailor snap-through loads in bistable variable stiffness laminates. *Int. J. Solids Struct.*, 2020;195:91–107.

[12] Santos, J. L. T., Choi, K. K. Shape design sensitivity analysis of nonlinear structural systems. *Struct. Optim.*, 1992;4(1):23–35.

[13] Cho, S., Jung, H. S. Design sensitivity analysis and topology optimization of displacement-loaded non-linear structures. *Comput. Methods Appl. Mech. Engg.*, 2003;192(22–23):2539–2553.

[14] Wu, X., Downes, M. S., Goktekin, T., Tendick, F. Adaptive nonlinear finite elements for deformable body simulation using dynamic progressive meshes. *Comput. Graph. Forum*, 2001;20(30):349–358.

[15] Bletzinger, K. U., Firl, M, Linhard, J., Wüchner, R. Optimal shapes of mechanically motivated surfaces. *Comput. Methods Appl. Mech. Engg.*, 2010;199(5–8):324–333.

[16] Kumar, A., Haldar, A., Weaver, P. M. An adaptive metastructure concept using bistable composite laminates. *Compos. Struct.*, 2024;330:117767.

17 Predicting the effects of volume fraction on mechanical behaviour of the polymer composites

Sristi Gupta[1,a], *Rahul Raghavendra Choudhury*[1], *Tanmoy Mukhopadhyay*[2], *Divyesh Varade*[1] *and Vinod Kushvaha*[3]

[1]Civil Engineering Department, Indian Institute of Technology Jammu, Jammu, Jammu and Kashmir, India

[2]Department of Aeronautics and Astronautics, University of Southampton, Southampton, UK

[3]Independent Researcher, Boise, USA

Abstract

This study investigates the application of artificial intelligence (AI) to analyse the mechanical behaviour of short fibre reinforced polymer (FRP) composites, aimed on predicting stress-field distribution in particulate polymer composites (PPC). Traditional experimental and computational approaches often face challenges to accurately quantify and optimise the behaviour of such complex materials. This research utilises a paired image-to-image translation (Pix2Pix) deep learning algorithm, which employs a conditional generative adversarial network (cGAN) to predict stress patterns, particularly in the principal direction (S_{yy}) under tensile loading. Key input parameters for the AI model include fibre volume fraction, Young's modulus, and time steps (frame numbers). The study models a 2D representative area element (RAE) for glass fibre-filled epoxy specimens using ABAQUS/CAE v2021. The corresponding stress patterns are used as reference for the AI model. The PPC specimens consist of rod shape glass fibres with an aspect ratio of 80, randomly oriented and positioned (at 1%, 2%, and 3% volume fractions) within epoxy matrices. Tensorflow, with GPU acceleration, is utilised to optimise model training and improve computational efficiency. A train-test split of 80:20 is used, and the model is trained with 2000 and 5000 iterations to analyse the impact of volume fraction on stress distribution. The stress field is partitioned into four distinct spectra based on the upper and lower bound stresses in the glass fibre and epoxy. This partitioning scheme improves the model's predictive accuracy for stress distributions across varying material properties, fibre orientations, and volume fractions. The results are visualised through histograms of *L2* norm and correlation metrics, with improved model performance observed as *L2* norm values approach zero and correlation values approach one. The results demonstrate that the proposed approach provides a more efficient and less labour-intensive alternative to conventional experimental and numerical methods in analysing the mechanical behaviour of polymer composite materials.

Keywords: Fibre-reinforced polymer (FRP), PPC, Quasi-static tension, Pix2Pix deep learning, conditional generative adversarial network

Introduction

Polymer composites, known for corrosion resistance, lightweight, and, high strength-to-weight ratio are widely used across various industries [1]. For example, in aerospace engineering, polymer composites reduce fuel consumption and enhance performance by forming key components like fuselags and wings [2]. In the automotive industry, polymer composites are employed in body panels, chassis components, and interior parts, offering improved fuel efficiency, durability, and crashworthiness [3]. Furthermore, these materials excel in marine applications due to their resistance to moisture, saltwater corrosion, and biofouling, making them ideal for marine infrastructure [4]. These versatile materials also extend their influence into other sectors, including construction, renewable energy, sports equipment, and healthcare [5].

The synthesis of polymer composites, achieved by combining materials with diverse mechanical properties, is transforming the manufacturing field in

[a]sristi.gupta@iitjammu.ac.in

DOI: 10.1201/9781003770442-17

devices and structures, offering superior mechanical performance [6]. In the fields of mechanical, aerospace, and civil engineering, material binders like thermoplastics, elastomers, or thermosetting agents are frequently blended with reinforcements/fillers to create complex and high-performing structures. Recent studies have provided key insights into the behaviour and characterisation of composite materials. For instance, Sharma et al. [7] performed an in-depth study on the fatigue responses of rod-shaped glass fibre reinforced polymer composites, analysing the influence of different volume fractions on mechanical behaviour and fatigue life. They have explored how varying fibre volume fractions affect the ultimate tensile strength (UTS) and fatigue life of glass-fibre reinforced composites, finding that a 10% volume fraction exhibited the longest fatigue life. Similarly, Capela et al. [8] focused on the fatigue resistance of short fibre reinforced polymer composites (SFRPCs), providing critical data on how varying carbon fibre volume fraction affects the mechanical performance of composite. Fatigue tests conducted with a constant stress amplitude demonstrated that fatigue strength increased with increasing fibre volume fraction, reaching a maximum at 17.5% volume fraction. Additionally, Singh et al. [9] investigated the effects of reinforcement by varying volume fraction and, particle size on the toughening characteristics of brittle thermosetting polymer composites, revealing critical insights into fracture toughness improvement. They demonstrated how smaller reinforcement particles and higher volume fractions significantly enhance fracture toughness in brittle thermosetting polymer composites.

In another key study, Fu et al. [10] addressed PPCs and the impact of filler matrix adhesion, particle size, and loading on mechanical characteristics such as strength and stiffness. Cebeci et al. [11] investigated the properties of polymer nanocomposites (PNCs) in contrast to conventional composites used in aerospace applications. They observed that the PNCs employ low-volume fraction filler-like concepts with poorly controlled morphologies with random orientation of fibres, posing challenges in the dispersion and alignment of nanostructures. These studies underscore the dynamic advancements in composite material synthesis and emphasise the pivotal role of interdisciplinary research in shaping the future of manufacturing and engineering practices [4, 9].

However, the analysis of newly developed advanced materials presents significant challenges that include costly nature of fabrication and testing processes. The numerical modelling approaches like the finite element method (FEM) tend to be often cumbersome, requiring fresh modelling and long simulation times for each new composite configuration. However, leveraging AI-driven computational solutions offer a promising alternative by learning from existing data and providing tailored solutions to complex material behaviour problems [12]. Bhaduri et al. [13] addressed computational challenges in analysing fibre-reinforced composites and proposed a deep learning approach for stress-field prediction, offering a novel perspective on material characterisation. Gupta et al. [14] introduced a machine learning (ML) algorithm to capture stress maps in polymer composites, highlighting the potential of AI-driven methodologies in materials science.

AI has proven highly effective in optimising composite designs based on FEM calculations, addressing diverse loading conditions in both static and dynamic analysis. However, these methods typically rely on numerical datasets and struggle to capture intricate microstructural details. To bridge this gap, recent advancements in image-based predictive models, which integrate traditional numerical simulations with AI, offer a solution by establishing a direct connection between material microstructure and physical field predictions such as stress distribution. Convolutional neural networks (CNNs), in particular, have been used to analyse microstructural images and extract relevant features that influence material behaviour, improving the accuracy of predictive models [5, 6].

Pattnaik et al. [15] addressed the transformative impact of AI on redefining the design and optimisation processes of fibre reinforcement in polymer composites, yielding material with different properties. Other studies on AI models demonstrated the effectiveness of integrating AI algorithms with experimental and computational methods to optimise material processing and characterisation, thereby reducing costs and enhancing efficiency [6, 15–17]. AI integration has proven to be useful in simulating material performance as a solution to finite element modelling, a computationally costly procedure primarily due to the complexity of composite materials [18]. The AI models reduce the reliance on traditional

FEM simulations, which are computationally intensive and limited by material complexity [19].

The research work employs a ML model to predict stress distribution in fibre reinforced matrix composite under mechanical loads. The model is trained for geometries with 1%, 2%, and 3% volume fractions of glass fibres, and incorporates varying Young's modulus to develop a robust predictor for stress fields. The adversarial network architecture, based on U-Net, is used to demonstrate the model's ability to predict stress maps accurately.

Methodology

Neural networks have greatly simplified the estimation of stress fields in PPC by offering a data-driven approach capable of learning from extensive datasets with varying parametric combinations. Unlike traditional methods that rely on explicit mathematical models and assumptions, neural networks effectively capture the intricate relationships between input parameters and output predictions [18]. By analysing large volumes of image training data, these networks identify specific correlations among images. The Pix2Pix algorithm, in particular, has demonstrated high effectiveness in segmenting stress fields in short fibre-reinforced composites across a diverse range of material elastic properties [20]. Figure 17.1 shows a typical workflow utilising AI technique.

Integrating AI models into the FEM involves detailed specifications of material attributes, geometric configurations, and boundary conditions (Specific parameters in Figure 17.1). Each parameter must be precisely defined to represent the specific physical behaviour of the composite under study. This approach results in an accurate material response, providing a viable alternative to extensive FEM simulations as depicted in Figure 17.1. Conditional generative adversarial network (cGAN)-based image-to-image translation algorithms facilitate the prediction of stress patterns in fibre reinforced polymer composites by learning the specific relationships between microstructure and material properties [14], [19]. The AI model generates stress field predictions based on training data from an experimentally and computationally validated model of PPC [14].

FEM simulations are carried out a priori to analyse the mechanical behaviour of composites across selected parametric combinations, generating a reference dataset for training the AI model. This trained

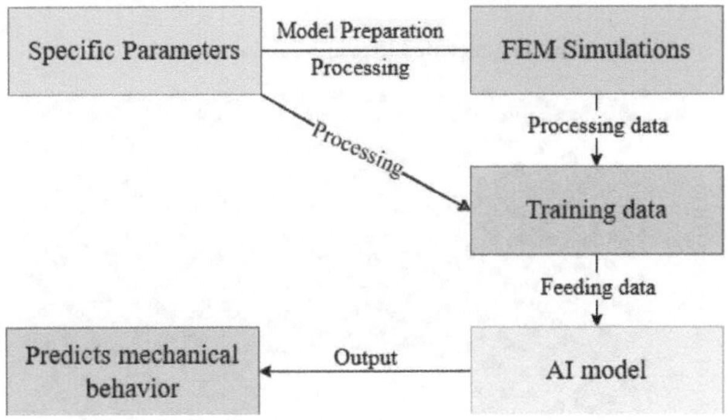

Figure 17.1 Schematic illustration of AI integration with FEM simulations of PPC material
Source: Authors

AI model then predict the mechanical behaviour of composites over a broader range of parametric variations, effectively reducing both computational time and resource consumption compared to traditional FEM simulations.

Composite modelling-based training data generation

The main objective of this methodology is to thoroughly analyse the stress distribution in a composite material, considering different mechanical properties within the elastic regime. To achieve this, data points are generated using ABAQUS/CAE v2021 by systematically varying Young's modulus of both the glass fibres and matrix (epoxy) within a range of ±30% of nominal values [14]. Specifically, the Sobol distributed nominal values are 1.25–2.33 GPa for neat epoxy and 49–91 GPa for glass fibres.

A rod-shaped glass fibre-filled epoxy specimen is simulated under static tension using displacement control, with a cross-head speed of 0.144 mm/min based on the specimen size [7]. For simplification, the modelling is limited to 2D representations using a representative area element (RAE) of the composite. In 2D, the glass rods are approximated as rectangular cross sections measuring 800 × 10 μm^2. The RAE, with dimensions of 8000 × 8000 μm^2, contains random points corresponding to various area fractions (Af), which were generated using custom Python code. Specifically, the code assigns 80 points for 1%, 160 points for 2%, and 320 points for 3% area fractions. These random points are obtained in a

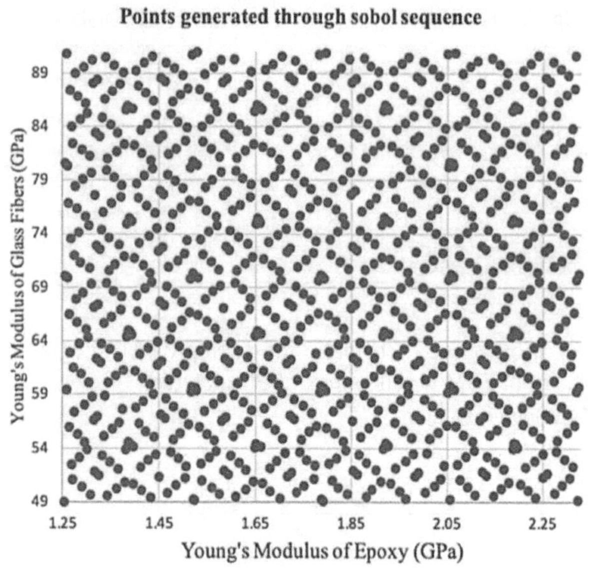

Figure 17.2 Visual representation of data extraction showcasing the sampled points spanning the ranges of Young's modulus values for both glass fibre and epoxy, generated using Sobol sequence
Source: Authors

Table 17.1 Elastic properties of constituent materials.

Material	Elastic Modulus (GPa)	Poisson's ratio
Glass fibre	70	0.2
Neat epoxy	1.79	0.35

Source: Authors

well-distributed manner using the simple sequential inhibition (SSI) process [21] (Refer to Figure 17.2).

The points represent positions of glass fibres in PPC, with generated angles ranging from −180° to 180°, corresponding to the randomly distributed points in RAE. A Python script run through ABAQUS/CAE v2021 models the RAE of the glass fibre-filled epoxy composite. The specimen is restrained in the y-direction at the bottom, with the centre node at the bottom constrained in the x-direction. The upper part is displaced at a velocity of 0.14 mm/min, simulating the cross-head speed of tension tests. CPS4R, a 4-node bilinear quadrilateral with reduced integration, is used for the plane stress elements [22]. The initial elastic properties of the neat epoxy and glass fibres are detailed in Table 17.1.

A total of 256 data values of Young's modulus are generated for both epoxy and glass fibre, utilising sobol sequences which are quasi-random, low-discrepancy sequences that ensure well-distributed sampling across the parameter space. After generating the Young's modulus values using the Sobol sequence, the automated Python code executes Abaqus/CAE v2021 simulations for each corresponding set of material properties. This results in the stored output database (odb) files for each pair which are subsequently processed, resized, and optimised for performance using other Python scripts to create training data for the AI model. Hence, the dataset consists of 256 Abaqus simulations, which serve as the sample size for training the AI model. In order to improve training process, different sets of output images (4 sets) are incorporated based on the maximum and minimum stresses, separately for epoxy and glass fibre.

An important variable during this simulation is the timestep, represented by a frame number in ABAQUS/CAE v2021. These time steps are varied from 0 to 102, where, the stress complexity increases as the timesteps increase in the simulation. To effectively capture stress development in the composite microstructure, images are extracted at six regular intervals, 15, 30, 45, 60, 75, and 90, representing the overall evolution from 0 to 102 time steps [20].

The extracted training data from ABAQUS/CAE v2021 forms the basis for creating the engineered dataset needed for the image-to-image translation task. Within this framework, three variables i.e., Young's modulus of epoxy, Young's modulus of glass fibre, and timesteps are precisely structured to form the supporting elements of input images. During training, labelled images (constituting the variables as mentioned earlier) of 512 × 512 pixels are used to generate synthetic counterparts while incorporating actual stress-field images for accurate loss computation and backpropagation. These labelled images are organised into an 8 by 8 grid of patches, with a series of 3 patches along rows that encode the fibre orientation, denoting the varying volume fractions. This detailed structuring ensures a comprehensive representation of stress distribution and optimised data utilisation. The sample input image depicted in Figure 17.3 exemplifies this approach, enabling robust data-driven learning and stress prediction in PPC.

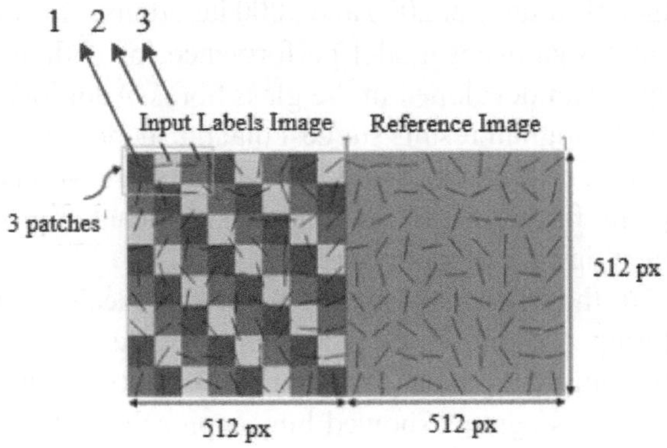

Figure 17.3 An 8 × 8 grid input image sample for 1% volume fraction where the values of variables are linearly scaled in 8-bit representation in grey scale. Each of the 3 patch series signifies 1. Young's modulus of epoxy; 2. Young's modulus of glass fibre and 3. timesteps
Source: Authors

The Pix2Pix model is trained for 1%, 2%, and 3% volume fraction with a train-test split of 80% for training and 20% for testing. Results are generated for 2000 and 5000 iterations for each case, emphasising the importance of iterations requirement based on the volume fraction of fibres in the neat epoxy.

2.2 AI model description

The cGAN model is employed as a predictive tool to estimate the stress field within the 2D model of PPC [16]. It consists of two primary components: the generator and the discriminator. The generator generates new instances within the problem domain, aiming to produce plausible examples while the discriminator's role is to differentiate between real and generated examples. Through iterative training using backpropagation, the generator learns to approximate the physical and statistical characteristics of the dataset. As the generator works to minimise its loss function, the discriminator seeks to maximise its own, fostering a competitive dynamic between the two [17]. This adversarial relationship encourages a strong correlation coefficient (~1) and minimal *L2* norm (mean square error, ~0). The cGAN functions as a supervised model, forming a specialised subclass of the original unsupervised GAN model. Unlike many generative models, the cGAN specifically predicts and generates data that matches a target distribution by incorporating label information, rather than relying solely on internal patterns within the data. This targeted approach allows the cGAN to achieve greater accuracy in tasks where alignment with specific labels is essential [23].

In the proposed implementation of the cGAN, the model is composed of two essential components: a PatchGAN discriminator, and U-Net generator. The U-Net architecture comprises two distinct paths: the encoder and the decoder. The encoder captures the context of the image while symmetrically expanding decoder performs precise localisation through transposed convolutions. The encoder transforms grey scale images into latent representations, and the decoder enlarges these representations to generate RGB images. With its symmetrically arranged layers, the generator functions within an encoder-decoder structure, aiming to increase error rates for the discriminator, which in turn optimally identifies fake images. During training, the generator and discriminator neural networks alternate execution. With each iteration, both components are refined to improve overall performance. In each training step, the generator processes grey scale images and generates RGB outputs.

Model evaluation

The *L2* norm is applied to quantitatively evaluate the similarity between generated images and reference images [3]. For two images, the *L2* norm between two images, I_1 and I_2, is defined as

$$L2 = \sum_{l,m,n} | I_1(l,m,n) I_2(l,m,n) | \qquad (1)$$

where, *l*, *m* refers to the pixels in the images, and n = 1, 2, 3 represent the three channels red, green, and blue of the image colours.

As the images produced by the generator successfully deceive the discriminator, the correlation coefficient is expected to reach the maximum. This mechanism assesses the degree of similarity between the two images (predicted and reference), providing a quantitative measure of their relationship. The correlation coefficient (Corr) serves as a valuable metric for evaluating the effectiveness of the generator in processing realistic outputs that deceive the discriminator. Mathematically, the correlation coefficient between the two images I_1 and I_2 is expressed as follows:

$$Corr = \frac{\sum_{l,m,n} I_1(l,m,n) \times I_2(l,m,n)}{\sqrt{\sum_{l,m,n} I_1^2(l,m,n)}\sqrt{\sum_{l,m,n} {I_2}^2(l,m,n)}} \qquad (2)$$

Results and discussion

To establish a robust correlation between reference images and predicted outcomes, it is essential for the discriminator's loss function to accurately identify real images from the training data as "real" and generated images from the generator as "fake." This study emphasises use of the TensorFlow library, which requires a graphical processing unit (GPU) for efficient and optimised execution of computational systems.

The Pix2Pix algorithm for predicting the mechanical behaviour of the PPCs is trained with the settings described in Section 2.1. The findings are presented as the magnitudes of *L2* norm and correlation coefficient histograms, where the *x*-axis represents the metric values, and the *y*-axis denotes the number of samples. Each bin has a lower bound, upper bound, and a ceiling edge; the *y*-coordinate corresponding to the ceiling edge of the bin gives the number of samples where the model obtained the *L2* norm and correlation coefficient values within the specified range. A well-performing model is characterised by *L2* norm values close to 0 and correlation values approaching 1. An analysis of the histograms on test datasets reveals that a greater number of bins close to zero in *L2* histogram and a greater number of bins near one in the correlation histogram signify good model performance.

Volume fraction – 1%

For the 1% volume fraction, four models are trained for each fibres, and matrix, corresponding to the maximum and minimum spectrum of stresses as detailed in Table 17.2. The stress patterns across both the minimum and maximum spectrum for the glass fibres are effectively captured, with the *L2* norm histograms displaying bins closer to 0 and the correlation histograms showing bins nearer to 1 for the models trained at 2000 and 5000 iterations. This indicates significant model performance for both stress spectrum developed in the glass fibres. Additionally, the histogram results suggest that the model's performance converges around 2000 iterations; beyond this point, further iterations yield no significant improvement in case of glass fibres.

In the case of epoxy, the AI model excels in capturing the minimum spectrum of stresses. After training sessions of 2000 and 5000 iterations, the *L2* norm histograms showed bins approaching 0, while the correlation histogram bins nearing 1. However, the model demonstrates less precision in predicting stress fields within the maximum stress spectrum at 2000 iterations, indicating the higher complexity of stress patterns. Predictions of stress fields at 5000 iterations suggest that a higher iteration count is necessary for the model to effectively learn and capture the features of FEM model.

Volume Fraction – 2% & 3%

At the 2% volume fraction, the AI model efficiently captures the minimum spectrum stresses in both the fibres and the epoxy. However, the maximum stress spectrum is more accurately captured in fibres than in the epoxy.

As the volume fraction increases from 1% to 2%, the stresses in the interfacial region between the fibres and the epoxy also rise, suggesting a higher range of stress values within the stress spectrum. The predictive capacity of the model remains strong at a 2% volume fraction, particularly in predicting stress fields across various critical regions of the microstructure. Training the model for 2000 iterations reveals an *L2* norm range spanning from 0 to 0.5, while the correlation coefficient ranges from 0.6 to 1, with 120 samples falling between 0.95 and 1. Extending the training session to 5000 iterations results in a notable enhancement in predicted patterns; the *L2* norm then

Table 17.2 Range of stress limits in colour spectrum in neat epoxy, and glass fibres.

Spectrum	Material	Lower limit	Upper limit	Range
1	Glass fibre	5.011×10^{-6}	1.39×10^{-5}	Min
2	Glass fibre	1.39×10^{-5}	1.63×10^{-4}	Max
3	Neat epoxy	-2.6×10^{-19}	0	Min
4	Neat epoxy	0	7.24×10^{-5}	Max

Source: Authors

ranges from 0 to 0.35, and the correlation histogram displays bins spanning from 0.65 to 1. These observations align with those observed at a 1% volume fraction, where increased iterations provide the AI model with more opportunities to refine its predictions, converging more accurately toward the desired outcomes.

In the subsequent analysis at 3% volume fraction, the results indicate a decline in accuracy for capturing the maximum stress values in spectrum in the epoxy compared to a case with a lower volume fraction of fibres in neat epoxy. This highlights the need for improved predictions within the maximum stress spectrum for the epoxy system with increasing volume fraction.

For the 3% volume fraction, an increase in stress values is observed for an increased population of fibres. AI model results show lower accuracy compared to the 2% volume fraction, as the stress distribution exhibits greater variability. Tables 17.3 and 17.4 summarise the ranges for *L2* norm and correlation coefficient, revealing a slight reduction in the correlation of predicted fibres with the increased volume fraction of fibres. In the case of images corresponding to maximum stress spectrum, the maximum samples of 140 are falling closer to range of 0, observed for 1% V_f. Conversely, the samples that depict strong correlation reduce to 120 and below for 2% and 3% V_f of fibres.

The accuracy observed at 2000 iterations is insufficient and fails to reliably predict stress patterns under loading direction for the case of images denoting the maximum stress spectrum in epoxy. Additionally, a higher fibre volume fraction within the epoxy matrix introduces interfacial complexities between fibres and epoxy, resulting in elevated stress magnitudes in the epoxy. It is observed that fine-tuning is necessary for these critical cases, where stress patterns are complex and demand higher precision.

Initially, a large-scale model covering an area of 8000 × 8000 μm^2 is developed in ABAQUS/CAE v2021, with the glass fibres sized at 800 × 10 μm^2. However, to analyse the stress fields in the fibres is challenging in 8000 × 8000 μm^2 RAE. Consequently, as shown in Figure 17.4 a reduced RAE of 2000 × 2000 μm^2 is considered for taking a clear visualisation of stresses in fibres, maintaining the same fibre dimensions (800 × 10 μm^2), and modelling for 1% fibre volume fraction. The boundary conditions and mesh convergence study are kept consistent as in the previous 8000 × 8000 μm^2 model. Subsequently, simulations are conducted, resulting in the extraction of 256 odb files based on varying ranges of Young's modulus for the glass fibre and neat epoxy. Finally, stress-field images are extracted based on the upper and lower bounds of the stresses developed in both the fibre and epoxy systems.

Overall, the cGAN effectively learns to predict the complete stress distribution within PPC, accounting for specific geometries, boundaries, and loading conditions. This approach efficiently captures all stress-strain components and different failure criteria.

Table 17.3 *L2* norm values for the maximum and minimum stress spectrum in glass fibres.

Volume fraction (V_f)	L2 norm			
	Minimum		Maximum	
	2000	5000	2000	5000
1%	0–0.006	0–0.016	0–0.014	0–0.012
2%	0–0.008	0–0.019	0–0.028	0–0.02
3%	0–0.010	0–0.018	0–0.03	0–0.038

Source: Authors

Table 17.4 Correlation coefficient values for the maximum and minimum stress spectrum in glass fibres.

Volume fraction (V_f)	Correlation coefficient			
	Minimum		Maximum	
	2000	5000	2000	5000
1%	0.980–0.998	0.992–0.999	0.9–0.999	0.99–0.999
2%	0.98–0.99	0.98–0.99	0.89–0.9	0.90–0.98
3%	0.82–0.88	0.87–0.89	0.8–0.9	0.84–0.99

Source: Authors

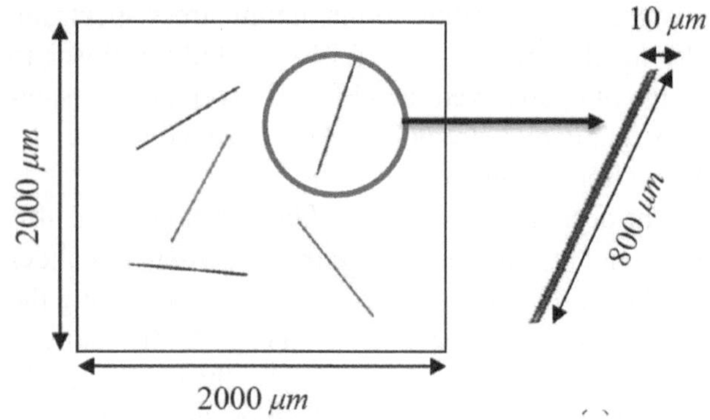

Figure 17.4 RAE (size 2000 × 2000 μm²) showing the stress field in glass fibres, fibres in the RAE correspond to 1% V_f
Source: Authors

3.3. Overall performance at varying volume fractions

At a 1% volume fraction, the Pix2Pix model achieved an impressive correlation value of approximately 0.999 and *L2* norm of less than 0.002, accurately predicting stress distributions in both minimum and maximum values spectrums for glass fibres. The results show convergence at 2000 iterations, indicating the model's stability in predicting stress fields in this context. However, the maximum stress spectrum in the epoxy corresponding to higher stress values exhibited complex stress patterns resulting in unsatisfactory histogram outcomes at 2000 iterations; as a result, predicted stress values diverged from the reference data.

With a 2% volume fraction, the model's predictive capability is reduced, particularly in predicting stress patterns aligned with loading direction. It is observed that the AI model struggles to replicate consistent results across varying volume fractions due to complexities inherent in the mechanical responses of the PPC.

On extending the analysis to a 3% volume fraction, an increase in the stress concentration around the fibres is observed accompanied by a decreasing correlation coefficient at the same iteration counts – 2000 and 5000. This decline in correlation can be attributed to the heightened density of fibres within the composite, resulting in an increased number of fibre-epoxy interfaces per unit volume. Consequently, the mechanical interactions between the fibres and the epoxy matrix lead to intensified stress concentrations at these interfaces.

Conclusion

The study demonstrates the significance of considering volumetric composition when analysing the mechanical behaviour of composite materials. It also highlights the efficacy of the developed deep learning model in accurately capturing and predicting complex stress distributions across varying volume fractions, particularly the stresses in fibres and minimum stress spectra in epoxy. Although the model struggled to replicate stresses in epoxy at higher volume fractions, in general, this efficient algorithm represents a substantial advancement in reducing the time and cost associated with designing new composite materials, bypassing traditional numerical inputs, and directly utilising microstructural images. The study effectively demonstrates a significant step towards image-to-image data-driven predictions, expediting microstructural analysis in composite materials and encouraging innovative strategies for material design.

References

[1] Oladele, I. O., Omotosho, T. F., Adediran, A. A. Polymer-based composites: An indispensable material for present and future applications. *Internat. J. Polymer Sci.*, 2020;2020(8834518):1–12.

[2] Mansor, M. R., Nurfaizey, A. H., Tamaldin, N., Nordin, M. N. A. Natural fibre polymer composites: Utilization in aerospace engineering. *Biomass Biopoly. Mater. Bioener.*, Verma, D., Fortunati, E., Jain, S., Zhang, X., (Eds.). Woodhead Publishing Series in Composites Science and Engineering. Woodhead Publishing, 2019:203–224.

[3] Akampumuza, O., Wambua, P., Ahmed, A., Wei, L., Qin, X.-H. A review of the applications of bio composites in the automotive industry. *Polymer Compos.*, 2017;38(11):2553–2569.

[4] Muthukumar, T., Aravinthan, A., Lakshmi, K., Venkatesan, R., Vedaprakash, L., Doble, M. Fouling and stability of polymers and composites in marine environment. *Internat. Biodeter. Biodegr.*, 2011;65(2):276–284.

[5] Oladele, I., Onuh, N. L., Siengchin, S., S. M. R., Adelani, S. Modern applications of polymer composites in structural industries: A review of philosophies, product development, and graphical applications. *Appl. Sci. Engg. Prog.*, 2024;17:6884.

[6] Hsissou, R., Seghiri, R., Benzekri, Z., Hilali, M., Rafik, M., Elharfi, A. Polymer composite materials: A comprehensive review. *Compos. Struct.*, 2021;262:113640.

[7] Arora, A., Sharma, A., Singh, A., Mahajan, D. K., Kushvaha, V. Fatigue response of glass-filled epoxy composites: A crack initiation and propagation study. *Internat. J. Fatigue*, 2023;170:107542.

[8] Capela, C., Oliveira, S. E., Ferreira, J. A. M. Fatigue behaviour of short carbon fibre reinforced epoxy composites. *Compos. Part B Engg.*, 2019;164:191–197.

[9] Singh, R. P., Zhang, M., Chan, D. Toughening of a brittle thermosetting polymer: Effects of reinforcement particle size and volume fraction. *J. Mater. Sci.*, 2002;37(4):781–788.

[10] Fu, S.-Y., Feng, X.-Q., Lauke, B., Mai, Y.-W. Effects of particle size, particle/matrix interface adhesion and particle loading on mechanical properties of particulate–polymer composites. *Compos. Part B Engg.*, 2008;39(6):933–961.

[11] Cebeci, H., de Villoria, R. G., Hart, A. J., Wardle, B. L. Multifunctional properties of high volume fraction aligned carbon nanotube polymer composites with controlled morphology. *Compos. Sci. Technol.*, 2009;69(15):2649–2656.

[12] Mukhopadhyay, T., Adhikari, S. Effective in-plane elastic moduli of quasi-random spatially irregular hexagonal lattices. *Internat. J. Engg. Sci.*, 2017;119:142–179.

[13] Bhaduri, A., Gupta, A., Graham-Brady, L. Stress field prediction in fibre-reinforced composite materials using a deep learning approach. *Compos. Part B Engg.*, 2022;238:109879.

[14] Gupta, S., Mukhopadhyay, T., Kushvaha, V. Microstructural image based convolutional neural networks for efficient prediction of full-field stress maps in short fibre polymer composites. *Defence Technol.*, 2023;24:58–82.

[15] Pattnaik, P., Sharma, A., Choudhary, M., Singh, V., Agarwal, P., Kukshal, V. Role of machine learning in the field of Fibre reinforced polymer composites: A preliminary discussion. *Mater. Today Proc.*, 2021;44:4703–4708.

[16] Gupta, S., Mukhopadhyay, T., Varade, D., Kushvaha, V. Investigating deep learning-based stress prediction in particulate polymer composites using multiple quality measures. *Proc. Internat. Conf. Fundamen. Indus. Res. Mater.* Tiwari, A., Ray, P. K., Sardana, N., Kumar, R. (Eds.). Singapore: Springer Nature, 2024:101–107.

[17] Hui, X., Xu, Y., Niu, J., Zhang, W. Rapid evaluation and prediction of cure-induced residual stress of composites based on cGAN deep learning model. *Compos. Struct.*, 2024;330:117827.

[18] Gupta, S., Mukhopadhyay, T., Varade, D., Kushvaha, V. Prediction of stress fields in particulate polymer composites using micromechanics-based artificial intelligence model. *Recent Dev. Struct. Engg.* Goel, M. D., Kumar, R., Gadve, S. S. (Eds.). Singapore: Springer Nature, 2024;1:113–123.

[19] Gupta, S., Choudhury, R. R., Kushvaha, V. Deep learning model to predict the stresses in polymer composite materials. *Cong. Ind. Soc. Theoret. Appl. Mech. (ISTAM)*, 2022;67.

[20] Gupta, S., Kushvaha, V., Varade, D. Stress field prediction in particulate polymer composite materials using paired image-to-image translation approach. *Adv. Theor. Appl. Mech.* Ray, R. K., Bora, S. N., Maiti, D. K. (Eds.). Singapore: Springer Nature, 2024:285–294.

[21] Kushvaha, V. Synthesis, processing and dynamic fracture behavior of particulate epoxy composites with conventional and hierarchical micro-/nano-fillers. Elec. Theses Disser., 2016;334:844–4000.

[22] Kawecki, B., Podgórski, J. Numerical results quality in dependence on abaqus plane stress elements type in big displacements compression test. *Appl. Comp. Sci.*, 2017;13(4):56–64.

[23] Isola, P., Zhu, J.-Y., Zhou, T., Efros, A. A. Image-to-image translation with conditional adversarial networks. *2017 IEEE Conference on Computer Vision and Pattern Recognition (CVPR),* 5967–5976, 2017, doi:10.1109/CVPR.2017.632

18 Linear thermal bending analysis of functionally graded square plate under different boundary conditions and geometric effects

Smruti Ranjan Sahoo[1,a], Surendra Verma[2] and B. N. Singh[1]

[1]Research Scholar, Department of Aerospace Engineering, Indian Institute of Technology Kharagpur, Kharagpur, West Bengal, India.

[2]Toronto Metropolitan University, Toronto, Canada

Abstract

A C^0 FE framework is developed in this study through the application of a penalty method on a C^1 five-variable "higher-order shear deformation theory". The method is tailored for analysing the linear-thermal-bending response of "functionally graded material" (FGM) plates. The properties of FGMs are estimated employing rule of mixtures, while the governing relations are obtained by implementing virtual work approach. For discretisation, a 9-noded Lagrange element incorporating 7 degrees of freedom are employed to each node. To assess performance, numerical studies are conducted by comparing results with Reddy's TSDT theory and a "non-polynomial higher-order shear deformation theory" (NPHSDT). finite element method (FEM) method is applied for analysing thermal flexural response of "FGM-plates" under diverse environments. The proposed formulation's reliability is established through validation against reference solutions reported in prior studies. Furthermore, a detailed parametric study explores the influence of material gradation, span-to-thickness ratio, thermal expansion coefficient ratio, and boundary constraints. Overall, the research provides meaningful insights into thermal bending characteristics of "FGM-plates", supporting the efficient design and structural evaluation of advanced engineering systems.

Keywords: FGM, FEM, HSDT, thermal bending

Introduction

Composite structures, especially laminated composites and sandwich structures, are distinguished for their remarkable "stiffness-to-weight" and "strength-to-weight" ratios, rendering those optimal materials for aeronautical, mechanical, and various engineering applications. These characteristics have ensured their extensive employment in essential structural functions for decades. Nevertheless, these materials possess intrinsic limitations in extremely high-temperature settings, as they are susceptible to thermal residual stresses resulting from rapid alterations in material properties at the interfaces. These forces may cause delamination, ultimately resulting in the catastrophic failure of multilayer structures. To rectify these deficiencies, a recently developed material category termed as "functionally graded materials" (FGMs) was proposed and has since been effectively embraced by the scientific community [1]. FGMs were developed as a solution for a space plane project that necessitated a thermal barrier material capable of enduring severe surface temperatures. The initial FGMs were engineered to withstand a temperature differential of 1000 K over a thickness of less than 10 mm [2]. This breakthrough represented a substantial progression in materials engineering, as FGMs successfully eradicated abrupt interface transitions by facilitating a progressive variation in material properties. Since its origin, FGMs have been thoroughly researched and developed for several engineering applications beyond aerospace. Their capacity to function efficiently in elevated temperature and thermomechanical conditions has rendered them essential in sectors such as automotive, maritime, and biomedical engineering. These materials

[a]smrutisahoo500@gmail.com

DOI: 10.1201/9781003770442-18

persist in transforming structural design, providing improved durability, dependability, and performance in rigorous applications [3].

FGMs are an innovative type of inhomogeneous advance-composite structure consisting smooth transition in material constituents across a defined spatial dimension, usually along the thickness, accomplished through a gradual distribution of two or more constituent phases in the desired directions. This gradual transition in properties, obtained through adjusting the volume fractions of constituent materials, eliminates abrupt interfaces and provides tailored performance advantages. Such unique material behaviour makes FGMs highly suitable for advanced applications in automotive, aerospace, marine, and biomedical engineering, where optimised performance under diverse thermal and mechanical conditions is critical. The analysis of linear thermal bending behaviour in FGM plates is particularly important for understanding how these materials respond to thermal loading. This understanding is crucial for designing structural components capable of withstanding extreme environments while maintaining efficiency and reliability. Accurate characterisation enables engineers to predict deflection, stresses, and failure modes, which is essential for ensuring the structural integrity of components exposed to high-temperature gradients, such as turbine blades, thermal barriers, and biomedical implants. This research lays a foundation for the development of advanced FGM-based systems, offering insights that can be directly applied to optimise their performance. Incorporating advanced modelling techniques, such as higher-order deformation theories, coupled with precise boundary condition analyses, enhances the predictive capabilities of these studies [4]. The outcomes have far-reaching implications for the design and deployment of FGM components in high-performance applications, contributing to innovations in materials engineering and structural optimisation.

In order to fabricate many types of FGMs, which are each tuned to attain certain material qualities, a variety of procedures are utilised. These techniques include, amongst others, centrifugal casting, plasma spraying, chemical vapour deposition (CVD) [5]. The criteria of the application and the type of FGM that is being produced both have a role in determining the specific advantages that each approach offers. Aluminium/Alumina (Al/Al_2O_3) FGMs, for example, are perfect for applications such as rocket nozzles and engine casings, which require a higher level of heat and corrosion resistance than other materials. In a similar manner, FGMs made of Aluminium/Silicon Carbide (Al/SiC) are utilised in settings that require a combination of high hardness, toughness, and heat resistance [6]. FGMs are designed to fulfil the particular needs of high-performance engineering applications thanks to the adaptability of these material combinations, which highlight their versatility. When it comes to structural components in the aerospace industry, FGMs have become indispensable. These components include wings, rotor blades, aircraft frames, and fuselages. In order to improve both performance and fuel efficiency, these components need to have a proportion of strength to weight that is high. FGMs are able to deliver optimum solutions that are capable of meeting these stringent standards since they are created by meticulously mixing metals and ceramics in varied compositions. An example of this would be the utilisation of titanium-based FGMs in conjunction with ceramic reinforcements in order to establish lightweight structures that possess excellent thermal and mechanical qualities. The versatility of FGM fabrication processes and the capability to customise material properties for individual applications highlight the revolutionary potential of these techniques in a variety of advanced engineering sectors, including aerospace, automotive, and others.

For advanced engineering applications to completely capitalise on the benefits of FGMs, it is imperative to possess a thorough comprehension of their thermo-mechanical behaviour. Several works have been performed over the years in order to analyse the performance of FGM flat panels under static and dynamic bearing conditions, utilising a variety of theoretical approaches. These analyses are derived from the "first-order shear deformation theory" (FSDT) and four-variable simplified HSDT [7]. These theoretical models offer a variety of viewpoints on the reaction of FGM plates to various loading. Many researchers have implemented equivalent shear-deformation theory in regards to the thermomechanical flexural study of FGM panels. The implementation of refined and unified theories in the study of FGM plates has been facilitated by recent advancements. For example, Zenkour and Radwan [8] adopted a refined exponential displacement approach to understand the hygrothermal effects on the FGM structures situated on Winkler foundations. In another literature, the thermo-mechanical behaviour of FGM sandwich structures is investigated by Hamidi et al. [9] incorporating stretching effects. These are enhanced theories for multi-layered and graded materials. Similarly, Houari et al. [10] suggested a HSDT in order to analyse the flexural behaviour of FGM structures under thermo-elastic loading showing its efficacy in capturing transverse stresses. To explore the bending performance of FGM structures with simultaneous thermo-mechanical loading on non-uniform elastic foundations, Bouderba and Madjid [11] implemented a simplified HSDT. Belkhodja et al. [12] conducted a comprehensive study of the flexural and stability behaviours of FGM sandwich panels employing a novel technique. In the same vein, Han et al. [13] examined the flexural characteristics of asymmetric sandwich FGM plates with the help of a refined HSDT. The numerical results were obtained using Navier's approach. Reddy and Chin [15] this study adopted FEM for evaluating the performance of FG cylinders and plates under coupled loading conditions. Emphasis was placed on the significance of material gradation on the thermal stress distribution. They found FGM plates reduce stress concentrations, enhancing structural stability. In addition to analytical methods, numerical approaches have been extensively employed to evaluate the performance of FGM plates. The flexural exploration of FGM plates under temperature effect using FEM and five-variable HSDT is still not well-studied. More research is needed in this area despite recent advancements. In order to address existing gaps and establish reliable methodologies for accurately anticipating the behaviour of FGMs under intricate thermo-mechanical conditions, additional research in this field is required.

The present work proposes a C^0 finite element formulation to implement a C^1 continuous HSDT for investigating linear bending behaviour of FGM structures considering temperature change. The proposed formulation addresses the challenges associated with achieving C^1 continuity in conventional finite element methods, enabling an accurate and computationally efficient analysis of FG plates. Linear thermal analysis, as conducted in this study, is of critical importance as it provides fundamental insights into the deflection and stress distribution in FG plates subjected to temperature variations. This understanding is vital for the design of thermally resilient structures in advanced applications, such as space technology, automotive, defence, and power generation fields, where precise thermal performance is essential for reliability and safety. The organisation of this paper is set as follows. Section 2 provides a comprehensive explanation of the penalty-based FE framework. In Section 3, the effectiveness of the proposed approach is evaluated by comparing its outcomes with findings reported in earlier open research studies. Following validation, a parametric analysis is performed to investigate how different geometric configurations and material gradations influence the thermal bending response of FGM plates. Finally, the key findings and implications of the study are summarised in Section 4.

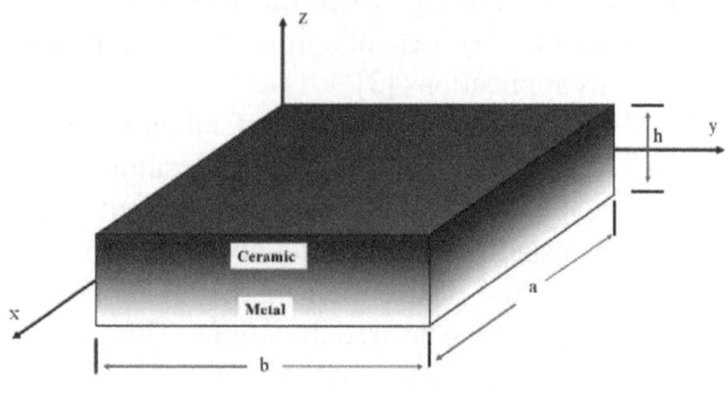

Figure 18.1 Representation of FGM flat panel
Source: Authors

Mathematical model

Mathematical model of NPHSDT

The mathematical model of non-polynomial-higher order shear deformation theory (NPHSDT) can be modelled and implemented by the means of mathematical modelling shown below.

A cartesian co-ordinate system is established at the corner of an FGM panel with length "a", width "b", and thickness "h", mentioned in Figure 18.1.

The effective elastic constants of the FGM structure are determined, following a "power-law distribution" considering the volume-fractions of material constituents. The material variation is mathematically defined by:

$$\bar{P}(z) = (\bar{P}_c - \bar{P}_m)\left(\frac{z}{h} + \frac{1}{2}\right)^n + \bar{P}_m \quad (1)$$

The kinematics of the FG plate is approximated by taking HSDT is represented as

$$U_1(x,y,z) = u_0 - z\frac{\partial w_0}{\partial x} + f(z)\theta_x(x,y) \quad (2)$$

$$V_1(x, y, z) = v_0 - z\frac{\partial w_0}{\partial y} + f(z)\theta_y(x, y) \quad (3)$$

$$W_1(x, y, z) = w_0 \quad (4)$$

In this context, the mid-plane stretching is represented by (u_0), (v_0), and (w_0), corresponding to the in-plane deformations along the x- and y-axes, together with the transverse displacement along the z-axis. Additionally, the rotations of the mid-plane are represented by θ_x and θ_y, which denote the angular displacements. Additionally, a transverse shear function $f_{(z)}$ is also introduced. This function captures the variation of shear deformation along the plate thickness and ensures an accurate representation of the non-linear distribution of shear stresses, thereby improving the fidelity of HSDT in modelling complex deformation patterns.

The strain vector ϵ derived from the displacement equation, involves the first derivative of the transverse displacement w. This necessitates C^1 continuity of the assumed displacement field across the entire plate domain to accurately capture the strain behaviour. However, conventional numerical techniques like the FEM predominantly employ Lagrangian elements, which inherently ensure only C^0 continuity of the field variables. To address this discrepancy, the formulation introduces independent degrees of freedom for the transverse displacement, thereby conforming to the C^0-continuity requirement of Lagrange FEM. This adjustment effectively simplifies the practical implementation of the method, as higher-order continuity requirements are challenging to achieve directly using standard finite elements. The introduction of these independent degrees of freedom, however, creates an artificial assumption that does not inherently satisfy the original governing equations. To mitigate the potential inaccuracies introduced by this assumption, a penalty stiffness matrix is incorporated into the formulation. This matrix acts as a corrective mechanism, ensuring that the enforced C^0 continuity does not compromise the physical fidelity of the solution while maintaining computational efficiency. This approach bridges the gap between the theoretical requirements of C^1 continuity and the practical implementation constraints of standard FEM.

Strain–displacement relation

The strain–displacement relation is a fundamental aspect of the analysis, connecting the displacement field to the induced strains in the structure. For the modified field variables, the strain vector $\{\epsilon\}$ is expressed using the derivatives of the displacement components. This relation captures the deformation behaviour of the FGM plate, ensuring that both in-plane and transverse deformations are accurately accounted for.

Mathematically, the strain vector $\{\epsilon\}$ is given by:

$$\{\epsilon_l\} = \begin{Bmatrix} \frac{\partial u}{\partial x} \\ \frac{\partial v}{\partial y} \\ \frac{\partial u}{\partial y} + \frac{\partial v}{\partial x} \\ \frac{\partial v}{\partial z} + \frac{\partial w}{\partial y} \\ \frac{\partial u}{\partial z} + \frac{\partial w}{\partial x} \end{Bmatrix}$$

The generalised thermoelastic stress-strain relation under plane stress conditions for the FGM plate is expressed by the Equation 5.

$$\begin{Bmatrix} \sigma_{xx} \\ \sigma_{yy} \\ \sigma_{xy} \\ \sigma_{yz} \\ \sigma_{xz} \end{Bmatrix} = \begin{bmatrix} Q_{11} & Q_{12} & 0 & 0 & 0 \\ Q_{12} & Q_{22} & 0 & 0 & 0 \\ 0 & 0 & Q_{66} & 0 & 0 \\ 0 & 0 & 0 & Q_{44} & 0 \\ 0 & 0 & 0 & 0 & Q_{55} \end{bmatrix} \left(\begin{Bmatrix} \epsilon_x \\ \epsilon_y \\ \gamma_{xy} \\ \gamma_{yz} \\ \gamma_{xz} \end{Bmatrix} - \Delta T \begin{Bmatrix} \alpha_x \\ \alpha_y \\ \alpha_{xy} \\ 0 \\ 0 \end{Bmatrix} \right) \quad (5)$$

In this formulation, $\{\bar{\sigma}\}$, $\{\bar{\epsilon}\}$ and $[\bar{Q}]$ represent the stress vector, strain-vector, and material stiffness matrix, respectively. $\{\alpha\}$ corresponds to the thermal expansion coefficients, while ΔT represents the temperature increment.

Within the FE framework, the plate structure is modelled using nine-noded isoparametric Lagrange element. This type of element improves computational precision by employing quadratic shape functions, which allow higher-order representations of both geometric configuration and field variables.

$$\{q\} = \sum_{i=1}^{9} N_i\{q_i\}$$

The equations governing static equilibrium are determined through the implementation of the "principle of virtual work", where the displacement vector $\{\delta q\}$ is eliminated. This leads to a set of algebraic relations that characterise the equilibrium state of the structure. The final expression is given as:

$$([K^*] + [K_\gamma])\{d\} - \{F_{th}\} = \{F_m\} \quad (6)$$

where:

- [K*]: Structural stiffness matrix, representing the "material and geometric properties" of the plate.
- [Kγ]: Stiffness matrix due to artificial constraints, introduced to satisfy C^0-continuity.
- $\{F_m\}$: Vector of applied mechanical loads, representing forces acting on the plate.
- $\{F_{th}\}$: Vector of thermal loads, accounting for the effects of temperature changes on the plate.
- {d}: Elemental displacement vector, containing the unknown nodal displacements to be determined.

Results and discussion

Multiple numerical problems are solved to check the efficacy of the TSDT and NPHSDT for several boundary condition configurations. Since the formulation employs a displacement-based approach, the system is constrained only in terms of displacement DOF in this model. The edge support conditions considered for the analysis are taken as:

- Simply supported condition (*SSSS*1)
  For Edge parallel to the x-axis:
  $u_o = w_o = \phi_x = \theta_x = 0$
  For Edge parallel to the y-axis:
  $v_o = w_o = \phi_y = \theta_y = 0$
- Clamped condition (*CCCC*)
  All sides:
  $u_o = v_0 = w_o = \phi_x = \phi_y = \theta_x = \theta_y = 0$

The different material properties of functionally graded materials are listed in Table 18.1.

Table 18.1 Properties of the FGM components.

Material	Properties		
	E	ρ	ν
Zirconia (ZrO_2)	151×10^9	3000	0.30
Alumina (Al_2O_3)	380×10^9	3800	0.30
Silicon carbide (SiC)	427×10^9	3210	0.17
Aluminium (Al)	70×10^9	2707	0.30

Source: Authors

Table 18.2 Validation of central deflection response of FGM plates under linear thermal loading.

n	*(a/h)*	Xiaohui Zhen [14]	FEM-TSDT (present)
0	5	3.227	3.292
	10	13.11	13.172
0.5	5	5.286	5.394
	10	21.48	21.586
1	5	5.842	5.990
	10	23.86	24.01

Source: Authors

Linear thermal bending

The effect of the volume fraction exponent (n) and the thickness-to-span ratio (a/h) on the non-dimensional central deflection ω of a SSSS1-FGM plate is investigated under thermal loading, expressed as $T(x,y,z) = [T1(2z/h)sin(\pi x/a)\ sin(\pi y/b)]$ The plate consists of aluminum (Al) and alumina (Al/Al_2O_3), characterised by a thermal expansion coefficient $\alpha c = 7.4 \times 10 - 6/k$. The results, presented in Tables 18.2 and 3, illustrate how these parameters influence central deflection, offering a clear reference for evaluating the thermo-mechanical performance of FGM plates under defined thermal environments with varying geometry and material gradation.

Validation of the proposed model against benchmark data obtained from HSDT, as shown in Table 18.2, confirms its reliability. A detailed parametric study further examines the influence of thickness ratios and volume fraction indices, with results compiled in Table 18.3. The findings highlight that deflection increases with larger values of n and higher (a/h) ratios. These outcomes establish a solid basis for future analyses of the thermal bending response of FGM plates under diverse thickness conditions.

The detailed assessment of the outcomes highlights that the central displacement is minimal for the plate composed entirely of alumina and maximal for the plate made entirely of aluminium. This behaviour is attributed to the significant difference in material stiffness; alumina possesses a much higher stiffness compared to aluminium. Consequently, alumina-dominant compositions resist deformation more effectively under thermal loads, while aluminium-dominant compositions exhibit greater flexibility and larger displacements. The findings demonstrate the essential contribution of material composition in determining the "thermo-mechanical" behaviour of FGM plates.

Effect of temperature ratio (α_m/α_c)

Figure 18.2 describes the effect of temperature ratio (α_m/α_c) on normalised deflection of FGM plate (a/h=10) under thermal loading. It is observed that with gradual increase in temperature ratio (α_m/α_c) the normalised deflection increases linearly for all boundary conditions.

Table 18.3 Linear thermal bending analysis of a square FGM (SSSS1-type) under thermal loading.

n	*(a/h)*	**FEM-TSDT (present)**	**FEM-NPHSDT (present)**
0	4	2.10751	2.10751
	10	13.172	13.172
	25	82.3247	82.3247
0.5	4	3.45096	3.45084
	10	21.5859	21.5858
	25	134.93	134.93
1	4	3.82851	3.82745
	10	24.0101	24.0089
	25	150.147	150.146
1.5	4	4.02882	4.02672
	10	25.3133	25.311
	25	158.345	158.342
2	4	4.17053	4.16737
	10	26.2441	26.2407
	25	164.209	164.205

Source: Authors

Influences of edge support conditions

Table 18.4 illustrates how different edge conditions affect the normalised deflection of a square FGM (Al/Al_2O_3) plate exposed to various temperatures T. The findings reveal that an increase in the volume fraction exponent (n) leads to a corresponding rise in normalised deflection. Among the cases considered, the maximum deflection occurs under the SSSS3 boundary condition, while the minimum deflection is observed for the CCCC1 condition.

Conclusion

This research introduces a penalty-based C^0 FE-formulation designed for C^1 continuous HSDT to analyse functionally graded plates exposed to sinusoidal thermal loading. The material characteristics of the "FGM plates" are investigated through "rule

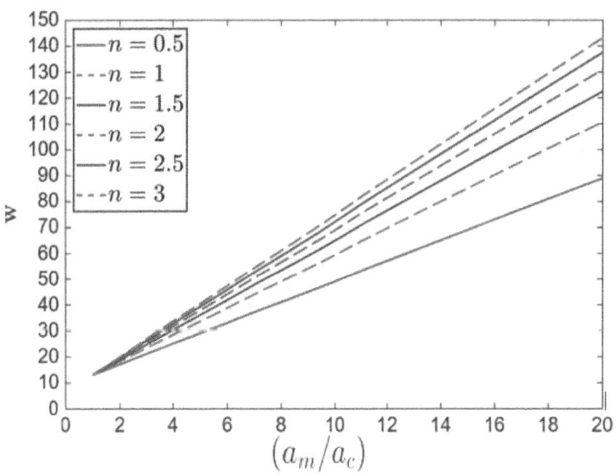

Figure 18.2 The effect of temperature ratio (α_m/α_c) on normalised deflection $\hat{w}$

Source: Authors

Table 18.4 Effect of edge conditions on normalised deflection of FGM (Al/Al_2O_3) plate (a/h = 10) under thermal loading.

	***n* = 0**	***n* = 0.5**	***n* = 1**	***n* = 1.5**	***n* = 2**
SSSS1	13.172	21.5859	24.0101	25.3133	26.2441
SSSS2	13.172	20.7607	22.6385	23.7985	24.7471
SSSS3	14.2115	22.1814	24.0013	25.1548	26.1603
SSSS4	13.172	20.6576	22.3455	23.36	24.2187
CCCC1	4.78701	7.82806	8.69474	9.16342	9.50231

Source: Authors

of mixtures", providing a reliable representation of material gradation. A nine-noded Lagrange element is adopted for mesh construction, ensuring both flexibility and accuracy in capturing deformation characteristics. The model's precision is thoroughly verified by comparing with established numerical benchmarks, demonstrating excellent consistency and confirming its reliability for FG plate analysis within the HSDT framework.

The developed FEM-HSDT methodology shows strong potential for future studies and serves as an effective validation tool for alternative thermal analysis approaches of FG plates. Furthermore, its adaptable nature allows integration with advanced micromechanical models, including Mori-Tanaka and self-consistent schemes, thereby improving the accuracy of material modelling for complex FG structures. This adaptability positions the framework as a versatile and robust tool for exploring FG plate behaviour under thermal environments, supporting its application in advanced simulations and real-world engineering practices.

References

[1] Koizumi, M., Niino, M. Overview of fgm research in Japan. *Mrs Bulletin*, 1995;20(1):19–21.

[2] Koizumi, M. FGM activities in Japan. *Compos. Part B Engg.*, 1997;28(1–2):1–4.

[3] Saleh, B., Jiang, J., Fathi, R., Al-hababi, T., Xu, Q., Wang, L., Song, D., Ma, A. 30 years of functionally graded materials: An overview of manufacturing methods, applications and future challenges. *Compos. Part B Engg.*, 2020;201:108376.

[4] Reddy, J. N. Mechanics of laminated composite plates and shells: theory and analysis. CRC press, 2003.

[5] Gupta, A., Talha, M. Recent development in modelling and analysis of functionally graded materials and structures. *Prog. Aerospace Sci.*, 2015;79:1–14.

[6] Kharghani, N., Guedes Soares, C. Application of layerwise hsdt and fracture analysis in the ultimate strength of composite plates with delamination in bending. *Internat. J. Solids Struct.*, 2022;234:111263.

[7] Daouadji, T. H., Benferhat, R., et al. Bending analysis of an imperfect fgm plates under hygro-thermo-mechanical loading with analytical validation. *Adv. Mater. Res.*, 2016;5(1):035.

[8] Zenkour, A. M., Radwan, A. F. Bending and buckling analysis of fgm plates resting on elastic foundations in hygrothermal environment. *Arch. Civil Mech. Engg.*, 2020;20(4):1–23.

[9] Hamidi, A., et al. A sinusoidal plate theory with 5-unknowns and stretching effect for thermomechanical bending of functionally graded sandwich plates. *Steel Compos. Struct.*, 2015;18(1):235–253.

[10] Houari, M. S. A., Tounsi, A., Anwar Bég, O. Thermoelastic bending analysis of functionally graded sandwich plates using a new higher order shear and normal deformation theory. *Internat. J. Mech. Sci.*, 2013;76:102–111.

[11] Bouderba, B., Houari M. S. A., Tounsi, A. Thermomechanical bending response of FGM thick plates resting on Winkler–Pasternak elastic foundations. *Steel Compos. Struct.*, 2013;14(1):85–104.

[12] Benbakhti, A., et al. A new five unknown quasi-3D type HSDT for thermomechanical bending analysis of FGM sandwich plates. *Steel Compos. Struct.*, 2016;22(5):975–999.

[13] Han, M., et al. Bending analysis of asymmetric functionally graded material sandwich plates in thermal environments. *Materials*, 2023;16(13):4682.

[14] Xiaohui, R., Zhen, W. A refined sinusoidal model for functionally graded plates subjected to thermomechanical loading. *J. Compos. Mater.*, 2019;53(14):1883–1896.

[15] Reddy, J. N., Chin, C. D. Thermomechanical analysis of functionally graded cylinders and plates. *J. Thermal Stress.*, 1998;21(6):593–626.

19 Stochastic dynamic response of tunnels subjected to asynchronous ground motion using domain reduction method

Bhavesh Banjare[a] *and G. R. Dodagoudar*[b]

Department of Civil Engineering, Indian Institute of Technology Madras, Chennai, Tamil Nadu, India

Abstract

Seismic response analysis of soil-fault-tunnel (SFT) systems involves numerical simulations that consider factors such as seismic sources, wave propagation paths, local site effects, and the spatial variation of soil properties. Past earthquakes, especially the 2008 Wenchuan (China) and 2011 Tohoku (Japan) events, have shown the vulnerability of tunnels. Traditional fault rupture models require simulating large areas, which is computationally intensive. This paper employs an efficient modelling technique known as the domain reduction method (DRM) to analyse the seismic response of SFT systems. The DRM works in two steps: first, calculating the nodal effective forces from fault rupture (free-field analysis), and then, limiting the analysis to a smaller domain that includes the tunnel, with an artificial boundary layer that uses the nodal forces from the first step. Monte Carlo simulations were performed to understand the overall response of the tunnel, considering the shear wave velocity (V_s) as a spatially varying random variable in the model. The study examined a circular tunnel cross-section subjected to inclined earthquake faults using a three-layer DRM approach, with 210 realisations for stochastic dynamic analysis of the SFT system. Numerical simulations were performed using Multi-hazard Analysis for STOchastic time-DOmaiN phenomena (MASTODON), based on the multi-physics object-oriented simulation environment (MOOSE) framework. The results show how fault inclination affects the tunnel, the generation of surface waves, and local site effects such as frequency shifts and the amplification or attenuation of seismic waves caused by the tunnel. These findings are useful for designing tunnels more efficiently and reliably, as well as developing engineered materials to reduce seismic wave impacts on critical infrastructure.

Keywords: Stochastic response, tunnel system, fault-rupture mechanism, asynchronous ground motion, domain reduction method

Introduction

Tunnels are critical components of underground infrastructure and are often located in areas prone to seismic activity. Earthquakes pose a significant risk to these underground structures due to the dynamic interaction between the surrounding soil and the tunnel lining. The seismic response of tunnels is typically influenced by various factors, including the magnitude of the earthquake, the soil properties, and the characteristics of the ground motion [1–5]. In particular, tunnels can experience asynchronous ground motion [6], which occurs when seismic waves reach different parts of the tunnel at different times. This can result in complex dynamic responses, making it essential to account for these effects in seismic analysis and design.

The conventional methods of seismic analysis often assume uniform ground motion across the entire structure. However, asynchronous ground motion, which arises from spatial variations in seismic wave propagation, can lead to significantly different tunnel responses. This complexity requires advanced modelling techniques to accurately capture the stochastic dynamic behaviour of tunnels subjected to such conditions. One such approach is the domain reduction method (DRM), a powerful numerical technique that allows for efficient simulation of seismic wave propagation in large-scale models, such as those involving tunnels and surrounding soil. By employing DRM [7–11], it is possible to reduce the computational domain to a manageable size while accurately representing the boundary conditions and wave propagation effects.

This study examines the stochastic dynamic response of tunnels subjected to asynchronous ground motion using the DRM. The goal is to

[a]ce23d035@smail.iitm.ac.in, [b]goudar@iitm.ac.in

DOI: 10.1201/9781003770442-19

develop a thorough understanding of how spatially varying seismic waves affect tunnel behaviour, with an emphasis on the probabilistic nature of earthquake loading. The research aims to fill the knowledge gap concerning the dynamic interaction between tunnels and asynchronous seismic waves, which is vital for the safe design of underground infrastructure in earthquake-prone areas. Through numerical simulations, the study seeks to provide insights into the impact of asynchronous ground motion on tunnel performance, helping to create more resilient tunnel designs.

Methodology

Domain reduction method

Domain reduction method was developed by Bielak et al. [12, 13]. It is a two-step computational approach used to analyse the effects of seismic activity on localised structural systems, such as tunnels. In the first step, known as the free-field analysis (Ω^0), nodal effective forces are calculated to represent the seismic effects caused by the fault-rupture mechanism. This is performed in a large, semi-infinite domain that includes the earthquake source, and geological features (Ω^+). The free-field analysis captures the natural wave propagation in a homogeneous medium without incorporating the structure itself (as shown in Figure 19.1a). These nodal forces, which encapsulate the earthquake's effects, are then transferred to the second step. In the second step, the focus shifts to a smaller, localised computational domain that contains the region of interest, such as a tunnel, and an artificial boundary layer (presented in Figure 19.1b). This reduced domain ensures a detailed analysis of the structural response to seismic forces while maintaining computational efficiency. The nodal forces from the first step are applied to the artificial boundary layer, allowing the seismic waves to propagate through the smaller domain accurately, without introducing artificial reflections. Figure 19.1c provides an overarching schematic, showing the semi-infinite domain with the earthquake fault, geological features, and the tunnel (Ω). DRM effectively bridges large-scale seismic wave modelling with localised structural analysis, enabling precise assessments of structural behaviour under earthquake-induced forces.

Governing Equation of the problem

$$M_G\ddot{u} + C_G\dot{u} + K_G u = F_{ext} \tag{1}$$

where M_G, C_G, K_G and F_{ext} denote global mass, damping, stiffness, and external force terms, respectively. The regions Ω and Ω^+ are discretised using the finite element method (FEM) as follows:

$$\begin{bmatrix} M_{ii}^{\Omega} & M_{ib}^{\Omega} \\ M_{bi}^{\Omega} & M_{bb}^{\Omega} \end{bmatrix} \begin{bmatrix} \ddot{u}_i \\ \ddot{u}_b \end{bmatrix} + \begin{bmatrix} C_{ii}^{\Omega} & C_{ib}^{\Omega} \\ C_{bi}^{\Omega} & C_{bb}^{\Omega} \end{bmatrix} \begin{bmatrix} \dot{u}_i \\ \dot{u}_b \end{bmatrix} + \begin{bmatrix} K_{ii}^{\Omega} & K_{ib}^{\Omega} \\ K_{bi}^{\Omega} & K_{bb}^{\Omega} \end{bmatrix} \begin{bmatrix} u_i \\ u_b \end{bmatrix} = \begin{bmatrix} 0 \\ F_b \end{bmatrix} \quad in\ \Omega \tag{2}$$

$$\begin{bmatrix} M_{bb}^{\Omega^+} & M_{be}^{\Omega^+} \\ M_{eb}^{\Omega^+} & M_{ee}^{\Omega^+} \end{bmatrix} \begin{bmatrix} \ddot{u}_b \\ \ddot{u}_e \end{bmatrix} + \begin{bmatrix} C_{bb}^{\Omega^+} & C_{be}^{\Omega^+} \\ C_{eb}^{\Omega^+} & C_{ee}^{\Omega^+} \end{bmatrix} \begin{bmatrix} \dot{u}_b \\ \dot{u}_e \end{bmatrix} + \begin{bmatrix} K_{bb}^{\Omega^+} & K_{be}^{\Omega^+} \\ K_{eb}^{\Omega^+} & K_{ee}^{\Omega^+} \end{bmatrix} \begin{bmatrix} u_b \\ u_e \end{bmatrix} = \begin{bmatrix} -F_b \\ F_e \end{bmatrix} \quad in\ \Omega^+ \tag{3}$$

where i, b, and e indices resemble the interior, boundary, and exterior nodes, superscripts Ω and Ω^+ define the region to which the block matrix corresponds. The global matrix equation for the entire domain is obtained by combining Equations (2) and (3):

The global stiffness matrix of the above problem (Figure 19.1b) is represented by Equation (4)

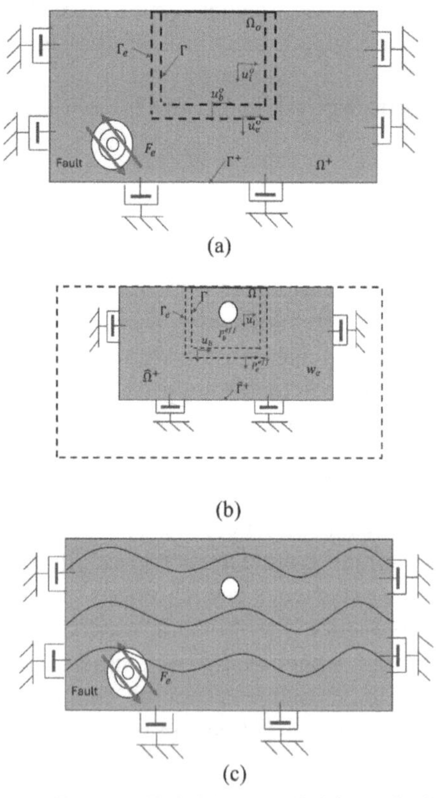

Figure 19.1 (a) Free-field analysis, (b) Reduced domain analysis using the DRM, and (c) Semi-infinite domain with earthquake fault and structure
Source: Authors

$$\begin{bmatrix} M_{ii}^{\Omega} & M_{ib}^{\Omega} & 0 \\ M_{bi}^{\Omega} & M_{bb}^{\Omega} + M_{bb}^{\Omega^+} & M_{be}^{\Omega^+} \\ 0 & M_{eb}^{\Omega^+} & M_{ee}^{\Omega^+} \end{bmatrix} \begin{bmatrix} \ddot{u}_i \\ \ddot{u}_b \\ \ddot{w}_e \end{bmatrix} + \begin{bmatrix} C_{ii}^{\Omega} & C_{ib}^{\Omega} & 0 \\ C_{bi}^{\Omega} & C_{bb}^{\Omega} + C_{bb}^{\Omega^+} & C_{be}^{\Omega^+} \\ 0 & C_{eb}^{\Omega^+} & C_{ee}^{\Omega^+} \end{bmatrix} \begin{bmatrix} \dot{u}_i \\ \dot{u}_b \\ \dot{w}_e \end{bmatrix} + \begin{bmatrix} K_{ii}^{\Omega} & K_{ib}^{\Omega} & 0 \\ K_{bi}^{\Omega} & K_{bb}^{\Omega} + K_{bb}^{\Omega^+} & K_{be}^{\Omega^+} \\ 0 & K_{eb}^{\Omega^+} & K_{ee}^{\Omega^+} \end{bmatrix} \begin{bmatrix} u_i \\ u_b \\ w_e \end{bmatrix} = \begin{bmatrix} 0 \\ -M_{be}^{\Omega^+}\ddot{u}_e^o - C_{be}^{\Omega^+}\dot{u}_e^o - K_{be}^{\Omega^+}u_e^o \\ M_{eb}^{\Omega^+}\ddot{u}_b^o + C_{eb}^{\Omega^+}\dot{u}_b^o + K_{eb}^{\Omega^+}u_b^o \end{bmatrix} \tag{4}$$

$$F^{eff} = \begin{bmatrix} F_i^{eff} \\ F_b^{eff} \\ F_e^{eff} \end{bmatrix} = \begin{bmatrix} 0 \\ -M_{be}^{\Omega^+}\ddot{u}_e^o - C_{be}^{\Omega^+}\dot{u}_e^o - K_{be}^{\Omega^+}u_e^o \\ M_{eb}^{\Omega^+}\ddot{u}_b^o + C_{eb}^{\Omega^+}\dot{u}_b^o + K_{eb}^{\Omega^+}u_b^o \end{bmatrix} \tag{5}$$

The sub-matrices M_{be}, K_{be}, C_{be}, M_{eb}, C_{eb}, and K_{eb} vanishes at all points in the sub-region Ω^+, except for the boundary nodes located within Γ region. The effective nodal forces of the DRM layer in the auxiliary problem are given in the form of Equation (5).

Numerical modelling and verification of DRM

In this study, numerical simulations are conducted using an open-source program, MASTODON [14], built on the MOOSE framework. The DRM follows a two-step approach. In the first step (free-field analysis), a model with dimensions 500 m × 700 m × 1000 m (Figure 19.2a) is used. This step calculates the DRM layer forces, which involve inner and outer layers. The process alternates between fixing nodal displacements of the inner layer to compute outer layer forces, and vice versa.

In the second step (Step II), a reduced domain model of dimensions 150 m × 300 m × 1000 m is employed, including the structure (tunnel) and the calculated DRM layer forces (Figure 19.2b). The fault is modelled with a 45° inclination, a planar fault dimension of 425 m × 300 m, and a shear modulus of 1.996 GPa. The material in the model is assumed to be a homogeneous and linearly elastic material model. The Newmark-β method 15, 16], known for its unconditional stability, is used in an implicit time integration scheme with parameters β = 0.25 and γ = 0.5. A ten-nodes-per-wavelength scheme is adopted for optimal mesh size [14]. To minimise wave reflections and scattering from the tunnel and free surface, viscous absorbing boundaries [17] are applied to all sides except the top boundary. This ensures effective dissipation of outgoing waves and maintains the accuracy of the seismic response simulation.

Figure 19.3 presents the acceleration time histories for the Y and Z components at the coordinates [X = 500 m, Y = 700 m, Z = 500 m]. The results demonstrate a strong agreement between the reduced domain models (Figure 19.2b), incorporating effective nodal forces through the DRM layer, and the comprehensive model encompassing the full fault-tunnel system (Figure 19.2c). This consistency verifies the accuracy and reliability of the DRM approach. The findings underscore the method's efficacy and robustness in capturing key dynamic responses while significantly reducing computational effort. Based on these results, all subsequent numerical simulations in this study will employ the DRM methodology.

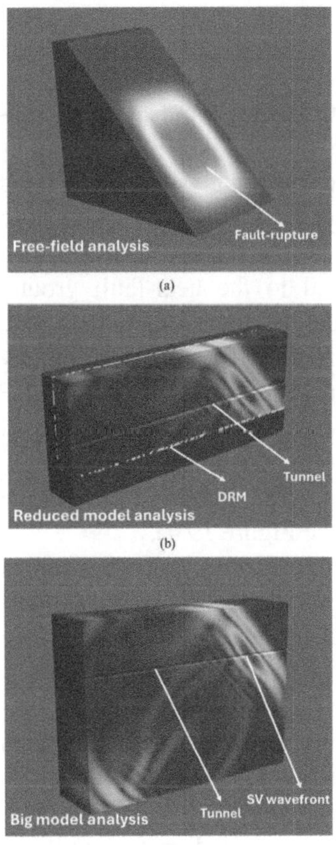

Figure 19.2 (a) Free-field Analysis, (b) Reduced model analysis using DRM, and (c) Big model analysis considering earthquake fault, and structure
Source: Authors

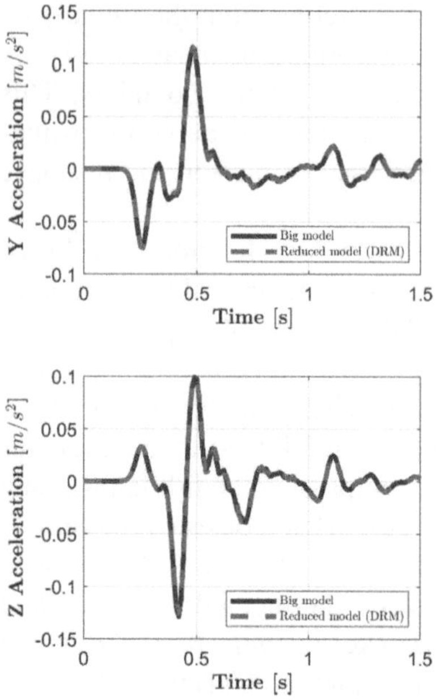

Figure 19.3 Acceleration comparison at [X = 500 m, Y = 700 m and Z = 500 m]; Y component, and Z component acceleration, respectively
Source: Authors

Algorithm 1 Stochastic Modeling of 3D Tunnel Response under Seismic Input with uncertainty in soil Properties

Input: Shear wave velocity bounds Vs_{min} and Vs_{max} for each soil type based on NEHRP 2020 Provision(soft, medium, hard); Number of realizations $N = 210$; Fault inclination angle $\theta = 45°$
Output: Statistical distribution of tunnel dynamic responses
for each soil medium $\mathcal{R}$ in {soft, medium, hard} **do**
  Define uniform distribution: $Vs \sim \mathcal{U}(Vs^{(\mathcal{R})}_{min}, Vs^{(\mathcal{R})}_{max})$
  for $i = 1$ to N **do**
    Sample a realization: $Vs^{(i)} \leftarrow$ Random sample from $\mathcal{U}(Vs^{(\mathcal{R})}_{min}, Vs^{(\mathcal{R})}_{max})$
    Assign $Vs^{(i)}$ to the numerical model of the tunnel domain
    Generate incident wavefield corresponding to $\theta = 45°$ seismic fault
    Compute the dynamic response of the circular tunnel subjected to the sampled wavefield and rock properties
    Store response quantities (e.g., displacement, stress, acceleration) for realization i
  end for
  Aggregate all realizations $\{1, \ldots, N\}$ for rock medium $\mathcal{R}$
  Compute statistical metrics (mean, standard deviation, percentiles) of tunnel response
end for
End

Figure 19.4 Flowchart of stochastic modeling of 3D tunnel response under seismic input with uncertainty in soil properties
Source: Authors

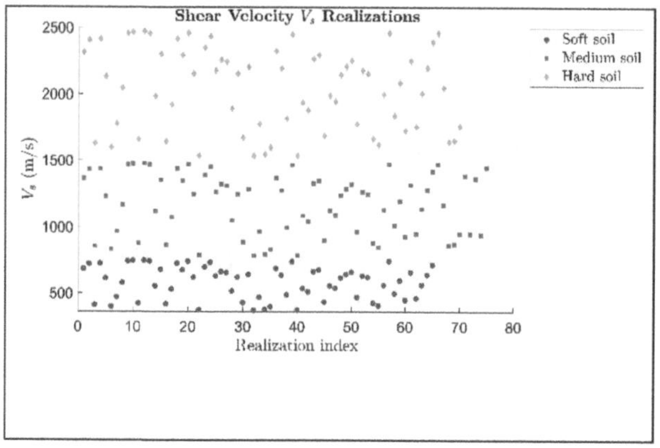

Figure 19.5 Shear velocity realisations
Source: Authors

Stochastic modelling

Uncertainties in both the seismic input and soil properties can influence the site response results, requiring the inclusion of stochastic analysis to understand the tunnel response. Here, we employ Monte Carlo simulation to perform the stochastic analysis of a circular tunnel subjected to the near-fault ground motion. Three soil media, namely, soft, medium, and hard, are chosen, wherein the shear wave velocity (V_s) is considered as uncertain. Based on NEHRP 2020 [18] provisions, N realisations of $V_s \sim U[V_{s,min}, V_{s,max}]$ are randomly sampled, where $V_{s,min}$ and $V_{s,max}$ are the minimum and maximum values of V_s for the respective soil media (see Figure 19.4).

Each realisation represents a unique variation in the soil properties, capturing the inherent randomness and variability that influence the seismic wave propagation and response of the tunnel. By incorporating these scenarios, the analysis provides a comprehensive understanding of how the uncertainties in V_s affect the dynamic behaviour of the tunnel [5, 19–21]. Specifically, 210 realisations of the shear wave velocity (V_s) were generated using a uniform sampling technique, as presented in Figure 19.5.

The stochastic framework ensures that the results reflect a broad range of possible outcomes rather than a single deterministic scenario, offering more robust insights into the tunnel's performance under varying geological conditions. This methodology is particularly valuable for assessing risk and designing tunnels with enhanced resilience to seismic activity. The uniform sampling approach ensures that all potential values of V_s within the specified range are equally represented, enabling an unbiased and systematic evaluation of the impact of material variability.

Results and discussion

The response spectra plots for the Y- and Z-directions across three types of geological media: soft, medium, and hard soils are shown in Figures 19.6–19.9 for circular tunnel section at 0° and 90° locations,

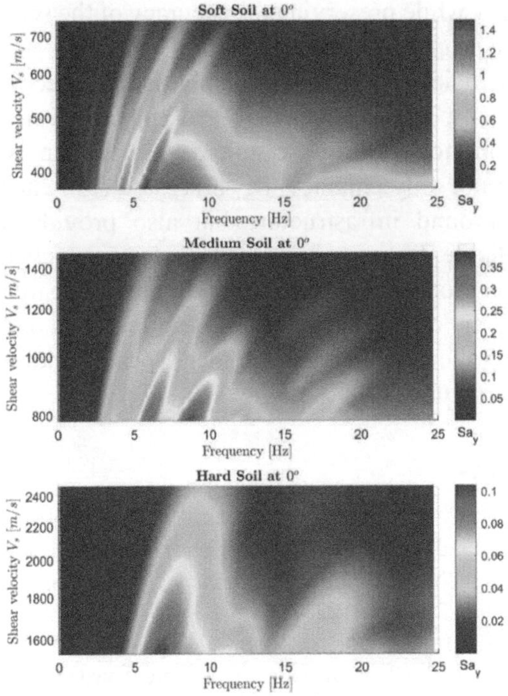

Figure 19.6 Response spectra for circular tunnel section for different soil media for Y-direction at 0° location
Source: Authors

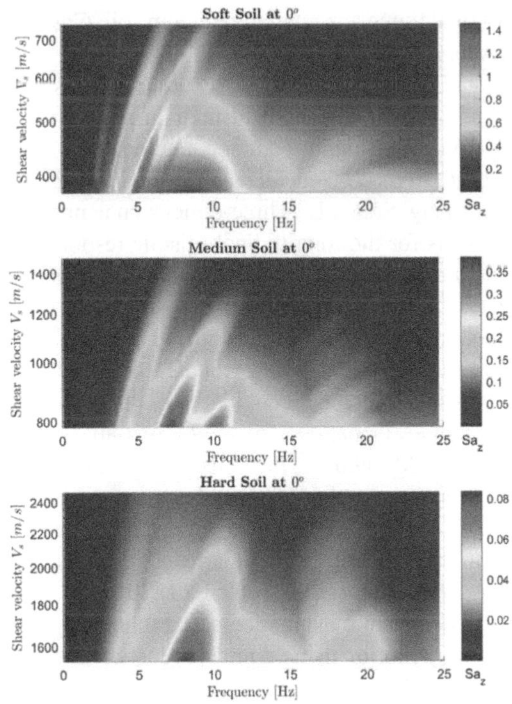

Figure 19.7 Response spectra for circular tunnel section for different soil media for Z-direction at 0° location
Source: Authors

respectively. These plots demonstrate how the dynamic response changes with variations in shear wave velocity (V_s), emphasising the impact of material stiffness on seismic behaviour. As V_s increases, multimodal dynamics become clearer. For soft soil, the peak broadband frequency decreases with rising V_s, indicating a shift of energy toward lower frequencies. Although higher peaks diminish, lower frequency peaks are preserved but shift noticeably. Similar patterns are seen in medium and hard soils, where the frequency peaks gradually decrease as the soil becomes stiffer. This trend reflects how increasing stiffness reduces the number of dominant vibrational modes.

At higher frequencies, energy dissipation is pronounced, leading to the disappearance of peaks in these regions. The overall response spectra exhibit a shift of broadband frequencies toward the lower end of the spectrum. Over time, the peak frequencies tend to merge, emphasising the concentration of seismic energy in a narrower range of lower frequencies.

This pattern highlights the interplay between material properties and seismic wave behaviour. Softer soils support broader frequency ranges with distinct multimodal peaks, while stiffer soils have fewer, more concentrated peaks at lower frequencies. These

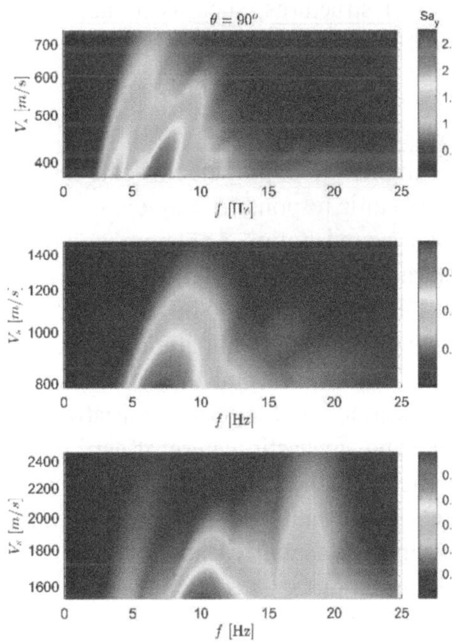

Figure 19.8 Response spectra for circular tunnel section for different soil media for Y-direction at 90° location
Source: Authors

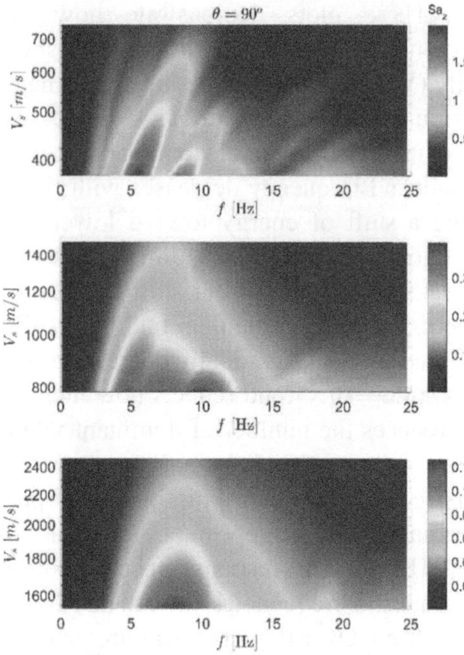

Figure 19.9 Response spectra for circular tunnel section for different soil media for Z-direction at 90° location
Source: Authors

findings underline the importance of material stiffness in shaping the seismic response of geological media and provide valuable insights into the design and assessment of structures, such as tunnels, subjected to seismic ground motions.

Conclusion

This study highlights the importance of considering the stochastic dynamic response of tunnels subjected to asynchronous ground motion, a phenomenon often overlooked in traditional seismic design approaches. The use of the DRM has proven to be a highly effective technique for accurately simulating the interaction between seismic waves and tunnel structures, particularly in accounting for spatial variations in ground motion. The stochastic nature of earthquake loading, combined with the spatial variability of seismic waves, can significantly influence the dynamic behaviour of tunnels, potentially leading to amplified stresses and deformations.

Through numerical simulations, this research has demonstrated that ignoring asynchronous ground motion can lead to underestimation of seismic demands on tunnels. The DRM approach enables a more precise analysis, reducing computational complexity while preserving the accuracy of the seismic wave propagation effects.

In conclusion, the findings emphasise the necessity of incorporating stochastic and spatially varying ground motion in the seismic design of tunnels. This will not only enhance the safety and resilience of underground infrastructure but also provide a more reliable framework for engineering practice in regions prone to seismic activity. Future research should focus on further refining these methods and expanding their application to other types of underground structures.

References

[1] Yu, H., Chen, J., Bobet, A., Yuan, Y. Damage observation and assessment of the Longxi tunnel during the Wenchuan earthquake. *Tunnel. Undergr. Space Technol.*, 2016;54:102–116.

[2] Lai, J., He, S., Qiu, J., Chen, J., Wang, L., Wang, K., Wang, J. Characteristics of seismic disasters and aseismic measures of tunnels in Wenchuan earthquake. *Environ. Earth Sci.*, 2017;76:1–19.

[3] Shen, Y., Gao, B., Yang, X., Tao, S. Seismic damage mechanism and dynamic deformation characteristic analysis of mountain tunnel after Wenchuan earthquake. *Engg. Geol.*, 2014;180:85–98.

[4] Wu, G., Zhao, H., Zhao, M., Zhu, Z. Stochastic analysis of dual tunnels in spatially random soil. *Comput. Geotech.*, 2021;129:103861.

[5] Banjare, B., Avatar, G. R. K. C.Stochastic assessment of 3-D tunnels in near-fault ground motion using modified domain reduction method. *Int. J. Numer. Anal. Methods Geomech.*, 2024;48:2259–2284.

[6] Li, P., Xiang Song, E. Three-dimensional numerical analysis for the longitudinal seismic response of tunnels under an asynchronous wave input. *Comput. Geotech.*, 2015;63:229–243.

[7] Kontoe, S., Zdravkovic, L., Potts, D. The domain reduction method for dynamic coupled consolidation problems in geotechnical engineering. *Int. J. Numer. Anal. Methods Geomech.*, 2008;32:659–680.

[8] Kontoe, S., Zdravkovic, L., Potts, D. M. An assessment of the domain reduction method as an advanced boundary condition and some pitfalls in the use of conventional absorbing boundaries. *Int. J. Numer. Anal. Methods Geomech.*, 2009;33(3):309–330.

[9] Banjare, B. Seismic evaluation of 3D tunnels in near fault ground motion using domain reduction method. *Elec. Theses Disser. Ind. Instit. Sci.*, 2023.

[10] Banjare, B., Dodagoudar, G. R. Effect of combined P- and S-waves on seismic response of 3D tunnels using domain reduction method. In: *Found. Dynam.*, 2025:235–248.

[11] Banjare, B., Avatar, G. R. K. C., Dodagoudar, G. R. Three-dimensional seismic response analysis of tunnels in layered media using modified domain reduction method. *Int. J. Numer. Anal. Methods Geomech.*, 2025;49(13):3027–3047.

[12] Bielak, J., Loukakis, K., Hisada, Y., Yoshimura, C. Domain reduction method for three-dimensional earthquake modelling in localized regions. *Part I Theory Bull. Seismol. Soc. Am.*, 2003;93:817–824.

[13] Yoshimura, C., Bielak, J., Hisada, Y., Fernández, A. Domain reduction method for three-dimensional earthquake modelling in localized regions. *Part II Verif. Appl. Bull. Seismol. Soc. Am.*, 2003;93:825–841.

[14] Coleman, J., Slaughter, A., Veeraraghavan, S., Bolisetti, C., Numanoglu, O. A., Spears, R., Hoffman, W., Hurt, E. *MASTODON Theory Manual*, 2017.

[15] Newmark, N. M. A method of computation for structural dynamics. *J. Engg. Mech. Div.*, 1959;85:67–94.

[16] Ahrabi, B. R., Mavriplis, D. J. An implicit block ILU smoother for preconditioning of Newton–Krylov solvers with application in high-order stabilized finite-element methods. *Comput. Methods Appl. Mech. Engg.*, 2020;358:112637.

[17] Lysmer, J., Asce, M., Kuhlemeyer, R. L., Asce, A. M. Finite dynamic model for infinite media. *J. Engg. Mech. Div.*, 1969;95:859–877.

[18] Fema. FEMA P-2082-1: NEHRP Recommended Seismic Provisions for New Buildings and Other Structures, n.d.

[19] Wu, G., Zhao, H., Zhao, M., Zhu, Z. Stochastic analysis of dual tunnels in spatially random soil. *Comput. Geotech.*, 2021;129:103861.

[20] Pang, R., Chen, K., Fan, Q., Xu, B. Stochastic ground motion simulation and seismic damage performance assessment of a 3-D subway station structure based on stochastic dynamic and probabilistic analysis. *Tunnel. Undergr. Space Technol.*, 2022;126:104568.

[21] Wang, F., Wang, H., Yang, H., Feng, Y., Jeremić, B. A modular methodology for time-domain stochastic seismic wave propagation. *Comput. Geotech.*, 2021;139:104409.

20 Impact of microgrooves on the surface of SS304 steel on wettability and corrosion

Abdul Rahman Afzal[1,a], *A. Vivek Anand*[2,b], *V. Hariprasad*[3,c], *K. Loganathan*[4,d], *B. Deepanraj*[5,e] *and Abdul Aziz Afzal*[5,f]

[1]Department of Industrial Engineering, University of Business and Technology (UBT) Dahban, North Jeddah, Saudi Arabia

[2]Department of Aeronautical Engineering, MLR Institute of Technology, Hyderabad, Telangana, India

[3]Department of Aerospace Engineering, IIAEM, Jain (Deemed-to-be) University, Bengaluru, Karnataka, India

[4]Department of Mathematics and Statistics, Manipal University Jaipur, India

[5]Department of Mechanical Engineering, Prince Mohammad Bin Fahd University, Al-Khobar, Saudi Arabia

Abstract

A novel biomimetic surface was proposed to minimise wetting and eradicate corrosion in SS304 steel. The microgrooves were formed on SS304 steel using a wire-cut electrical discharge machining process. The wetting and corrosion study were conducted using the standard sessile drop method and potentiodynamic polarisation experiment in saline water (pH = 8 and duration = 7 days). The contact angle, corrosion current density and corrosion rate were used to quantify the effect of wetting and corrosion in a micro-patterned surface. The wetting and corrosion parameters were compared with the variation in size of the microgrooves. The micro-patterns decrease the interface between the solid and liquid medium, isolating the substrate from the corrosive medium. The microgroove surface shows excellent corrosion resistance and minimum wetting when compared to the plain surface.

Keywords: Hydrophobicity, microgrooves, roughness, corrosion and contact angle

Introduction

Nowadays, researchers are focusing on increasing the performance and service life cycle of the flight vehicles, road vehicles and steel pipelines because of the necessity associated with present-day technologies. In humid environments, component failures in vehicles primarily result from corrosion, which threatens the lifespan of components [1, 2]. Coatings are one of the possible ways to eradicate corrosion, but their lifetime and durability are limited; after a certain time period, they will peel off. Nature itself provides a possible solution, namely the Lotus effect. The surface has a lotus effect, resulting in reduced wetting, which in turn prevents the surface from undergoing corrosion easily. Young's equation, which describes the balance at the three-phase contact of solid, liquid, and vapour, can be used to measure contact angle and quantify wetting. The surface's hydrophobic or hydrophilic character is determined by the three-phase interaction. The textured surface, such as unidirectional grooves and pillars, has better hydrophobic characteristics, with a contact angle of more than 120°C when compared to the conventional smooth surface [3–5].

Earlier researchers believed that a super hydrophobic surface would lose its wetting property and would not eradicate corrosion when it is completely immersed in water. Even if the sample is dipped in the water, the same effect remains up to a certain height due to the equilibrium between capillary pressure and the pressure generated due to the liquid gravity force. Capillary pressure supports the air pocket, which opposes the penetration of liquid. The surface micro asperities determine the immersion height up to which hydrophobicity remains. Corrosion rate depends upon the surface area exposed to a corrosive environment. Patterning the surface leads to variation in surface area, and heterogeneous wetting will occur due to air or vapour pockets accumulating in the valleys of the groove, as stated by Cassie Baxter.

[a]a.afzal@ubt.edu.sa, [b]vivekanand7adr@gmail.com, [c]aero007@gmail.com, [d]cloganathankaruppusamy304@gmail.com, [e]dbalakrishnan@pmu.edu.sa, [f]aafzal@pmu.edu.sa

DOI: 10.1201/9781003770442-20

The air pockets are formed in between the channels due to hydrophobicity which opposes the penetration of particles responsible for corrosion deep inside the surface [6–8].

The patterned surface can be formed by roughening the smooth surface. Suppose the roughness is induced on the surface in the form of unidirectional grooves or pillars. In that case, it will lead to a hydrophobic surface, whereas forming micro pits will lead to a hydrophilic surface. The surface with micro pits will show poor corrosion resistance when compared to the smooth surface. Once the surface is made hydrophilic, it will not be turned into hydrophobic. The patterned surface can be fabricated either mechanically or using chemical processes. In some cases, a passive layer is formed on the surface during the fabrication process, and it will eradicate corrosion. This will not work in peculiar cases of corrosion like fretting, pitting or stress corrosion cracking. The passive layer will break due to external loading, accelerating corrosion compared to the smooth surface. The possibility for crack initiation at the root of the roughness asperity will be higher due to hardness [9–11]. To highlight these problems, in our previous work, the wetting behaviour of grooved and pillar surfaces were investigated under the influence of uniaxial stress [12]. The grooved surface exhibits its superior wetting properties compared to the pillar surface due to its isotropic wetting character.

Most of the components of aircraft and ground vehicles are fabricated using stainless steel S304 wherever high strength is required. In our work, wetting and corrosion characteristics are controlled by mimicking the biological surface, like paddy on the surface of stainless steel S304.

Experimental

Material and surface characterisation

The micro grooves were formed on the SS304 rod with a 12 mm diameter and a height of 30 mm. A tungsten electrode with a 0.18 mm diameter was used as the slicing wire, and to limit wear, the constant anode supply of the terminal wire was kept up by using a nourished framework. The wire and the specimen were kept at 0.02 mm to limit the unevenness on the sides of the microgrooves due to the spark. The groove and channel width are not designed with equal dimensions to avoid the Wenzel model, where wettability is higher. The fabricated microgroove patterned surface is shown in Figure 20.1.

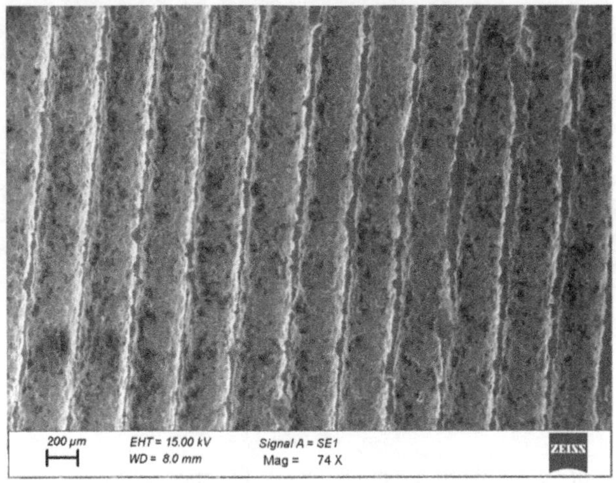

Figure 20.1 SEM image of microgroove patterned surface
Source: Authors

Table 20.1 Geometrical dimensions of the specimens.

S. No.	Specimen	Groove width (GW) (mm)	Channel width (CW) (mm)	GW/CW
1	G1	0.4	0.3	1.3
2	G2	0.6	0.3	2
3	G3	0.8	0.3	2.7
4	G4	1	0.3	3.3

Source: Authors

Table 20.1 lists the geometrical dimensions of the microgrooves as determined by a 3D optical profilometer. The channel and groove are kept at 0.1 mm in height.

Wetting experiment

The standard sessile drop method was preferred to conduct a wetting study on the test specimens for both the plane and the patterned surface. The samples were thoroughly cleaned before each test, and a customised setup was fabricated to conduct a wetting study. A water droplet of 8 µl volume was placed on the specimen's surface, and the image was captured after 5–10 seconds, once the droplet had settled, using a high-speed camera (Motion Xtra N3) [13, 14]. Figure 20.2 shows the water droplet image that developed on the plain and microgroove surfaces.

The wetting parameters i.e., static contact angle (perpendicular and parallel) of the water droplet was measured at three different places on the specimen

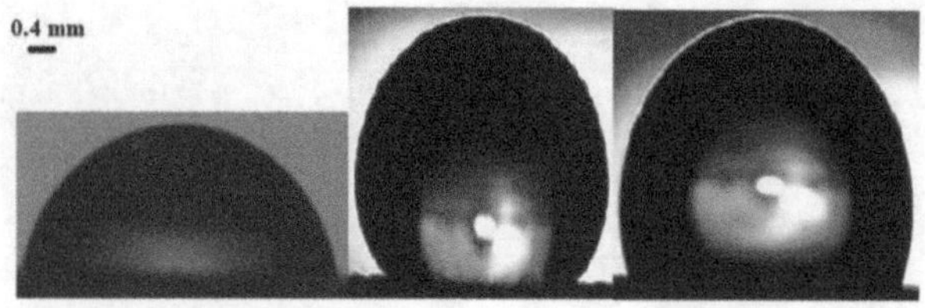

Figure 20.2 Water droplet image formed on (a) Plain, microgroove patterned surface, (b) Perpendicular and (c) Parallel
Source: Authors

Table 20.2 Wetting parameters of the specimens.

S. No.	Specimen	Contact angle θ_{Per} (Deg)	Contact angle θ_{Par} (Deg)
1	Plain	74	
2	G1	125	103
3	G2	133	106
4	G3	131	104
5	G4	121	101

Source: Authors

using ImageJ software to quantify the effect of hydrophobicity and listed in Table 20.2.

Corrosion experiment

The commonly used method to eradicate corrosion is the application of coatings. The primary function of any coating is to prevent water absorption and repel water from the surface. The microgroove surface also has the same water-repellent characteristics, which encourages us to believe that the micro-patterned surface can eradicate corrosion and withstand for a prolonged time period when compared to coatings. Corrosion tests were conducted on the specimens, according to ASTM standard G31 for electrochemical corrosion studies in saline water with a pH value of 8. The polarisation resistance technique was chosen for the corrosion test due to its accurate measurement even at very low corrosion rate.

The electrochemical workstation (Metrohm, Netherlands) was used in conjunction with a system to generate a TAFEL plot, and fitting parameters were obtained using NOVA software. The cyclic voltammetry test was conducted to find the corrosion rate. The workstation comprises three electrodes: a working electrode (specimen), a reference electrode

Table 20.3 Corrosion fitting parameters extracted from the TAFEL plot.

S. No.	Specimen	Ecorr (V)	Icorr (A/cm^2)	Rate of corrosion (mm/year)
1	Plain	-0.29	2.76E-7	0.00394
2	G1	-0.19	3.31E-8	0.00043
3	G2	-0.14	3.07E-8	0.00031
4	G3	-0.21	3.49E-8	0.00039
5	G4	-0.26	3.73E-8	0.00047

Source: Authors

(silver dipped in KCl solution), and a counter electrode (platinum). The fluctuation of corrosion potential is controlled by dipping the specimen in saline water for a saturation time period of 60 minutes before conducting the test. The potential between the electrodes is maintained constant at -0.8V and the scan rate is 1 mV/S. The fitting parameters used in this study, extracted from TAFEL plot for micro-patterned surfaces, are shown in Table 20.3.

Results and discussion

Wetting study

The plain surface undergoes more wetting when compared to the micro-patterned surface due to its heterogeneous nature, following the Casie Baxter model. A high-resolution camera captures the sessile water droplet on the substrate at both perpendicular and parallel angles to the grooves, owing to its anisotropic wetting behaviour. The morphology of the liquid drop varies for all cases due to variation in the surface area caused by texturing. The variation in wetting parameter with respect to the micro-patterns is shown in Figure 20.3.

The contact angle of a water droplet on the micro-patterned surface rises with an increase in surface area, regardless of whether it is oriented perpendicular or parallel to the substrate. The anisotropic wetting characteristic of the grooves causes variations in the contact angle measured both parallel and perpendicular to them. The pinning between the three phases (solid, liquid and air) contact line of the water droplet was not uniform in all directions, and changes with respect to the direction of the groove. The contact angle formed on the perpendicular to the grooves shows a higher value when compared to the parallel

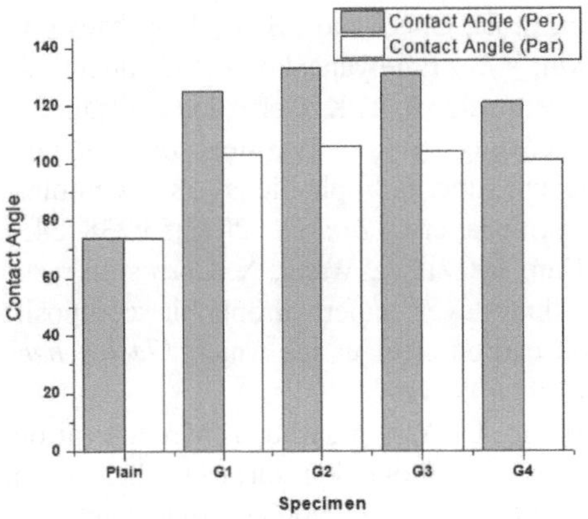

Figure 20.3 Variation in contact angle for plain and microgroove patterned surface
Source: Authors

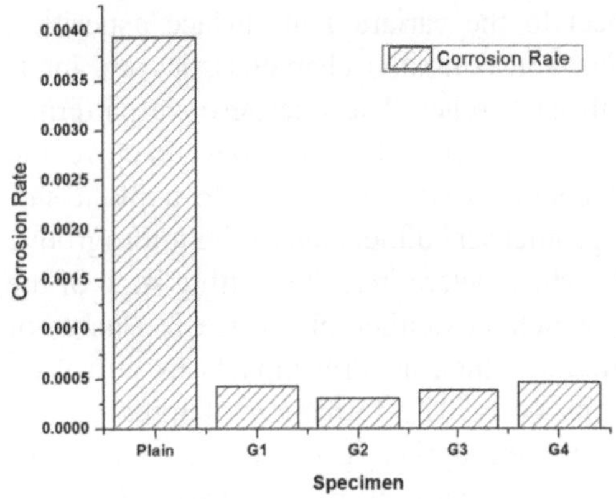

Figure 20.4 Variation in corrosion rate for plain and microgroove patterned surface
Source: Authors

case and its controversy for the contact diameter. The reason for this variation is that the spreading of the water droplet across the surface was restricted due to the presence of the groove. Meanwhile, a water droplet can spread easily in another direction along channels.

Corrosion study

A hydrophobic surface, characterised by its better wetting properties, such as a higher contact angle, exhibits better corrosion resistance compared to a smooth surface. A one-dimensional grooved surface will come under a hydrophobic surface having a higher contact angle when compared to a bare surface. The corrosion inhibition property of a grooved textured surface with different surface areas was studied using the polarisation resistance technique in a saline water environment. The fitting parameters obtained from the TAFEL plot used for this study are shown in Table 20.3.

Corrosion potential (Ecorr) values obtained in TAFEL plot for different surface areas are shown in Figure 20.4. The corrosion potential (Ecorr) value varies for different surface areas. The variation in corrosion potential leads to a decrease in the dissolution of Fe in saline water. Corrosion current density is responsible for the formation of iron (II) chloride. Ferrous chloride, which is easily soluble in water, accelerates the corrosion process by dissolution of Fe. The variation in corrosion rate for different specimens is shown in Figure 20.4. The plain surface exhibits more corrosion compared to the microgroove patterned surface. The hydrophobicity can be improved by means of a one-dimensional groove textured surface, which in turn increases the corrosion inhibition property of the material. The surface morphology plays the major role; depending upon the size of roughness, the surface will act as either hydrophobic or hydrophilic. If it is hydrophobic it inhibits corrosion otherwise, it accelerates the process of corrosion.

The correlation between wetting and corrosion reveals the mechanism behind corrosion in microgroove-patterned SS304 steel surface under a saline water environment. The water droplet sitting on the groove was not penetrating inside the channels as long as the wetting was in the Casie-Baxter state. Once the wetting shifts from Casie-Baxter to Wenzel state due to a change in surface morphology, the water droplet can penetrate easily inside the channels. Anodic reactions occur at the surface of the channels, whereas cathodic reactions occur at the surface of the grooves. Due to these reactions the electrons will pass from anode to cathode i.e., to the grooves. The loss of electrons in the channels turns the electrolyte (saline water) into a positive charge. The positively charged electrolyte will attract the negatively charged Cl-ions, which increases the acidity of the corrosive medium in contact with the channel. Hence, corrosion will be more aggressive on the channels, not on grooves.

Conclusion

The microgroove pattern was formed on the surface of SS304 steel with different geometrical dimensions. The wetting and corrosion behaviour of plain and microgroove patterned SS304 steel was investigated with respect to the variation in surface asperities. The wetting and corrosion characteristics are interlinked with each other. The microgroove-patterned surface exhibits excellent hydrophobic and corrosion resistance characteristics compared to a plain surface. The geometrical dimension of the microgroove patterned surface determines the wetting state of the substrate, which lies either in the Casie-Baxter or Wenzel model, which in turn impacts its corrosion resistance property. The formation of microgroove patterns with appropriate geometrical dimensions will be able to combat wetting and corrosion.

References

[1] Liu, Y., Zhang, B., Zhang, Y., Ma, L., Yang, P. Electrochemical polarization study on crude oil pipeline corrosion by the produced water with high salinity. *Engg. Fail. Anal.*, 2016;60:307–315.

[2] Sengottaiyan, S., Shaisundaram, V. S., Basha, M. S., Deepanraj, B., Senthilkumar, N. Effect of duplex and austenite stainless steel fillers on mechanical, microstructural, and corrosion characteristics of dissimilar welds using cold metal transfer welding. *Results Engg.*, 2025;26:104818.

[3] Goswami, A., Rahman, M. A. Numerical study of energetics and wetting syability of liquid droplets on microtextured surfaces. *Coll. Poly. Sci.*, 2017;295(10):1787–1796.

[4] Salem, T. K., Budakli, M., Sahan, O., Arik, M. An experimental and analytical study on the influence of superhydrophobic micro-textured surfaces on liquid wetting phenomena. *Coll. Surf. A Physiochem. Engg. Aspects*, 2018;555:191–200.

[5] Dhinakarraj, C. K., Senthilkumar, N., Palanikumar, K., Deepanraj, B. Machinability evaluation of magnesium composite using response surface methodology and nature-inspired metaheuristic algorithms. *Internat. J. Interac. Design Manuf.*, 19(4):2607–2627.

[6] Boinovich, L. B., Gnedenkov, S. V., Alpysbaeva, D. A., Egorkin, V. S., Emelyanenko, A. M., Sinebryukhov, S. L., Zaretskaya, A. K. Corrosion resistance of composite coatings on lo-carbon steel containing hydrophobic and superhydrophobic layers in combination with oxide layers. *Corr. Sci.*, 2012;55:238–245.

[7] Yu, D., Tian, J., Dai, J., Wang, X. Corrosion resistance of three-layer superhydrophobic composite coating on carbon steel in sea water. *Electrochem. Acta*, 2013;97:409–419.

[8] Zhi, J., Zhang, L., Yan, Y., Zhu, J. Mechanical durability of superhydrophobic surfaces: the role of surface modification technologies. *Appl. Surf. Sci.*, 2017;392:286–296.

[9] Wang, P., Zhang, D., Qiu, R. Liquid/solid contact mode of super-hydrophobic film in aqueous solution and its effect on corrosion resistance. *Corr. Sci.*, 2012;54:77–84.

[10] Ishizaki, T., Masuda, Y., Sakamoto, M. Corrosion resistance and durability of superhydrophobic surface formed on magnesium alloy coated with nanostructured cerium oxide film and fluroalkylsilane molecules in corrosive NaCl aqueous solution. *Langmuir*, 2011;27:4780–4788.

[11] Guo, M., Toloei, A., Rotermund, H. The effect of surface patterning on corrosion resistance of biomedical devices. *J. Mater. Engg. Perform.*, 2016;25(25):4190–4198.

[12] Anand, A. V., Arumugam, V., Kannan, R. Static wetting characteristics of micro-textured stainless-steel surfaces under uniaxial loading condition. *Internat. J. Surf. Sci. Engg.*, 2017;11(2):74–190.

[13] Anand, A. V, Arumugam. V., Jayalakshmi, S., Arvind Singh, R. Innovative approach for suppressing corrosion of SS304 steel in saline water environment. *Anti-Corr. Methods Mater.*, 2018;65(5):484–491.

[14] Annakodi, V. A., Arvind Singh, R., Jayalakshmi, S., Zhang, Y., Srinivas Rao, K., Shabadi, R. Anticorrosion behaviour of SS304 microgroove surfaces in saline water. *Metals*, 2021;11:1543.

21 Low-velocity impact on fibre composites: A review of damage mechanisms and material behaviour

Shilpa R. S.[a] *and Jyothilakshmi R.*[b]

Department of Mechanical Engineering, M. S. Ramaiah Institute of Technology, Bangalore, Karnataka, India

Abstract

Glass fibre reinforced polymer (GFRP) and carbon fibre reinforced polymer (CFRP) composites are extensively utilised in the aerospace and automotive sectors owing to their high performance-to-weight ratio. These materials are often exposed to impact loading during service, which leads to a reduction in both stiffness and strength of the structural components. The behaviour of composite laminates under low-velocity impact is governed by multiple interacting parameters that critically influence their damage mechanisms and overall performance. In this study GFRP and CFRP composite laminates were fabricated using Vacuum bagging technique with the staking sequence of 8 layers, 10 layers and 12 layers. These laminates are subjected to impact energies 8 joules, 16 joules and 24 joules and their performances were analysed based on the effect of thickness of composite laminates with absorbed energy, elastic energy, peak force, displacement and contact duration. Damage areas were also found and analysed. Based on the study it was found that Carbon epoxy laminates demonstrate superior peak load capacity, improved damage tolerance, and reduced displacement, making them more appropriate for high-performance applications where stiffness and strength are critical. In contrast, glass epoxy laminates show higher energy absorption through deformation, making them more suitable for scenarios where cost-effectiveness and ductile damage tolerance are prioritised. An increase in laminate thickness consistently improved impact resistance in both materials, leading to higher peak loads and lower displacement. However, the improvement was more significant in carbon epoxy laminates, which displayed a sharper rise in mechanical performance with additional layers.

Keywords: Low velocity impact, absorbed energy, peak force, effect of thickness

Introduction

Composite materials are extensively employed in various industries, including aerospace, automotive, marine, and biomedical sectors, due to their outstanding specific mechanical properties and lightweight characteristics [1]. The utilisation of advanced composite materials, particularly fibre-reinforced polymer composites (FRPs), has significantly increased in recent years because of their superior attributes. FRPs are highly regarded for their excellent strength-to-weight ratio and making them ideal for the applications where it is essential to minimising the weight without sacrificing the integrity of the structure. Furthermore, FRPs exhibit resistance in the corrosion, ease of installation, low density, and long-term durability, making them suitable for diverse fields such as the automotive industries, aerospace industries, architectural industries, and marine industries. In civil engineering, fibre-reinforced polymers are commonly applied in reinforcement bars, sandwich panels, and structural retrofitting, thereby improving the performance and longevity of construction projects [2]. Carbon fibre reinforced plastic (CFRP) and glass fibre reinforced plastic (GFRP) are particularly favoured in aerospace applications, automotive applications, and marine applications due to their remarkable specific strengths, light weight properties, and resistance to impact energies. Over their life time, composite laminates may come across low or high velocity impacts from external objects [3].

Glass fibre reinforced polymers are used commonly in composite materials because of outstanding strength-to-weight ratio, corrosion resistance, and versatility in design. These attributes make them ideal for use in the automotive and aerospace sectors, where lightweight and dependable materials are crucial. However, during their operational lifespan, composite materials can be exposed to impact loads, which may cause structural damage [4].

Glass-based composites, such as fibre glass, are commonly used for their cost efficiency, corrosion

[a]shilpars1519@gmail.com, [b]jyothirswamy@gmail.com

DOI: 10.1201/9781003770442-21

resistance, and reliable mechanical strength. Similarly, carbon-based materials, including CFRPs, are valued for their outstanding strength-to-weight ratio and superior fatigue resistance. However, differences in thickness can alter the mechanical properties of these materials, affecting their performance under varying loading conditions [5].

Composite structures are prone to impacts from foreign objects, such as dropped tools during maintenance or low-velocity airborne debris. These impacts can take place during manufacturing, routine operations, or maintenance procedures. Due to the lack of through-thickness reinforcement and weak interfacial bonding in composites, structural damage frequently occurs under these circumstances. Consequently, impact events can alter the mechanical characteristics of composites, reducing the structural integrity and stability. In composite laminates, four primary failure modes are observed: matrix cracking, delamination, fibre breakage, and debonding. Among these, delamination and matrix cracking are the most prevalent types of damage caused by low-velocity impacts. Such defects significantly degrade the mechanical performance of the structure and shorten the lifespan of composite materials. Since impact damage is inevitable in real-world applications, developing strategies to enhance the impact resistance of composite structures is crucial.

During their operational lifespan, composite structures may experience impacts from foreign objects at both low and high velocities. The impact energies are absorbed through structural failures such as delamination, fibre breakage and matrix cracking. The inherently low toughness in the carbon fibre often results in less impact resistance in laminates. Even when surface damage is barely visible (BVID), internal defects can still lead to significant deterioration in mechanical properties. The impact resistance of composite laminates is largely influenced by factors including types of fibre, resin composition, layup design, thickness, impact velocity, and projectile shape [6].

CFRP laminates in practical applications may inevitably experience impact forces, such as accidentally dropping of tool during maintenance or collisions with ground service vehicles. These impacts fall under the category of low velocity impact (LVI) and can lead to barely visible impact damage (BVID) on the surface. Although these damages may seem insignificant, internal failures such as fibre breakage, in-plane ply cracks, and delamination can still occur within the laminate [7, 8].

Unintended low impacts energies, like dropped tools during the maintenance work, or debris during transportation, or hailstone strikes, can result in considerable reduction in the strength and localised damage. The extent of damage within the composites laminates can different from minor surfaces defects to complete penetration, which is dependent on the intensity of the impacts. Common failure mechanisms in the composite laminates include fibre breakages, matrix cracks, matrix debonding, and delaminations. Among all these failures, delaminations have the most severe impact on the stiffness and strength of composites [9].

Composite are always subjected to different in plane, out of plane loading conditions throughout the life time. Several literatures highlights that impact events, particularly low velocity impacts, are among the most crucial factors to consider when designing and applying composite structures [10–12]. Researchers have emphasised that a primary concern in the design of fibre reinforced composite laminates for impact loading is the damage tolerance of both the fibre and matrix system. Additionally, it has been stated that the toughness of fibre glass epoxy composites is significantly influenced by strain rate damage and the matrix material behaviour [13].

The impact energy resistance in the composite laminates depends on different factors, including resin composition, layup configurations, thickness of laminates, velocity, and shape of the projectile [14–16]. In terms of fibrous reinforcement, research has demonstrated that fabric structures offer enhanced performance under impact loading. Despite limitations in velocity range and damage measurement accuracy, the work is thorough, well-presented, and provides both experimental data and analytical modelling [17].

Materials and methods

1. Different types of materials such as carbon fibres, glass fibres, Kevlar, are used based on their woven patterns like plain woven, twill woven, unidirectional and bidirectional as per the requirement of the objective to be tested.
2. The above said materials are reinforced using epoxy resin mixed with hardener. Some of the materials are reinforced with flame retardant resins.

3. The above materials are fabricated by vacuum bagging techniques and some of the authors have used vacuum-assisted resin transfer moulding process.
4. Varying layering techniques were used in order to obtain varying thickness.
5. All the testing specimens were prepared based ASTM standards and required dimensions.
6. Drop weight testing methods were used to test the specimens.
7. The impact energy loads varies from 10 J to 70J.
8. Some of the glass carbon composites were subjected to low velocity testing at low temperature.
9. Specimens were pneumatically fixed in the testing chamber and at −25°C and −50°C temperatures by using liquid nitrogen. Impact tests were carried out.

Results and discussions

The specimens exhibit consistent stiffness as impact energy increases, with corresponding rises in maximum force and displacement as shown in the Figure 21.1. As expected, these values also scale with the thickness of the specimens. Across all tested energy levels, the curves demonstrate a stiffening trend near the point of maximum force. Interestingly, while significant stiffness changes—typically indicative of damage—are generally observed after specific force and displacement thresholds, this behaviour is largely absent in the current findings, with only a little exceptions., e.g., kevlar with 8 layers [K8] and glass with 15 layers [G15] for the impact energy at 115 J, at which a small rising curve is observed when the force reaches the maximum. Comparing the thickness and density, i.e., kevlar with 8 layers [K8] and glass with 13 layers [G15] and kevlar with 13 layers [K13] and glass with 20 layers [G20], no distinct curve is evident in the maximum force and displacement, with the glass specimens exhibiting lower energy absorption compared to the previously reported values for the same reinforcements using an epoxy resin for example the laminae absorbs on average 14.9% and 4.4% consumes less energy when compared with [K8]–[G15] and [K13]–[G20], respectively [17–19].

The glass kevlar hybrid specimens [G6/K5] and [G8/K4] absorb noticeably less energy when compared with their symmetric counterparts. The energy absorbed by the non-symmetric specimens is approximately 15% lower at 115 J as shown in the Figure 21.2. Nevertheless, the non-symmetric specimens surpass the symmetric specimens in terms of energy absorption capacity, as they can endure greater loads with in the failure condition. This behaviour contrasts with prior findings for epoxy composites, where symmetric and non-symmetric configurations exhibited comparable performance. Despite variations in aramid ratios, the two non-symmetric specimens show similar behaviour, with energy absorption comparable to that of the glass specimen. This similarity, despite a lower areal density, underscores a beneficial hybridisation effect [17–19].

Figure 21.3 shows that the impacted side shows the radial plasticisation, similar to the other glass specimen's lamina on the impacted side, but the plasticised area is smaller. The **[G4/K2]S** specimen exhibits a cross-shaped fracture on its back, with damage characteristics resembling those observed in other specimens within the same subsection. Crack extends outward from centre of the specimen's section. This specimen absorbs the full impact energies,

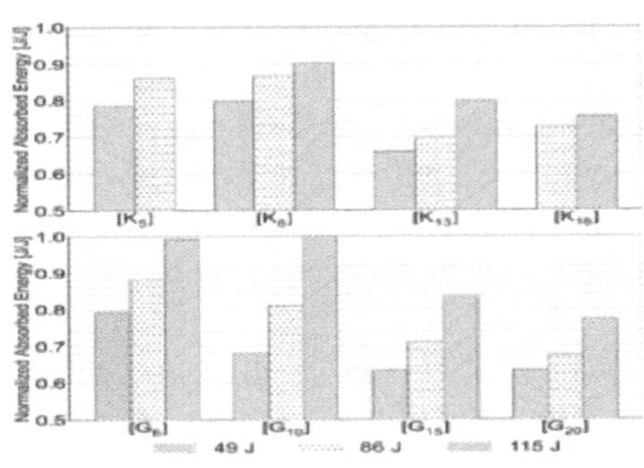

Figure 21.1 Normalized absorbed energy of Kevlar and glass specimens
Source: Authors

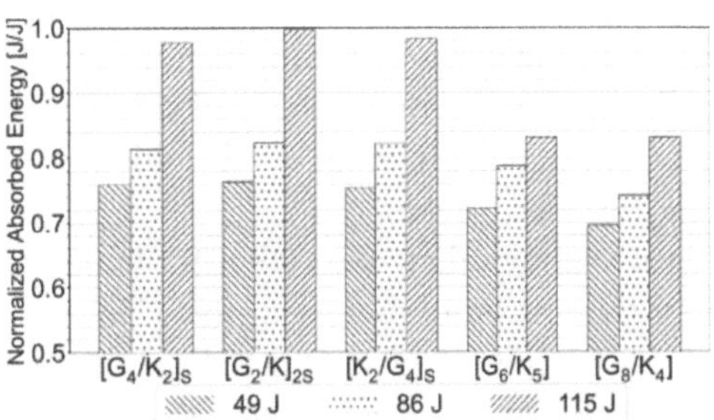

Figure 21.2 Normalized absorbed energy of hybrid specimens
Source: Authors

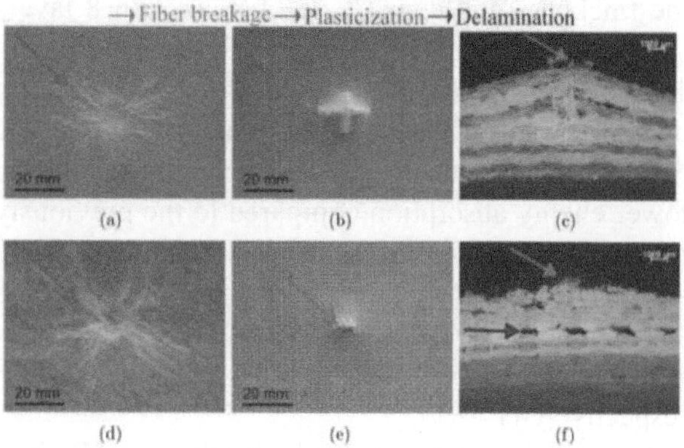

Figure 21.3 Views of $[G_2/K]_{25}$ and $[G_6/K_5]$ specimens impacted at 115 J: (a) front, (b) back, (c) cross-section views of the $[G_2/K]_{25}$ specimen, and (d) front, (e) back, (f) cross-section views of the $[G_6/K_5]$ specimen
Source: Authors

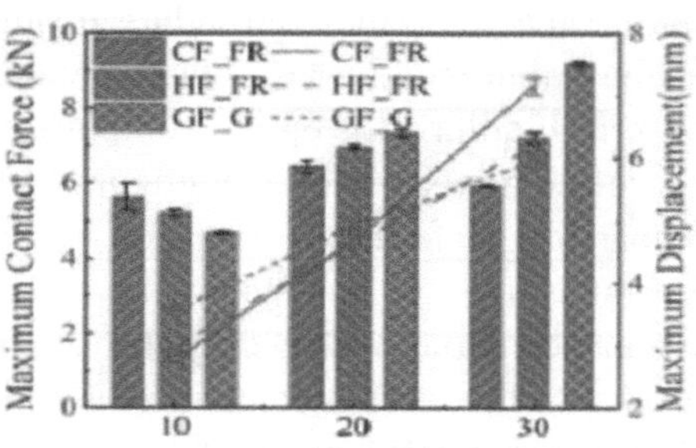

Figure 21.4 Impact Energy (J)
Source: Authors

which is 2% more than the energy absorbed by the other specimens which are symmetric.

In case of hybrid specimens **[G6/K5]**, the front glass layers display damage similar to the impact side, characterised by a significant deformed area around the impact zone. Fibre pull-outs are evident on the back side of the laminates.

According to **Figure 21.3**, the front layers of the glass and Kevlar show no signs of cracking or delamination. However, the back side exhibits fibre breakages and pull-outs. Furthermore, the backside layers of the Kevlar show delaminations and void formation between the fabric layers.

Figure 21.4 shows that the CF_FR laminate demonstrates the maximum highest force in contact but the maximum The smallest displacement observed suggests that the structure possesses higher flexural stiffness, indicating its enhanced resistance to bending under applied loads, resulting from its greater modulus and greater thickness, allows it to

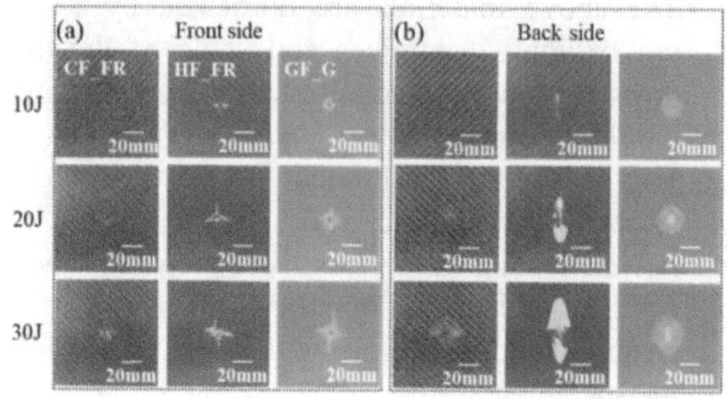

Figure 21.5 Damage morphology of CF_FR, HR_FR, and GF_G under 10 J, 20 J, and 30 J impact energy
Source: Authors

absorb an equivalent quantity of energies with less out of plane deformation. The none of the 3 laminate types exhibit visible at this energy level, widespread surface failure was observed. On the other hand, the GF_G laminate exhibits the greatest maximum displacement, which is likely because of its comparatively thickness will be smaller. Furthermore, the CF_FR sample shows a rapid decline in contact force after reaching its maximum, indicating the possibility of significant damage [20].

It is observed that at this point a cross-shaped brittle fracture occurs. When the contact force is at maximum were compared between CF_FR and HF_FR is higher and can sustain a longer displacement period for a greater contact force. This shows that HF_FR is less prone to a very high brittle fracture. A comparable result is also seen at the incident energy of 30 J as shown in the Figure 21.5.

Figure 21.6 shows that for the composite specimens with thickness of 2.72 mm, the maximum contact force increases by 12.13% when the temperature drops from 20°C to −25°C and by 13.24% when it decreases further to −50°C and by 13.24% when it decreasesthickness of 3.52 mm, the contact force which is maximum raised by 16.48% and 5.83% for the same temperature intervals, respectively. In all tests, an increase in specimen thickness leads to a higher maximum contact force, with a 20.06% rise attributed to thickness alone. In comparison to the effects of temperature and effect of thickness, impact energy has a minimal influence on the maximum contact force [21, 22].

Impact energy leads to perforation around the penetration area due to greater breakage in the fibre, while specimens with thickness of 2.72 mm did

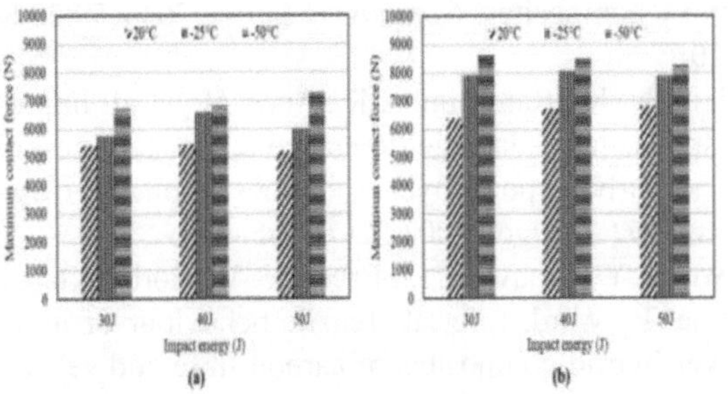

Figure 21.6 Maximum contact force change due to temperature change of composites with (a) 2.72 mm and (b) 3.52 mm thicknesses
Source: Authors

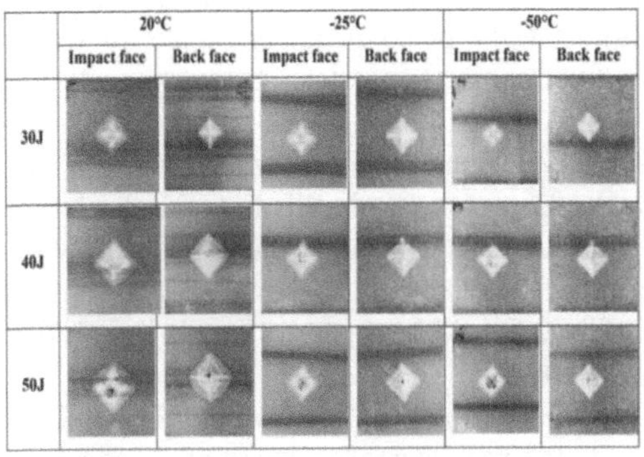

Figure 21.7 Impact damages of S2-glass/epoxy specimens with 2.72 mm thickness at 20°C, -25°C, and -50°C and different impact energies
Source: Authors

not show increase delaminations. The damage area which is influenced by the temperature is smaller compared to the effect of impact energy or thickness of the specimen, ref Figure 21.7.

Conclusion

1. The glass and Kevlar coupons displayed considerable matrix yielding and plasticisation on the front surface. On the back surface, the glass specimens showed fibre breakage, while the Kevlar specimens exhibited fibre pull-outs. The hybrid specimens demonstrated characteristics of significant plastic deformation on the front surface. The specimens with glass fabric on the back side layers showed a fibre breakage pattern as a cross-shaped structure, whereas Kevlar shows pull-outs in the fibre.
2. The woven carbon fibre laminate containing flame-retardant epoxy resin demonstrates a greater damage area, more significant fibre breakage, and reduced CAI strength compared to the laminate using standard epoxy resin under identical impact energy, largely due to its lower toughness. This demonstrates comparatively weaker resistance to damages related to impact energy. Moreover, it displays a 26% reduction in the initial peak force and a 25% decrease in the associated displacement compared to the carbon fibre laminate containing standard epoxy resin the maximum contact force raises by 19% and 29%. When there is a decrease in temperature from -25°C to -50°C, respectively. Furthermore, as the sample thickness becomes higher when the maximum contact force increases to 25%. However, impact loads has little to no significant impact on the maximum contact force.
3. As impact loads increases, the impact damage area expands, where as it shrinks when the temperature decreases. For thin specimens, failure mode shifts rebounding to perforation when the penetration site as the impact loads rises from 40 J to 50 J.

In conclusion, the study highlights the significant impact on the specimen thickness, temperature, and resin type on the impact performance on composites. As specimen increases the thickness, the maximum contact force rises, demonstrating better resistance to impact damage. Temperature variations, particularly decreases, lead to a substantial increase in maximum contact force, with lower temperatures enhancing the material's resistance to deformation. However, impact energy has a relatively minor effect compared to temperature and thickness. The hybrid specimens, combining different reinforcement types, showed promising results, with a higher maximum contact force and reduced damage compared to single-material specimens. Glass fibres, in particular, exhibited less fibre pull-out and breakage than Kevlar fibres, leading to improved performance under impact. The flame-retardant epoxy resin showed poorer impact resistance, with higher damage areas and more fibre breakage, compared to standard epoxy resin. Despite this, the hybrid composites demonstrated a positive hybridisation effect, with better overall performance than the individual materials. Delamination was not significantly affected by temperature, while the impact energy played a more prominent role in fibre damage and failure modes. Finally, the study underscores the significance of considering material properties and environmental factors, such as temperature and resin type, to optimise composite performance for specific applications.

References

[1] Ogunleye, R. O., Rusnakova, S., Zaludek, M., Emebu, S. The influence of ply stacking sequence on mechanical properties of carbon/epoxy composite laminates. *Polymers*, 2022;14:5566.

[2] Seifoori, S., Mahdian Parrany, A., Mirzarahmani, S. Impact damage detection in CFRP and GFRP curved composite laminates subjected to low-velocity impacts. *Compos. Struct.*, 2021;261:113278.

[3] Luo, R. K., Green, E. R., Morrison, C. J. Impact damage analysis of composite plates. *Int. J. Impact Engg.*, 1999;22(4):435–447.

[4] Schoeppner, G. A., Abrate, S. Delamination threshold loads for low velocity impact on composite laminates. *Compos. Part A-Appl. S.*, 2000;31(9):903–915.

[5] Belingardi, G., Vadori, R. Low velocity impact tests of laminate glass-fiber-epoxy matrix composite material plates. *Int. J. Impact Engg.*, 2002;27(2):213–229.

[6] Sinmazçelik, T., Arici, A. A., Günay, V. Impact-fatigue behaviour of unidirectional carbon fibre reinforced polyetherimide (PEI) composites. *J. Mater. Sci.*, 2006;41(19):6237–6244.

[7] Chen, D., Luo, Q., Meng, M., Li, Q., Sun, G. Low velocity impact behaviour of interlayer hybrid composite laminates with carbon/glass/basalt fibres. *Compos. Part B Engg.*, 2019;176:107191.

[8] Lei, Z. X., Ma, J., Sun, W. K., Yin, B. B., Liew, K. M. Low-velocity impact and compression-after-impact behaviors of twill woven carbon fibre/glass fibre hybrid composite laminates with flame retardant epoxy resin. *Compos. Struct.*, 2023;321:117253.

[9] Kaware, K., Kotambkar, M. Experimental investigation of hybridization effect of Kevlar and glass fibres on CFRP composite under low velocity impact. *Internat. J. Crashworth.* 2023. 1–2.

[10] Lopes, C. S., Seresta, O., Coquet, Y., et al. Low-velocity impact damage on dispersed stacking sequence laminates. *Part I Exper. Compos. Sci. Technol.*, 2009;69(7–8):926–936.

[11] Mars, J., Chebbi, E., Wali, M., et al. Numerical and experimental investigations of low velocity impact on glass fibre-reinforced polyamide. *Compos. B Engg.*, 2018;146:116–123.

[12] Sarasini, F., Tirill o, J., Ferrante, L., et al. Drop-weight impact behaviour of woven hybrid basalt-carbon/epoxy composites. *Compos. B Engg.*, 2014;59:204–220.

[13] Naik, N. K., Ramasimha, R., Arya, H., et al. Impact response and damage tolerance characteristics of glass-carbon/epoxy hybrid composite plates. *Compos. Part B Engg.*, 2001;32(7):565–574.

[14] Swolfs, Y., Crauwels, L., Breda, E. V., Gorbatikh, L., Hine, P., Ward, I., et al. Tensile behaviour of intralayer hybrid composites of carbon fibre and self-reinforced polypropylene. Compos A Appl Sci Manuf 2014;59:78–84.

[15] Swolfs, Y., Gorbatikh, L., Verpoest, I. Fibre hybridisation in polymer composites: A review. *Compos. A Appl. Sci. Manuf.*, 2014;67:181–200.

[16] Enfedaque, A., Molina-Aldareguía, J. M., G´alvez, F., Gonz´alez, C., LLorca, J. Effect of glass fiber hybridization on the behavior under impact of woven carbon fiber/epoxy laminates. *J. Compos. Mater.*, 2010;44(25):3051–3068.

[17] Rezasefat, M., Gonzalez-Jimenez, A., Ma, D., Vescovini, A., Lomazzi, L., da Silva, A. A., Amico, S. C., Manes, A. Experimental study on the low-velocity impact response of inter-ply S2-glass/aramid woven fabric hybrid laminates. *Thin Walled Struct.*, 2022;177:109458.

[18] Abdallah, E. A., Bouvet, C., Rivallant, S., Broll, B., Barrau, J.-J. Experimental analysis of damage creation and permanent indentation on highly oriented plates. *Compos. Sci. Technol.*, 2009;69(7):1238–1245.

[19] Vescovini, A., Cruz, J. A., Ma, D., Colombo, C., Salerno, A., Bianchi, O., Amico, S. C., Manes, A. Experimental investigation on low-velocity impact behaviour of glass, Kevlar, and hybrid composites with an elastomeric polyurethane matrix. *Compos. Part C Open Acc.*, 2024;13:100426.

[20] Lei, Z. X., Ma, J., Sun, W. K., Yin, B. B., Liew, K. M. Low-velocity impact and compression-after-impact behaviors of twill woven carbon fibre/glass fibre hybrid composite laminates with flame retardant epoxy resin. *Compos. Struct.*, 2023;321:117253.

[21] Reddy, T. S., Reddy, P. R. S., Madhu, V. Low velocity impact studies of E-glass/epoxy composite laminates at different thicknesses and temperatures. *Def. Technol.*, 2019;15(6):897–904.

[22] Karakuzu, R., Kıyık, S., Kandas, H. Low-velocity impact and compression-after-impact behaviors of S2-glass/epoxy laminated composites under impact loading at low temperatures. *Mech. Adv. Mater. Struct.*, 2023;31(30):1–11.

22 Buckling and post-buckling of FG-CNTRC plate subjected to in-plane triangular load

Rakesh Kumar V. G.[a], *Shuvendu Narayan Patel*[b] *and Rajesh Kumar*[c]

Birla Institute of Technology and Science, Pilani, Rajasthan, India

Abstract

Buckling and post-buckling behaviour of functionally graded carbon nanotube reinforced composite (FG-CNTRC) plate under in-plane triangular loading is presented in this article by finite element method (FEM). Present investigation includes uniform distribution (UD), FG-X and FG-A distribution of carbon nanotubes (CNTs) within plate thickness. Extended rule of mixture is adopted to compute net mechanical properties. Plate element is modelled by 8-noded degenerated shell element using FSDT and isoparametric formulation. Buckling and post-buckling analyses are performed by solving eigenvalue equation and non-linear equilibrium equation. To ensure accuracy of formulation, obtained results are verified with previous results. Later focused on effect of volume fraction, distribution of CNTs, and various triangular loads on post-buckling behaviour.

Keywords: Buckling, post-buckling, isoparametric formulation, arc-length method, carbon nanotubes, green-Lagrangian strain

Introduction

Composite materials are used widely across industries such as aviation, marine, construction, and automotive sectors because of their superior performance and lightweight characteristics. They are commonly used as structural components in form of plates and shells. These plates are assembled together to form complex structure and they experience several random loads. These plates are exposed to various compressive loads, and it leads to buckling before material failure initiates. Thus, study of buckling behaviour is essential in the design process. Buckling docs not always imply plate failure. Even after initiation of buckling, the plates can withstand a significant amount of load before failure. This additional load-carrying capacity beyond buckling represents post-buckling. In post-buckling region, large deformations cause the plate to stretch, exhibiting geometric non-linear behaviour. However, beyond certain range of stretching the material non-linearity start to couple with geometric non-linearity [1]. This additional capacity can be considered in design process to reduce material consumption. Thus, investigating post-buckling behaviour is important for efficient design of structures. Post-buckling behaviour of moderately thick plates with uniform load is studied [2] and concluded that post-buckling strength of moderately thick plate is higher than thin plate. Liew et al. [3] adopted Ritz method to plot post-critical paths of composite plates. Previous works show post-buckling strength is two to three times the buckling load. To improve strength of plates, various methods are adopted such as increasing the thickness of plate, providing stiffeners or reinforcing the plate with different nano-materials. One such technique is to enhance post-buckling response by reinforcing the plate with CNTs.

CNTs were first reported by Iijima [4], and their exceptional mechanical properties of CNTs, make them a promising reinforcing material as alternative for conventional fibres. The addition of CNTs improves mechanical strength by improving elastic modulus and tensile properties of plate [5]. To optimise the reinforcement, functional gradation of CNTs is adopted. CNTs are dispersed across the plate thickness to achieve the functional gradation by powder metallurgy process [6]. Also, it is essential to discuss post-buckling capacity of functionally graded composite plates with various non-uniform loads.

Determining buckling and post-buckling of FG-CNTRC plates is necessary, an appreciable amount of work has carried out earlier. Kiani [7] discussed significance of prebuckling BC's in determining critical loads of FG-CNTRC plates. Adhikari

[a]p20200407@pilani.bits-pilani.ac.in, [b]shuvendu@pilani.bits-pilani.ac.in, [c]rajesh.kr@pilani.bits-pilani.ac.in

DOI: 10.1201/9781003770442-22

et al. [8] adopted HSDT to discuss buckling and explained diminishing of fibre volume fraction with increase in aspect ratio. Nguyen et al. [9] presented post-buckling response under axial compression by analytical method. Watts et al. [10] analytically predicted post-buckling strength.

In all of these investigations, the researchers have focused mainly to present buckling behaviour of FG-CNTRC plates under uniform and parabolic loads. The study of post-buckling response of FG-CNTRC plate by FEM is very limited. However, a lot of scope to report post-buckling strength under non-uniform loads. In present investigation, finite element code is developed in FORTRAN 90 and post-buckling behaviour of FG-CNTRC plates under triangular loads is reported. Effect of CNT configuration, volume fraction, in-plane triangular loads on post-buckling behaviour are reported.

Theory and formulation

Shell element is modelled discretely based on assumptions of Ahmad et al. [11]. Here a three dimensional solid element is degenerated by assuming its shell thickness is considerably lesser. Isoparametric shell element consists of 8-nodes with five dof's per node, where θ_x and θ_y represent rotation about y and x axes respectively and θ_z is neglected. All details about formulation can be referred in [1]. Governing equations for buckling and post-buckling are given in Equations 1 and 2, respectively.

$$[[K] - P_{cr}[K_G]]\{d\} = \{0\} \tag{1}$$

$$[[K_L] + [K_{NL}](\delta)]\{\delta\} = \{F\} \tag{2}$$

Results and discussion

Accuracy of current formulation is verified by considering previous examples. Afterwards details of present investigation are provided.

Convergence and validation

The accuracy of current formulation and mesh size required to carry out analysis is discussed. For this purpose simply supported $(45/-45)_2$ plate under bidirectional compression is considered (Figure 22.1). Geometric properties are $a = b$, $b/h = 10$, $E_{11} = 25\ E_{22}$, $G_{12} = G_{13} = 0.5\ E_{22}$, $G_{23} = 0.2\ E_{22}$ and $v_{12} = 0.25$ [2, 3].

The plate is discretised into various mesh sizes to perform buckling and post-buckling analysis. Equation (1) is solved to report non-dimensional critical loads ($\lambda = N_x b^2 / E_{22} h^3$) The out-of-plane displacement is initiated by applying negligible load at plate centre. Post-buckling paths are traced by solving Equation (2) by arc-length method. It is noticed that, results gets converged for 8 × 8 mesh and also post-buckling paths show close agreement with available results as given in Figure 22.2(b). Thus, all the further analyses are carried out by considering 8 × 8 mesh size.

Material properties

FG-CNTRC plates are made by reinforcing single walled CNTs (SWCNTs) with poly methyl methacrylate (PMMA) matrix. Mechanical properties are provided as in Table 22.1.

FG-CNTRC plate geometry is represented in Figure 22.3. Linear and functional gradation of CNTs across the plate thickness are considered (Figure 22.4). Distribution of CNTs includes linear (UD) and functional gradation (FG-X, FG-A) configurations. CNTs are more at exterior surfaces and lesser at plate centre in FG-X configuration and in FG-A distribution CNTs are varied linearly from top to bottom surface.

Net mechanical properties are computed by extended rule of mixture [11]. Initially, number of imaginary layers across thickness is considered

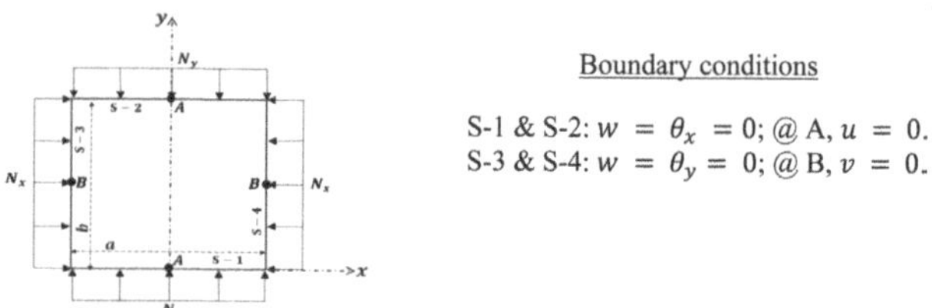

Figure 22.1 Simply supported plate under biaxial uniform compression
Source: Authors

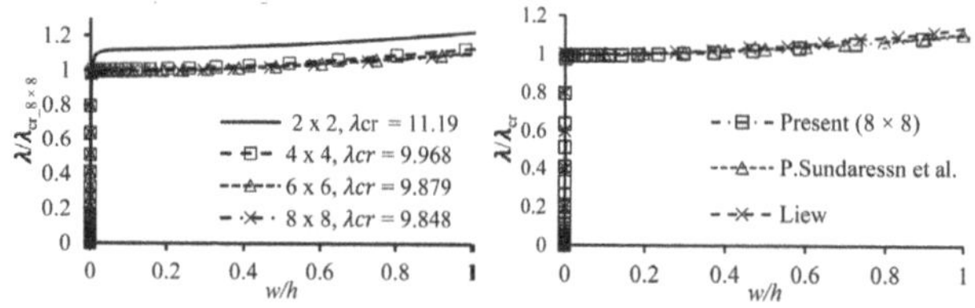

Figure 22.2 (a) Buckling and post-buckling results for different mesh sizes (b) Convergence and validation of post-buckling paths
Source: Authors

Table 22.1 Elastic properties.

SWCNTs				PMMA matrix	
E_{11}^{CNT} (TP_a)	E_{22}^{CNT} (TP_a)	G_{12}^{CNT} (TP_a)	Poisson's ratio v_{12}^{CNT}	E_m GP_a	Poisson's ratio v^m
5.6466	7.08	1.9445	0.175	2.5	0.34

Source: Authors

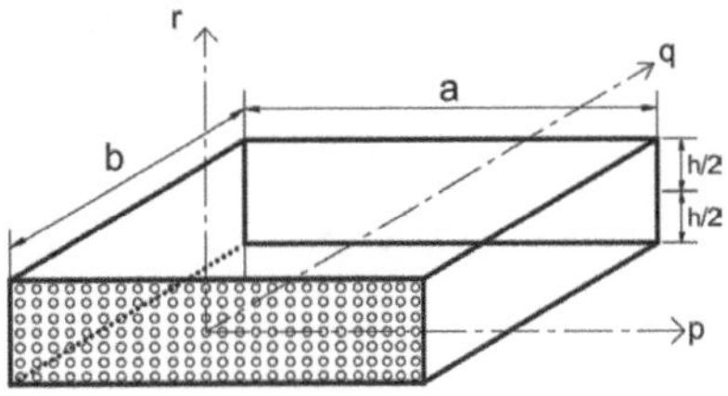

Figure 22.3 Geometry of FG-CNTRC plate
Source: Authors

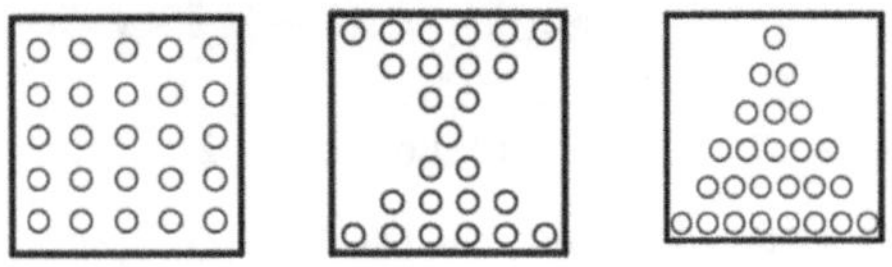

$V_{CNT}^* = V_{cnt}$ $\quad V_{CNT}^* = 4V_{cnt}\frac{|r|}{h}$ $\quad V_{CNT}^* = V_{cnt}\left(1 - 2\frac{|r|}{h}\right)$

Figure 22.4 CNT configurations
Source: Authors

Table 22.2 Non-dimensional buckling loads for different number of layers.

No. of layer	2	4	6	8	10	12	14	16	18
Buckling load	59.5960	78.3114	81.7874	82.9022	83.4813	83.7393	83.9280	84.0446	84.1243

Source: Authors

Table 22.3 Non-dimensional buckling loads of FG-CNT plate ($a = b$, $b/h = 50$, $V_{cnt} = 0.17$).

CNT distribution	Kiani [7]	Adhikari et al. [8]	Pham et al. [9]		Watts et al. [11]	Present (FEA)
			Analytical	ABAQUS		
UD	59.1758	59.1919	60.0992	60.3467	59.0174	59.5960
FG-X	84.3354	84.3870	86.3828	84.9354	83.4438	83.7393

Source: Authors

and effective material properties at top and bottom of every layer is computed. Average properties at upper and lower surface of each layer represent corresponding layer properties. For this purpose buckling of FG-X plate under parabolic load [7] is considered and results are presented in Table 22.2 and noticed that buckling loads are converged at 12th layer.

Also obtained results have close agreement with available results as given in Table 22.3. Further, all analyses are performed by considering 12-imaginary layers.

Post-buckling of FG-CNTRC plates

This section discusses post-buckling of single layered FG-CNTRC plates ($a = b = 1$, $h = 0.01$) under in-plane triangular compression. Total load magnitude acting on plate remains same in all load conditions. Uniformly distributed and different triangular loads considered in the study and boundary conditions are presented in Figure 22.5. Non-dimensional/dimensionless buckling loads are calculated by, $k_{cr} = \lambda_{cr} b^2/\pi^2 D_m$, where $D_m = E_m h^3/12 (1 - v^2)$. Dimensionless central displacement is $\overline{w} = w/h$. Dimensionless load = (*Applid load* × b^2)/$\pi^2 D_m$.

Firstly, buckling analysis is performed and buckling loads are given in Table 22.4 for different CNT configurations, volume fraction and loadings. Buckling strength improves with higher CNT volume fraction because of significant improvement in effective mechanical properties and thereby the stiffness. This results in higher buckling load as the

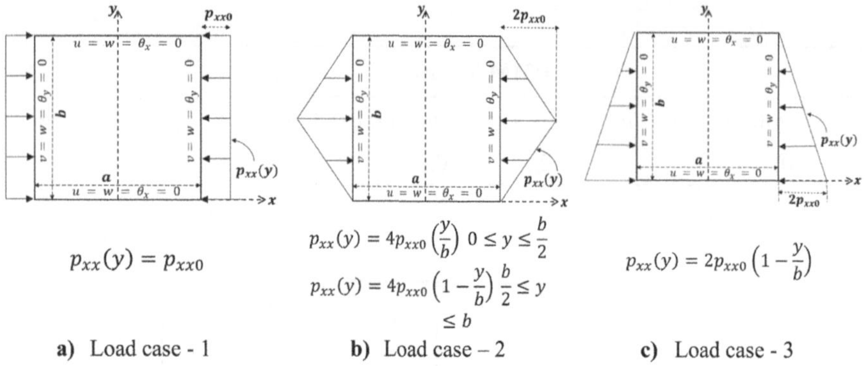

Figure 22.5 Loads and load equations
Source: Authors

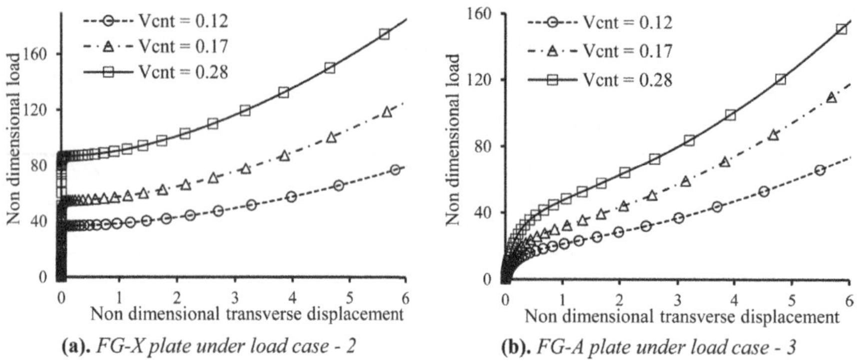

Figure 22.6 Post-buckling paths for FG-CNTRC plate with varied CNT volume fraction. (a) FG-X plate under load case-2. (b) FG-A plate under load case-3
Source: Authors

Table 22.4 Non-dimensional buckling loads.

Load case		UD	FG-X	FG-A
Load case-1	0.12	36.1391	51.8392	17.9945
	0.17	53.3734	76.8119	26.3834
	0.28	83.6947	122.0970	40.1861
Load case-2	0.12	25.9922	36.5444	13.1904
	0.17	38.5733	54.6232	19.4502
	0.28	59.7098	86.4937	29.3367
Load case-3	0.12	31.2106	43.5750	16.2461
	0.17	46.6940	65.6939	24.0906
	0.28	70.7737	102.0443	35.7227

Source: Authors

plate can able to carry higher compressive loads. The post-buckling path is given in Figure 22.6 and it shows the post-buckling load carrying capacity improves for higher CNT volume fraction. The bifurcation point does not exist in FG-A distribution because of unsymmetrical dispersion of CNTs across plate thickness.

Table 22.4 shows that, buckling strength of FG-X plate is better than UD and FG-A patterns. In FG-X, concentration of CNTs is higher towards the outer surface of plate, which contributes to enhance mechanical properties of plate which results in higher bending rigidity. Therefore, FG-X plate has highest buckling capacity. The post-buckling paths for all three CNT distributions under different loads are presented in Figure 22.7. Post-buckling strength is considerably higher and it is found to be two times the critical load.

The post-buckling paths for FG-CNT plates with varying load cases are reported (Figure 22.8). Table 22.4 show critical load for load case-1 (uniform load) is higher followed by load case-3 and load case-2. As the load magnitude is concentrated more towards plate centre, higher stresses will develop which reduces overall stiffness and diminishes load

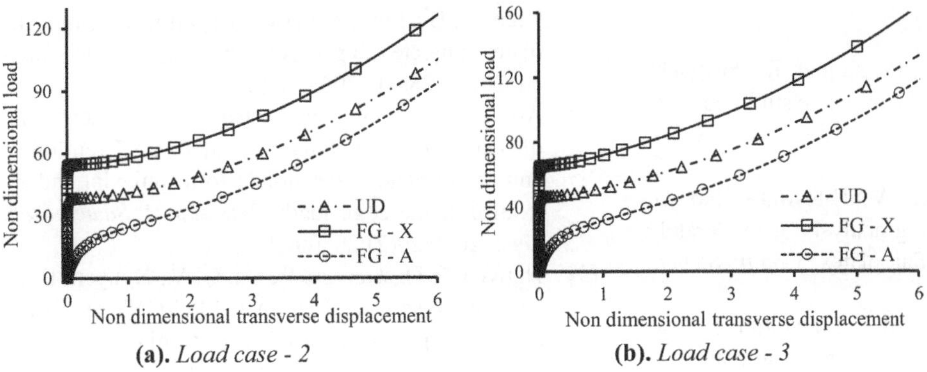

Figure 22.7 Post-buckling equilibrium path for different CNT distribution (. (a) Load case-2. (b) Load case-3
Source: Authors

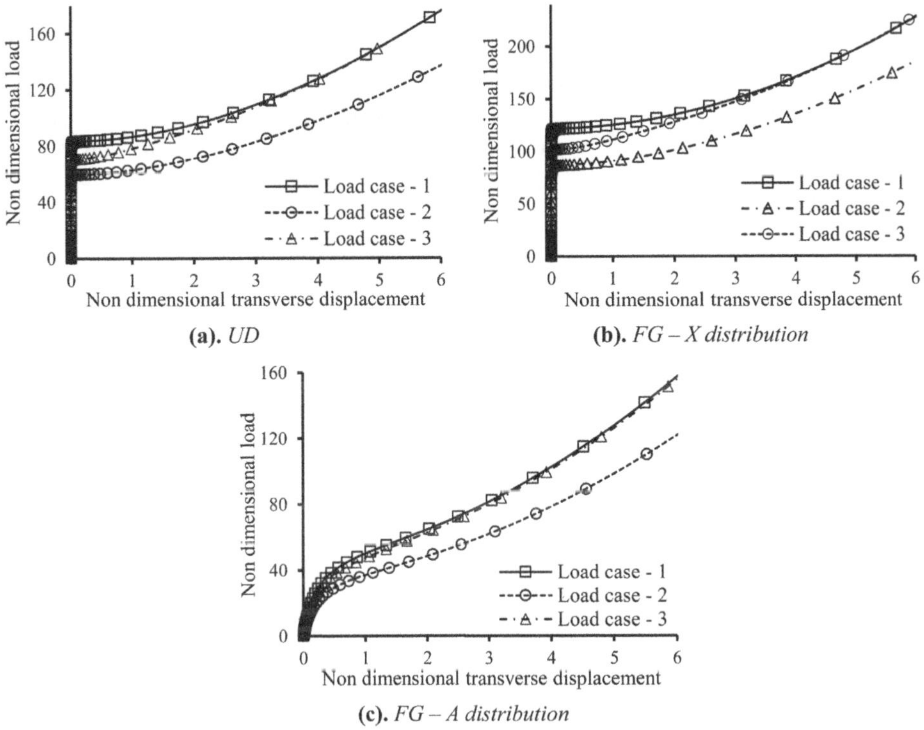

Figure 22.8 Post-buckling equilibrium paths with varying load cases
Source: Authors

carrying capacity. Additionally, it is noticed that, load case-3 does not exhibit bifurcation point due to unsymmetrical load distribution.

Summary

This article discusses buckling and post-buckling response of FG-CNTRC plates with triangular loads. Firstly, correctness of current formulation is checked with available results and later post-buckling results are presented.

1. Buckling and post-buckling strength of FG-X plate under various non-uniform load is higher followed by UD and FG-A plate.
2. Higher CNT volume fraction improves post-buckling capacity due to improved stiffness.
3. Post-buckling strength of FG-CNTRC plate with uniform load is higher than that of triangular loads. FG-X and FG-A distributions are more suitable for load case-2 and load case-3, respectively.

References

[1] Patel, S. N., Kumar, V. G. R., Kumar, R. Postbuckling study of the laminated composite stiffened plate subjecetd to parabolic in-plane loading. *Compos. Sci. Technol.*, 2022:341–361.

[2] Sundaresan, P., Singh, G., Venkateswara Rao, G. Buckling and post-buckling analysis of moderately thick laminated rectangular plates. *Comp. Struct.*, 1996;61(1):79–86.

[3] Liew, K. M., Wang, J., Tan, M. J., Rajendran, S. Postbuckling analysis of laminated composite plates using the mesh-free kp-Ritz method. *Comp. Methods Appl. Mech. Engg.*, 2006;195:551–570.

[4] Iijima, S. Helical microtubules of graphitic carbon. *Nature Publishing Group*, 1991;354:56–58.

[5] Aodkeng, S., Sinthupinyo, S., Chamnankid, B., Hanpongpun, W. Effect of carbon nanotubes/clay hybrid composite on mechanical properties, hydration heat and thermal analysis of cement-based materials. *Constr. Build. Mater.*, 2022;320:126652.

[6] Kwon, H., Christopher, R., Bradbury, Leparoux, M. Fabrication of functionally graded carbon nanotube-reinforced aluminum matrix composite. *Adv. Engg. Mater.*, 2011;13(4):325–329.

[7] Kiani, Y. Buckling of FG-CNT-reinforced composite plates subjected to parabolic loading. *Acta Mechanica*, 2017;228(4):1303–1319.

[8] Adhikari, B., Singh, B. N. Buckling characteristics of laminated functionally graded CNT-reinforced composite plate under nonuniform uniaxial and biaxial in-plane edge loads. *Internat. J. Struct. Stab. Dyn.*, 2020;20(2):2050022.

[9] Nguyen, P. D., George, P., Vu, Q. V., Nguyen, D. D. Buckling response of laminated FG-CNT reinforced composite plates - Analytical and finite element approach. *Aerospace Sci. Technol.*, 2022;121:107368.

[10] Watts, G., Kumar, R., Singh, S., Sengar, V., Reddy, G. R., Patel, S. N. Postbuckling and postbuckled vibration behaviour of imperfect trapezoidal sandwich plates with FG-CNTRC face sheets under nonuniform loadings. *Aerospace Sci. Technol.*, 2022;127:107716.

[11] Ahmad, S., Iropns, B. M., Zienkiewicz, O. C. Analysis of thick and thin shell structures by curved finite elemenets. *Internat. J. Num. Methods Engg.*, 1970;2:419–451.

23 Effect of void location on crack-void interaction in plastically compressible hardening solids

Satyabrat Pandey[a], Sandeep Kumar[b] and Debashis Khan[c]

Mechanical Engineering Department, IIT (BHU) Varanasi, India

Abstract

This study examines the interaction between crack and void in materials that undergo plastic deformation and volume changes under stress. It focuses on two specific void locations: one in the line of crack tip and the other at 45° with respect to the crack direction. The research involves two types of bilinear hardening solids subjected to mode I loading. Finite element finite deformation formulation along with a viscoplastic material model is used to simulate the deformation. To account for change in volume during deformation, a plastic compressibility parameter is introduced. The study investigates how factors such as crack-tip opening displacement, crack extension, void growth, and ligament length vary on application of load. The results show that the location of the void, plastic compressibility and hardening slope significantly influence the deformation of both the crack and the void. A void placed at an angle from the crack direction, such as at 45°, reduces the crack's extension more as compared to the same when the void is located just in line of the crack-tip. The plastic compressibility also slows down the growth of crack and void. The hardening slope of material as well plays a key role in deformation behaviour under loading.

Keywords: Void growth, crack-void interaction, plastically compressible solid, and linear hardening

Introduction

Cracks, voids, cavities, and other defects play a critical role in the failure behaviour of engineering components. These defects are common and can originate from various sources such as the manufacturing or fabrication process, plastic deformation, or exposure to different environmental conditions. Understanding how these defects influence deformation behaviour is essential to predicting how components might fail. Studying the interaction between different types of defects, like cracks and voids, offers valuable insights for improving the design of engineering components. While many studies have focused on analysing crack and void growth separately [1–4], several researchers have also explored their interaction and the resulting deformation behaviour. These studies use different models and parameters to simulate how cracks and voids grow and to explain how they deform under stress.

With the consideration of plastic compressibility in the solids which is significantly shown by structural polymers, transformation toughened ceramics, metallic foams, soils and plastics, there are many researchers recently have worked in this field for studying growth of the crack and plastic deformation under monotonic and fatigue loadings. Some studies have been conducted for small-scale deformation [5] while the others are for large or finite scale deformation [3, 6]. From these studies, it is seen that the plastic compressibility with plastic dilatancy have significant effect on the deformation of the material.

To date, there has been only limited research on void growth, the characterisation of crack–void interactions, and the corresponding near-tip field parameters in pressure-sensitive materials [7, 8].

Although, there are many studies to explore the crack-void interaction, but there is not much exploration for plastically compressible solids. These plastically compressible materials offer many attractive qualities such as energy absorption, compatibility with some of body-parts, highly desirable thermal and electrical interfaces, etc.

This study focuses on the understanding of the effect of plastic compressibility, hardening slope and void location simultaneously on the deformation behaviour of a mode I crack by taking two different strain hardening materials. In this investigation, one void is modelled in line of the crack-tip and in other case is at 45° angle from the crack direction. The study includes the results for the response

[a]satyabratpandey.rs.mec20@iitbhu.ac.in, [b]sandeep.mec@iitbhu.ac.in, [c]dkhan.mec@iitbhu.ac.in

DOI: 10.1201/9781003770442-23

of crack-tip opening displacement (CTOD), crack growth, void growth and ligament length with the applied load. For the comparison purpose both plastically compressible and plastically incompressible materials are taken into the consideration.

Mathematical model

Constitutive model

This investigation utilises the constitutive model reported by Khan et al. [2]. The deformation tensor, **d** is aggregate of isotropic elastic component, $\mathbf{d}^e = \mathbf{L}^{-1}: \hat{\boldsymbol{\tau}}$ (**L** described as elastic constants) and visco-plastic component, $\mathbf{d}^p$. In this formulation, $\hat{\boldsymbol{\tau}}$ denotes the Jaumann rate of the Kirchhoff stress, while **L** represents the elastic moduli tensor. Cauchy stress $\boldsymbol{\sigma}$ and Kirchhoff stress $\boldsymbol{\tau}$ are related through = $J\boldsymbol{\sigma}$, where $J = \det(\mathbf{F})$ and, **F** is tensor of deformation gradient. Assuming small elastic strains, the plastic contribution, ($\mathbf{d}^p$), as:

$$\mathbf{d}^{\mathrm{p}} = \frac{3}{2}\frac{\dot{\epsilon}_{\mathrm{p}}}{\sigma_{\mathrm{e}}}\mathbf{p} \qquad (1)$$

Where, $\mathbf{p} = -\alpha\mathrm{tr}(\boldsymbol{\tau})\mathbf{I} + \boldsymbol{\tau}$

And $\dot{\epsilon}_{\mathrm{p}} = \left(\frac{\sigma_{\mathrm{e}}}{\mathrm{g}}\right)^{1/m} \dot{\epsilon}_0$

where α is the parameter associated with plastic compressibility, the parameters $\dot{\epsilon}_0$ is reference strain rate and m is hardening exponent, and the hardness function g is defined as:

$$\mathrm{g}(\epsilon_{\mathrm{p}}) = \sigma_0 \begin{cases} 1 + \epsilon_{\mathrm{p}} h_1, & \epsilon_{\mathrm{p}} < \epsilon_1 \\ 1 + \epsilon_1 h_1 + (\epsilon_{\mathrm{p}} - \epsilon_1) h_2, & \epsilon_{\mathrm{p}} > \epsilon_1 \end{cases} \qquad (2)$$

Here, σ_0 and σ_e represents the reference stress and the effective stress takes the form

$$\sigma_{\mathrm{e}}^2 = \frac{3}{2}\boldsymbol{\tau}:\mathbf{p} = \frac{3}{2}\left[-\alpha\left(\mathrm{tr}(\boldsymbol{\tau})\right)^2 + \boldsymbol{\tau}:\boldsymbol{\tau}\right] \qquad (3)$$

The square bracketed term of the above Equation (3) is non-negative when $\alpha \leq 1/3$. In case of $\alpha = 1/3$, the equation resembles to that of a plastically Mises material model under the constraint of plastic incompressibility.

The plastic dissipation rate per unit volume is formulated as:

$$\boldsymbol{\tau}:\mathbf{d}^{\mathrm{p}} = \sigma_{\mathrm{e}}\dot{\epsilon}_{\mathrm{p}} \qquad (4)$$

Problem formulation

In this study, a semi-circular geometry with outer radius $R_0 = 2.0$ in arbitrary units (see Figure 23.1a) has been considered for the finite element simulation. A blunted notch with an initial radius b_0 (measured in the same units as R_0), as shown in Figure 23.1b. In front of the crack tip, a void of the same initial radius b_0 is introduced, with a crack–void separation of d_0 (Figure 23.1b). Another void is considered at 45° from the crack direction with same radius and crack-void distance as previous one, shown in Figure 23.1c. Here $b_0 = 0.001$ and $d_0 = 0.01$ taken. During loading, the plastic zone size is kept small in comparison to to assure the small-scale yielding

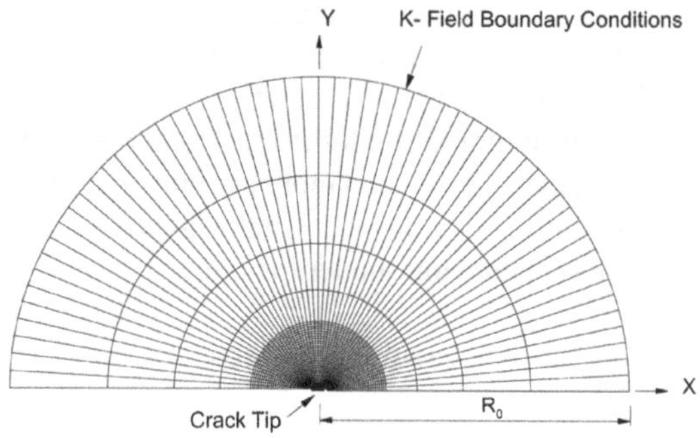

(a)

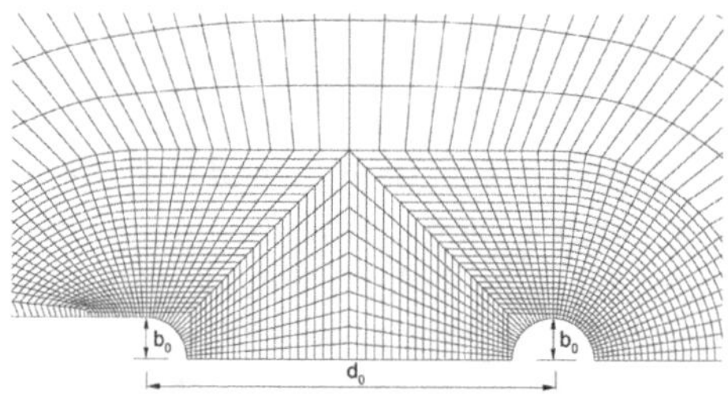

(b)

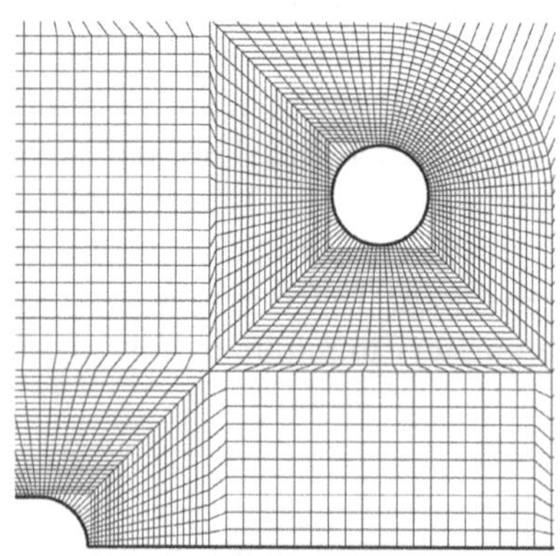

(c)

Figure 23.1 (a) Complete model, (b) near the crack tip with a void ahead of the crack, and (c) near the crack tip with a void positioned at 45° from the crack direction

Source: Authors

(SSY) conditions. Due to symmetry, it is sufficient to model only half of the specimen.

As described in the work of Mohan et al., two different materials (B and C) are used [9], have been selected for generating the numerical results. The function which governs hardness, $g(\varepsilon_p)$ of those mentioned materials are portrayed in Figure 23.2. Materials reported as B and C are bilinear hardening materials in which B has relatively higher hardening slope than that of C [9–11]. The parameters $h_1 = 24$ and $\varepsilon_1 = 0.085$ have been used for both the materials. Also, for the in this study, $E/\sigma_0 = 100$, strain rate for reference, $\dot{\varepsilon}_0 = 1$ and $m = 0.02$.

Using Equations (5) and (6), isotropic linear elastic mode I displacement field is specified along the outside boundary of the semi-circular domain, Figure 23.1a.

$$\dot{u}_1 = \frac{2(1+\nu)\dot{K}_I}{E}\sqrt{\frac{R}{2\pi}}\cos\left(\frac{\theta}{2}\right)\left[1 - 2\nu + \sin^2\left(\frac{\theta}{2}\right)\right] \quad (5)$$

and

$$\dot{u}_2 = \frac{2(1+\nu)\dot{K}_I}{E}\sqrt{\frac{R}{2\pi}}\sin\left(\frac{\theta}{2}\right)\left[2 - 2\nu - \cos^2\left(\frac{\theta}{2}\right)\right] \quad (6)$$

Where, K_I represents the stress intensity factor, valid under SSY, the applied J value i.e., J_{app} is

$$J_{app} = K_I^2(1 - \nu^2)/E \quad (7)$$

The formulation for the present finite element finite deformation problem has widely been used in the past [12], and consequently not included here for brevity. A constant value of $\dot{K_I}/\sigma_0\dot{\varepsilon}_0\sqrt{b} = 31.62$

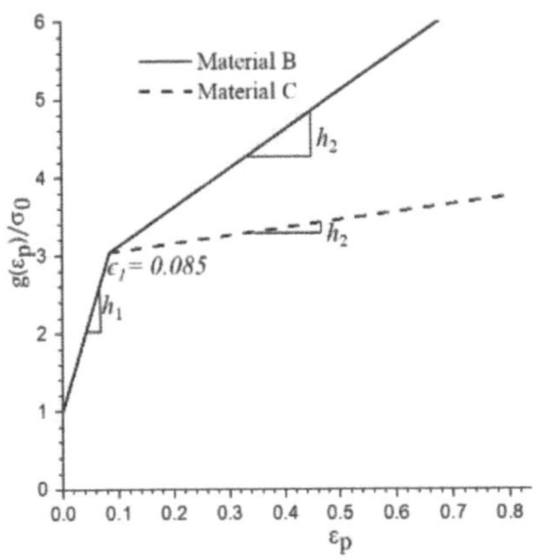

Figure 23.2 Hardness function $g(\varepsilon_p)$ with ε_p for B ($h_2 = 5.0$) and C ($h_2 = 1.0$), [10]
Source: Authors

is specified and corresponds to $\dot{K}_I = 1\, MPa\sqrt{m}s^{-1}$.. Numerical results have been generated for both $\alpha = 0.333333$ and $\alpha = 0.28$, which corresponds to plastically incompressible and compressible solids.

Linear incremental scheme employed for deformation history calculation with a $\Delta t = 0.0002$s, while constitutive update utilises the rate tangent method developed by Peirce [13] has been used. Finite element meshing is performed using quadrilateral elements, in which each subdivided into four crossed constant-strain triangular elements. These crossed quadrilateral elements, when properly oriented and with an appropriate aspect ratio, are widely employed to accurately capture localised deformation under large strain conditions [8].

Results and discussion

Initially mesh convergence study has been carried out to fix the number of finite elements in the simulation model. Maximum ϵ_p observed in vicinity of crack-tip and void has been observed at several mesh density levels for the B with $\alpha = 0.333333$ with $J_{app}/(\sigma_0 b_0) = 19$. It is seen that there is a decline in the relative error with increase in the number of elements, and the maximum value of ε_p for 3553 and 4285 elements are the same for the case where void is in line of the crack-tip. Looking at the correctness of the solution as well as the simulation time, the mesh density is fixed at 3553 elements for the case where void is in line of the crack-tip. Following similar approach, the number of elements for the case where void present at 45° from the crack direction is fixed to 6273. In vicinity of crack-tip and void, sufficient finer mesh is considered. The radial size of such a finite element is approximately $b_0/20$.

The verification of the current numerical model is further tested by comparing the near-tip ϵ_p distribution with the corresponding result of Aravas and McMeeking [14]. In their study, Aravas and McMeeking considered a plastically incompressible solid with $E/\sigma_0 = 300$, $\nu = 0.3$ and power-law material without hardening at $J_{app}/(\sigma_0 b_0) = 4$. Considering same simulation setting, the near-tip distribution of plastic strain is generated and illustrated in Figure 23.3. The distribution and the highest value of the plastic strain are nearly matching with those of Aravas and McMeeking [14]. The deformed crack-tip, void growth nature and crack-void ligament configuration are too similar with those of Aravas and McMeeking [14]. Thus, the present numerical scheme gives us

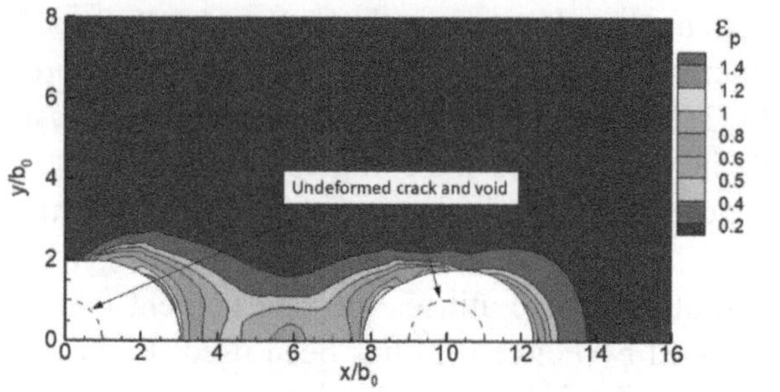

Figure 23.3 Distributions of plastic strain for non-hardening power law material at load of $J_{app}/(\sigma_0 b_0) = 4$ with $E/\sigma_0 = 300$ and $v = 0.3$
Source: Authors

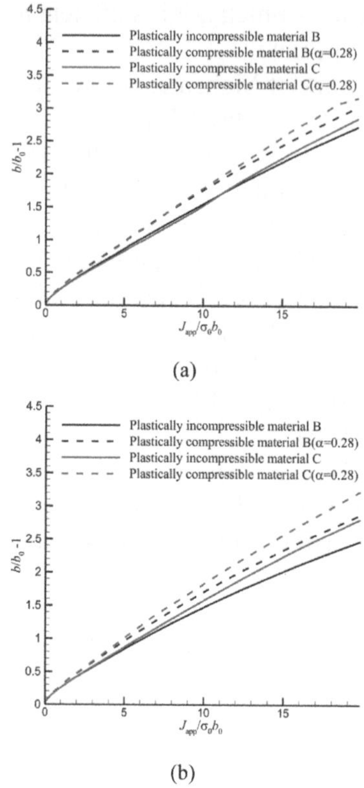

Figure 23.4 Plot for the CTOD ($b/b_0 - 1$) with applied J-integral ($J_{app}/\sigma_0 b_0$) (a) void in line of the crack-tip, and (b) void at 45° with the crack direction
Source: Authors

sufficient confidence to move forward for further studies.

Variation of CTOD and crack extension

The plot of CTOD has been shown in Figure 23.4, in which it can be observed that the CTOD is higher in plastically compressible solid relative to the corresponding of plastically incompressible solid. Difference in the CTOD values of plastically incompressible and compressible solids is more when the solid is having lower hardening slope, i.e., here it is reflected for material C. It is observed in the plastically incompressible material B (i.e., material with higher hardening slope), that CTOD is minimum in the case of void at 45° from crack direction. This result is qualitatively matching with the results of Aravas and McMeeking [14]. Here it is reflected that the CTOD is not significantly affected by the location of void, however, later it will be shown that several other parameters under investigation are very much affected by the location of void. For material C with $\alpha = 0.28$, slight oscillation in the CTOD curve is visible when the void is in line of the crack tip but not when the void is at 45° with the crack direction. It is also visible here that the CTOD plot with normalised J-integral is almost linear suggesting that fully developed HRR solution exists.

In Figure 23.5, ϵ_p distribution as well as deformed crack tip and void expansion for the plastically incompressible material C is shown for both void locations. When the void is in line of the crack-tip, the deformed void is like an oblate void shape whereas when the void is at 45°, the void is deformed into a prolated void shape with reduced void expansion. Further discussions on void growth are followed later. The contour shows that the CTOD is similar in

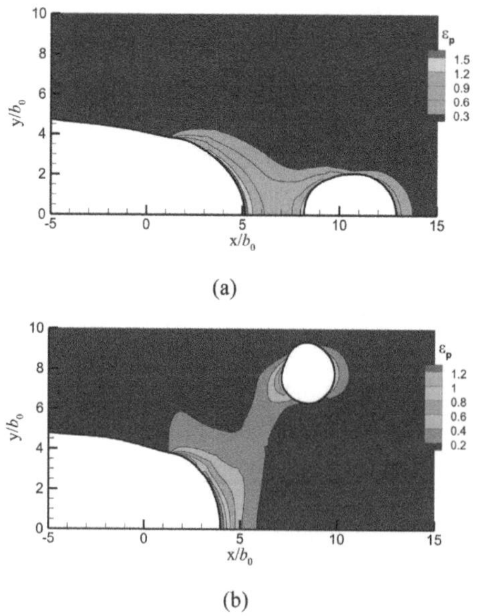

Figure 23.5 Distribution of ε_p in material C (with $\alpha = 0.333333$), $J_{app}/\sigma_0 b_0 = 19.74$ (a) void is in line of the crack tip, and (b) void is at 45° with the crack direction
Source: Authors

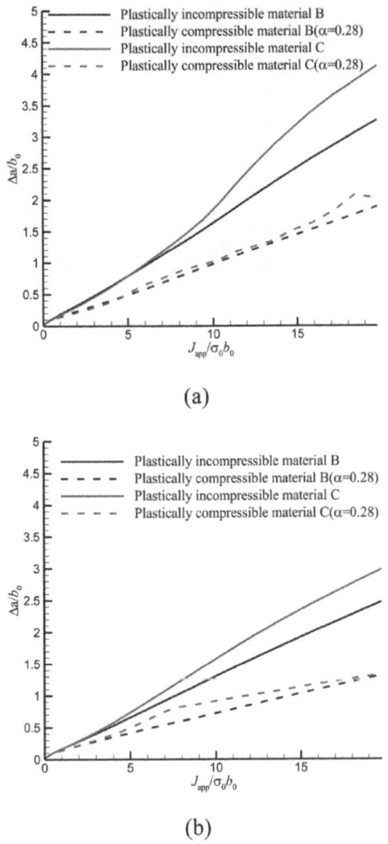

Figure 23.6 Plot for the crack extension with applied J-integral ($J_{app}/\sigma_0 b_0$) (a) void in line of the crack tip, and (b) void at 45° with the crack direction
Source: Authors

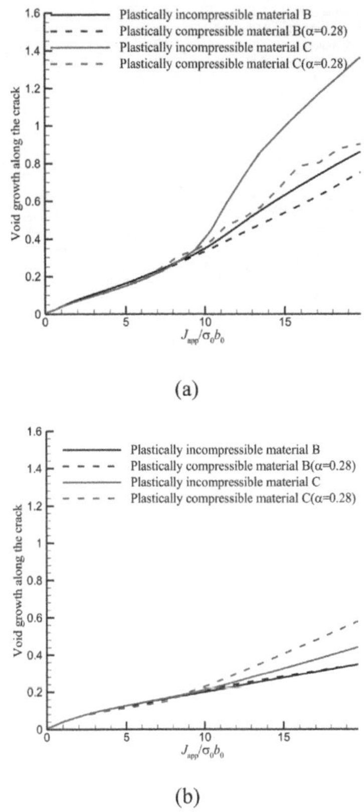

Figure 23.7 Plot for the growth of void along the crack ($a_2 / 2b_0 - 1$) with applied J-integral ($J_{app}/\sigma_0 b_0$) (a) void in line of the crack tip, and (b) void at 45° with the crack direction
Source: Authors

both cases, however, the plastic strain generated is lower when the void is at a 45° angle relative to the plastic strain when the void is in line of the crack-tip (same load level). For C with $\alpha = 0.28$, it has been it was noted that the void expansion is more in both the directions as compared to Figure 23.5b for the void location at 45°, but, when the void is in line of the crack tip, growth of void is more in the vertical direction.

In Figure 23.6, crack tip extension is plotted against the $J_{app}/\sigma_0 b_0$. Figure 23.6 shows that the crack extension is higher in plastically incompressible solids relative to same in plastically compressible solids and this behaviour is matching with our earlier studies, Pandey et al. [7]. Further, it is reflected that crack growth is less when the void is at 45° as compared to the corresponding of the case with other void location. This reduction in crack growth may be because of the fact that the crack-tip blunting is more when the void is at 45°, Figure 23.5. In plastically compressible solids, there is significant reduction in crack extension, Figure 23.6. In material C with $\alpha = 0.333333$, when the void is just in front of the crack tip, it is visible that there is little jump in the crack growth while with $\alpha = 0.28$, some fluctuations are observed in the crack extension curve.

Variation of void growth and ligament length

Theoretical calculation of void growth is an important parameter to study as the growth of void has critical influence in the ligament reduction and ultimately on failure. In this study, growth of void is characterised by calculating the growth in two directions, one along the crack by expression $a_2/2b_0-1$, and other in normal direction to the crack by expression $a_1/2b_0-1$, a_1 and a_2 are the deformed diameters of the void in normal and along the crack direction, respectively (Figure 23.7).

Figure 23.8 represents void growth ($a_2/2b_0-1$) along the crack direction while Figure 23.9 illustrates

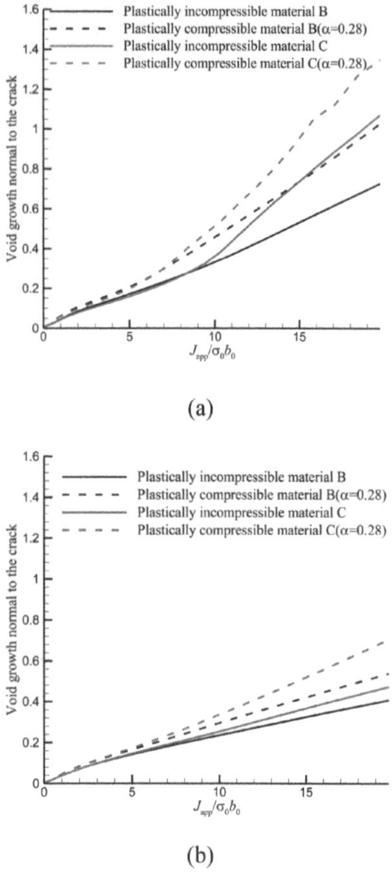

Figure 23.8 Plot for the void growth normal to the crack (a_1 / $2b_0$−1) with applied *J*-integral (J_{app}/ $\sigma_0 b_0$) (a) void in front of the crack tip, and (b) void at 45° with the crack direction
Source: Authors

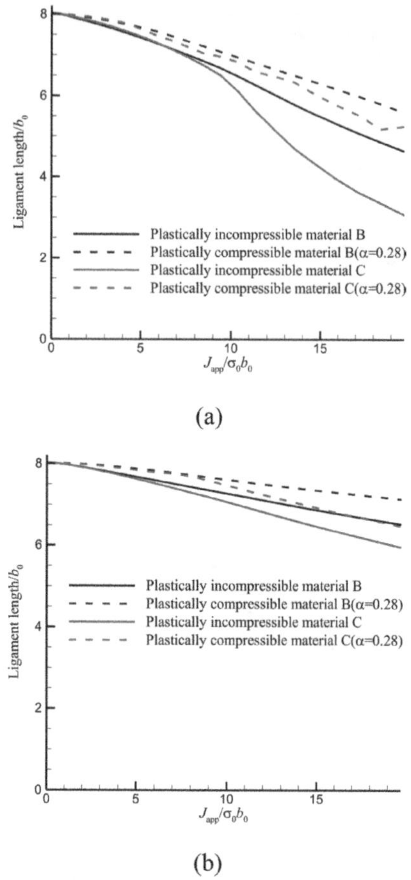

Figure 23.9 Plot for the ligament length with applied *J*-integral (J_{app}/ $\sigma_0 b_0$) (a) void in line of the crack tip, and (b) void at 45° with the crack direction
Source: Authors

void growth (a_1/$2b_0$−1) normal to the crack direction for both void locations. When the void is located at 45°, the void growth is appeared to be less for both the materials in comparison to the void growth for the geometry with void in line of the crack-tip. When void located in line of the crack-tip (Figures 23.7a and 8a), there is a sudden change in the void growth for material C with α = 0.333333 and this behaviour is observed in both Figures 23.7a and 23.8a. There are more or less oscillations observed in the void growth curve of material C with α = 0.28 and this is because of the plastic strain localisation as observed in Figure 23.6. With increase in load, this fluctuation is increased. Void growth is more along the crack direction when the material has α = 0.333333 and initial void is located in line of the crack tip. When the void is located at an angle of 45° as described, plastic compressibility dominates the void growth. When material have plastically compressibility and the initial void location is in line of the crack tip, growth of void is more in perpendicular to the crack-tip line and the deformed void is like a prolated void shape. Therefore, failure mechanism in such cases is not because of void coalescence (Figure 23.7a) but by a tearing mechanism (Figure 23.8a). In Figures 23.7b and 23.8b, when void is at 45° from crack direction, plastically compressible material C shows higher void growth in both directions (along and perpendicular to the crack). From Figures 23.7 and 23.8, it can be said that material with lower hardening slope (i.e., material C) experiences higher void growth.

Figure 23.9 shows the reduction in ligament length during loading. From both the Figures 23.9a,b, it is visible that plastic compressibility offers resistance in ligament length reduction. When the void is in line of the crack tip, for plastically incompressible C, the ligament reduction is maximum, Figure 23.9a.

On the other hand, when the void is at 45° angle, the reduction in ligament length is significantly less and this is because of the reduced crack and void growth. Minimum ligament reduction takes place for plastically compressible material B with void location at 45° angle with the crack direction.

Conclusion

This study investigates the crack-void interaction in plastically compressible and incompressible solids under mode I loading. The results highlight that the void's location significantly impacts crack extension, void growth, and ligament reduction. Specifically, when the void is at 45° from the crack direction, there is reduction in crack and void growth as compared to the same when void is directly in line of the crack tip. Plastically compressible solids exhibit higher CTOD and lower crack extension. Void in plastically compressible solid provides resistance to crack-void ligament reduction and failure mechanism in such cases is not because of void coalescence but by tearing mechanism. On the other hand, for a plastically incompressible solid, material with lower hardening slope exhibits higher crack-void ligament reduction.

References

[1] Rudnicki, J. W., Rice, J. R. Conditions for the localization of deformation in pressure-sensitive dilatant materials. *J. Mech. Phy. Solids*, 1975;23(6):371–394.

[2] Mittal, Y., Khan, D., Pandey, S., Gupta, G. C. Fatigue crack growth in plastically compressible solids: Role of negative stress ratio, plastic compressibility and strain softening. *J. Solid Mech.*, 2020;12(4):902–911.

[3] McMeeking, R. M. Finite deformation analysis of crack-tip opening in elastic-plastic materials and implications for fracture. *J. Mech. Phy. Solids*, 1977;25(5):357–381.

[4] Needleman, A., Tvergaard, V. A numerical study of void distribution effects on dynamic, ductile crack growth. *Engg. Frac. Mech.*, 1991;38(2–3):157–173.

[5] Spitzig, W. A., Richmond, O. Effect of hydrostatic pressure on the deformation behaviour of polyethylene and polycarbonate in tension and in compression. *Poly. Engg. Sci.*, 1979;19(16):1129–1139.

[6] Khan, D., Singh, S., Needleman, A. Finite deformation analysis of crack tip fields in plastically compressible hardening–softening–hardening solids. *Acta Mech. Sinica*, 2017;33:148–158.

[7] Pandey, S., Kumar, S., Khan, D. A detailed study on void growth and crack-void interaction in plastically compressible bilinear hardening and trilinear hardening-softening-hardening solids. *Mech. Adv. Mater. Struct.*, 2024:1–17.

[8] Bandil, P., Khan, D., Shah, P., Kaul, S., Goswami, R. Numerical simulation of void growth in front of a blunting crack-tip in plastically compressible solids. *J. Brazil. Soc. Mech. Sci. Engg.*, 2021;43(3):152.

[9] Mohan, N., Cheng, J., Greer, J. R., Needleman, A. Uniaxial tension of a class of compressible solids with plastic non-normality. *J. Appl. Mech.*, 2013;80(4):040912.

[10] Needleman, A., Hutchens, S. B., Mohan, N., Greer, J. R. Deformation of plastically compressible hardening-softening-hardening solids. *Acta Mech. Sinica*, 2012;28:1115–1124.

[11] Hutchens, S. B., Needleman, A., Greer, J. R. Analysis of uniaxial compression of vertically aligned carbon nanotubes. *J. Mech. Phy. Solids*, 2011;59(10):2227–2237.

[12] Pandey, S., Khan, D., Alam, I. Distribution of crack-tip stresses during fatigue loading with an overload event: Role of initial crack-tip shape, plastic compressibility and material softening. *J. Theor. Appl. Mech.*, 2021;59(2):239–250.

[13] Peirce, D., Shih, C. F., Needleman, A. A tangent modulus method for rate dependent solids. *Comp. Struct.*, 1984;18(5):875–887.

[14] Aravas, N., McMeeking, R. M. Finite element analysis of void growth near a blunting crack tip. *J. Mech. Phy. Solids*, 1985;33(1):25–49.

24 Non-linear dynamics of functionally graded electro-magneto-active elastomer-based actuators

Anurag Harsh[1], *Aman Khurana*[1,a] *and Deepak Kumar*[2]

[1]Department of Mechanical Engineering, Indian Institute of Technology Indore, Madhya Pradesh, India

[2]Department of Mechanical Engineering, Maulana Azad National Institute of Technology Bhopal, Madhya Pradesh, India

Abstract

In soft robotics, precisely controlling localised point loads is made possible by electro-magneto-active polymers, including a thickness gradient. Understanding their non-linear oscillation patterns is crucial for optimising actuator design during time-dependent motion. The present work investigates the non-linear dynamic characteristics of soft actuators that are composed of functionally graded electro-magneto-active elastomers. This dynamic behaviour is examined while considering the Gent model and characterising the thickness gradient by a power-law distribution. The governing equation of motion of the actuator is derived using the Euler-Lagrange framework, allowing us to analyse the non-linear behaviour of the system. The equilibrium states of the actuator over a range of defined functionally graded parameters are obtained along with the time-history responses, Poincaré plots, and phase diagrams to examine the stability, periodic behaviour, beating phenomena, and resonance characteristics of the system.

Keywords: Functionally graded elastomers, electro-magneto-active, non-linear oscillations, natural frequency

Introduction

Soft material-based smart actuators are increasingly pivotal in advancing modern technologies [1–4]. However, their development is often challenged by the need to meet the multifaceted requirements of modern intelligent system architectures. One of the primary concerns for innovators is selecting the most suitable material for a specific system design. This has spurred interest in innovative materials like electro-magneto-active (EMA) polymers, capable of responding rapidly to electrical, magnetic, and mechanical inputs [5–8]. EMA polymers are preferable because they can be subjected to significant deformation and having properties such as quick reaction, high strength to weight ratio and economically affordable [9]. Because of such special characteristics, EMA materials have proven to be a great choice for application in artificial muscles [10], tuneable lenses for vision [11], and minimum energy structures [12], etc. In most of the above-discussed applications, the non-linear dynamics of EMA materials have been a distinguishing feature. Consequently, there are various studies on non-linear dynamic behaviour of EMA-based actuators in the existing published literature [13–15].

Existing research on non-linear vibrations and dynamics has predominantly focused on homogeneous smart elastomers. Significant opportunities exist to enhance such elastomer performance by adopting functionally graded designs and providing tailored material properties to better adapt to varying conditions. Despite this potential, studies on the stability and modelling of functionally graded soft elastomers (FGSEs) still need to be explored, leaving much of this promising field unexplored. Presently, only a few papers [16–18] have addressed the stability and modelling of FGSEs. However, the dynamic behaviour analysis of FGSEs subjected to combined electro-magneto-mechanical loadings remains unexplored.

The advancements in actuator designs using EMA and electro-graded materials have sparked interest in developing innovative actuators based on functionally graded EMA polymers. This research examines the behaviour of electro-magneto-mechanical instability (EMMI) in functionally graded EMA polymeric actuators, with an emphasis on their

[a]amankhurana@iiti.ac.in

DOI: 10.1201/9781003770442-24

performance under static and DC dynamic actuation, aiming to support advancements in contemporary technology. To the authors' best understanding, the impact of functionally graded parameters on the EMMI behaviour of smart elastomers has yet to be explored in the existing literature. The potential impact of this study on the field of materials science and engineering, particularly in the development of smart materials, is significant.

Problem description and material model

Figure 24.1 presents an illustration of FGSE membrane subjected to an electrical load V, a magnetic field H, and an equal biaxial mechanical tensile load $P_1=P_2=P$. The schematic illustrates the deformation mapping of the system from the initial configuration (B_i) to the final configuration (B_f). The associated coordinate systems for these configurations are (X_1, X_2, X_3) and (x_1, x_2, x_3), respectively. The dimensions of the FGSE membrane in the initial and final configurations are represented as (L_1, L_2, L_3) and (l_1, l_2, l_3) in their respective coordinate systems. Let λ_1, λ_2, and λ_3 are the three principal stretches in the X_1, X_2, and X_3 directions, given by:

$$\lambda_1 = \frac{x_1}{X_1} \; ; \lambda_2 = \frac{x_2}{X_2} \; ; \text{and } \lambda_3 = \frac{x_3}{X_3} \tag{1}$$

And assuming equal biaxial ($L_1 = L_2 = L$) case with an applied Equi-biaxial loads yields $l_1 = l_2 = l$.

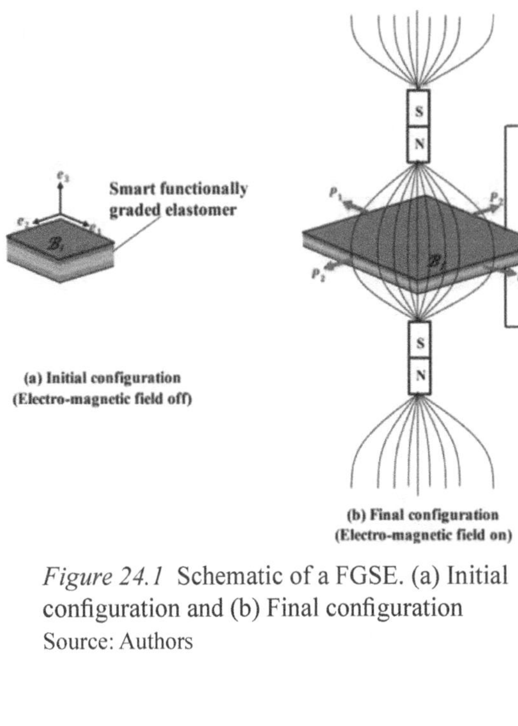

Figure 24.1 Schematic of a FGSE. (a) Initial configuration and (b) Final configuration
Source: Authors

Considering the membrane material is incompressible, which expressed as:

$$\lambda_1\lambda_2\lambda_3 = 1 \; ; \lambda_1 = \lambda_2 = \lambda \; ; \text{and } \lambda_3 = \lambda^{-2} \tag{2}$$

Assuming the mass density (ρ), shear modulus (μ), electric permittivity (ε), and magnetic permeability (α) vary smoothly with respect to thickness following the power law [19] and all of such properties vary with the same law gradient index K. Such as assumption is mathematically expressed as:

$$\rho(x_3) = \rho_2 + (\rho_1 - \rho_2)\left(\frac{X_3}{L_3}\right)^K \tag{3}$$

where, $\rho = \rho_1$ at $X_3 = L_3$, and $\rho = \rho_2$ at $X_3 = 0$.

$$\text{Similarly, } \varepsilon(X_3) = \varepsilon_2 + (\varepsilon_1 - \varepsilon_2)\left(\frac{X_3}{L_3}\right)^K \tag{4}$$

$$\mu(X_3) = \mu_2 + (\mu_1 - \mu_2)\left(\frac{X_3}{L_3}\right)^K \tag{5}$$

$$\text{and, } \alpha(X_3) = \alpha_2 + (\alpha_1 - \alpha_2)\left(\frac{X_3}{L_3}\right)^K \tag{6}$$

The associated kinetic energy of the system is given by

$$KE = \int_0^{l_3}\int_0^{l_2}\int_0^{l_1} \frac{1}{2}\left\{\rho_2 + (\rho_1 - \rho_2)\left(\frac{X_3}{L_3}\right)^K\right\} (\dot{x}_1^2 + \dot{x}_2^2 + \dot{x}_3^2)dx_1 dx_2 dx_3 \tag{7}$$

The solutions of the above equation yields

$$KE = \frac{1}{3}\frac{(\rho_1 + \rho_2 K)}{(1+K)} L_3 L^2 \dot{\lambda}^2 \tag{8}$$

As the considered smart elastomeric membrane is too thin as compare to lateral dimensions, so that $\left(\frac{L_3}{L}\right)^2$ or higher order terms are neglected in Equation 8. Based on the stated condition, the total potential energy of the system is given by:

$$PE = \int_0^{l_3}\int_0^{l_2}\int_0^{l_1} (\xi_h + \xi_e + \xi_m)dx_1 dx_2 dx_3 - W_d \tag{9}$$

Where,

$\xi_h = \frac{\mu J_m}{2} In\left(1 - \frac{\lambda_1^2 + \lambda_2^2 + \lambda_3^2 - 3}{J_m}\right)$ (given by Gent strain energy model [20])

$\xi_e = \frac{1}{2}\varepsilon\left(\frac{V}{L_3}\right)^2 \lambda^4 \; ; \xi_m = \alpha H^2 \lambda^{-4}$ and

$W_d = \int_{X_1}^{x_1} P dx_1 + \int_{X_2}^{x_2} P dx_2$

Upon invoking the definition of parameters from Equations (4–6) in Equation (9), we get:

$$PE = \begin{cases} \frac{(\mu_1+\mu_2 K)J_m L_3 L^2}{2(K+1)} \ln\left(1 - \frac{2\lambda^2+\lambda^{-4}-3}{J_m}\right) - \frac{(\varepsilon_1+\varepsilon_2 K)L_3 L^2}{2(1+K)}\left(\frac{V}{L_3}\right)^2 \lambda^4 \\ -\frac{(\alpha_1+\alpha_2 K)L_3 L^2}{2(1+K)} H^2\lambda^{-4} - 2\left(\frac{P}{L_3 L}\right) L_3 L^2(\lambda - 1) \end{cases} \quad (10)$$

The Lagrangian of the system is given as:

$$L = KE - PE \quad (11)$$

Further, the equation of motion of the system, is given by Euler-Lagrange method is expressed as:

$$\frac{d}{dt}\left(\frac{\partial L}{\partial \lambda}\right) - \frac{\partial L}{\partial \lambda} = 0 \quad (12)$$

After incorporating the definition of Lagrangian (L) and implementing various non-dimensional parameters in Equation 12, the non-dimensional form of the governing equation of motion is given by:

$$(1 + aK)\frac{d^2\lambda}{d\tau^2} + (1 + bK)\frac{J_m(\lambda-\lambda^{-5})}{(J_m-2\lambda^2-\lambda^{-4}+3)} - (1 + cK)\bar{V}^2\lambda^3$$
$$- (1 + dK)\bar{H}^2\lambda^{-5} - (1 + K)\bar{\sigma} = 0 \ (13) \quad (13)$$

where,

$$a = \frac{\mu_2}{\mu_1}; b = \frac{\rho_2}{\rho_1}; c = \frac{\varepsilon_2}{\varepsilon_1}; d = \frac{\alpha_2}{\alpha_1}; \tau = \frac{t}{L\sqrt{\rho_1/3\mu_1}};$$
$$\bar{V} = \sqrt{\varepsilon_1/\mu_1}\left(\frac{V}{L_3}\right); \bar{H} = H\sqrt{\alpha_1/\mu_1}\text{ ; and } \bar{\sigma} = \frac{P}{L_3 L\mu_1}.$$

Results and discussions

The proposed model developed in the previous section is now applied to investigate the influence of functionally graded parameters on the EMMI behaviour of EMA-based polymer actuators, considering both static and DC dynamic actuation modes.

This study utilises energy-based technique [21] to calculate the static response. Using this approach, the potential energy is derived in a non-dimensional form as follows:

$$\overline{PE} = -\frac{(1+bK)J_m}{2(K+1)}\ln\left(1 - \frac{2\lambda^2+\lambda^{-4}-3}{J_m}\right) - \frac{(1+cK)}{2(1+K)}\bar{V}^2\lambda^4$$
$$- \frac{(1+dK)}{2(1+K)}\bar{H}^2\lambda^{-4} - 2\bar{\sigma}(\lambda - 1) \quad (14)$$

The derivative of the preceding equation, gives the static equilibrium equation as:

$$\frac{d\overline{PE}}{d\lambda} = (1 + bK)\frac{J_m(\lambda-\lambda^{-5})}{(J_m-2\lambda^2-\lambda^{-4}+3)} - (1 + cK)\bar{V}^2\lambda^3$$
$$+ (1 + dK)\bar{H}^2\lambda^{-5} - (1 + K)\bar{\sigma} = 0 \quad (15)$$

The above equation provides the stretch for a given mechanical, electrical, and magnetic load. The double derivative of the Equation 15 with respect to stretch is written as:

$$\frac{d^2\overline{PE}}{d\lambda^2} = \begin{Bmatrix} (1 + bK)J_m\left(\frac{(1+5\lambda^{-6})}{(J_m-2\lambda^2-\lambda^{-4}+3)} + \frac{4(\lambda-\lambda^{-5})^2}{(J_m-2\lambda^2-\lambda^{-4}+3)^2}\right) \\ -3(1 + cK)\bar{V}^2\lambda^2 - 5(1 + dK)\bar{H}^2\lambda^{-6} \end{Bmatrix} = 0 \ (16)$$

The critical values and the instability of the membrane are determined by a numerical solution to the preceding Equations 15–16, for obtaining the static critical dimensionless voltage (V_S^C) and corresponding static critical stretch (λ_S^C), or static critical dimensionless magnetic field (M_S^C) and corresponding static critical stretch (λ_S^C) values as local maxima when providing the constant dimensionless magnetic field or dimensionless voltage value. The actuated stretch (λ^a) is given as:

$$\lambda_S^a = \frac{\lambda_S}{\lambda_{ps}}\text{ (for static actuation) and,}$$
$$\lambda_D^a = \frac{\lambda_D}{\lambda_{ps}}\text{ (for dynamic actuation)}$$

To get the initial stretch (λ_{ps}) that is caused by the prestress only is obtained by putting the values of $\bar{V} = 0$ and $\bar{H} = 0$ in Equation 15. After invoking the values, we get

$$(1 + bK)\frac{J_m(\lambda_{ps}-\lambda_{ps}^{-5})}{(J_m-2\lambda_{ps}^2-\lambda_{ps}^{-4}+3)} - (1 + K)\bar{\sigma} = 0 \quad (17)$$

The solution of the Equation 17 provides the value of λ_{ps}. Energy-based approach is used to get the dynamic actuation. The non-dimensional Hamiltonian is given as:

$$\bar{h} = \overline{KE} + \overline{PE} \quad (18)$$

$$\bar{h} = \frac{(1+aK)}{(1+K)}\dot{\lambda}^2 - \frac{(1+bK)J_m}{2(K+1)}\ln\left(1 - \frac{2\lambda^2+\lambda^{-4}-3}{J_m}\right)$$
$$- \frac{(1+cK)}{2(1+K)}\bar{V}^2\lambda^4 - \frac{(1+dK)}{2(1+K)}\bar{H}^2\lambda^{-4} - 2\bar{\sigma}(\lambda - 1) \quad (19)$$

Due to the conservative nature of the system, we can write

$$\Delta\bar{h} = \bar{h}_2 - \bar{h}_1 = 0 \quad (20)$$

$$\Delta\bar{h} = \begin{Bmatrix} -\frac{(1+bK)J_m}{2(1+K)}\ln\left(\frac{J_m-2\lambda_D^2-\lambda_D^{-4}+3}{J_m-2\lambda_{ps}^2-\lambda_{ps}^{-4}+3}\right) - \frac{(1+cK)}{2(1+K)}\bar{V}^2(\lambda_D^4 - \lambda_{ps}^4) \\ -\frac{(1+dK)}{2(1+K)}\bar{H}^2(\lambda_D^{-4} - \lambda_{ps}^{-4}) - 2\bar{\sigma}(\lambda_D - \lambda_{ps}) \end{Bmatrix} = 0 \ (21)$$

Where, at time $\tau = 0$, $\lambda = \lambda_{ps}$ and $\dot{\lambda} = 0$ and at time $\tau = \tau_2$, $\lambda = \lambda_D$ and $\dot{\lambda} = 0$.

The solution of Equation 21 provides the λ_D vs. $\bar{V}$ and λ_D vs. $\bar{H}$ graph and the solution of $\partial \bar{h}/\partial \lambda = 0$ gives the corresponding critical values on specific given conditions. Plots between stretch vs. dimensionless electrical load ($\bar{V}$) and stretch vs. magnetic field ($\bar{H}$) of static and dynamic actuation of the functionally graded electro-magneto-active (FGEMA) membrane have been generated for $J_m = 100$, a = b = c = d = 0.5, various values of gradient index (*K*), prestress ($\bar{\sigma}$) and the constant value of $\bar{H} = 0.4$ and $\bar{V} = 0.4$. All the critical values obtained, are listed in Tables 24.1–24.3. The data in these tables reveal that if prestress condition has been varied for the above-described constant electric and magnetic loading condition of FGEMA membrane, the static and dynamic instability increases for the lateral stretch as the functional grading of the membrane becomes more pronounced. As K increased from 0 to 2.5 with $\bar{\sigma}$ increasing from 0.1 to 0.5. Accordingly, critical stretch for unit critical dimensionless voltage (λ^C/V^C) and for unit critical dimensionless magnetic field (λ^C/M^C) also increased with both static and dynamic actuation. λ^C/V^C increased from 15.5% to 45.2% (static) and 4.4% to 32.2% (dynamic), whereas λ^C/M^C increased from 4.8% to 98.7% (static) and 1.9% to 141.

Figure 24.2a–c depicts how prestress and gradient index impact the stretch response to the electric field in both scenarios. As the value of prestress ($\bar{\sigma}$) and gradient index (*K)* increase the critical value of $\bar{V}$ decreases and the critical value of λ increases. Figure 24.2d–f shows the effect of prestress and gradient index on the magnetic field-stretch response in both modes of actuation. As the value of $\bar{\sigma}$ and *K*

Table 24.1 Effect of parameter K (gradient index) on critical electrical and magnetic load (when $\bar{\sigma} = 0.1$, $J_m = 100$ and a = b = c = d = 0.5).

	$\bar{H} = 0$, $\bar{V} = 0$	$\bar{H} = 0.4$				$\bar{V} = 0.4$			
		Static		Dynamic		Static		Dynamic	
	λ_{ps}	λ_S^C	V_S^C	λ_D^C	V_D^C	λ_S^C	M_S^C	λ_D^C	M_D^C
K = 0.0	1.0174	1.3390	0.6215	1.5790	0.5946	2.248	5.2745	2.572	1.9216
K = 0.5	1.0211	1.3570	0.6075	1.5895	0.5885	2.239	5.1656	2.560	1.9006
K = 1.0	1.0235	1.3735	0.5961	1.5940	0.5845	2.233	5.0930	2.554	1.8863
K = 1.5	1.0253	1.3900	0.5863	1.5985	0.5816	2.227	5.0411	2.548	1.8759
K = 2.0	1.0267	1.4035	0.5776	1.6000	0.5794	2.224	5.0022	2.542	1.8681
K = 2.5	1.0277	1.4185	0.5696	1.6015	0.5777	2.221	4.9719	2.539	1.8619

Source: Authors

Table 24.2 Effect of parameter K (gradient index) on critical electrical and magnetic load (when $\bar{\sigma} = 0.25$, $J_m = 100$ and a = b = c = d = 0.5).

	$\bar{H} = 0$, $\bar{V} = 0$	$\bar{H} = 0.4$				$\bar{V} = 0.4$			
		Static		Dynamic		Static		Dynamic	
	λ_{ps}	λ_S^C	V_S^C	λ_D^C	V_D^C	λ_S^C	M_S^C	λ_D^C	M_D^C
K = 0.0	1.0465	1.3870	0.5718	1.6405	0.5493	2.176	4.4570	2.479	1.7503
K = 0.5	1.0572	1.4200	0.5492	1.6630	0.5345	2.149	4.1836	2.446	1.6852
K = 1.0	1.0646	1.4485	0.5328	1.6795	0.5247	2.131	4.0009	2.425	1.6391
K = 1.5	1.0700	1.4725	0.5198	1.6915	0.5177	2.119	3.8701	2.407	1.6045
K = 2.0	1.0741	1.4935	0.5091	1.7005	0.5125	2.107	3.7718	2.395	1.5778
K = 2.5	1.0774	1.5130	0.4999	1.7080	0.5085	2.101	3.6952	2.383	1.5564

Source: Authors

Table 24.3 Effect of parameter K (gradient index) on critical electrical and magnetic load (when $\bar{\sigma} = 0.5$, $J_m = 100$ and a = b = c = d = 0.5).

	$\bar{H} = 0$, $\bar{V} = 0$	$\bar{H} = 0.4$				$\bar{V} = 0.4$			
		Static		Dynamic		Static		Dynamic	
	λ_{ps}	λ_S^C	V_S^C	λ_D^C	V_D^C	λ_S^C	M_S^C	λ_D^C	M_D^C
K = 0.0	1.1054	1.4980	0.4932	1.7710	0.4768	2.032	3.0764	2.296	1.3641
K = 0.5	1.1335	1.5685	0.4590	1.8370	0.4493	1.963	2.5040	2.209	1.1465
K = 1.0	1.1539	1.6270	0.4364	1.8880	0.4316	1.915	2.1071	2.143	0.9632
K = 1.5	1.1695	1.6750	0.4203	1.9255	0.4193	1.876	1.8104	2.092	0.8005
K = 2.0	1.1817	1.7185	0.4080	1.9570	0.4102	1.843	1.5763	2.050	0.6481
K = 2.5	1.1915	1.7560	0.3983	1.9810	0.4033	1.816	1.3838	2.014	0.4955

Source: Authors

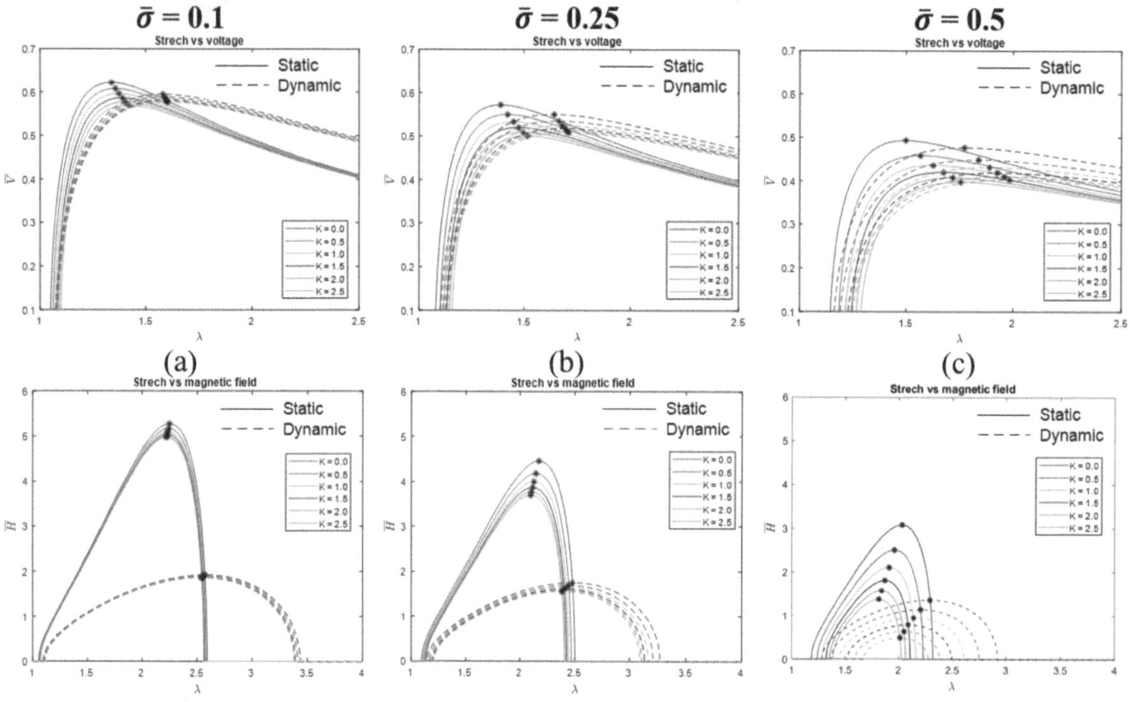

Figure 24.2 Stretch of the membrane for various values of K and $\bar{\sigma}$, $J_m = 100$, a = b = c = d = 0.5. (a–c) λ vs. $\bar{V}$ when $\bar{H} = 0.4$ and, (d–f) λ vs. $\bar{H}$ when $\bar{V} = 0.4$

Source: Authors

increase the critical value of λ decreases and the critical value of also decreases.

Figure 24.3a illustrates how changes in the magnetic load strength ($\bar{H} = 0.4$, 1.0, and 2.0) influence both the static and DC dynamic stretch, along with the electric voltage applied to the membrane. It inferred that as the value of $\bar{H}$ increases the value of limit point electric load decreases and stretches at the beginning of instability increases. Likewise, Figure 24.3b demonstrates the influence on $\bar{H}$ and corresponding stretch due to variation in electric loads, $\bar{V} = 0.30$, 0.35 and 0.40, when $\bar{\sigma} = 0.5$, $J_m = 100$, K = 1, and a = b = c = d = 0.5. It is concluded that as the value of the electric load increases, the stretch at

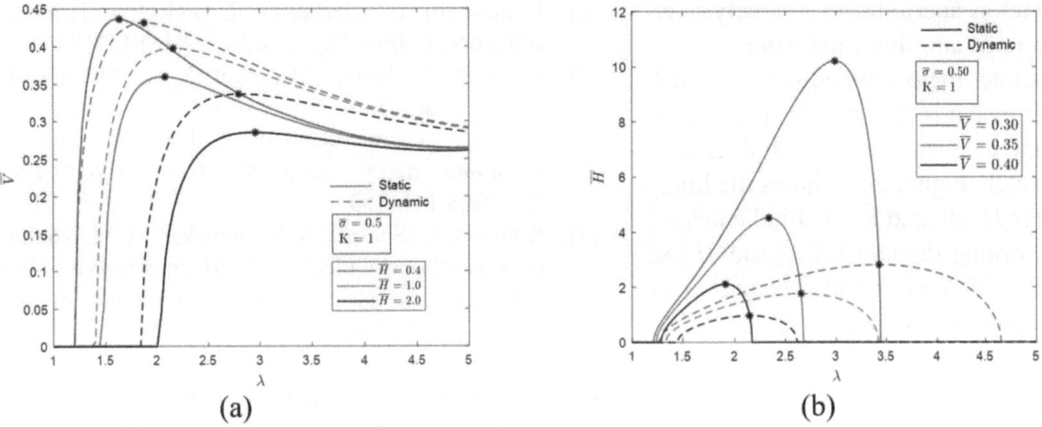

Figure 24.3 Stretch of the membrane for $J_m = 100$, K = 1, $\bar{\sigma} = 0.5$, a = b = c = d = 0.5 (a) λ vs. $\bar{V}$ for $\bar{H} = 0.4$, 1, 2 and, (b) λ vs. $\bar{H}$ for $\bar{V} = 0.30$, 0.35 and 0.40
Source: Authors

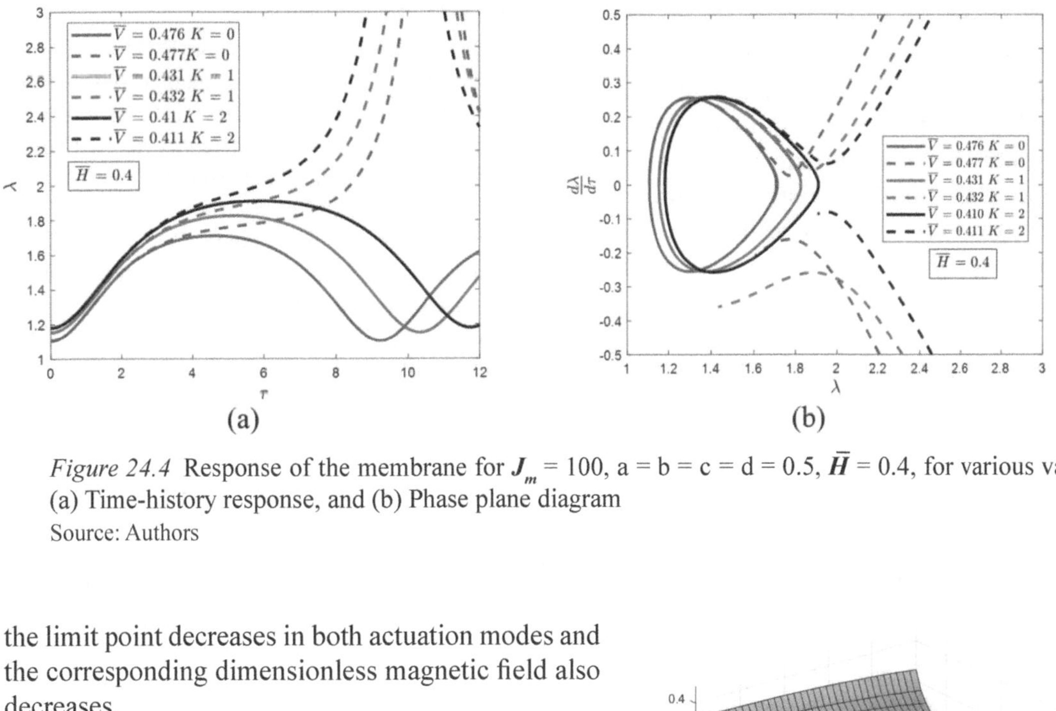

Figure 24.4 Response of the membrane for $J_m = 100$, a = b = c = d = 0.5, $\bar{H} = 0.4$, for various values of K and $\bar{V}$, (a) Time-history response, and (b) Phase plane diagram
Source: Authors

the limit point decreases in both actuation modes and the corresponding dimensionless magnetic field also decreases.

Figures 24.4a, b demonstrate the time-history response of the dynamic stretch actuation and phase-plane plots respectively. The required plots are obtained by solving Equation 13 through numerical methods using the Runge-Kutta technique. The plots show the DC dynamic response of the membrane for the value of just below and above of limit points dimensionless electrical load at different values of gradient index when $\bar{H} = 0.4$. It is inferred from the solution that as the value of $\bar{V}$ just below or approaching the limit point, the stretch is smooth and periodic and as the value of $\bar{V}$ above the limit point,

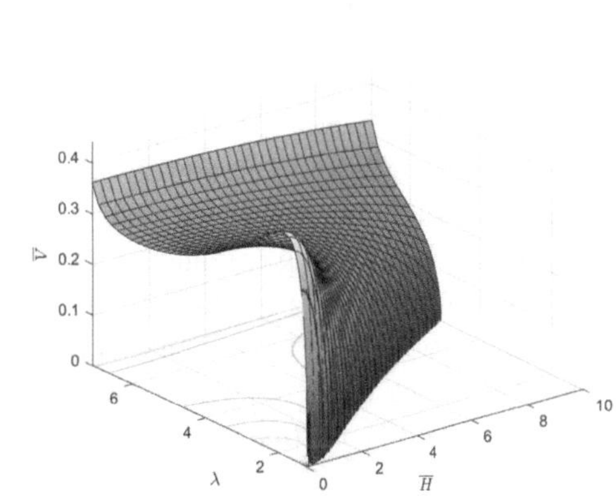

Figure 24.5 Dynamic lateral stretch (λ) vs. the dimensionless electric load ($\bar{V}$) and the dimensionless magnetic field variable ($\bar{H}$) for $J_m = 100$, K = 1, $\bar{\sigma} = 0.5$, a = b = c = d = 0.5
Source: Authors

the membrane stretch is aperiodic, extensively overshoot and undergo large amplitude actuation.

Figure 24.5 illustrates the combined variation of $\bar{V}$ and $\bar{H}$ on the values of λ. It is observed from the curve as the value of $\bar{H}$ increases the values of $\bar{V}$ decreases for the constant stretch. Figure 24.5 shows the hump at around $\lambda = 1.5$ for $\bar{H} = 0$, and $\bar{V} = 0.4$ indicates the limit point. The working domain for $\bar{V}$ and $\bar{H}$ can be determined for stable and optimise actuation by using the derived equation for specific stretch.

Concluding remarks

The present study presents a theoretical framework to investigate the static and DC dynamic response of a FGEMA elastomer-based actuator. To incorporate the functionality in the smart elastomer, the critical influencing parameters, i.e., mass density, shear modulus, electric permittivity, and magnetic permeability, are varied by the power law with the same law gradient index in the thickness direction. It is assumed that membrane is thin and deforms homogeneously. This study presents the deformation of the membrane under static and DC dynamic electric loading under varied prestress, magnetic field, and power law gradient index. The following are the critical findings of the present study:

i. Gradient index and prestress significantly govern the static and dynamic response of the smart EMA membrane.
ii. As the values of the gradient index and prestress increase, the instability of the membrane is enhanced.
iii. Increment in gradient index results in the enhancement of voltage actuation range and stretch.

By tuning the gradient index, the response of the smart electro-magneto-active membrane can be more precisely tailored, enabling greater control and versatility. This optimisation not only enhances its performance across a wide range of engineering applications but also opens up new possibilities for precise control and versatility in smart materials.

References

[1] Guo, Y., Liu, L., Liu, Y., Leng, J. Review of dielectric elastomer actuators and their applications in soft robots. *Adv. Intell. Sys.*, 2021:2000282.

[2] Hajiesmaili, E., Clarke, D. R. Dielectric elastomer actuators. *J. Appl. Phys.*, 2021;129(15):151102.

[3] Khurana, A., Kumar, A., Raut, S. K., Sharma, A. K., Joglekar, M. M. Effect of viscoelasticity on the nonlinear dynamic behaviour of dielectric elastomer minimum energy structures. *Int. J. Solids Struct.*, 2021;208:141–153.

[4] Khurana, A., Sharma, A. K., Joglekar, M. M. Nonlinear oscillations of electrically driven aniso-visco-hyperelastic dielectric elastomer minimum energy structures. *Nonlinear Dynam.*, 2021;104(3):1991–2013.

[5] Kumar, D., Ghosh, S., Roy, S., Santapuri, S. Modeling and analysis of an electropneumatic braided muscle actuator. *J. Intell. Mater. Syst. Struct.*, 2021;32(4):399–409.

[6] Alameh, Z., Yang, S., Deng, Q., Sharma, P. Emergent magnetoelectricity in soft materials, instability, and wireless energy harvesting. *Soft Matter.*, 2018;14(28):5856–5868.

[7] Patra, A. K., Khurana, A., Kumar, D. Modeling and analysis of a thermo-electro-magneto-viscoelastic actuator. *Int. J. Appl. Mech.*, 2023;16(02):2450015.

[8] Kumar, D., Sarangi, S., Bhattacharyya, R. Universal relations in nonlinear electro-magneto-elasticity. *Arch. Appl. Mech.*, 2020;90(7):1643–1657.

[9] Kumar, D., Sarangi, S., Saxena, P. Universal relations in coupled electro-magnetoelasticity. *Mech. Mater.*, 2020;143:103308.

[10] Bar-Cohen, Y. Electroactive polymers as artificial muscles-reality and challenges. *19th AIAA Appl. Aerodyn. Conf.*, 2001:1492.

[11] Carpi, F., Frediani, G., Turco, S., De Rossi, D. Bioinspired tunable lens with muscle-like electroactive elastomers. *Adv. Funct. Mater.*, 2011;21(21):4152–4158.

[12] Khurana, A., Naskar, S., Varma, R. K., Mukhopadhyay, T. Smart electro-magneto-viscoelastomer minimum energy structures with particle-reinforcements: Theoretical equilibrium and nonlinear dynamics of actuated configurations. *Int. J. Eng. Sci.*, 2024;194:103974.

[13] Agrawal, A., Khurana, A. Unveiling the dynamics of particle-reinforced electro-magneto-active circular membrane. *Int. J. Non Lin. Mech.*, 2025;174:105064.

[14] Patra, A. K., Khurana, A., Kumar, D., Saxena, P. Impact of compliant electrodes on the dynamics of electromagnetoactive membranes. *Int. J. Non Lin. Mech.*, 2024;167:104906.

[15] Khurana, A., Kumar, D., Sharma, A. K., Joglekar, M. M. Nonlinear oscillations of particle-reinforced electro-magneto-viscoelastomer actuators. *J. Appl. Mech.*, 2021;88(12):121002.

[16] Su, Y., Ogden, R. W., Destrade, M. Bending control and stability of functionally graded dielectric elastomers. *Ext. Mec. Lett.*, 2021;43:101162.

[17] Zhou, W., Chen, Y., Su, Y. Bifurcation of a finitely deformed functionally graded dielectric elastomeric tube. *Int. J. Non Lin. Mech.*, 2020;127:103593.

[18] Alibakhshi, A., Rahmanian, S., Destrade, M., Zurlo, G. Local and global dynamics of a functionally graded dielectric elastomer plate. *Int. J. Eng. Sci.*, 2024;195:103987.

[19] Pascon, J. P. Large deformation analysis of functionally graded visco-hyperelastic materials. *Comput. Struct.*, 2018;206:90–108.

[20] Gent, A. N. A new constitutive relation for rubber. *Rub. Chem. Tech.*, 1996;69(1):59–61.

[21] Khurana, A., Kumar, D., Sharma, A. K., Joglekar, M. M. Static and dynamic instability modelling of electro-magneto-active polymers with various entanglements and crosslinks. *Int. J. Non Lin. Mech.*, 2022; 139:103865.

25 Effect of wind load directions on floating photovoltaic arrays

Safwan N. K.[a], *Arun C. O.*[b] *and Gokulnath C.*[c]

Department of Civil Engineering, Indian Institute of Technology, Palakkad, Kerala, India

Abstract

Floating photovoltaic (FPV) systems represent a suitable green alternative to capture solar energy, especially where land space is scarce for land-based solar plants. Like land-based solar systems, FPV systems are predominantly subjected to wind. In this study, the effect of wind flow on floating solar panels has been studied numerically using computational fluid dynamics (CFD) analysis in ANSYS fluent. The drag and lift coefficients were estimated on the FPV array for different wind directions. The geometry and boundary conditions for the model were adopted from the existing literature. The simulations are performed with an inlet wind speed of 35.9 m/s, replicating site conditions while the outlet pressure is set to atmospheric levels. The results show that the wind directions 0° and 180° showed higher drag and lift forces than the others. The gap between the panels and the bottom of the boundary was found to have a significant impact on force variations along the array of panels.

Keywords: Floating photovoltaic (FPV), wind load analysis, floating solar panels, computational fluid dynamics (CFD), drag coefficient, lift coefficient, structural stability, renewable energy systems

Introduction

Floating photovoltaic (FPV) systems have become a promising opportunity for harvesting solar power, mainly in water-scarce regions. The market need for FPV systems is escalating due to the growing necessity for alternative energy sources. These systems offer advantages such as land conservation and enhanced cooling effects from water bodies [1]. The rapid development of FPV technology has also brought advancements in the materials and efficiency of solar panels. Various FPV technologies are now being applied to FPV systems, with recent studies comparing the performance of different PV module types in diverse environmental conditions [2]. Innovations in materials, anchoring mechanisms, and array configurations have promoted resilience and adaptability to water-based environments such as reservoirs and lakes. The above developments bring new challenges, such as the evaluation of external forces, including wind and waves, which will be required to determine their strength and stability check on the system [3].

Studies have shown that wind loading significantly affects the strength and stability of the FPV system, with variations in drag and lift forces depending on panel orientation and wind direction [4, 5]. Existing design codes lack provisions for quantifying wind forces on solar panels since they were primarily developed for the wind load estimation of building systems. So, one must rely on the experimental method or numerical analysis, primarily through computational fluid dynamics (CFD). As a method for assessing drag and lift forces acting on FPV arrays, CFD plays a crucial role in understanding fluid-structure interactions. By quantifying these forces, researchers can identify optimal design strategies to reduce structural stress and enhance the structural stability of FPV systems.

This paper presents a detailed CFD analysis to quantify these forces on an FPV array. The analysis will aid in identifying the critical areas that may lead to further understanding of force distribution over the panels. The primary objective of the current study is to understand the effect of wind direction on the drag and lift experienced by the FPV system.

CFD analysis of FPV

A panel array with boundary conditions taken from ref [6], as shown in Figures 25.1 and 25.2, is considered for the study. The configuration includes four rows with eight panels each, totalling 32 panels. Each solar panel is of length *(l)* 2 m, width (b) 1 m,

[a]safvankodur7@gmail.com, [b]arunco@iitpkd.ac.in, [c]gokul@iitpkd.ac.in

DOI: 10.1201/9781003770442-25

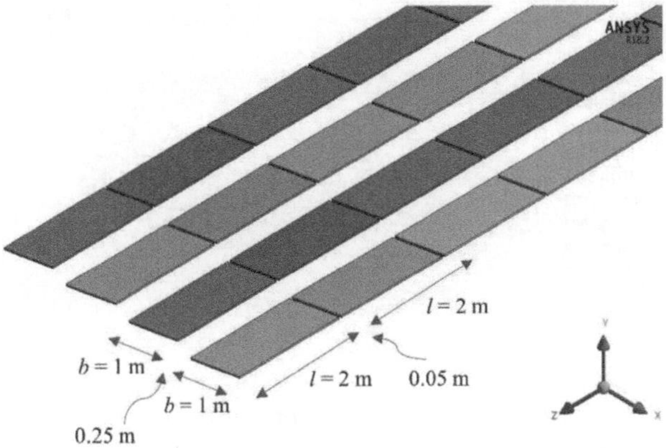

(a) Enlarged view

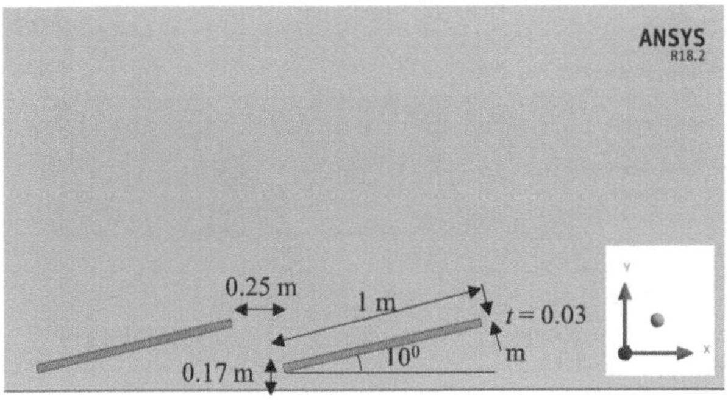

(b) Side view

Figure 25.1 Geometry of the panel array. (a) Enlarged view. (b) Side view
Source: Authors

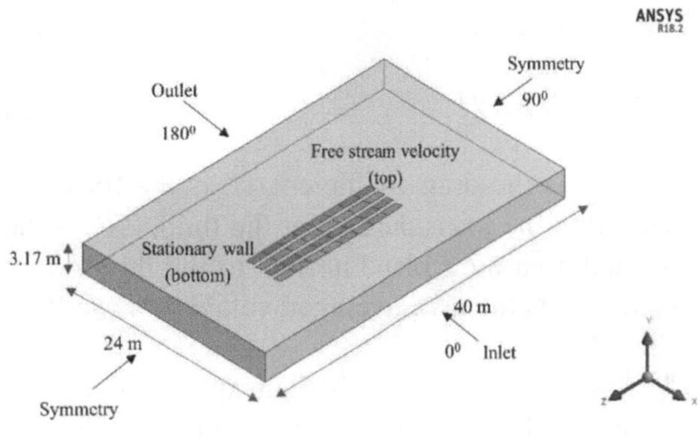

Figure 25.2 Boundary conditions
Source: Authors

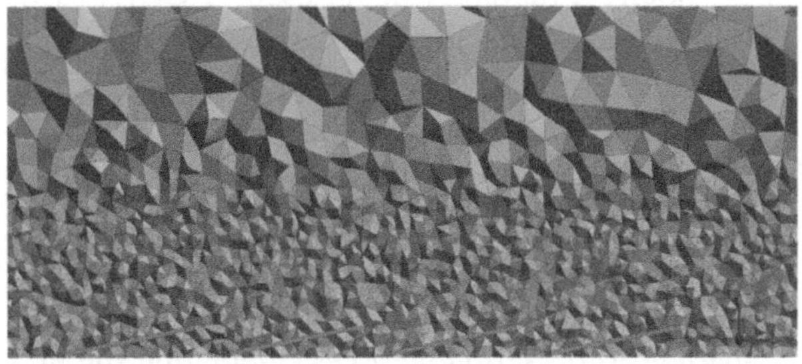

Figure 25.3 Discretisation of the air domain
Source: Authors

and thickness *(t)* 0.03 m, with an inclination angle of 10°. The spacing between the adjacent panels in a row is 0.05 m, while the spacing between the panels in different columns is 0.25 m.

The inlet boundary condition is given based on IS 875 Part-3, with a wind velocity of 35.9 m/s, the designated design wind speed for the project site in Madhya Pradesh, India. The outlet pressure is set to atmospheric pressure (0 Pascal). This ensures that the flow behaves realistically as it leaves the simulation domain. The working fluid is air at 15°C, close to standard ambient temperature, making the results more relevant for real-world scenarios where the floating solar panels operate. For the current study, the liquid part below the panel is not considered. Hence, the bottom of the air domain is modelled as a stationary wall. The top boundary is defined with free stream velocity and zero shear to simulate an open environment where the flow is undisturbed and not influenced by surface interactions. The sides are treated as symmetrical boundaries, and the panels are modelled as stationary walls.

The air domain for the simulation is assumed to be 40 m in length, 24 m in width, and 3.17 m in height. The entire panel array is centrally located within the fluid domain. FLUENT is used as the numerical tool for CFD analysis. The air domain is discretised with a tetrahedral grid. The mesh uses linear elements and applies the curvature option with the advancing front method for element sizing [7]. From a convergence study, it was decided to discretise the air domain using a mesh consisting of 7.55 million nodes and 5.43 million elements. The discretisation of the air domain is shown in Figure 25.3.

Estimation of the drag and lift forces

Wind-induced drag and lift forces on the panels are estimated using ANSYS FLUENT's CFD approach. The drag force (F_D) and lift force (F_L) are evaluated using the following formula:

$$F_D = \frac{1}{2}\, C_D\, \rho\, A\, V^2 \qquad (1)$$

$$F_L = \frac{1}{2}\, C_L\, \rho\, A\, V^2 \qquad (2)$$

where C_D is the drag coefficient, C_L is the lift coefficient, ρ is the mass density of the fluid, "*A*" is the projected area on a plane normal to the flow direction, and V is the area-averaged velocity of the fluid at the inlet.

The analysis uses a steady-state approach, assuming flow conditions do not change with time. The viscous model used is the shear stress transport (SST) k-ω model to effectively capture the effects of the near-wall region and free shear flow [7]. The pressure-velocity coupling scheme used is COUPLED. This scheme allows for the simultaneous solution of the pressure and velocity fields, which helps to improve convergence and accuracy in resolving the interactions between these two variables [8]. It is particularly beneficial in turbulent flows, making it suitable for analysing the wind loading on solar panel structures.

Spatial discretisation involves calculating the gradient via the least squares cell-based method, while pressure, momentum, turbulent kinetic energy, and specific dissipation rate are discretised using second-order accurate schemes. The convergence condition for the drag coefficient C_D is established at 0.001. The solution is initialised using the hybrid initialisation method, which minimises the likelihood of numerical instability and divergence, thus leading to a more efficient and accurate simulation process [8]. The reference area for computing the drag and lift forces is the summation of the frontal-projected area for the whole panel array, which is 13.004 m^2. Additionally, the pseudo-transient option and high-order term relaxation are enabled to capture the transient nature of fluid dynamics while still aiming for a steady-state solution, ensuring more reliable and accurate results.

Results and discussion

Figure 25.4 shows the pressure distribution across the panels when in inlet velocity is from –ve X direction (0°). The pressure is highest on the panels in the front row and gradually decreases on subsequent rows. This reduction in pressure is primarily due to the shielding effect created by the panels in front, which block and divert the wind, reducing its impact on the panels positioned behind them.

Figure 25.4 Variation of pressure over the panels when the wind is acting in -ve X direction
Source: Authors

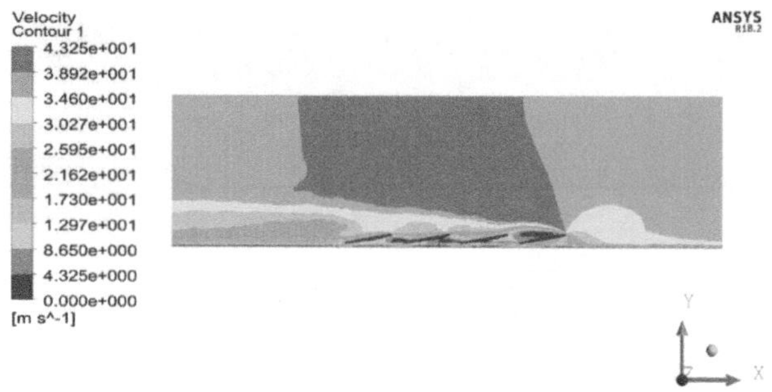

Figure 25.5 Variation of velocity over the panels when the wind is acting in -ve X direction
Source: Authors

The velocity reaches its lowest point when the wind meets the first panel and then gradually increases as it progresses through the subsequent panels, as shown in Figure 25.5. The highest velocities are observed above the panel arrays, near the top of the fluid domain. This occurs due to the drop in pressure and the corresponding increase in velocity as the wind flows over the inclined surface of the panels, consistent with Bernoulli's principle.

Furthermore, the interaction of wind with the sharp edges of the panels results in complex flow patterns with localised regions of higher velocity compared to the free-stream wind. This can be visualised in Figures 25.4 and 25.5.

When the wind acts in 180°, it directly impacts the top of the solar panels in the back row, as shown in Figures 25.6 and 25.7. The shielding effect is evident in this scenario as well. Vortices can form along the trailing edge and downstream of the panels as the wind flows over the inclined solar panels.

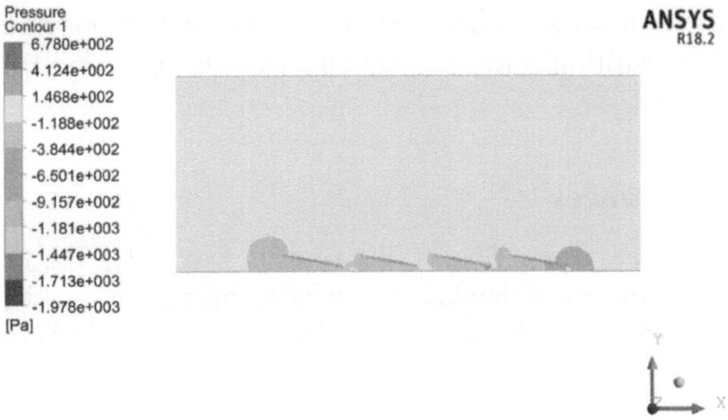

Figure 25.6 Variation of pressure over the panels when the wind is acting in +ve X direction
Source: Authors

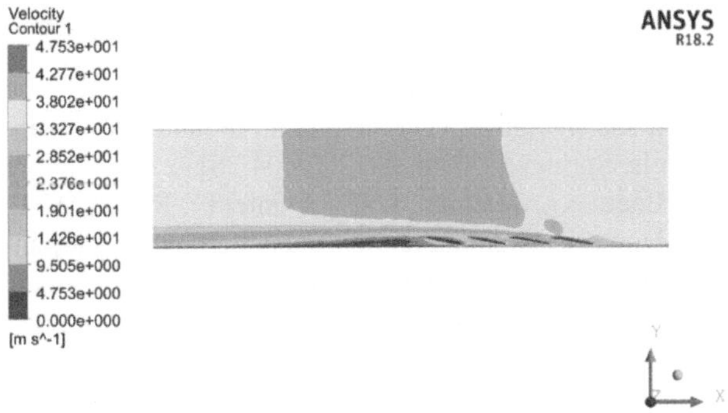

Figure 25.7 Variation of velocity over the panels when the wind is acting in +ve X direction
Source: Authors

Factors such as the panel's inclination angle, the wind speed, and the spacing between panels in the array vary the frequency and intensity of the vortex shedding. This vortex shedding can reduce the effective wind velocity hitting the downstream panels, leading to lower drag forces. However, it may also introduce more turbulent fluctuations in the flow.

Figure 25.8 illustrates the variation in wind-induced drag force on the solar panel array as a function of wind direction. The drag force reaches its peak when the wind hits the panels from the bottom, corresponding to 0° orientation. In this case the total drag force is observed as 5.12 kN. Conversely, when the wind is perpendicular to the panels at a 90° orientation, the drag force is found to be minimal (0.375 kN), due to the much smaller-exposed area of 0.96 m^2.

When the wind hits the panels from the top, corresponding to 180° cases, drag is expected to be less compared to 0° cases. This is more evident because

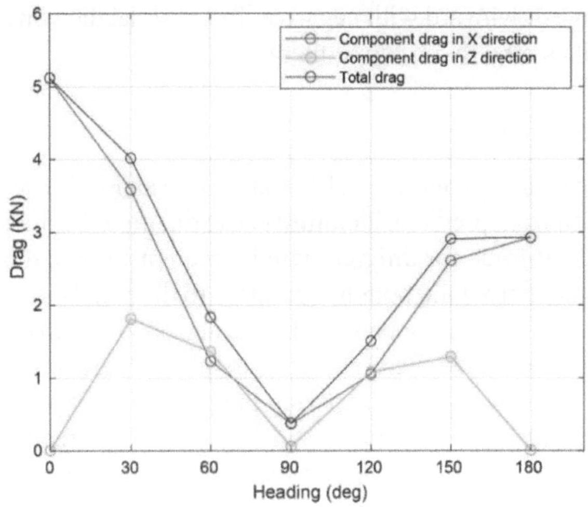

Figure 25.8 Variation in the drag force on the panel array for different wind directions
Source: Authors

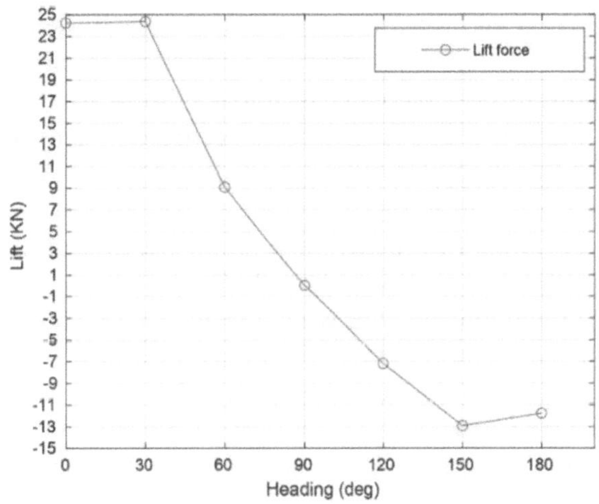

Figure 25.9 Variation in the lift force on the panel array for different wind directions
Source: Authors

the panel will experience a smoother flow over the panels compared to the first case. In this case, the total drag force is observed as 2.927 kN.

The variation in lift force on the panel array against the direction of the wind is shown in Figure 25.9. When the wind is in the -ve X direction (i.e., 0°), the panels experience upward movement due to positive lift forces as the wind strikes directly underneath them. A maximum lift force of 24.25 kN is observed in this case. As the wind heading angle changes from 0 to 180°, the lift force decreases. For wind-heading angles greater than 90° degrees, the panels tend to move downward with negative lift forces as the wind strikes directly on top of them.

Conclusion

The study conducted a CFD analysis on an FPV array taken from previous literature [7], estimating the drag and lift force for different wind directions. Since the distance from the bottom boundary to the panels is as low as 0.17 m, the panel acts like a ground-mounted setup. As such, there is a considerable difference in performance when comparing the performance at 0° and 180° wind directions, and with those reported by Arora et al. [7], If we had increased the separation between the panels and the bottom of the air domain, we would have expected symmetric variation in the drag and lift forces around the 90° angle. The reason for such an unsymmetrical variation of about 90° in this case is stated below:

When the wind comes from the -ve X direction, the lower surfaces of the panels directly expose the airflow. Hence, great pressure is experienced on the windward face, with a possibility of flow separation on the leeward side. On the other hand, when the wind is coming from the +ve X direction, the wind impacts the upper surfaces of the panels. This results in a different flow pattern due to the presence of the panels and associated wake influence.

Furthermore, with the wind flow in the -ve X direction, the leading edges of the panels cause flow separation along the edges of the panels. However, when the wind is from the +ve X direction, a different type of flow separation and wake behaviour occurs at the rear edges. This change in wake behaviour and the resulting turbulence create varying drag and lift forces.

References

[1] Lee, G.-H., Choi, J.-W., Kim, J., Seo, J.-H., Ha, H. Numerical simulations of wind loading on the floating photovoltaic systems. *J. Visualiz.*, 2021;24(3):471–484.

[2] Kumar, M., Niyaz, H. M., Gupta, R. Challenges and opportunities towards the development of floating photovoltaic systems. *Solar Energy Mater. Solar Cells*, 2021;233:111408.

[3] Ahmed, A., Elsakka, M., Elminshawy, N., Mohamed, A., Sundaram, S. Recent advances in floating photovoltaic systems. *Chem. Record*, 2023;23(12):e202300229.

[4] Reina, G. P., Stefano, G. D. A CFD study of wind loads on high aspect ratio ground-mounted solar panels. Springer, Cham, 2017;10404:261–272.

[5] Baetu, G., Teleman, C.-E., Axinte, E., Ro¸sca, V.-E. Numerical simulation of wind action on a solar panels array for different wind directions. *Bull. Polytech. Instit. Jassy Constr. Architect. Sec. (4)*, 2013:9–20.

[6] Irtaza, H., Agarwal, A. CFD simulation of turbulent wind effect on an array of ground-mounted solar PV panels. *J. Instit. Engrs Series A*, 2018;99(2):205–218.

[7] Arora, P., Prasad Patel, B., Ahmad, S., Jain, A. K., Joe Joe, C., Aggarwal, P. CFD-based simulation for the stability of floating solar arrays. *Internat. Conf. Offshore Mech. Arctic Engg.*, 2023;86908:V008T09A017.

[8] Ansys Fluent Theory Guide. 2021.

26 Influence of pre-existing damage on crack evolution using bond-based peridynamics model

Amish Kumar Aman[a] *and Arun C. O.*[b]

Department of Civil Engineering, Indian Institute of Technology Palakkad, Kerala, India

Abstract

The peridynamics models are capable of capturing long-range interactions as well as the damage growth processes that most of the other local approaches are unable to account. This study examines whether the bond-based peridynamics (BB-PD) model can account for the presence of initial damage and studies its effect on further crack growth. A plate with a circular hole at its centre, which is under the action of quasi-static loading along its horizontal edges, is considered as an example problem. The creation of such a hole can lead to damage around it. The current study proposes a simple method to introduce initial damage by reducing the bond stiffness. The reduction is applied to the bonds between particles within their respective horizons for a small region around the hole. For the present study, the initial damage is considered uniform. The results indicate that the existence of damage and its subsequent growth contribute to crack propagation and elevated damage levels near the hole. The study proves that initial damage can be incorporated into the BB-PD model using the proposed method, and it can successfully simulate further progress of the damage.

Keywords: Pre-existing damage, crack initiation, crack propagation, bond-based peridynamics

Introduction

The investigation of damage and fracture is significant in scientific research. Due to its complexity, researchers continue to pay attention to developing appropriate numerical methods for analysing damage and fracture processes. There are various finite element method (FEM)-based approaches for dealing with damage and fracture, such as the extended finite element method (XFEM) and cohesive zone model (CZM). However, each of these methods continues to encounter issues. CZM simulates fractures by assuming that a cohesive zone would exist near the crack tip where energy dissipation and elastic properties are preserved through an opening displacement function [1]. However, the path of the crack has to be predefined. XFEM incorporates discontinuities into the displacement fields of elements [2] that model fractures without any initial assumption for the crack paths. Although it has been successfully used to simulate the majority of applications involving cracks. There are problems in trying to capture other complex behaviours for cracks involving events such as branching, coalescence, and interactions with material heterogeneities, particularly in dynamic conditions or multi-material systems. The computational cost and the required precision for a large number of simulations are rather huge; hence, in some cases, its applicability to real-world scenarios is rather limited. One of the best alternatives in this situation is peridynamics.

Peridynamics emerged as a creative extension of continuum mechanics, effectively addressing the difficulties in modelling discontinuities, such as cracks. Classical continuum mechanics loses its mathematical relevance in the presence of cracks and other singularities because of its spatial derivatives. However, peridynamics which is based on non-local continuum mechanics [3, 4], inherently possess the capability to simulate fracture processes due to its integral formulation. One of the initial formulations in this direction is bond-based peridynamics (BB-PD), wherein interactions among material particles are shown by paired forces, or bonds, that are elongated and broken when stretch exceeds a specified threshold. This capability enables the seamless modelling of crack initiation, propagation, and branching [5–8], eliminating the need for remeshing or other complex approaches related to finite element methods.

A key area requiring enhancement in BB-PD is the modelling of pre-existing damage. Damages often pre-exist in materials, and under loading conditions, they influence the behaviour of materials.

[a]amish.civilstructure@gmail.com, [b]arunco@iitpkd.ac.in

DOI: 10.1201/9781003770442-26

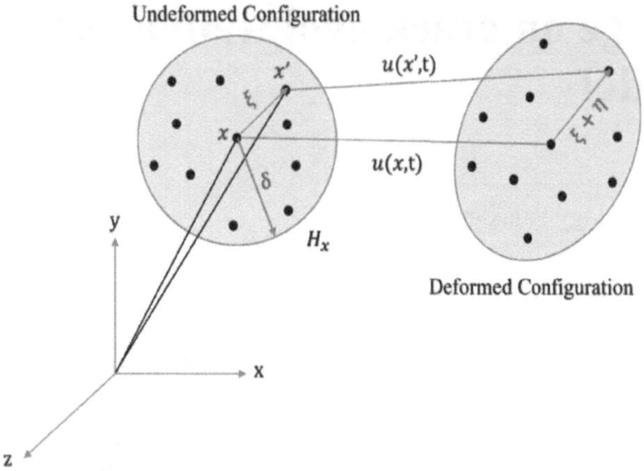

Figure 26.1 BB-PD model
Source: Authors

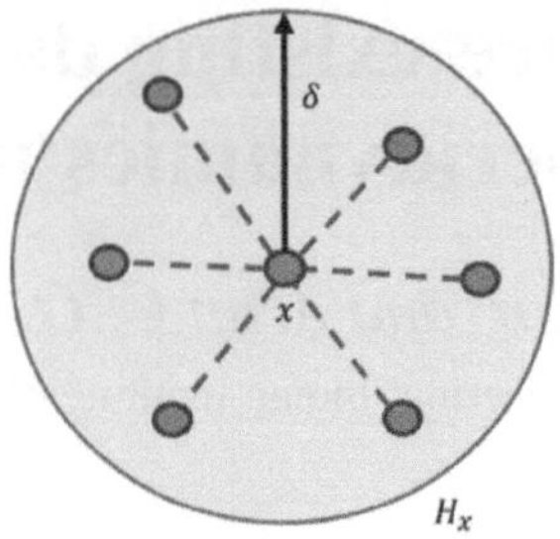

Figure 26.2 Horizon of material particle x
Source: Authors

The existence of cracks or defects in the structure can significantly impact its functionality. Integrating pre-existing damage into the criteria for bond failure can improve the predictions of crack propagation. The improved model needs to incorporate pre-existing defects, microcracks, and material heterogeneities, providing a more accurate representation of fracture processes. These modifications would provide peridynamics a more adaptable and powerful instrument in fracture mechanics, particularly in systems where pre-existing damage substantially influences the damage evolution. The current work proposes a new and simple approach to incorporate the presence of initial damages into peridynamic simulation.

Bond-based peridynamics theory

The BB-PD is based on non-local continuum theory inspired by molecular dynamics where a material particle interacts with neighbouring particles over a constrained range known as the horizon. It is the simplest peridynamics model in which material particles x and x′ exchange forces through spring like interaction shown in Figure 26.1. The relative position in the reference configuration of the material particles is denoted by $\boldsymbol{\xi}$ and the relative displacement in the deformed configuration by $\boldsymbol{\eta}$. The equation of motion for BB-PD model [5] is given as follows:

$$\rho(x)\ddot{\mathbf{u}}(x,t) = \int_{H_x} \mathbf{f}(\boldsymbol{\eta},\boldsymbol{\xi},t)dV_x' + \mathbf{b}(\mathbf{x},t) \quad (1),$$

where $\rho(x)$ is the density of the material, $\ddot{\mathbf{u}}(x, t)$ is the acceleration of the particle x, H_x is the horizon, $\mathbf{f}(\boldsymbol{\eta}, \xi, t)$ is the bond force per unit squared volume between particle x and x′, dV_x' is the volume linked to particle x′ and $\mathbf{b}(\mathbf{x}, t)$ as the body force in Equation 1. Further the bond force per unit squared volume is obtained [5].

$$\mathbf{f} = cs(\boldsymbol{\xi},\boldsymbol{\eta})\frac{\xi}{|\xi|} \quad (2),$$

where,

$$s = \frac{|\xi+\eta|-|\xi|}{|\xi|} \quad (3).$$

Here, c and s are the bond stiffness and stretch between two material particles x and x′, respectively. Bond stiffness can be evaluated by equating strain energy density derived from continuum mechanics as well as peridynamics theory [5]. Strain becomes equal to the bond stretch for uniform dilatation. For a bond, the strain energy per unit squared volume $w(\xi, \eta)$ stored is given as:

$$w(\xi,\eta) = \frac{1}{2}cs^2|\xi| \quad (4)$$

For a material particle x, the strain energy density W_{PD} is computed as half of $w(\xi, \eta)$ stored [5] in the bond between the particles within its horizon as shown in Figure 26.2.

$$W_{\text{PD}} = \frac{1}{2}\int_{H_x} w(\xi,\eta)\text{d}V_{x'} \quad (5)$$

The expression for the bond stiffness for different cases can be found considering δ as horizon radius shown in Figure 26.2, h as thickness and E as elastic modulus [5, 9].

$$c = \begin{cases} \frac{9E}{\pi h\delta^3} & \text{plane stress} \\ \frac{48E}{5\pi h\delta^3} & \text{plane strain} \\ \frac{12E}{\pi\delta^4} & \text{3D} \end{cases} \quad (6)$$

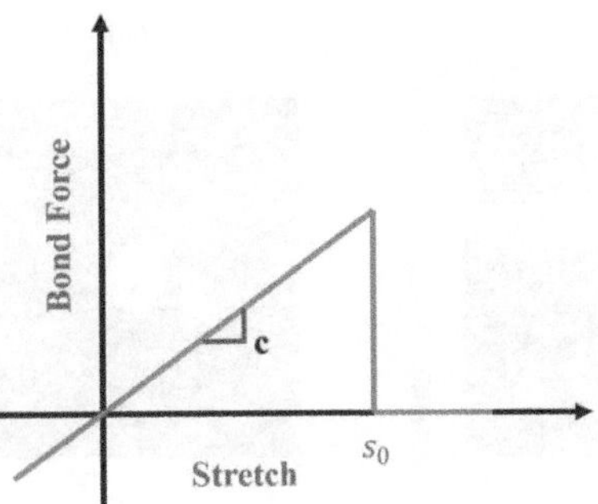

Figure 26.3 PMB damage model
Source: Authors

The bond stiffness is not independent of Poisson's ratio; however, it is limited to 1/4 for plane strain and 3D problem, and 1/3 for plane stress problem, in the BB-PD model [8]. For the damage criterion, the prototype micro-elastic brittle (PMB) model [5] is used. When the bond between any two material particles breaks, the bond force drops to zero and remains the same until the simulation ends in the PMB model as illustrated in Figure 26.3.

A valued scalar function $d(\xi, t)$ describes the status of bond failure [5] by taking a value of 1 if the bond stretch has never been greater than the critical stretch value s_0, and 0 otherwise.

$$d(\xi,t) = \begin{cases} 1 & \text{if } s(\xi,\eta) < s_0 \\ 0 & \text{otherwise} \end{cases} \quad (7)$$

Critical stretch s_0 is determined by equating fracture energy G_0 and the energy released per unit area W_d from peridynamics theory. According to peridynamics, the crack propagation results in the breakage of bonds along the newly formed crack surface, subsequently releasing the energy stored in these bonds. The energy released per unit area W_d of the newly formed crack surface [5], is calculated as:

$$W_d = \begin{cases} 2\int_0^{\delta}\int_z^{\delta}\int_0^{\cos^{-1}(z/|\xi|)} (a)|\xi|\, d\varphi d\xi dz & 2D \\ \int_0^{\delta}\int_0^{2\pi}\int_z^{\delta}\int_0^{\cos^{-1}(z/|\xi|)} (a)|\xi|^2 \sin\varphi\, d\varphi d\xi dz & 3D \end{cases} \quad (8)$$

where "a" represents $\frac{1}{2}cs_0^2|\xi|$ in Equation 8. Thus, the expression for the critical stretch [5, 9] is derived as.

$$s_0 = \begin{cases} \sqrt{\dfrac{4G_0}{c\delta^4}} & 2D \\ \sqrt{\dfrac{10G_0}{\pi c\delta^5}} & 3D \end{cases} \quad (9)$$

The local damage $D(\xi, t)$ is defined as the ratio of bonds broken to the total bonds associated with material particle x [5], which is given as:

$$D(\xi,t) = 1 - \frac{\int_{H_x} d(\xi,t)dV}{\int_{H_x} dV} \quad (10)$$

If $D(\xi, t) = 0$, it means the material particle is undamaged and $D(\xi, t) = 1$, shows that the material particle is fully damaged.

Proposed modelling of initial damage in the BB-PD theory

It is proposed to use the reduction in bond stiffness in this study for incorporating the initial damage in the BB-PD model. This is achieved by coupling with the continuum damage model, which allows reducing the bond stiffness. The bond stiffness values in the material domain where the initial damage exists are multiplied by $(1 - \phi)$ where ϕ is the initial local damage for the material particle x in the domain. Hence, in the modified bond stiffness c' for the plane stress problem, it takes the reduced stiffness value as in Equation 11, if the particle is in the zone where initial damage presents, otherwise takes the original bond stiffness value, which is given in Equation 6.

$$c' = \frac{9(1-\phi)E}{\pi h\delta^3} \quad (11)$$

Modification of bond stiffness leads to the modification of the bond force. The expression for the modified bond force $\mathbf{f}'$ is presented as:

$$\mathbf{f}' = d(\xi,t)c's(\xi,\eta)\frac{\xi}{|\xi|} \quad (12)$$

The above expression takes care of the reduction in bond force for the particles within their respective horizons in the domain of initial local damage. This reduction in bond force allows the bond to break earlier than the case without initial damage. The local damage parameter $D(\xi, t)$ of the material particles remains the same as in Equation 10. However, the local damage calculation of the particle in the domain also accounts for initial damage through the scalar valued function $d(\xi, t)$. This function depends on the bond stretch of the particles, which gets affected by presence of initial local damage.

The number of interacting neighbours is less for surface particles than for interior particles, as shown in Figure 26.4. Surface correction adjusts force

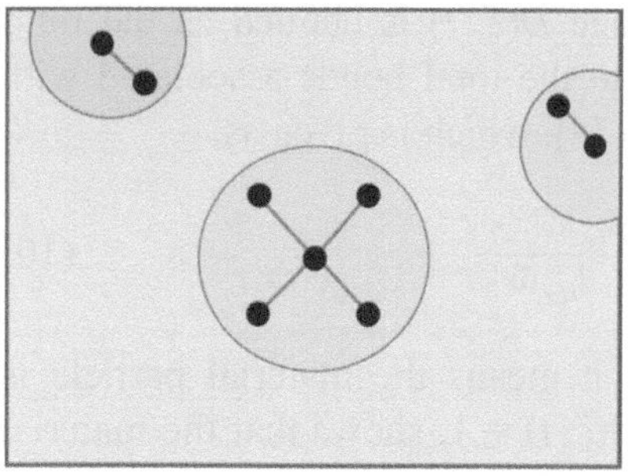

Figure 26.4 Horizon of material points inside the body and near to the boundaries
Source: Authors

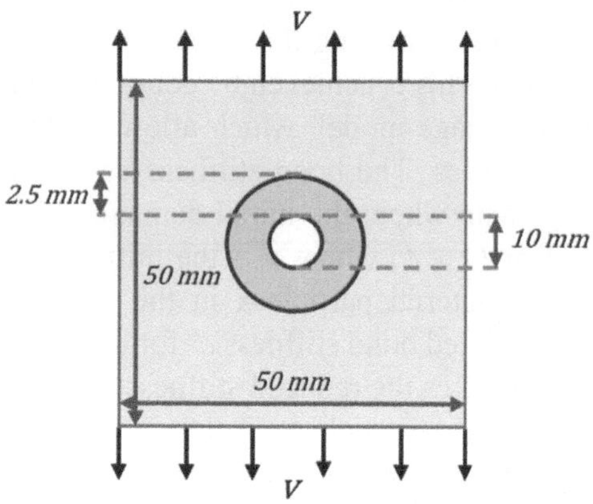

Figure 26.5 A plate with the central hole: Geometry and boundary conditions
Source: Authors

contributions to ensure that material behaviour at the boundary is the same as in the interior of the body. The energy method is used for the surface correction [10].

Numerical example

A plate with a central hole of diameter 10 mm is studied as illustrated in Figure 26.5 under the quasi-static loading along the y-direction on its horizontal edges is considered [11]. The length (l) and width (w) of the plate are 0.050 m, and the thickness (h) is 0.0005 m. The material properties taken for the study are elastic modulus (E) 192000 MPa, mass density (ρ) 8000 kg/m^3, and Poisson's ratio (v) 1/3. The particle spacing (Δx) is set at 0.0005 m, and the horizon (H_x) is equal to 3.015 times Δx. The critical stretch, s_0 is designated as 0.02. Two cases have been considered: Case I, where no initial damage exists in the domain, and case II, where 5% initial damage is considered around

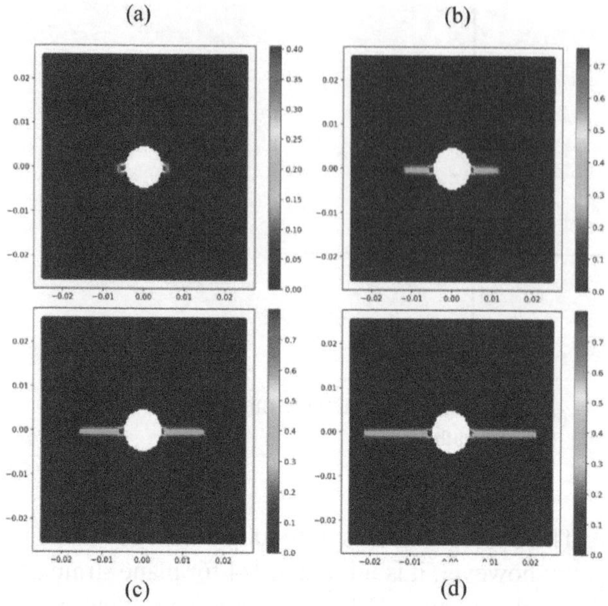

Figure 26.6 Damage contour obtained at various time steps for the case I, (a) 800, (b) 875, (c) 925, (d) 1000
Source: Authors

the hole for an annular width of 2.5 mm. At $y = \pm 25$ *mm* of the plate, a velocity (v) of $\pm 2.7541 \times 10^{-7}$ m/s in the y-direction is applied. The adaptive dynamic relaxation (ADR) method [12] is used to consider the quasi-static state of loading. This method considers the damping coefficient at each iterative step as an adaptive parameter which improves the rate of convergence toward the equilibrium solution. This method stabilises the simulation process in quasi-static loading conditions where the time-dependent effects are negligible. This keeps accuracy intact while adapting the damping coefficient for achieving the steady-state solution efficiently.

Initially, simulations are done assuming no initial damage in the domain. The damage growth at different load steps for case I is illustrated in Figure 26.6. Figure 26.7 illustrates the damage contour at various load steps for case II. The damage initiation begins at the location of high stress concentration, as illustrated in Figures 26.6a and 26.7a for case I and case II, respectively. Afterwards, the crack propagates towards the vertical boundaries, which is shown in Figure 26.6b–d as well as in Figure 26.7b–d at various time steps for both the cases. It is observed that certain particles within the domain sustained a local damage value exceeding 0.38, which should lead to self-similar crack branching according to Madenci and Oterkus [11], but due to the application of slow

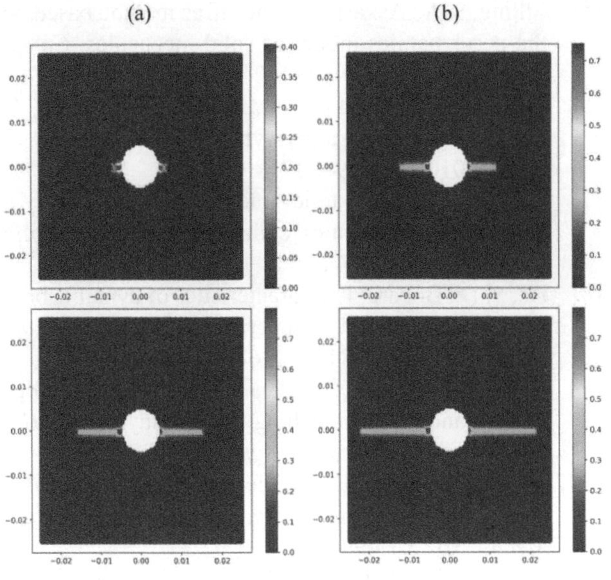

Figure 26.7 Damage contour obtained at various time steps for the case II, (a) 800, (b) 875, (c) 925, (d) 1000
Source: Authors

Table 26.1 Bond force and stretch at the 775th time step.

Model	Bond force (N/m³)	Stretch
Case I	1.19×10^{12}	1.99374×10^{-2}
Case II	1.13×10^{12}	1.99824×10^{-2}

Source: Authors

rate velocity, crack propagation advances towards the vertical boundaries.

In both cases, the first bond failure is found to have occurred between two material particles near the central hole located at [-0.00525, -0.00075] and [-0.00525, -0.00025]. The bond force between these two particles reaches to a value of zero at the 776th time step in case II, indicating that the bond has failed. The bond stretch and the relative bond force for both the cases at the 775th time step, which is just a step before the failure of bond are recorded and shown in Table. 26.1. Upon comparing the bond force, it is found that case II exhibits a lower bond force than case I. The bond force at this particular load step for case II is about 5% lower than case I as expected. This decrease in bond force indicates a decrease in bond stiffness, as it requires less bond force to reach the critical stretch. On the other hand, the bond stretch is higher in case II, which indicates that the simulation is able to capture the expected accelerated failure due to pre-existing damage. The bond stretch at this particular load step, for case II is 0.23% higher than case I.

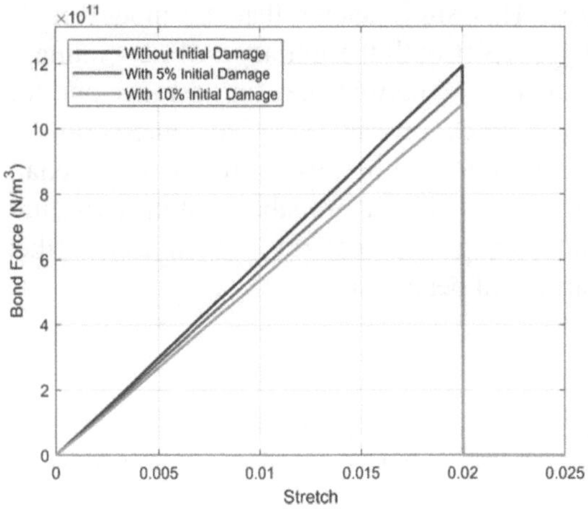

Figure 26.8 Bond force vs. stretch with and without the initial damage
Source: Authors

The variation of the bond force and stretch for the various percentages of initial damage in the annular domain around the circle is shown in Figure 26.8. The bond force increases linearly until the stretch attains its critical value, at which point the bond force abruptly diminishes to zero as anticipated. For a given bond force of 1.0×10^{12} N/m³, the stretch for case I is found to be 1.68×10^{-2}, whereas for case II, it is found to be 1.77×10^{-2}. It indicates that the presence of initial damage reduces the bond stiffness. Additionally, it indicates that for a given bond force, the bond stretch in case II is greater, which demonstrates accelerated crack growth. When initial damage is 10%, the reduction in the stiffness is even more evident as can be observed in Figure 26.8.

Conclusion

The current work proposes a method to incorporate pre-existing damage through a change in bond stiffness and further evaluates its impacts on the bond force, stretch, and crack propagation. This research demonstrated the efficacy of the BB-PD model in incorporating initial damage within the simulation. This makes the method of peridynamics a more versatile tool for the analysis of damage evolution in structures. This study shows that the model could accurately depict both the initiation of new damage and the interaction between existing damage and the subsequent crack propagation. This interaction is particularly important because in real-world scenarios, there always exists a probability of the existence of initial damages in material, which can further affect structural behaviour.

References

[1] Barenblatt, G. I. The formation of equilibrium cracks during brittle fracture. General ideas and hypotheses. Axially-symmetric cracks. *J. Appl. Mathem. Mech.*, 1959;23(3):622–636.

[2] Belytschko, T., Black, T. Elastic crack growth in finite elements with minimal remeshing. *Internat. J. Num. Methods Engg.*, 1999;45(5):601–620.

[3] Silling, S. A. Reformulation of elasticity theory for discontinuities and long-range forces. *J. Mech. Phys. Solids*, 2000;48(1):175–209.

[4] Silling, S. A., Epton, M., Weckner, O., Xu, J., Askari, E. Peridynamics states and constitutive modeling. *J. Elast.*, 2007;88:151–184.

[5] Silling, S. A., Askari, E. A meshfree method based on the peridynamics model of solid mechanics. *Comp. Struct.*, 2005;83(17–18):1526–1535.

[6] Silling, S. A., Weckner, O., Askari, E., Bobaru, F. Crack nucleation in a peridynamic solid. *Internat. J. Frac.*, 2010;162:219–227.

[7] Silling, S. A. Dynamic fracture modeling with a meshfree peridynamics code. *Comput. Fluid Solid Mech.*, 2003:641–644.

[8] Ha, Y. D., Bobaru, F. Characteristics of dynamic brittle fracture captured with peridynamics. *Engg. Frac. Mech.*, 2011;78(6):1156–1168.

[9] Ha, Y. D., Bobaru, F. Studies of dynamic crack propagation and crack branching with peridynamics. *IUTAM Symp. Dyn. Frac. Fragm.*, 2010:229–244.

[10] Le, Q. V., Bobaru, F. Surface corrections for peridynamics models in elasticity and fracture. *Comput. Mech.*, 2018;61:499–518.

[11] Madenci, E., Oterkus, E. Peridynamics theory, Peridynamics theory and its applications. Springer, 2013. 171–180.

[12] Kilic, B., Madenci, E. An adaptive dynamic relaxation method for quasi-static simulations using the peridynamics theory. *Theoret. Appl. Frac. Mech.*, 2010;53(3):194–204.

27 A discrete finite element analysis of coursed rubble masonry gravity dam

Sruthi Jayadeep[1,a], *Arun C. O.*[1,b] *and Robin Davis P.*[2,c]

[1]Department of Civil Engineering, Indian Institute of Technology, Palakkad, Kerala, India

[2]Department of Civil Engineering, National Institute of Technology, Calicut, Kerala, India

Abstract

Dams are the most critical structures constructed to suppress the floods as well as to supply water for irrigation purposes and energy generating infrastructure. Safety of the dam plays a big role as its failure leads to huge socio-economic losses. Rubble masonry gravity dams are very common in the southern part of India especially in the state of Kerala. The discrete nature of rubble masonry makes the numerical modelling of the masonry structure a difficult one. The current manuscript explores a technique which captures the discrete nature of coursed rubble masonry using deformable blocks. For the study, a masonry dam situated at Malampuzha of Palakkad District in Kerala is considered. The non-overflow section of the dam is considered for numerical analysis. The dam foundation numerical model is constructed using RS2 software, where the joint networks between the blocks are explicitly modelled using the Barton-Bandis slip model to account for joint behaviour and surface roughness. To capture the discrete nature of the masonry structure, each rubble block is modelled using a discrete joint element and which is further discretised using finite elements. Plane strain conditions are assumed and a static analysis is performed. The proposed method could predict location of yielded masonry joints. The analysis shows that the dam is safe and intact for the material and joint properties, loading and boundary conditions considered.

Keywords: Masonry gravity dam, discrete finite element, joint deformation

Introduction

Masonry gravity dams are widely constructed in the southern part of India due to availability of abundant natural materials and their durability and density. These dams have been integral to regional water management, irrigation and power generation. The design of these structures relies on their own weight to resist external forces such as water pressure, and it makes the stability of the dam a crucial consideration for safety and performance. Numerical modelling of masonry structures has shown challenges in accurately representing the heterogeneous nature of the structure, as the materials used consist of a combination of rubble and mortar [1]. One of the most extensively used tools for numerical modelling of the dam is the finite element method (FEM) [2]. Most of the FEM model that exists in the literature assumes the structure as a continuous homogeneous medium [2, 3]. Unfortunately, capturing the discrete nature of the masonry blocks and their interactions and failure patterns using direct FEM approach is very cumbersome [3], and alternate methods need to be developed.

This has led to the focus on using the discrete element method (DEM) which treats the masonry dam as a collection of distinct blocks. In the DEM approach, the domain is considered as an assembly of discrete blocks to obtain complex shapes [4]. This makes it highly suitable for analysing masonry structures [5, 6]. There are studies which have demonstrated that DEM can simulate failure mechanisms and joint behaviour more realistically in masonry structures [7]. Since heterogeneity affects the development of fractures and various microcracks in a dam construction, it is crucial to take into account the discrete nature of the masonry while carrying out the analysis [8]. However, in DEM, in general the domain is modelled as interactions between rigid element/blocks. DEM models are more suitable when full separation between the blocks is expected [9]. However, for structures where large displacements or slips are not expected, FEM can still be effectively employed by modifying the discretisation scheme to account for such interactions. One such method is the use of joint elements that behave as deformable

[a]sruthijayadeep98@gmail.com, [b]arunco@iitpkd.ac.in, [c]robin@nitc.ac.in

DOI: 10.1201/9781003770442-27

blocks capturing the joint behaviour. This approach can be computationally efficient as DEM often require a significantly large number of contact points to achieve stress computations [9].

This study focuses on modelling a gravity dam which is built using coursed masonry units. It aims to capture the discrete nature of the masonry dams by modelling the dam using different blocks, where each block is further discretised using finite elements. The contact between each of the blocks are explicitly modelled using Barton-Bandis (B-B) slip model, that captures effects of joint surface roughness of the masonry blocks [10]. Malampuzha dam which is a masonry gravity dam is taken as a case study to explore the application of this modelling approach.

The discrete-finite element approach

The idealisation of masonry structures in the finite element approach using joint elements is achieved by separately representing the mechanical behaviour of each unit and their interaction along the joints as contact surfaces between distinct elements as shown in Figure 27.1. A significant benefit of this method is its ability to manage deformable contacts. This enables to divide the blocks into finite elements as can be observed in Figure 27.1.

The contacts between the masonry blocks are modelled using an empirical slip model known as B-B model, which was put forth by Barton [11]. The surface roughness impact on joint strength along with the deformations that occurs at the discontinuities is captured by this model. Furthermore, it can also model the non-linear shear strength requirement for rocks [12]. The Mohr-Coulomb model generally over predicts the shear strength in normal stress

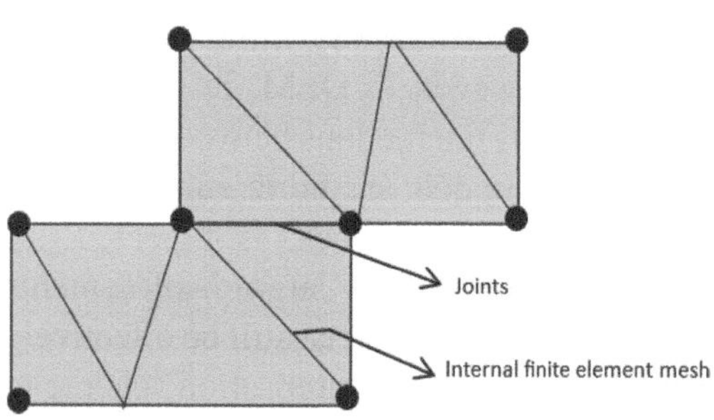

Figure 27.1 Representation of joint network internally discretised by finite elements matching the nodal points
Source: Authors

conditions compared to the B-B model [12]. The stress-displacement behaviour in normal direction is as follows [10],

$$\sigma_{n=} - \frac{u_n \cdot k_n}{1 - \frac{u_n}{v_m}} \tag{1}$$

where, k_n is the normal stiffness of the joint, u_n is the normal displacement of joint opening or closure and v_m is the maximum closure limit. k_n can be calculated as [10],

$$k_n = 0.0178\left[\frac{JCS}{a_{jn}}\right] + 1.748JRC_0 - 7.155 \tag{2}$$

where, *JCS* is the laboratory-scale joint wall compressive strength. JRC_0 is the laboratory-scale coefficient of roughness and a_{jn} is the aperture of joints at zero normal stress [10]. Mobilised roughness concept defines the shear resistance. The contact model uses the following limiting shear stress [10].

$$\tau_s = \sigma_n \cdot tan\left(JRC_{mob} \cdot \left(\frac{JCS}{\sigma_n}\right) + \phi_r\right) \tag{3}$$

where ϕ_r is the residual friction angle, *JCS* is the joint compressive strength, and JRC_{mob} is the mobilised joint roughness coefficient. All these parameters are taken from existing literature for this study [10].

Numerical modelling of the masonry gravity dam

Malampuzha dam is a rubble masonry dam, which is one of the largest irrigation projects in the State of Kerala which was commissioned in 1955. Figure 27.2 shows the aerial view of Malampuzha reservoir. The Malampuzha reservoir is the main source of water used for drinking and agricultural demands of people in and around Palakkad Taluk. The dam is constructed across the Malampuzha river. The main dam length is 1133.85 m and height is 38.45 m. The reservoir has two saddle dams and an earthen dam too. The storage capacity of the dam is 226.0 mm^3.

The current study uses a commercially available software package, Rocscience RS2 for numerical modelling of the dam. The highest non-overflow section of the dam is modelled using two-dimensional plane strain approximation. The section is discretised using deformable blocks with joint elements as shown in Figure 27.3. Each masonry block are modelled using discrete joint elements which

Figure 27.2 The aerial view of Malampuzha dam located in Palakkad
Source: Authors

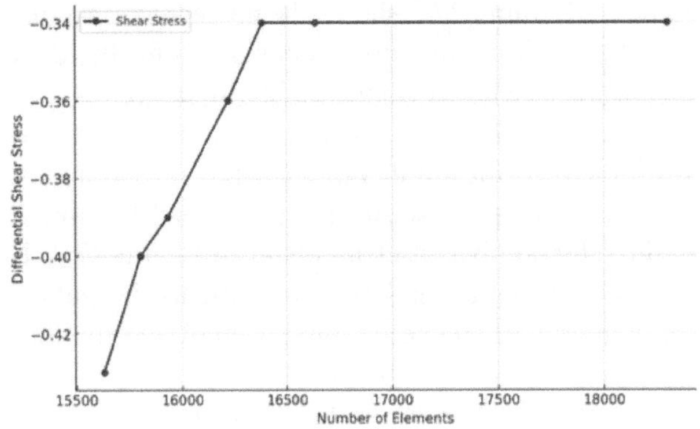

Figure 27.4 Mesh dependency study on the dam foundation system
Source: Authors

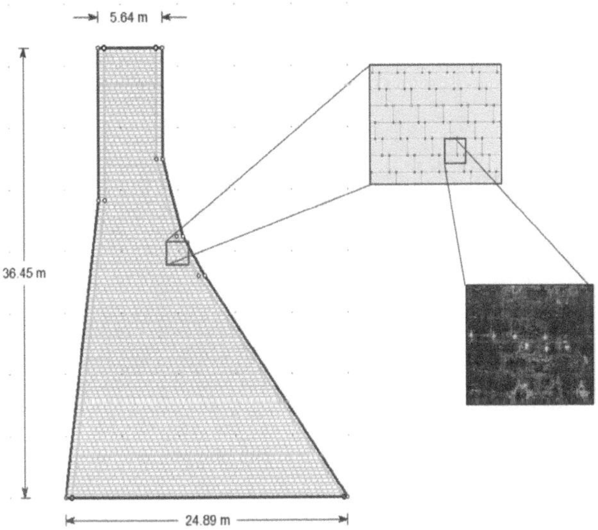

Figure 27.3 The discretisation scheme adopted for the non-overflow section of Malampuzha dam showing joint element network which replicates the rock alignment
Source: Authors

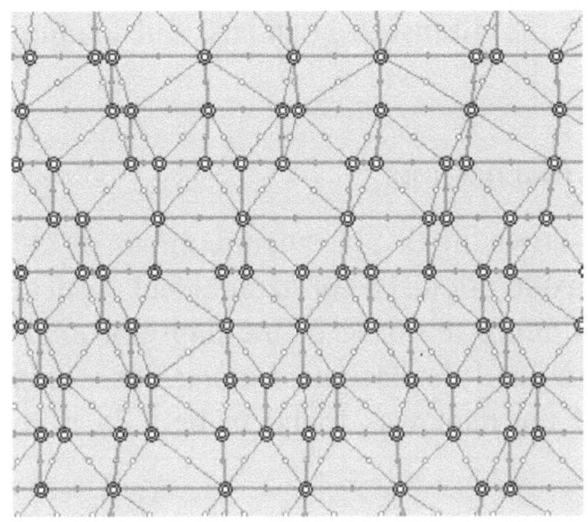

Figure 27.5 Jointed networks with open joints further discretised by 6 nodded triangular finite elements
Source: Authors

are connected to each other using the B-B model. A carefully selected coursed bonding network is used for masonry arrangement to mimic the same in actual dam as seen in Figure 27.3. Coursed rubble blocks are generated for the dam core with a cross joint length of 0.3 m to reflect the actual alignment of rocks in the structure. Moreover, each of these discrete elements is further discretised using 6 nodded triangular elements as shown in Figure 27.5. The numerical study is conducted by considering static loading that includes gravity loading and hydrostatic loadings. An upstream hydrostatic loading for a maximum water height of 31.21 m and downstream tail water pressure for height of 2.54 m is considered that has been taken from the details obtained from the site. In this study, a section without drainage gallery is considered and the effect of seepage is not included.

Here, a foundation rock up to a depth of 60 m is taken into consideration. Mohr-Coulomb failure model is applied to the foundation domain, to take into consideration the non-linearity of the problem. Displacements at the foundation base are constrained in every direction. The lateral boundaries are allowed to move vertically; however, the horizontal displacements are restrained. The foundation region is discretised using only finite elements by considering it as a continuum as shown in Figure 27.6.

The physical parameters of the rock material used for the dam core and that in foundation are given in

Table 27.1. Table 27.2 shows the mechanical parameters for the joint slip model considered for the dam core which are taken from existing literature.

A mesh convergence study is performed by gradually increasing the number of elements in the domain, and the differential shear stress obtained is plotted against the number of elements as in Figure 27.4. It is observed that around 16,300 elements the values are converging. Hence for further study the number of elements is fixed as 16,379.

The dam foundation system is discretised using 16,379 six-nodded triangular finite elements. A staged modelling approach is being used, first incorporating the joint networks and later applying the necessary loading conditions to simulate real-world conditions. A static analysis is performed using RS2 by considering the discussed boundary and loading conditions.

Results and discussions

Figure 27.7a shows the displacement contour of the dam under consideration. The horizontal and vertical displacements along the base of the dam are analysed. The horizontal displacement along the x axis at the base of the dam was found to be 0.24 mm,

Table 27.2 The parameters for joint model.

Properties	Unit	Rock
Normal stiffness	MPa/m	6.9×10^5
Shear stiffness	MPa/m	4.5×10^5
Friction angle	degree	35
Cohesion	MPa	6.0
Joint roughness coefficient	-	8
Joint compression strength	MPa	30

Source: Badakshtan et al. [10]

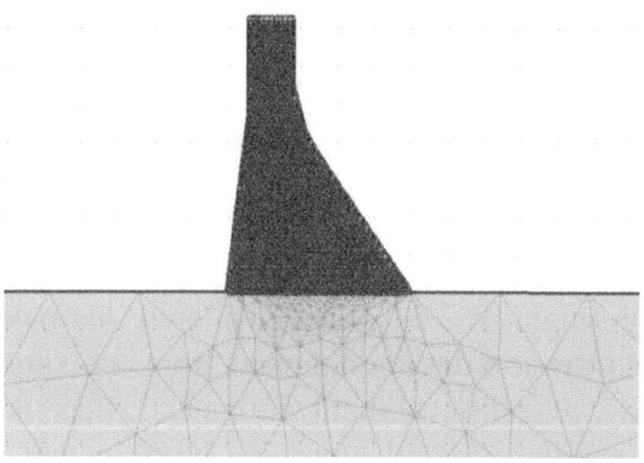

Figure 27.6 2D modelling of the dam-foundation system in RS2 where the dam modeled by jointed elements and foundation as continuum FE
Source: Authors

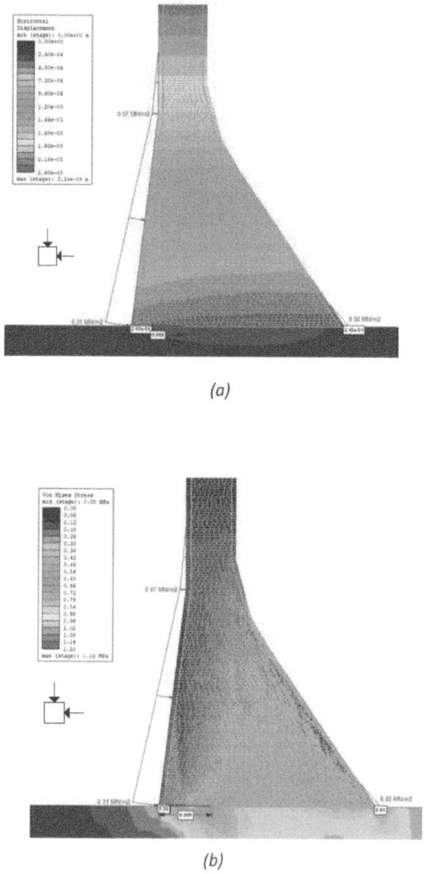

Figure 27.7 (a) Displacement along x-axis in mm (b) Von-Misses stress distribution in MPa within the dam body under static loadings
Source: Authors

Table 27.1 Material properties of Malampuzha dam.

Material	Young's modulus [GPa]	Poisson's ratio	Porosity	Density [kg/m³]
Rock	33.6	0.22	0.5	2277
Foundation	27.58	0.33	0.5	2643

Source: Badakshtan et al. [10] and Tidke et al. [13]

which is minimal and demonstrates the dam's lateral stability under the combined effects of the upstream and downstream pressures. Vertical displacements are observed at both the heel and toe of the dam. The heel exhibited an upward displacement of 0.10 mm, while the toe showed a downward displacement of -0.13 mm. The crest of the dam also experiences a total displacement of 2.36 mm.

The stresses acting in the x direction on the dam vary significantly where at the heel, the stress reaches a tensile value of -0.26 MPa, while at the toe, it increases to a compressive value of 0.30 MPa. This variation aligns with the expected behaviour of gravity dams, where tensile stresses occur at the upstream heel due to the water pressure and the bending effect of the structure while compressive stresses are developed at the downstream end. The tensile stress at the heel is well within the tensile strength of the masonry material. Therefore, the dam's structural integrity is not compromised by tensile stresses in this region. The vertical stresses at the toe of the dam were observed to be 0.18 MPa which closely coincides with the analytical calculation of 0.20 MPa at the heel indicating a higher load concentration at the downstream portion of the dam. The von-Mises stresses are found to be 0.18 MPa at the heel and 0.84 MPa at the toe in Figure 27.7b.

Figure 27.8a shows the predicted yielding joint patterns within the dam body that indicates the vulnerable joints, but do not signify joint failure. Yielding patterns of joints are appeared to be more concentrated near the toe of the dam, as expected due to the compressive forces. There is relatively less yielding near the heel, where tensile forces are more prominent, but within the material's tensile capacity. The central portion of the dam also shows a minor degree of yielded joints pattern. This could be due to internal redistribution of stresses caused by the pressure and the interaction between individual masonry blocks Figure 27.8b. However, despite the yielded patterns are observed, there was no slip of joints to be seen. Hence, the dam seems to maintain overall integrity, as the most critical yielding of joints is localised in regions that are expected to have higher stresses.

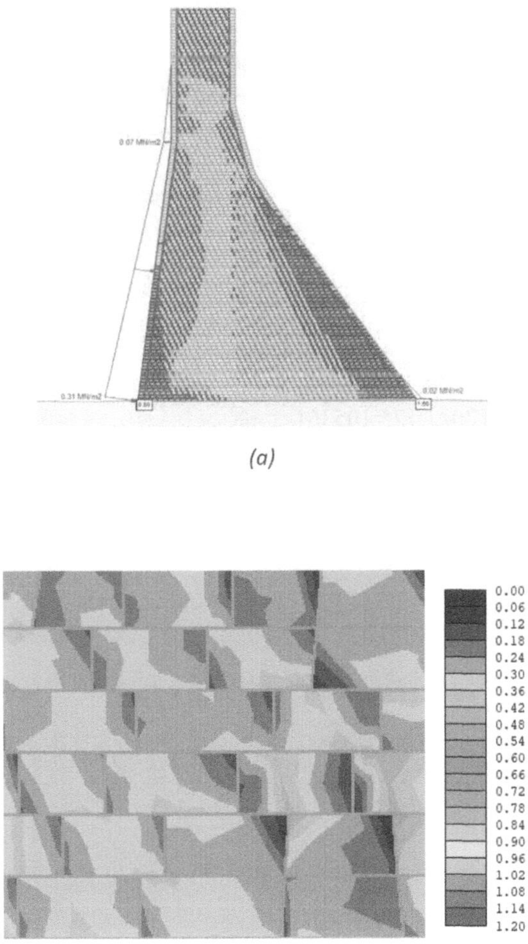

Figure 27.8 (a) Yielded joint trajectories within the dam (b) Variation of stress distribution within each of the discrete elements
Source: Authors

Conclusion

The study focuses on the discrete modelling of coursed rubble masonry gravity dam using cross jointed networks to mimic the distribution of rubble masonry blocks. A discrete finite element model using joint elements is used for discretising the rubble masonry domain of the dam while the foundation rock is discretised using finite elements. This model incorporates a B-B slip model to simulate the contacts between rock-rock interfaces of the rubble masonry structure. A plane strain analysis is conducted. The cross-jointed network enabled the capture of detailed, discontinuous stress behaviour and possible yielding within the joints. It can be noted that the yielding observed does not conclusively indicate joint failure, as the mechanical properties used in this study were referenced from literature rather than site-specific data. A more realistic study incorporating site-specific testing would yield more reliable and accurate insights into joint performance under static conditions and could serve as a basis for further in-depth analysis. However, within the analysed contours, no significant relative displacement is observed, indicating that the dam structure remains stable and intact under the current loading conditions.

Acknowledgement

The authors would like to acknowledge the financial support given by Kerala State Council for Technology, Science and Environment for the project. The authors also extend their thanks to the engineers and other staffs of Irrigation Design and Research Board, and Irrigation Malampuzha Division-Govt. of Kerala for supporting and sharing the available data.

References

[1] Azevedo, N. M., Pinho, F. F., Cisma¸siu, I., Souza, M. Prediction of rubble-stone masonry walls response under axial compression using 2d particle modelling. *Buildings*, 2022;12(8):1283.

[2] Zienkiewicz, O. C., Taylor, R. L. The finite element method fifth edition volume 1: The basis. Massachusetts: Butterworth-Heinemann. 2000. 143–153.

[3] Wittke, W., Wittke, M., Kiehl, J. R. Interaction of a masonry dam and the rock foundation. *Geotech. Geol. Engg.*, 2012;30(3):581–601.

[4] Bretas, E. M., Lemos, J. V., Louren¸co, P. B. Hydromechanical analysis of masonry gravity dams and their foundations. *Rock Mech. Rock Engg.*, 2013;46:327–339.

[5] Bretas, E. M., Lemos, J. V., Louren¸co, P. B. A dem based tool for the safety analysis of masonry gravity dams. *Engg. Struct.*, 2014;59:248–260.

[6] Hamp, E., Gerber, R., Pulatsu, B., Quintero, M. S., Erochko, J. Nonlinear seismic assessment of a historic rubble masonry building via simplified and advanced computational approaches. *Buildings*, 2022;12(8):1130.

[7] Villeneuve, M. C., Diederichs, M. S., Kaiser, P. K. Effects of grain scale heterogeneity on rock strength and the chipping process. *Internat. J. Geomech.*, 2012;6:632–647.

[8] Lan, H., Martin, C. D., Hu, B. Effect of heterogeneity of brittle rock on micromechanical extensile behaviour during compression loading. *J. Geophy. Res. Solid Earth*, 2010;115(B1).

[9] Lemos, J. V. Discrete element modelling of masonry structures. *Internat. J. Architect. Herit.*, 2007;1(2):190–213.

[10] Badakhshan, E., Veylon, G., Peyras, L., Vaunat, J. A simplified method for predicting overflow-induced crack propagation in gravity dams using genetic algorithm and material-based model. *Internat. J. Rock Mech. Min. Sci.*, 2024;181:105842.

[11] Barton, N., Bandis, S., Bakhtar, K. Strength, deformation and conductivity coupling of rock joints. *Internat. J. Rock Mech. Min. Sci. Geomech. Abstracts*, 1985;22:121–140.

[12] Barton, N. R., Bandis, S. C. Characterization and modelling of the shear strength, stiffness and hydraulic behaviour of rock joints for engineering purposes. *Rock Mech. Engg.*, 2017;1:3–40.

[13] Tidke, A. R., Adhikary, S. Seismic fragility analysis of the koyna gravity dam with layered rock foundation considering tensile crack failure. *Engg. Fail. Anal.*, 2021;125:105361.

28 Non-linear low velocity impact responses of composite conoidal panels

Jatin Garhekar[a] *and Kaustav Bakshi*[b]

Department of Civil Engineering, Indian Institute of Technology Indore, Madhya Pradesh, India

Abstract

In various industrial applications, composite curved panels are preferred as these are lightweight and stiffer than flat ones. Naturally, such panels are advantageous for marine, aerospace and civil structures. The vibration performances of curved structures can be significantly improved by integrating efficient structural designs with advanced composite materials. The high strength to weight ratio of composites leads to economic constructions. The composite curved panels can get impacted by water/wind borne debris, ice in cold climatic conditions and drop hammer conditions during service. Since moderately thin surfaces can experience large amplitudes of vibrations even during low velocity impact, it is essential to incorporate geometrically non-linear strains. The review of literature highlights that composite conoidal panels were studied for industrially important parameters but there is gap in research studies on relative impact performances using geometric non-linearity. The present paper aims to work on this area and proposes a numerical code which integrates von-Karman strains, constant shear deformation along thickness direction, and eight-nodded curved elements to ensure C^0 continuity. Newton-Raphson iteration solves the non-linear problems at each time. The contact forces are computed through modified Hertzian contact law. Time dependent equations are solved by Newmark's average acceleration algorithm. This study reports deflections of centrally impacted clamped panels for varying speed of impactor and laminations.

Keywords: Low velocity impact, conoidal panels, geometric non-linearity, laminated composites

Introduction

Laminated composite material has been developed and used because of the quest for smart materials. This material has become incredibly popular for having a high ratio of stiffness to weight and an excellent choice for structural applications where weight savings are critical without compromising structural integrity. Conoidal shells are characterised by their aesthetically appealing, doubly curved and easy to construct geometry which offers low weight for a given rigidity. The entry of ambient atmosphere through both ends is an additional factor for their popularity in architectural design, particularly in practical applications requiring natural airflow. Naturally, the application of composite conoids as roofs of long unsupported facilities are preferred among engineers as these structures can result in reduced foundation and seismic forces. Moreover, the material consumption for these structures can be optimised by aligning stiffness to the principal directions of load transfer, offering greater design flexibility. A group of researchers worked on composite conoids. Nayak and Bandyopadhyay [1–3] and Das and Chakravorty [4–6] studied bending and vibration of composite conoids with and without stiffeners. Delaminated conoids were studied by Kumari and Chakravorty [7, 8]. The authors of refs [1–8] adopted geometrically linear finite elements and Mindlin's theory. Vibration and stability of functionally graded shells were researched by Pradyumna and Bandyopadhyay [9, 10] using parabolic transverse shear and linear formulations. The first ply failure of laminated conoids and vibration of composite shells were studied by Bakshi and Chakravorty [11–16] using both geometric linearity and non-linearity. Chaubey et al. [17] worked on thick composite conoids while Rout and Hota [18] reported free vibration of CNT reinforced sandwiched shells. The authors Rout and Hota [18] adopted von-Karman strains in their paper.

The composite conoidal panels can get impacted by wind and water borne debris, snow in hilly areas and drop hammer conditions in industrial applications. Despite several advantages of laminated composites, the weakness in transverse shear makes these panels more vulnerable to get damaged by impact than the traditional steel and concrete counterparts. The low

[a]phd2401104009@iiti.ac.in, [b]kaustav.bakshi@iiti.ac.in

DOI: 10.1201/9781003770442-28

velocity impact may lead to latent damage which may culminate to complete collapse if not identified and rectified. Hence, the study of impact responses for composite conoids is one of the prerequisites to avoid the sudden catastrophe. Sun and Chen [19] proposed a contact law to compute impact induced forces on composite plates which accounted for permanent indentation. That law and geometric linearity was used by Ganapathy and Rao [20, 21] to study impact induced failure of composite cylindrical and spherical shells. The non-linear approach was adopted by Krishnamurthy et al. [22, 23] to report damages in composite cylindrical panels under low velocity impact. Changliang et al. [24] preferred non-linear formulation to predict delamination of composite filament wound vessel under low velocity impact. The stiffened plates for low velocity impact were researched by Li et al. [25] using progressive failure model based on layer wise/solid-elements method. Sharma et al. [26] conducted experiments on thin composite laminates to find out the contact forces initiated by low velocity impact. The effects of stacking sequences on intralaminar damage of semi cylindrical woven composite shells were researched by Ferreira et al. [27]. A few recent papers [28, 29] have worked on impact responses of stiffened composite conoids but those used the linear formulation only.

The literature clearly depicts that composite conoids were studied by several researchers using geometrically linear and non-linear approaches. The studies involved bending, vibration, stability and first ply failure by the researchers. The failure studies were conducted for static loads only. Similar studies on impact responses are also not very few in numbers, but researchers concentrated mostly on simple geometries like cylindrical and spherical forms. The complicated ones, like conoidal shells with cross curvature, received limited attention and were done using linear strains were only. The moderately thin panels may show large amplitudes of deflections under transverse impact. Hence, geometric non-linear formulation is essential to study impact responses of composite conoids which are missing in literature. The deficiency is aimed to be filled up by this paper where the clamped composite conoidal panels are studied for being impacted centrally. The impact is considered to act transversely to the panel surface. The responses in terms of dynamic deflections are computed using a C^0 continuous finite element code which is isoparametric and non-linear in nature.

Mathematical model

A laminated composite conoidal panel (Figure 28.1) is studied here which has thickness "h" and radii of curvatures "R_{yy}" and "R_{xy}". The thickness can be made of any number of layers where in each of those, the fibres can be at any arbitrary orientation.

Force-displacement relationship

The composite is considered as linearly elastic and the constitutive relationship is given below:

$$\{F\}=[D]\{e\} \tag{1}$$

The force vector $\{F\}$ and matrix $[D]$ are the same as reported by Bakshi and Chakravorty [11]. The strains $\{\varepsilon\}$ are the following,

$$\{\varepsilon^0\} = \{\varepsilon_x^0 \quad \varepsilon_y^0 \quad \gamma_{xy}^0 \quad k_x \quad k_y \quad k_{xy} \quad \gamma_{xz}^0 \quad \gamma_{yz}^0\}^T \tag{2}$$

Finite element formulation

The generalised displacements are assumed as the following:

$$u = u_0 + z\alpha,\ v = v_0 + z\beta,\ w = w_0 \tag{3}$$

u_0, v_0 and $_0$ = mid-surface translations. α and β are rotations (refer to Figure 28.1). An eight nodded element is adopted in this study which has the following degrees of freedom at each node:

$$\{\delta_i\} = \{u_i \quad v_i \quad w_i \quad \alpha_i/ \quad \beta_i/\}^T i=1 \text{ to } 8$$

$\{\varepsilon\}$ assumes von-Karman strains and constant transverse shear strain as given below:

$$\{\varepsilon^0\}=\{\varepsilon^0\}^L+\{\varepsilon^0\}^{NL} \tag{3}$$

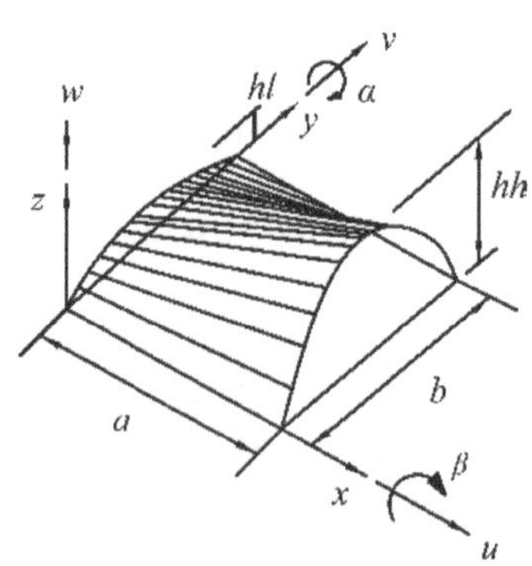

Figure 28.1 The conoidal panel
Source: Authors

$$\{\varepsilon^0\}^L = \begin{Bmatrix} \frac{\partial u_0}{\partial x} \\ \frac{\partial v_0}{\partial y} - \frac{w_0}{R_{yy}} \\ \frac{\partial u_0}{\partial y} + \frac{\partial v_0}{\partial x} + \frac{w_0}{R_{xy}} \\ \alpha + \frac{\partial w_0}{\partial x} \\ \beta + \frac{\partial w_0}{\partial y} \end{Bmatrix}, \{\varepsilon^0\}^{NL} = \begin{Bmatrix} \frac{1}{2}\left(\frac{\partial w_0}{\partial x}\right)^2 \\ \frac{1}{2}\left(\frac{\partial w_0}{\partial y} - \frac{v}{R_{yy}}\right)^2 \\ \left(\frac{\partial w_0}{\partial x}\right)\left(\frac{\partial w_0}{\partial y} - \frac{v}{R_{yy}}\right) \\ 0 \\ 0 \end{Bmatrix}$$

$$\text{and } \begin{Bmatrix} k_x \\ k_y \\ k_{xy} \\ k_{xz} \\ k_{yz} \end{Bmatrix} = \begin{Bmatrix} \frac{\partial \alpha}{\partial x} \\ \frac{\partial \beta}{\partial y} \\ \frac{\partial \alpha}{\partial y} + \frac{\partial \beta}{\partial x} \\ 0 \\ 0 \end{Bmatrix}$$

The strain- displacement relationship is as follows:

$$\{\varepsilon^0\} = \left([B]^L + \tfrac{1}{2}[B]^{NL}\right)\{\delta_i\}, \frac{\partial\{\varepsilon^0\}}{\partial\{\delta\}} = [\bar{B}], \quad (4)$$

$[\bar{B}] = [B]^L + [B]^{NL}$, $i = 1$–8 for the eight nodded element adopted here. $[B]^L$ and $[B]^{NL}$ are the same as given in reference [12].

Impact formulation

The equation of motion for the impactor is given as,

$m_i\ddot{\omega}_i + F_c = 0$, $m_i\ddot{\omega}_i$ represents inertia of impactor. (5)

F_c is contact force to be computed using indentation law given by Sun and Chen [19].

During loading phase, F_c is given as,

$$F_c = k\alpha^{1.5} \quad 0 < \alpha \leq \alpha_m \quad (6)$$

k designates contact stiffness. α = indentation which is equal to the difference of deflections of impactor and conoid at a constant time. α_m = maximum indentation.

F_c, during unloading and reloading are,

$$\text{Unloading:} \quad F_c = F_m\left[\frac{\alpha - \alpha_0}{\alpha_m - \alpha_0}\right]^{2.5} \quad (7)$$

$$\text{Reloading:} \quad F_c = F_m\left[\frac{\alpha - \alpha_0}{\alpha_m - \alpha_0}\right]^{1.5} \quad (8)$$

F_m = Maximum contact force obtained from loading phase.

α_0 = permanent indentation during loading – unloading phases. It is determined by following the guidelines of Sun and Chen [19].

$$\alpha_0 = \beta_c(\alpha_m - \alpha_p) \text{ if } \alpha_m < \alpha_{cr} \quad (9)$$

$$\alpha_0 = 0 \text{ if } \alpha_m > \alpha_{cr} \quad (10)$$

For graphite-epoxy composite, $\beta_c = 0.094$ and $\alpha_{cr} = 0.01667$ cm.

Governing equations

The equation of motion for the composite conoid under dynamic load is developed following Lagrange's equation of motion based on Hamilton's principle.

$$\frac{d}{dt}\left\{\frac{\partial U_2}{\partial \delta_e}\right\} - \left\{\frac{\partial U_2}{\partial \delta_e}\right\} + \left\{\frac{\partial}{\partial \delta_e}(U_1 + U_3)\right\} = 0 \quad (11)$$

where $U_1 = \frac{1}{2}\int_v \{\varepsilon\}^T\{\sigma\}dv$, $U_2 = \frac{1}{2}\int_A \{\dot{\delta}\}^T[m]\{\dot{\delta}\}\,dA$..

$[m]$ is defined by reference [11].

$U_3 = -\int\int_A \{\delta\}^T\{q\}\,dA$..

The steps given by Nanda and Bandyopadhyay [29] are followed to arrive at the following one from Equation (11),

$$[M_s]\{\ddot{\delta}_i\} + \int_A [\bar{B}]^T[D]\left([B]^L + \tfrac{1}{2}[B]^{NL}\right)dA\ \{\delta_i\} = \{F_e\} \quad (12)$$

where $[M_s]$ = mass of the conoid as given in reference [11].

$$\{F_e\} = \{0\ 0\ F_c\ 0\ 0\}^T$$

Newmark's constant-acceleration time integration algorithm solves Equations (5) and (11). Steps of the solution method are described in reference [29]. The Newton-Raphson method is adopted at each time step to solve the non-linear problem. The contact force for a given time is computed using the guidelines given by Sun and Chen [19]. The iteration procedure continues till the equilibrium condition is satisfied.

Problems

Two sets of problems are solved here. The first set of problems is selected with the objective of verifying correctness of the proposed non-linear impact formulation applicable for composite conoidal panels. The section is entitled as "benchmark problems". The present code is further used to solve impact problems of clamped conoids for which the details are given in the section entitled as "additional problems".

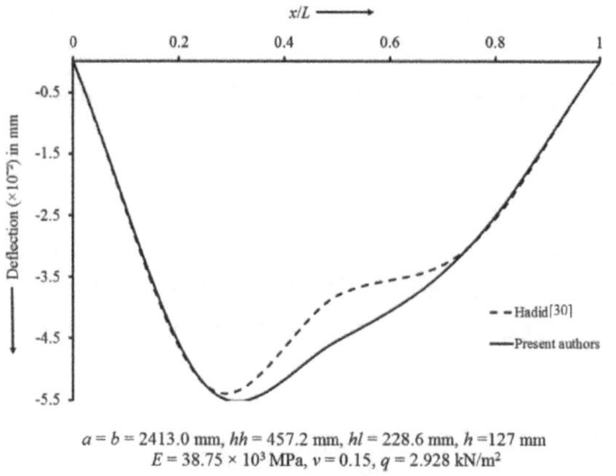

Figure 28.2 Deflections of uniformly loaded conoid along $\frac{y}{b} = 0.5$
Source: Authors

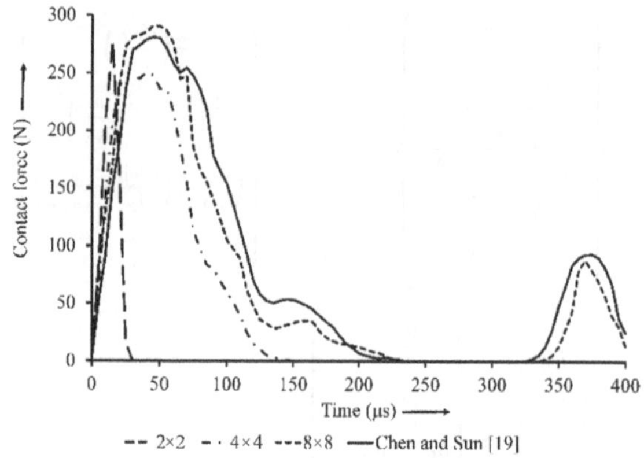

Figure 28.3 Contact forces of composite plate with simple supports
Source: Authors

Benchmark problems

The first problem solves displacements of a conoid made of isotropic material. The results obtained from the proposed code are reported in Figure 28.2 where material properties and dimensions of the panel are given. The figure compares present results with those given by Hadid [30]. The comparison shows that present results exhibit close match with what was given in literature. Hence, the figure confirms that conoidal panels are correctly formulated in proposed code.

The next problem reports contact forces for a simple supported laminated square flat panel of 20 cm length and 0.269 cm thickness. The lamination is $[0^0/90^0/0^0/90^0/0^0]_s$. A spherical steel ball of 1.27 cm diameter and velocity of 3.0 m/s impacted the plate at centre. Mass density of impactor is 7.96×10^{-5} N-sec^2/cm^4. The material properties of composite are: E_{11}=120 GPa, E_{22}= 7.9 GPa, G_{12}= G_{23}= G_{13}= 5.5 GPa, = 0.30, ρ = 1.58×10^{-5} N-sec^2/cm^4. The contact forces are computed for time step of 2.5 micro-second (μs) and three different finite element meshes which are 2×2, 4×4 and 8×8. The contact forces computed using proposed code are furnished in Figure 28.3 which compares the present results with those given by Sun and Chen [19].

The figure shows present results are in close match with published ones for 8×8 mesh. Hence, correctness of present code for dynamic solutions under low velocity impact is confirmed. For other meshes, the results do not match closely with published ones. Hence, the 8×8 mesh is utilised for additional problems solved further in this paper.

The correctness of proposed non-linear approach is additionally confirmed by Figure 28.4. The displacements of an isotropic cylindrical panel obtained using present approach are compared with those given by Wu et al. [31] and Carrera et al. [32]. R_{xy} = 10^{30} is adopted to simulate cylindrical panel using the code for conoids. The properties of cylindrical panels are: E = 3.10275 kN/mm^2, ν = 0.3, length of 0.508 m, R = 2.540 m and included angle = 0.1 radian. Earlier results were obtained beyond limit points using Carrera Unified Formulation based finite element approach. Since deflections at limit points are generally greater than serviceable limits of civil engineering and the proposed approach is formulated for civil engineering structures, the present results are obtained till first limit point only. The close agreement of results noted in Figure 28.4 confirms the correctness of proposed non-linear approach.

Additional problems

The impact responses of clamped conoidal panels are solved as additional problems. The panels are considered to have 45°/-45° and 0°/90° composites. The impact by a spherical impactor is considered at the centre of panels. The velocity of impactor is taken as 1, 2, 4, 8, 10 m/s. An 8×8 mesh is adopted for the panels with dimensions: a = 1.0 m, b =1.0 m, thickness = 0.01 m, hh = 0.2 m, hl = 0.05 m. The material properties of graphite-epoxy panels and those of impactor are the same as reported by Sun and Chen [19]. These properties are reported in the previous section for Figure 28.3. The contact forces and dynamic deflections of conoids are furnished in Table 28.1 and Figures 28.5 and 28.6.

Results and discussion

The contact forces and dynamic deflections of the composite are computed with converged time step.

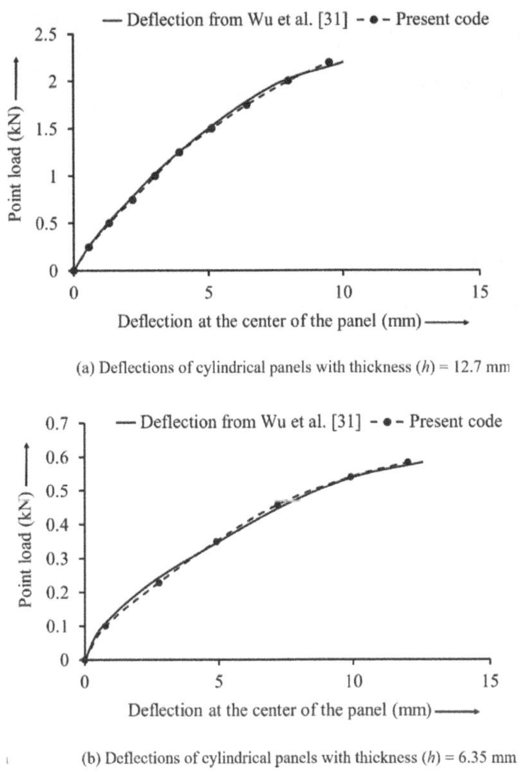

Figure 28.4 Non-linear deflections of cylindrical panel
Source: Authors

Table 28.1 reports the maximum values of contact forces and dynamic deflections for varying velocities of impactor. The table shows that the maximum ones increase as the speed of impactor varies. Naturally, the maximum dynamic deflections of the composite panels become higher with increasing speed of impactor.

The contact force and deflection histories for both 45°/-45° and 0°/90° composites are given in Figure 28.5. The figure clearly shows that for low velocities of central impact, the contact force histories for composite conoidal panels follow a nearly parabolic pattern with one peak only. The contact forces show null values for times which are 100 μs or less than that. The contact forces are higher when the impactor has higher velocity which meets the expectations. The maximum contact force for both composite panels is nearly the same as found in the figure. This should be interesting that for greater impactor speed, the convergence of contact forces towards null values takes place at an earlier time. The phenomenon can be explained as higher velocity of impactor leads to rapid elastic rebound which culminates to earlier detachment from the target panel surface. This is why the contact forces reach to null values earlier when the velocity of impactor increases. The downward dynamic deflections (given with negative sign) of 45°/-45° and 0°/90° composites for impact are reported in the same figure. Like the contact forces, the dynamic deflections for both composite panels are also very close as shown by Table 28.1 and Figure 28.5. This is important to note

Table 28.1 Maximum contact force and maximum dynamic deflections for varying velocities impactor.

Lamination	v (m/s)	Maximum contact force (N)	Maximum downward deflection (m)
0°/90°	0	0	0
	1	296.6435	-1.37×10^{-05}
	2	683.0495	-2.82×10^{-05}
	4	1573.766	-5.77×10^{-05}
	8	3633.271	-1.18×10^{-04}
	10	4789.018	-1.48×10^{-04}
45°/-45°	0	0	0
	1	293.7741	-1.31×10^{-05}
	2	676.4517	-2.71×10^{-05}
	4	1556.621	-5.59×10^{-05}
	8	3606.147	-1.15×10^{-04}
	10	4746.631	-1.45×10^{-04}

Source: Authors

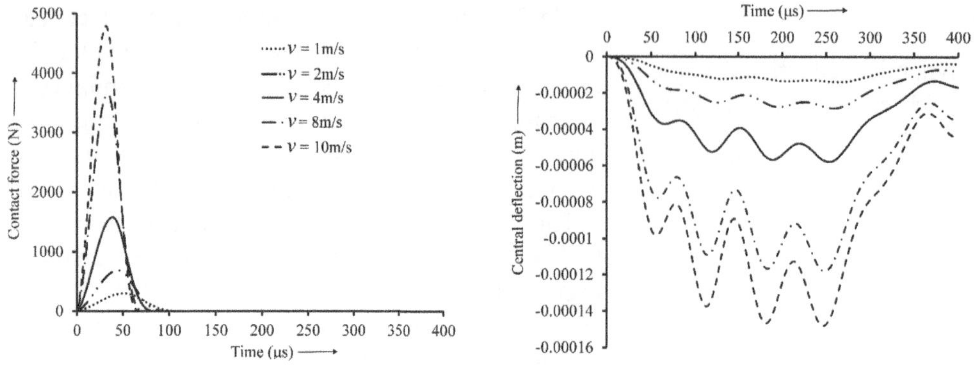

(a) Impact performances of 0°/90° conoid

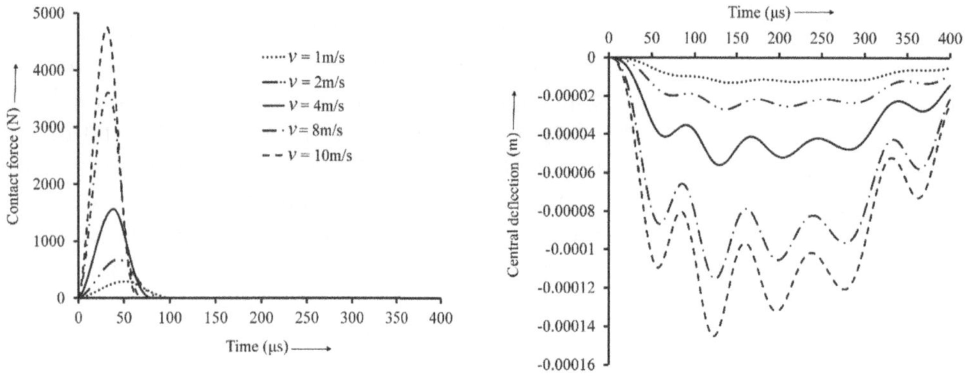

(b) Impact performances of 45°/-45° conoid

Figure 28.5 Contact force and deflection histories of clamped composite conoids for varying velocities of impactor
Source: Authors

that for any given velocity of impactor, the deflection history for a composite do not match with its contact force history. For greater velocity of impactor, the downward deflections are higher in magnitude and for 10 m/s velocity of impactor, the deflections are highest. An engineer must note that for both composites, the highest contact force and the corresponding peak deflections do not take place at identical time instance. A comparison of contact forces and deflections reveal that the peak deflections take place after the occurrence of highest contact force. It must be noted that the deflections of conoidal panels for any time is the cumulative effects of impact force at that time and inertia contributions of preceding time. This is why the peak deflection is delayed from the occurrence of highest contact force. Figure 28.5 further reveals that the dynamic deflections of composite conoidal panel continue even after 100 μs when the contact force is no longer acting. The deflections of 45°/-45° panel reveal that the transient deflections continue with successively occurring peaks where the deflections at those peaks are lower than the maximum value. However, for 0°/90° panel the peak showing maximum deflection is obtained after the contact forces attain the null values. It indicates that study of variations of dynamic deflections is important even after completion of impact load cycle. Hence, the effect of inertia should not be neglected by the engineering while studying impact responses of composite conoidal panels.

The variations of maximum dynamic deflections and maximum contact forces for varying velocities of impactor are depicted in Figure 28.6. Both maximum deflection and contact force show an increasing pattern with velocity. The figure shows that the variations follow a quadratic distribution with the velocity values. The figure reports representative equations also which can help the engineers to find out the maximum responses for any intermediate velocity of impactor. Hence, it is expected that these representative equations can be helpful to the engineers for practical industrial applications.

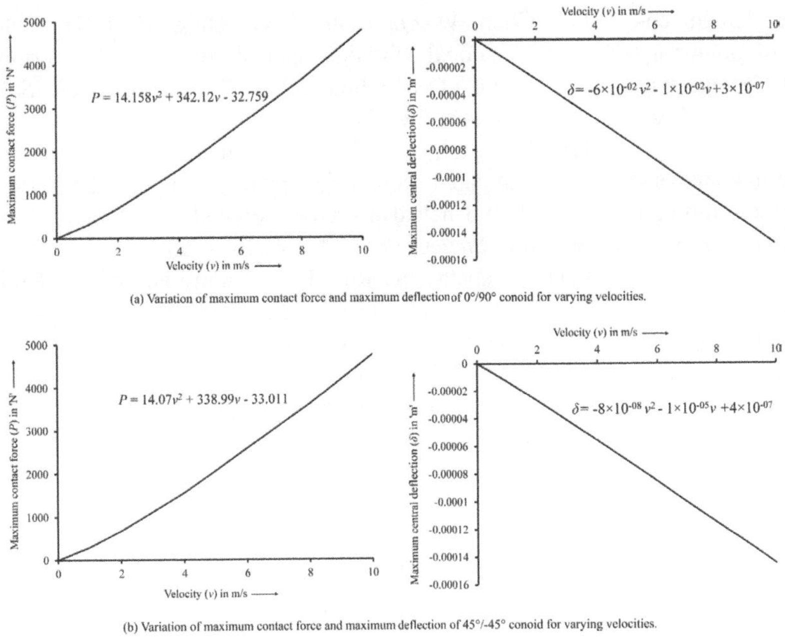

Figure 28.6 The variations of maximum contact force and maximum central deflections of clamped conoids for varying velocities of impactor

Source: Authors

Conclusion

The conclusions drawn from this study are as follows:

1. The results of benchmark problems clearly confirm that the proposed approach can correctly predict the contact forces and dynamic deflections of laminated conoids under central impact by a spherical impactor.
2. The low velocity impact on composite conoids generates parabolically distributed contact forces with only one peak. The impact forces increase as the velocity of impactor increases. This is important to highlight that the greater velocity leads to sharp elastic rebound and earlier detachment of impactor from the target panel surface. Naturally, the duration of contact force is lower when the speed of impactor is higher.
3. This study indicated that the peak contact force and peak deflection do not take place at the identical time instant for both composite conoids. Infact for cross-ply composite, the maximum force is identified after the contact force reached null value. Hence, the engineers must not that the maximum design stress can be obtained even after completion of contact force load cycle. Naturally, the dynamic investigation should be continued till the peak dynamic deflections are obtained.
4. The maximum force and dynamic deflections are found to increase as the speed of impactors increases. The relations between the maximum responses to the speed of impactor are not a linear one and this study suggests representative equations which can be used in real-life situations to interpolate the maximum responses for varying speed values.

Acknowledgement

The authors gratefully acknowledge the financial assistance by M/s Mazagaon Dock Shipbuilders to support this research.

References

[1] Nayak, A. N., Bandyopadhyay, J. N. Free vibration analysis and design aids of stiffened conoidal shells. *J. Engg. Mech.*, 2002;128(4):419–427.

[2] Nayak, A. N., Bandyopadhyay, J. N. Free vibration analysis of laminated stiffened shells. *J. Engg. Mech.*, 2005;131(1):100–110.

[3] Nayak, A. N., Bandyopadhyay, J. N. Dynamic response analysis of stiffened conoidal shells. *J. Sound Vib.*, 2006;291(3–5):1288–1297.

[4] Das, H. S., Chakravorty, D. Design aids and selection guidelines for composite conoidal shell roofs—A finite element application. *J. Reinf. Plast. Compos.*, 2007;26(17):1793–1819.

[5] Das, H. S., Chakravorty, D. A finite element application in the analysis and design of point-supported composite conoidal shell roofs: Suggesting selection guidelines. *J. Strain Anal. Engg. Des.*, 2010;45(3):165–177.

[6] Hari Sadhan Das, Chakravorty, D. Bending analysis of stiffened composite conoidal shell roofs through finite element application. *J. Compos. Mater.*, 2011;45(5):525–542.

[7] Kumari, S., Chakravorty, D. Finite element bending behaviour of discretely delaminated composite conoidal shell roofs under concentrated load. *J. Reinf. Plast. Compos.*, 2010;29(21):3287–3296.

[8] Kumari, S., Chakravorty, D. Bending of delaminated composite conoidal shells under uniformly distributed load. *J. Engg. Mech.*, 2011;137(10):660–668.

[9] Pradyumna, S., Bandyopadhyay, J. N. Free vibration and buckling of functionally graded shell panels in thermal environments. *Int. J. Struct. Stab. Dyn.*, 2010;10:1031–1053.

[10] Pradyumna, S., Bandyopadhyay, J. N. Dynamic instability behavior of functionally graded shells using a higher-order theory. *J. Engg. Mech.*, 2011;136(5):551–561.

[11] Bakshi, K., Chakravorty, D. Relative static and dynamic performances of composite conoidal shell roofs. *Steel Compos. Struct.*, 2013;15(4):379–397.

[12] Bakshi, K., Chakravorty, D. Numerical study on failure of thin composite conoidal shell roofs considering geometric nonlinearity. *KSCE J. Civ. Engg.*, 2020;24(3):913–921.

[13] Bakshi, K., Chakravorty, D. First ply failure study of thin composite conoidal shells subjected to uniformly distributed load. *Thin-Walled Struct.*, 2014;76:1–7.

[14] Bakshi, K., Chakravorty, D. First ply failure loads of composite conoidal shell roofs with varying lamination. *Mech. Adv. Mater. Struct.*, 2015;22(12):978–987.

[15] Bakshi, K., Chakravorty, D. A finite element prediction of first ply failure and delamination in composite conoidal shells using geometric nonlinearity. *Sādhanā*, 2020;45(1):218.

[16] Bakshi, K., Giri, A. Application of von-Karman model for laminated composite singly curved shells with non-uniform boundary conditions. *Mech. Adv. Mater. Struct.*, 2024:1–11.

[17] Chaubey, A. K., Kumar, A., Chakrabarti, A. Novel shear deformation model for moderately thick and deep laminated composite conoidal shell. *Mech. Based Des. Struct. Mach.*, 2018;46(5):650–668.

[18] Rout, M., Hota, S. S. Geometrically nonlinear free vibration of CNTs reinforced sandwich conoidal shell in thermal environment. *Acta Mech.*, 2023;234(6):2677–2694.

[19] Chen, W. Q., Bian, Z. G., Ding, H. J. Three-dimensional vibration analysis of fluid filled orthotropic FGM cylindrical shells. *Int. J. Mech. Sci.*, 2004;46:159–171.

[20] Ganapathy, S., Rao, K. P. Interlaminar stresses in laminated composite plates, cylindrical/spherical shell panels damaged by low-velocity impact. *Compos. Struct.*, 1997;38(1–4):157–168.

[21] Ganapathy, S., Rao, K. P. Failure analysis of laminated composite cylindrical/spherical shell panels subjected to low-velocity impact. *Comput. Struct.*, 1998;68(6):627–641.

[22] Krishnamurthy, K. S., Mahajan, P., Mittal, R. K. A parametric study of the impact response and damage of laminated cylindrical composite shells. *Compos. Sci. Technol.*, 2001;61(12):1655–1669.

[23] Krishnamurthy, K. S., Mahajan, P., Mittal, R. K. Impact response and damage in laminated composite cylindrical shells. *Compos. Struct.*, 2003;59(1):15–36.

[24] Changliang, Z., Mingfa, R., Wei, Z., Haoran, C. Delamination prediction of composite filament wound vessel with metal liner under low velocity impact. *Compos. Struct.*, 2006;75(1–4):387–392.

[25] Li, D. H., Liu, Y., Zhang, X. Low-velocity impact responses of the stiffened composite laminated plates based on the progressive failure model and the layerwise/solid-elements method. *Compos. Struct.*, 2014;110:249–275.

[26] Ferreira, L. M., Muñoz-Reja, M., Reis, P. N. B. Impact response of semicylindrical woven composite shells: The effect of stacking sequence. *Int. J. Impact Engg.*, 2024;189:104952.

[27] Sharma, A., Bhattacharjee, C., Jana, A., Asheequl, I., Neogi, S. Impact response of laminated composite simply supported stiffened conoidal sheel with cut out. *Adv. Engg. Mater.*, 2022; 2: 93–112.

[28] Irshad, A., Neogi, S. D., Chakravorty, D. Impact behaviour of laminated composite stiffened clamped conoidal shell roof by finite element. *IOP Conf. Series Environ. Earth Sci.*, 2024;1326(1):012073.

[29] Nanda, N., Bandyopadhyay, J. N. Nonlinear transient response of laminated composite shells. *J. Engg. Mech.*, 2024;134(11):983–990.

[30] Hadid, H. A. An analytical and experimental investigation into the bending theory of elastic conoidal shells. *Doc. Disser.*, 1964.

[31] Wu, B., Pagani, A., Chen, W. Q., Carrera, E. Geometrically nonlinear refined shell theories by Carrera Unified Formulation. *Mech. Adv. Mater. Struct.*, 2021;28:1721–1741.

[32] Carrera, E., Pagani, A., Augello, R., Wu, B. Popular benchmarks of nonlinear shell analysis solved by 1D and 2D CUF-based finite elements. *Mech. Adv. Mater. Struct.*, 2020;27(13):1098–1109.

29 Design of mechanism and kinematic analysis of robotic exoskeleton of hand for rehabilitation

Muzammil Bagewadi[a], *Nandeesha H. L. and Niranjana Nair*

Department of Mechanical Engineering, M S Ramaiah Institute of Technology, Bangalore, Karnataka, India

Abstract

Robotic hand exoskeletons are increasingly recognised as effective tools for rehabilitation yet replicating the human hand's intricate kinematics poses significant design challenges. This study presents the design and kinematic analysis of a robotic hand exoskeleton developed to enhance rehabilitation therapies for patients with impaired motor function. A design research methodology was employed to define key objectives, including adjustable mechanism design, safe force application, integration of surface EMG sensors for real-time feedback, and customisable grasping patterns. The process began with a detailed kinematic study of the metacarpophalangeal (MCP), proximal interphalangeal (PIP), and distal interphalangeal (DIP) joints to establish constraints and motion requirements. Building on these findings, a three-link, three-degree-of-freedom mechanism inspired by the extensor tendon system was developed to replicate natural finger movements. Iterative refinements ensured adaptability to different hand sizes, prevention of hyperextension, and compatibility with active, assistive, and resistive rehabilitation modes. Kinematic analysis using modelling and simulations verified the device's ability to reproduce natural range of motion, while dynamic analysis confirmed its structural integrity and safety under operational loads. The proposed exoskeleton demonstrates the potential of combining biomechanical insights, adaptive mechanisms, and user-centred design research methodologies to deliver a customisable, safe, and effective rehabilitation solution.

Keywords: Mechanism, robotic exoskeleton, prehension, actuator and rehabilitation

Introduction

Robotic exoskeletons for hand rehabilitation represent an innovative fusion of robotics, biomechanics, and medical technology, addressing the rising demand for advanced rehabilitative solutions. These devices aim to restore motor functionality by guiding and assisting the user's hand through controlled, repetitive movements, critical for patients recovering from hand injuries, strokes, or neurological conditions [1, 2]. As the prevalence of such conditions increases globally, there is a growing emphasis on integrating these technologies into therapeutic practices to enhance outcomes and reduce recovery times [3].

The human hand, with its 27 bones, over 20 degrees of freedom (DoFs), and complex musculotendinous structures, presents unique challenges for designing robotic exoskeletons [4, 5]. Its versatility stems from intricate kinematics and dynamics, allowing for a wide range of precise and powerful movements. Robotic exoskeletons aim to replicate this complexity, ensuring precise alignment with natural motion. For instance, studies have focused on detailed kinematic modelling to ensure devices mimic human hand movements accurately while accommodating inter- and intra-finger constraints [6]. This attention to design is critical, as even minor deviations can impact user comfort and therapeutic effectiveness [7, 8].

Among the key applications of robotic exoskeletons is stroke rehabilitation, a domain where upper-limb impairments are prevalent. Stroke-induced conditions, such as spasticity, abnormal synergies, and loss of voluntary control, hinder motor recovery. Exoskeletons offer a solution by enabling repetitive and task-specific training, a cornerstone of activity-dependent neuroplasticity [9, 15]. Bilateral devices that use movements of the unaffected hand to guide the impaired hand have shown significant promise in restoring motor function, leveraging the brain's capacity for reorganisation [10]. This bilateral approach aligns with findings on neural plasticity, where coordinated activities between both hands enhance recovery [11, 12].

[a]bagewadi.muzammil212@gmail.com

DOI: 10.1201/9781003770442-29

Recent technological advancements have further refined the design and functionality of robotic hand exoskeletons. The advent of lightweight materials, such as 3D-printed composites, has allowed the creation of highly customisable and ergonomic devices [13]. These innovations not only reduce manufacturing costs but also enable tailored solutions for individual patients. Additionally, the integration of artificial intelligence (AI) has revolutionised the adaptability of these devices. Machine learning algorithms analyse real-time patient data, optimising the exoskeleton's response and tailoring therapeutic interventions [1, 9].

Control mechanisms in robotic exoskeletons have also seen significant advancements. Traditional approaches have been augmented by systems that utilise electromyography (EMG) signals to interpret user intent, resulting in more intuitive operations [8, 10]. Additionally, mechanisms like timing belts and epicyclic gear trains have been incorporated to improve motion transmission and replicate natural hand movements more accurately [6, 7]. Such innovations ensure the devices are not only functional but also comfortable and easy to use, increasing their acceptability among patients [5].

Despite these advancements, challenges remain. Designing devices that balance flexibility, rigidity, and alignment with natural joint constraints requires meticulous engineering. Misalignments can lead to discomfort or even exacerbate existing injuries [4, 6]. Furthermore, the miniaturisation of components to maintain a lightweight profile while ensuring durability and functionality remains a critical area of focus [1, 7]. Interdisciplinary collaboration between engineers, clinicians, and therapists is essential to address these challenges and ensure the devices meet both technical and clinical requirements [2, 10].

Moreover, the potential for robotic exoskeletons extends beyond physical rehabilitation. By integrating gamification elements, these devices can enhance patient engagement and motivation, making therapy more interactive and enjoyable. For example, some systems incorporate virtual reality environments that simulate real-world tasks, enabling users to practice meaningful movements in a controlled setting [3, 9]. This holistic approach enhances both physical and psychological recovery, underscoring the transformative potential of these devices [12].

In conclusion, robotic hand exoskeletons signify a paradigm shift in rehabilitation, offering precise, adaptable, and patient-specific solutions. With continuous advancements in AI, materials science, and control systems, these devices are poised to become indispensable in modern therapeutic practices. Addressing existing challenges through multidisciplinary research and innovation will ensure their widespread adoption and success in improving patient outcomes [1, 2, 4, 6, 8, 12].

Methodology

Figure 29.1 shows the objectives as follows develop an indigenous and effective solution for rehabilitation: Reduce rehabilitation time by mimicking grasp motion with adjustable mechanism design and gripping strengths. Integrate surface EMG sensors: Provide real-time feedback and sensory simulation for touch and pressure, enhancing rehabilitation outcomes. Customise grasping patterns: Tailor motion sequences to individual rehabilitation needs, including pre-programmed sequences for passive rehabilitation. Ensure adjustable exoskeleton sizing: Accommodate various hand shapes while implementing safety features to prevent excessive force application, ensuring user safety.

Results and discussion

Kinematic study of the hand

Before delving into the intricate details of the mechanism design for the robotic hand exoskeleton, it's imperative to understand the underlying kinematics of the human hand. Our approach begins with a comprehensive kinematic study to replicate these movements accurately. The human hand boasts a remarkable range of motion (ROM) facilitated by the complex arrangement of bones, muscles, and tendons. To emulate this dexterity, our kinematic study focuses on dissecting the movements of key joints in the hand, including the metacarpophalangeal (MCP), proximal interphalangeal (PIP), and distal interphalangeal (DIP) joints.

By analysing the normal ROM of these joints, we establish static constraints on joint angles, delineating the limits within which our robotic hand exoskeleton must operate. These constraints serve as guiding principles in the design phase, ensuring that our mechanism closely mimics natural hand movements. Furthermore, dynamic constraints are identified to capture the intricate interplay between different finger joints during motion. These constraints elucidate

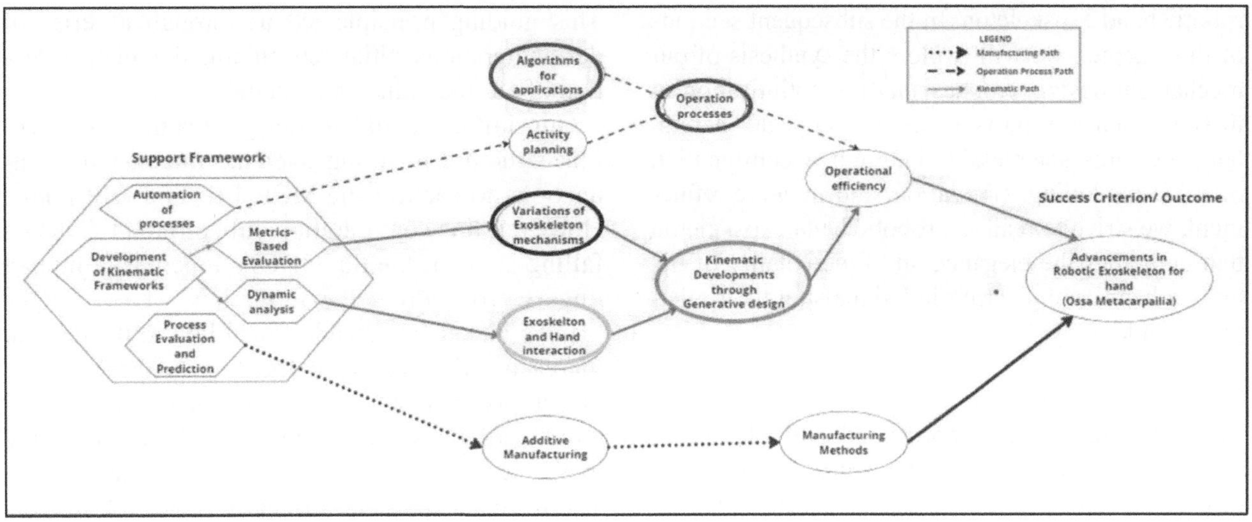

Figure 29.1 Design research methodology – Objectives defined
Source: Authors

the proportional relationships between joint angles, enabling coordinated movements across multiple fingers. Intrafinger constraints: The angle of the DIP joint is typically 2/3 of the PIP joint, and the angle of the PIP joint is typically 3/4 of the DIP joint during natural hand movements.

Interfinger constraints: As represented in Figure 29.2 certain joints of different fingers are interconnected, leading to coordinated movements. For example, bending the index finger at the MCP joint may automatically cause bending of the MCP joint in the middle finger, with certain proportionality [14]. Similarly, flexing the ring finger may cause slight flexion in the middle finger and little finger, with the same angle rotation Armed with this foundational knowledge, we proceed to translate these biomechanical principles into our mechanism design, ensuring that our robotic hand exoskeleton not only replicates natural hand movements but also enhances rehabilitation outcomes through precise and intuitive control mechanisms.

In our pursuit to model the hand's kinematics, we adopted a systematic approach, categorising movements into flexion-extension (pitch), rotation (roll), and abduction (yaw). Leveraging The Kutzbach's mobility criterion, we conducted an extensive analysis to determine the DOF of the hand's spatial mechanism, resulting in a total of 26 DOFs [4].

However, recognising the complexity inherent in capturing all DOFs, we strategically focused our attention on flexion-extension movements at each joint in the finger. This decision was guided by the essential role of grasp movements in physiotherapy, where precision and control are paramount. This selective approach not only simplifies the system's architecture but also aligns with our overarching goal of developing a user-friendly and effective

Finger	Flexion	Extension	Abd./add.
Thumb			
TMC	50°–90°	15°	45°–60°
MCP	75°–80°	0°	5°
IP	75°–80°	5°–10°	5°
Index			
CMC	5°	0°	0°
MCP	90°	30°–40°	60°
PIP	110°	0°	0°
DIP	80°–90°	5°	0°
Middle			
CMC	5°	0°	0°
MCP	90°	30°–40°	45°
PIP	110°	0°	0°
DIP	80°–90°	5°	0°
Ring			
CMC	10°	0°	0°
MCP	90°	30°–40°	45°
PIP	120°	0°	0°
DIP	80°–90°	5°	0°
Little			
CMC	15°	0°	0°
MCP	90°	30°–40°	50°
PIP	135°	0°	0°
DIP	90°	5°	0°

Figure 29.2 Static constraints [5]
Source: Authors

robotic hand exoskeleton. In the subsequent sections of this chapter, we will explore the synthesis of our mechanism design, emphasising the optimisation of flexion-extension movements to replicate natural hand gestures seamlessly. Through a combination of CAD modelling, simulation, and iterative refinement, we strive to realise a robotic hand exoskeleton that embodies the elegance and functionality of the human hand, empowering individuals on their rehabilitation journey.

Synthesis of mechanism

In developing a robust robotic hand exoskeleton, we explored kinematic models, refined design configurations, and adjusted link lengths to ensure functionality, adaptability, and user comfort. Selecting the appropriate mechanism involved addressing safety, functionality, and adaptability, focusing on rehabilitation effectiveness and user comfort. Key design criteria included mimicking natural hand movements, allowing a wide range of motion, preventing excessive force on joints, accommodating variations in hand sizes, preventing hyperextension with mechanical stops, scalability for all fingers, maintaining a palm-free design for object interaction, and supporting active, active-assistive, and resistive rehabilitation modes. Our analysis of various kinematic models presented unique benefits and challenges. Coinciding joint centres, where exoskeleton joints aligned with finger joint pivot points, offered simplicity but struggled with finger length variations. Models with remote centres of motion introduced non-fixed rotation points to enhance adaptability for diverse users. Redundant mechanisms, with additional degrees of freedom, allowed more natural and fluid finger movements. Tendon-driven mechanisms, using flexible actuators like tendons or artificial muscles, closely mimicked natural hand biomechanics, improving rehabilitation outcomes. Under actuated mechanisms provided simplicity and ease of control while maintaining functionality, making them practical for implementation. This approach ensured a design that balanced complexity and functionality, creating a scalable, adaptable, and effective rehabilitation device.

Kinematic model concept

In our pursuit of designing a robust robotic hand exoskeleton, our approach was inspired by the biomechanics of the extensor tendons in the human hand. This guiding principle led us through a series of design iterations, ultimately culminating in the selection of our final kinematic model.

Our initial candidate design featured a three-link structure hovering over the finger, with connections to each of the MCP, PIP, and DIP joints. However, this configuration only offered two DOF, falling short of the three DOFs inherent in human fingers. To address the limitations of two DOFs [4], we introduced an additional link to the design, increasing the degree of freedom to three. While this adjustment appeared promising, we encountered challenges related to the ROM, primarily due to the length restrictions imposed by the design. The need for disproportionately large link lengths to cover the desired ROM rendered this candidate impractical. In response to the limitations observed in candidate 2, we introduced further, introducing a third link to the design. This additional link provided the necessary flexibility to accommodate the full range of motion required for natural finger movements. Through this iterative process, our final candidate emerged as the optimal kinematic model for the robotic hand exoskeleton. This design, inspired by the extensor tendon mechanism, promises to deliver natural and fluid finger movements, enhancing the efficacy of rehabilitation interventions and empowering individuals on their path to recovery.

After this initial analysis, we engaged in a process of iterative design refinement, wherein multiple configurations were explored to ensure the faithful replication of natural finger motions. Inspired by the extensor mechanism, our preliminary designs emphasised contact points at three distinct locations along the finger's structure, while concurrently mitigating interference between adjacent digits (Figure 29.3).

Throughout this iterative process, we encountered various challenges and opportunities for optimisation. While simpler configurations with two degrees of freedom fell short in achieving the requisite ROM for natural finger movements, increasing the number of links in the final linkage to three degrees of freedom posed challenges in constraining motion within desired parameters. However, by dividing the final linkage into three loops, we were able to achieve the desired flexibility, particularly in facilitating MCP joint flexion, while maintaining three degrees of freedom (Figure 29.4).

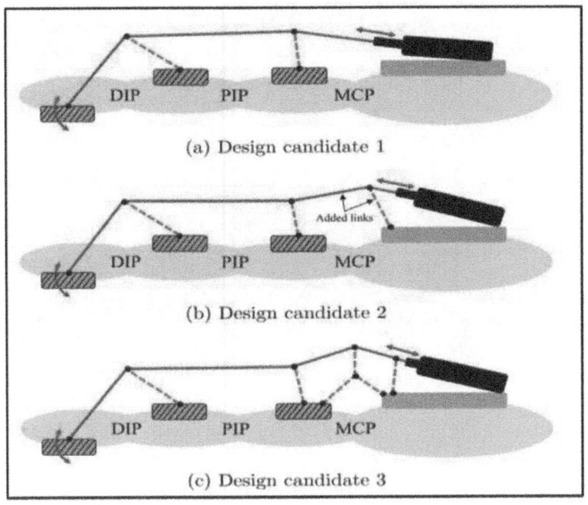

Figure 29.3 Iterative refinement
Source: Authors

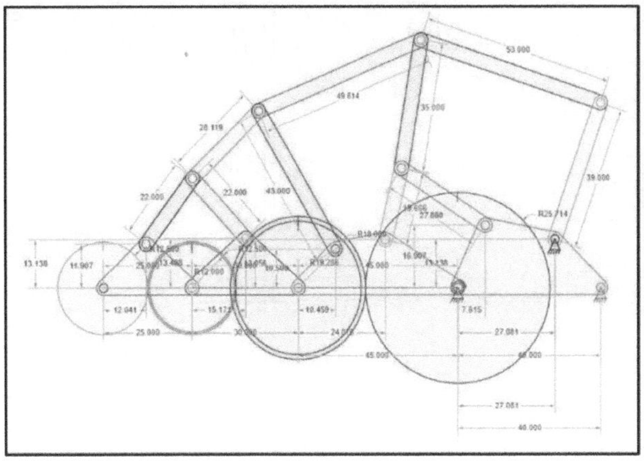

Figure 29.4 Final mechanism developed
Source: Authors

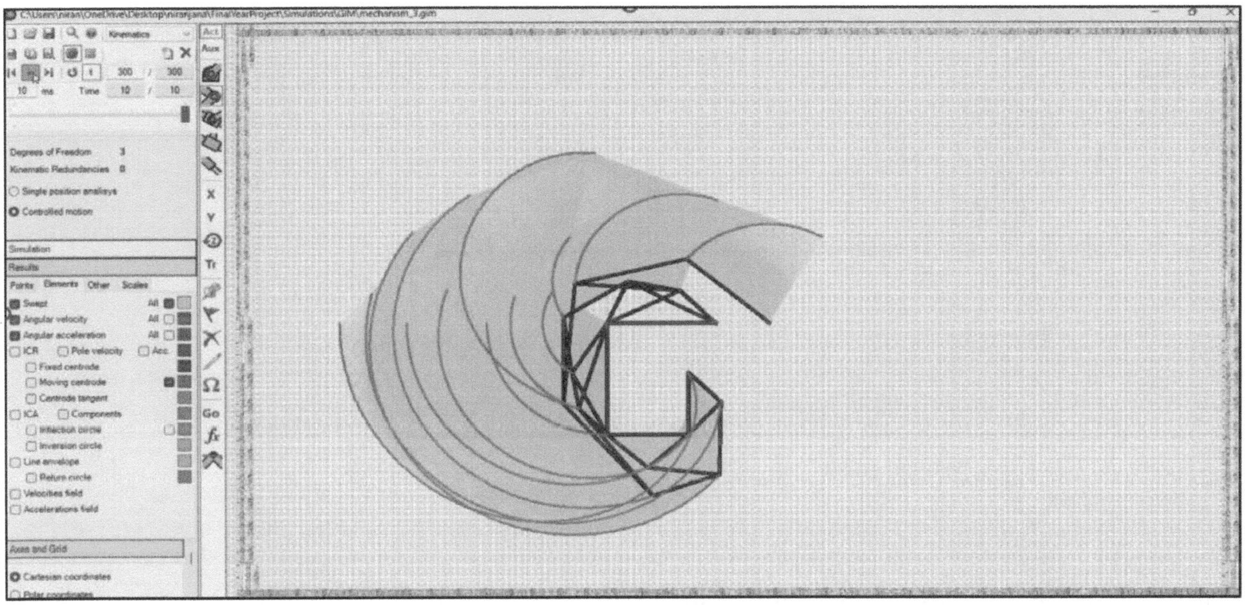

Figure 29.5 Range of motion
Source: Authors

Kinematic analysis

The kinematic analysis of the robotic hand exoskeleton began with the creation of a comprehensive CAD model using Fusion 360, a powerful design tool known for its precision and versatility. The CAD model served as the basis for subsequent simulations and analyses, ensuring that the virtual representation accurately mirrored the physical design of the exoskeleton. Utilising GIM software by Compmech Research Group, University of the Basque Country (UPV/EHU), Spain, the exoskeleton underwent rigorous kinematic studies. The primary objective was to verify the ROM of the device and assess its ability to replicate natural hand movements effectively. As shown in Figure 29.5, one of the key aspects of the kinematic analysis was the re-evaluation of the exoskeleton's ROM [16].

Through meticulous simulation and visualisation, the full extent of the exoskeleton's motion was verified, encompassing flexion-extension, abduction-adduction, and rotational movements at each finger joint. Visual representations of the ROM were generated, providing comprehensive documentation of the device's capabilities. Furthermore, kinematic

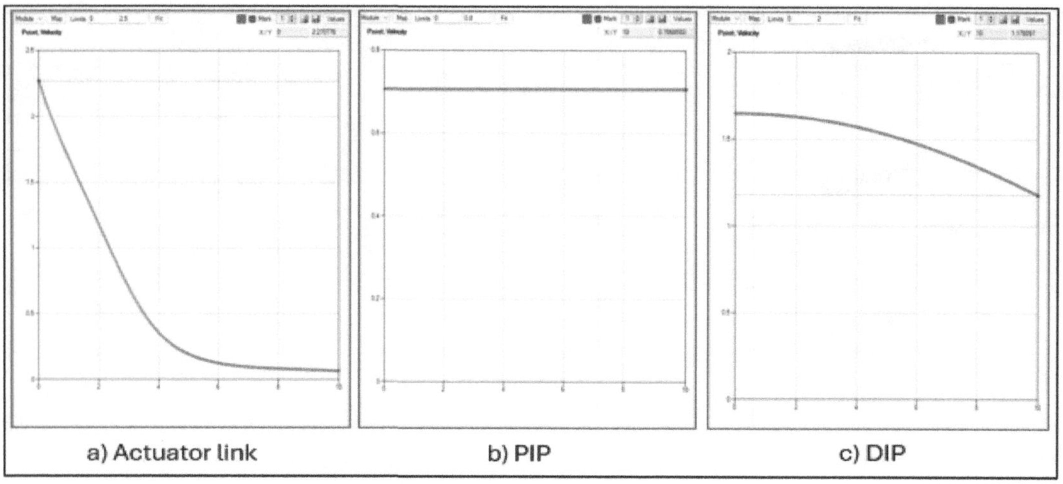

Figure 29.6 Kinematic analysis
Source: Authors

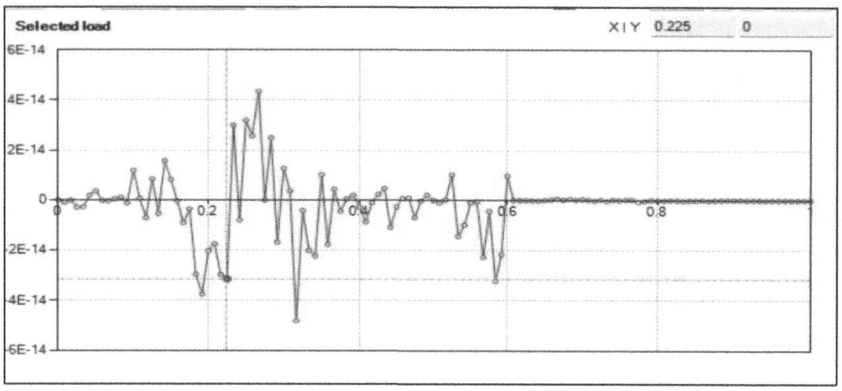

Figure 29.7 Load vs. position graph
Source: Authors

analysis delved into the specific motion at critical points of the exoskeleton, including the actuator end, PIP, and MCP joints. Constraints were applied to mimic natural hand movements, aiming to achieve a balance between natural motion and controlled speed. By accurately replicating human hand kinematics, the exoskeleton sought to enhance rehabilitation outcomes and assist individuals in performing activities of daily living with ease and precision.

Through this systematic process of synthesis and refinement, we arrived at a conceptual kinematic model that adeptly balances flexibility, ROM, and fidelity to natural finger movements. This refined design lays the groundwork for subsequent stages of development, promising the realisation of a robust and effective robotic hand exoskeleton tailored for rehabilitation applications (Figure 29.6).

Dynamic analysis

Dynamic analysis focused on evaluating the forces, torques, and accelerations experienced by the exoskeleton during operation. This analysis was crucial for assessing the structural integrity, actuator performance, and safety of the device under varying loading conditions. During dynamic analysis, the exoskeleton was subjected to servomotor torque input (10 kgfcm) to simulate real-world conditions. The maximum torque experienced by the MCP joint was found to be 150 N-mm as shown in Figure 29.7, consistent with literature findings, indicating the safety and reliability of the exoskeleton design [17]. This reassurance was crucial for ensuring user safety and in medical robotics, robot design must adapt to the challenges of the living world (Figure 29.8).

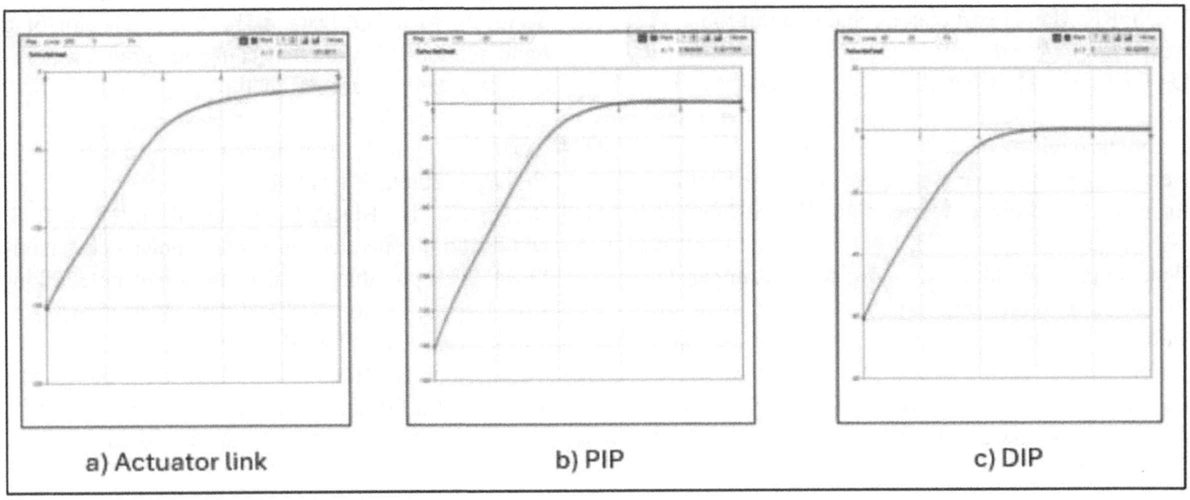

Figure 29.8 Dynamic analysis
Source: Authors

Conclusion

By developing an indigenous and effective solution that reduces rehabilitation time, the integration of adjustable mechanism designs and varying gripping strengths allows for a more natural and efficient recovery process. This approach is further enhanced by enabling natural wrist rotation within a customisable range, which caters to the specific needs of each user. This integration ensures that patients receive accurate and immediate data, which is vital for tracking progress and adjusting treatment plans accordingly. The customisation of grasping patterns to fit individual rehabilitation requirements, including pre-programmed sequences for passive rehabilitation, ensures that each patient receives personalised care. This tailored approach addresses the unique challenges faced by different individuals, promoting more effective and faster recovery outcomes. Lastly, ensuring adjustable exoskeleton sizing to accommodate various hand shapes while incorporating safety features to prevent excessive force application underscores commitment to user safety and comfort. This adaptability not only makes the technology accessible to a wider range of users but also guarantees that the rehabilitation process is both safe and effective. Collectively, these objectives form a comprehensive strategy for advancing rehabilitation technology, focusing on personalisation, safety, and effectiveness. This holistic approach promises significant improvements in rehabilitation outcomes, ultimately enhancing the quality of life for individuals undergoing rehabilitation.

Acknowledgement

This research was supported by M. S. Ramaiah Institute of Technology (MSRIT), Banglore–560054.

References

[1] Rojek, I., Dorożyński, J., Mikołajewski, D., Kotlarz, P. Overview of 3D printed exoskeleton materials and opportunities for their AI-based optimization. *Appl. Sci.*, 2023;13(14):8384.

[2] Urquhart, L., Petrakis, K., Hansen, J., Wodenhouse, A., Mariani, M., Lauer-Schmaltz, M., Loudon, B. Prototyping approaches for rehabilitation devices: From product embodiment to data management. *Comp.-Aided Design Appl.*, 2022:145–157.

[3] Proietti, T., Guigon, E., Roby-Brami, A., Jarrassé, N. Modifying upper-limb inter-joint coordination in healthy subjects by training with a robotic exoskeleton. *J. NeuroEngg. Rehab.*, 2017;14(1):55.

[4] ElKoura, G., Singh, K. Handrix: Animating the human hand. *Symp. Comp. Anim.*, 2003:110–119.

[5] Chen, F. C., Appendino, S., Battezzato, A., Favetto, A., Mousavi, M., Pescarmona, F. Constraint study for a hand exoskeleton: Human hand kinematics and dynamics. *J. Robotics*, 2013:1–17.

[6] Rahman, A., Al-Jumaily, A. Design and development of a bilateral therapeutic hand device for stroke rehabilitation. *Internat. J. Adv. Robotic Sys.*, 2013;10(12):405.

[7] Figliolini, G., Ceccarelli, M. A novel articulated mechanism mimicking the motion of index fingers. *Robotica*, 2002;20(1):13–22.

[8] Schabowsky, C. N., Godfrey, S. B., Holley, R. J., Lum, P. S. Development and pilot testing of

HEXORR: Hand exoskeleton rehabilitation robot. *J. Neuro Engg. Rehab.*, 2010;7(1):36.

[9] De Fazio, R., Mastronardi, V. M., Petruzzi, M., De Vittorio, M., Visconti, P. Human–machine interaction through advanced haptic sensors: A piezoelectric sensory glove with edge machine learning for gesture and object recognition. *Future Internet*, 2022;15(1):14.

[10] Worsnopp, T., Peshkin, M., Colgate, J., Kamper, D. An actuated finger exoskeleton for hand rehabilitation following stroke. *Proc. 10th IEEE Internat. Conf. Rehab. Robot. (ICORR '07),* 2007:1–6.

[11] Popovic, D., Stein, R., Jovanovic, K., Dai, R., Kostov, A., Armstrong, W. Sensory nerve recording for closed-loop control to restore motor functions. *IEEE Trans. Biomed. Engg.*, 1993;40(10):1024–1031.

[12] Chiri, A., Vitiello, N., Giovacchini, F., Roccella, S., Vecchi, F., Carrozza, M. C. Mechatronic design and characterization of the index finger module of a hand exoskeleton for post-stroke rehabilitation. *IEEE/ASME Trans. Mechatr.*, 2011;17(5):884–894.

[13] Yang, J., Peng, Q., Zhang, J., Gu, P. Design of a hand rehabilitation device using integrated axiomatic and benchmarking methods. *Procedia CIRP*, 2018;78:295–300.

[14] Curtis, L. Grant's atlas of anatomy (12th edition). *Ref. Rev.*, 2009;23(1):44–45.

[15] Brokaw, E. B., Black, I., Holley, R. J., Lum, P. S. Hand spring operated movement enhancer (HandSOME): A portable, passive hand exoskeleton for stroke rehabilitation. *IEEE Trans. Neural Sys. Rehab. Engg.*, 2011;19(4):391–399.

[16] Lu, J., Chen, W., Tomizuka, M. Kinematic design and analysis of a 6-DOF upper limb exoskeleton model for a brain-machine interface study. *IFAC Proc. Vol.*, 2013;46(5):293–300.

[17] Arroyave-Tobon, S., Drapin, J., Moretto, P., Linares, J. Design features of the kinematic chain of the ant exoskeleton. *Procedia CIRP*, 2023;119:855–860.

30 Multi-layer perceptron for damage detection in thin aluminium plate

Kamal Kishor Prajapati[a], Akshay Rai[b] and Mira Mitra[c]

Department of Aerospace Engineering, Indian Institute of Technology Kharagpur, West Bengal, India

Abstract

Structural health monitoring (SHM) is essential for maintaining the reliability and safety of various engineering structures. This study investigates the application of a multi-layer perceptron (MLP) model for detecting damage in a thin aluminium plate by analysing Lamb wave responses. A 1.6 mm thin square aluminium plate with side length of 1200 mm serves as the test structure under both healthy and damaged conditions. The dataset has been constructed using a combination of finite element (FE) simulations and experimental measurements. The problem of damage detection has been framed as a binary classification task, which has been tackled using a MLP, with the aim of distinguishing between the two states based on the Lamb wave responses. The MLP model, featuring multiple fully connected layers, has been trained to learn the complex relationships within the dataset and identify damage with high accuracy. The trained MLP model has been tested on a completely unseen test dataset and, has achieved an accuracy of 99% in classifying the plate's condition. The results underscore the MLP's effectiveness in accurately detecting damage in aluminium structures, offering a robust, data-driven approach for real-time structural monitoring and fault detection.

Keywords: Multi-layer perceptron, Lamb wave, SHM, finite element, safety hazards

Introduction

Structural health monitoring (SHM) facilitates the secure functioning and longevity of essential infrastructure across multiple engineering fields, including aerospace, mechanical, and civil engineering [1, 2]. The dependability of SHM techniques is crucial, particularly in cases where structural failures may result in substantial economic losses and safety hazards. Among various structural health monitoring methods, wave-based approaches have become increasingly favoured for their efficacy in detecting and localising structural anomalies. Lamb waves are widely employed in damage detection because they can propagate over extensive areas, allowing for the monitoring of large structural sections with a minimal number of sensors [3–5].

Recent progresses in artificial intelligence (AI) and machine learning (ML) have facilitated the creation of robust data-driven tools for SHM, enabling researchers to identify complex wave response patterns indicative of damage [6, 7]. Conventional SHM techniques, although efficient, frequently depend on predetermined damage characteristics, which may restrict adaptability to diverse structural configurations and damage types. Conversely, ML models, particularly neural networks, possess the potential to autonomously extract key characteristics from data without requiring manual feature selection. This capability renders them a promising choice for SHM applications that require flexible adaptation to varying conditions and support for real-time analysis [8, 9]. Conventional ML algorithms such as support vector machines (SVMs), k-nearest neighbours (k-NN), and random forests (RFs) have been effectively utilised in numerous SHM tasks [10, 11]. SVM are notably proficient in binary classification and have been extensively utilised for damage detection, where differentiating between damaged and undamaged conditions is essential [12]. Random forests have become popular because of their robustness and interpretability, effectively handling high-dimensional data and utilising an ensemble of decision trees to enhance accuracy and mitigate over fitting risks, which is crucial for analysing sensor data affected by environmental variables [11]. Notwithstanding its simplicity, k-NN has proven beneficial in SHM contexts with constrained labelled data, facilitating direct anomaly detection through proximity to established data points [13, 14].

[a]kkpkgp@gmail.com, [b]raia261990@gmail.com, [c]mira@aero.iitkgp.ac.in

DOI: 10.1201/9781003770442-30

Artificial neural networks (ANNs) represent a significant advancement over traditional methods, offering improved adaptability and feature-learning capabilities. ANNs have demonstrated efficacy in SHM because of their capacity to simulate intricate input-output relationships without reliance on manually defined features. An ANN can autonomously extract relevant damage-sensitive characteristics from data, eliminating the necessity for substantial feature engineering [15, 16]. Several researchers have applied ANN for various SHM purposes. For instance, Lu et al. [15] utilised ANN for detecting crack defects in aluminium plates, while Atashipour et al. [17] implemented ANN to identify defects in thick steel beams by analysing ultrasonic wave responses. These capabilities have enabled ANN to excel in classification and regression tasks in SHM, where understanding non-linear dependencies is essential. ANNs have been employed to assess damage severity and classify diverse damage states, exhibiting effectiveness in applications involving complex wave-based data [18, 19].

Multi-layer perceptrons (MLPs) are a specific variant of ANNs that have become prominent in SHM owing to their effective layered structure, which adeptly captures intricate relationships via non-linear transformations. In contrast to more complex ANN architectures that incorporate recurrent or convolutional components, MLPs are fully connected feed forward networks. This characteristic renders them especially suitable for tasks that necessitate the recognition of complex patterns devoid of spatial or temporal dependencies. Stacking multiple dense layers enables MLPs to capture hierarchical and non-linear data characteristics, facilitating more refined classifications than traditional linear methods [20, 21]. MLPs demonstrate strong performance in binary classification tasks, particularly in contexts where data directly correlates damage-sensitive features with output labels, positioning them as an optimal choice for SHM applications aimed at distinguishing between pristine and damaged states. Moreover, the straightforwardness and effectiveness of MLPs promote swift model training and evaluation, allowing for immediate implementation in SHM contexts [22, 23].

A variety of researchers have explored the application of deep learning (DL) methods for SHM, moving beyond simple ANNs to facilitate automated and precise analysis of intricate, high-dimensional data. DL algorithms, including convolutional neural networks (CNNs), recurrent neural networks (RNNs), and autoencoders have been effectively utilised in various SHM tasks, significantly enhancing the accuracy of damage identification and localisation [24, 25]. Despite the effectiveness of advanced DL architectures, the MLP remains a strong option for binary classification tasks in SHM. MLPs can reliably capture intricate, non-linear relationships in data, rendering them particularly useful when feature selection is challenging or when data lacks clear spatial or sequential patterns. Their simplicity, adaptability, and ability to provide accurate damage classification between intact and damage structures make MLPs an appealing choice for SHM applications that require quick and precise assessments.

This study employs an MLP model to examine Lamb wave responses for defect detection in a thin aluminium plate. Thin aluminium plates are susceptible to localised damage, including cracks or corrosion, which can severely undermine structural integrity if undetected. Lamb waves are utilised in this study because of their appropriateness for thin plate-like structures, facilitating efficient propagation and interaction with damage locations. These waves are produced and detected by piezoelectric (PZT) transducers acting as an actuator and sensors. The experimental configuration used in this study consists of a square aluminium plate, 1.6 mm in thickness and 1200 mm per side, evaluated under both undamaged and artificially damaged conditions. A comprehensive dataset has been developed by integrating finite element (FE) simulations with experimental measurements to represent realistic variations in Lamb wave propagation resulting from structural modifications. The damage detection problem has been framed as a binary task: distinguishing between undamaged and damaged states. The MLP model has been selected for its capability to manage extensive datasets and intricate input-output relationships, particularly due to the distinctive propagation and scattering properties of Lamb waves in thin plates. The MLP model utilises supervised learning to discern nuanced variations in wave response patterns between the two states. To enhance model robustness, the architecture integrates several hidden layers with non-linear activation functions, enabling it to capture complex patterns in the dataset. Moreover, dropout layers are added to mitigate over fitting and enhance the network's generalisation capability.

This study is structured in the following manner: Section 2 discusses the dataset, covering its architecture and the formulation of the training and testing datasets. Section 3 outlines the architecture of the proposed MLP model, as well as its fundamental principles. Section 4 evaluates the performance of the proposed MLP model. Ultimately, Section 5 encapsulates the conclusions derived from this study.

Data overview and preparation

This study evaluates the effectiveness of MLP architecture for identifying damage in a thin aluminium plate. To achieve this, a dataset is essential for both training and testing the MLP model. The dataset utilised in this research closely resembles the one employed in the author's previous work [26].

Layout of the dataset

To create the dataset, a thin square aluminium plate (Al-5052) measuring 1200 mm in length and 1.6 mm in thickness has been utilised. Lamb wave responses were gathered through both FE simulations and experimental methods. To generate diagnostic Lamb waves and capture those responses, PZT wafer transducers are affixed to the plate in a pitch-catch arrangement, as shown in Figure 30.1a. In this setup, one transducer acts as an actuator while the other serves as a sensor. The actuator receives a 3.5-cycle Hanning windowed tone burst signal at a predetermined centre frequency, which produces diagnostic Lamb waves that scan the plate. Upon encountering structural deterioration, these waves underwent significant alterations, including scattering and mode conversions, which can be identified by the PZT sensor, yielding critical information regarding potential damage.

For simulations involving damage, a rectangular cavity measuring 0.8 mm in depth and 2 mm in width has been introduced into the plate. In the experimental setup, two types of damages have been created:

(1) a notch in the plate and (2) an added mass (nut-bolt) weighing 14 grams that was bonded to the surface. An example of a Lamb wave response for the nut-bolt damage along with pristine condition at 100 kHz has been illustrated in Figure 30.1b.

The dataset comprises a total of 264 samples, with 132 samples collected from pristine (undamaged) conditions and the remaining 132 samples obtained from damaged conditions. The dataset consists of three primary categories of samples. The initial set comprises 216 samples produced via FE simulations, constituting the bulk of the dataset. The second and third sets comprise experimental samples, each containing 24 specimens. The second set was designed with a nut-bolt configuration to replicate damage, whereas the third set incorporated a notch into the plate. Collectively, these samples constitute a thorough dataset for evaluating damage detection.

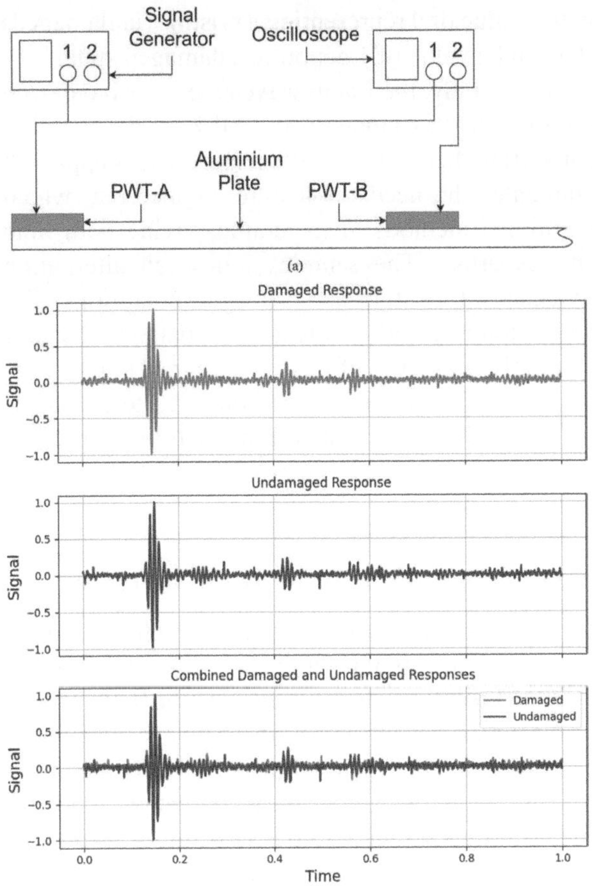

(b)

Figure 30.1 (a) Layout of the setup used in experiments; (b) Experimental Lamb wave data comparing responses for damaged and undamaged states
Source: Authors

Dataset formation for model training and testing

This study utilises a dataset comprising 264 samples, each featuring 1-D time-series sensor data with a total of 1925 data points. Each sample undergoes normalisation of time relative to the overall duration of the experiment, while the sensor data is adjusted based on its peak absolute value. Every sample is given a distinct identifier (sampleID) in conjunction with a condition label. This condition label is binary, with a value of 0 representing a pristine (undamaged) state and a value of 1 denoting a damaged state.

In this study, the Lamb wave responses have been used directly as inputs to the MLP model, without converting them into specific features. This approach eliminates the need for feature engineering, which traditional methods often require, saving both time and expertise. The samples follow an alternating sequence of pristine and damaged conditions. For model training and evaluation, the dataset was split randomly, with 70% of samples assigned to the training set and 30% held out for testing. The testing dataset was entirely segregated from the training process to guarantee a fair assessment of the model's efficacy on novel samples.

Architecture of the proposed MLP model

An MLP comprises an input layer, multiple hidden layers, and an output layer that are linked together by weighted connections between neurons. In this architecture, data moves across the network, with each neuron calculating a weighted sum of its incoming signals, applying a non-linear activation function, and sending the output to the subsequent layer. This design enables MLPs to identify intricate, non-linear patterns, rendering them effective for classification and regression tasks. MLPs are trained through back propagation, an algorithm that modifies weights according to the discrepancies between predicted and actual outputs. This iterative process helps the model to improve its accuracy over time.

This study presents the development of an MLP model for damage identification in a thin aluminium plate. The problem has been framed as a binary classification task. The proposed MLP model contains an input layer with 32 dense neurons and ReLU activation, which effectively captures non-linear relationships in the data. This is followed by a batch normalisation layer to mitigate internal covariate shift, facilitating stable and efficient training. To further ensure model robustness, a dropout layer with a rate of 0.2 is included to avoid over fitting. An additional hidden layer of 8 dense neurons with ReLU activation further refines the feature representation, followed by batch normalisation and another 0.2 dropout layer to enhance generalisation. The output layer features 2 neurons utilising sigmoid activation, making it ideal for binary classification by generating probability values for each class. To improve model performance in binary classification scenarios, the Adam optimiser is combined with a binary cross-entropy loss function. Model accuracy is tracked as a key metric. Early stopping is applied with a patience of 20 epochs, ensuring training halts if no progress is observed in validation loss. The model has been configured to train for 200 epochs with a batch size of 32, however, early halting ended training at 185. Additionally, 40% of the training data has been allocated for validation to balance training efficiency with validation accuracy.

Performance evaluation of the proposed MLP model

This section addresses the generalisation and performance assessment of the developed MLP architecture. During training, 40% of the training dataset is set aside as a validation dataset. While some researchers use the testing dataset for validation during training, this study keeps the testing dataset completely unexposed. This method offers a precise assessment of the model's performance on new, previously unencountered samples.

The training curve shown in Figure 30.2 clearly illustrates the MLP model's learning progression over time. Initially, the training and validation accuracies show rapid improvements, with occasional fluctuations, as the model begins to capture the data's underlying patterns. Around the middle of the training process, both accuracy curves start to stabilise and converge. Simultaneously, the loss for both the training and validation datasets falls steadily, indicating that the model is reducing errors. The convergence of

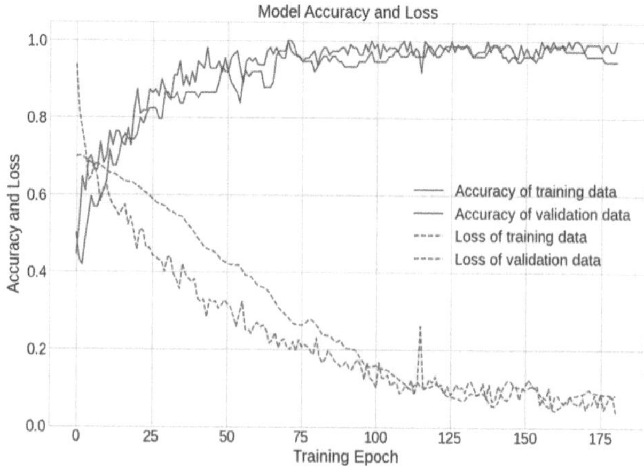

Figure 30.2 Loss and accuracy plots versus training epochs for proposed MLP model
Source: Authors

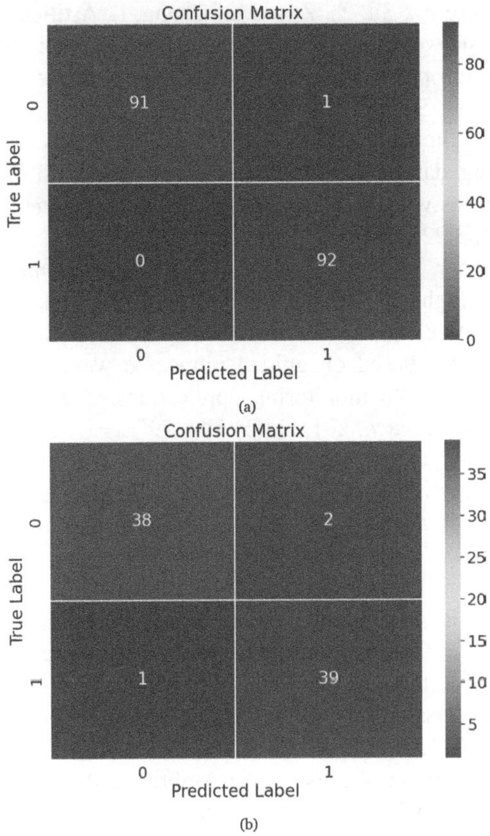

Figure 30.3 Confusion matrix for (a) training dataset; (b) testing dataset
Source: Authors

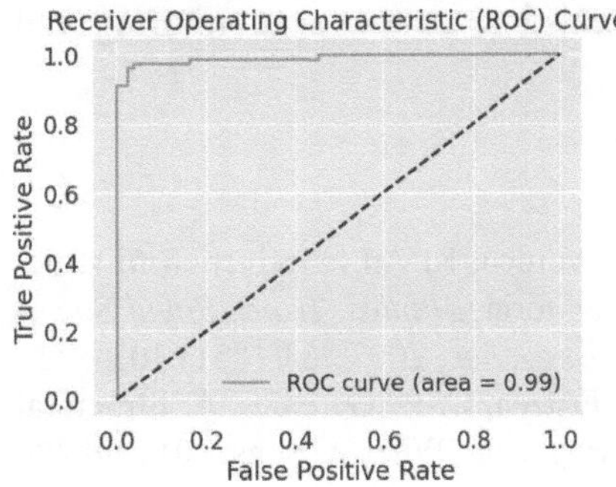

Figure 30.4 ROC curve for the proposed MLP model
Source: Authors

the accuracy and loss curves suggests effective learning with minimal over fitting. The model has been trained for 185 epochs.

Model's performance evaluation has been conducted using a confusion matrix. As illustrated in Figure 30.3, the results indicate that the MLP achieved an accuracy of 99% on the training dataset and 96% on the testing dataset, demonstrating strong generalisation capability. The confusion matrix (Figure 30.3b) indicates the high performance of the developed model by displaying the correct and incorrect classifications for both classes (pristine and damaged) in the testing dataset. The matrix reveals a high number of true positives and true negatives, with only a few misclassifications. This distribution confirms that the model has successfully learned to distinguish between the two conditions with strong accuracy on unseen data. Another practical way to assess the efficacy of the established model is to plot the receiver operating characteristic (ROC) curve. The ROC curve depicts the correlation between the true positive rate and the false positive rate at different thresholds, demonstrating the model's classification efficacy. A model exhibiting an ideal ROC curve attains the top-left corner, signifying accurate identification of positives while reducing false positives. The area under the curve (AUC) offers a single score for evaluating model efficacy with a score of 1 signifying flawless accuracy and 0.5 representing random guessing.

The ROC plot for the developed model (Figure 30.4) indicates that the model possesses an AUC of 0.99, demonstrating its near-perfect ability to distinguish between the two classes. The elevated AUC score validates the model's proficiency in differentiating between intact and damaged samples. The ROC curve, in conjunction with the training curve and confusion matrix data, indicates that the developed MLP architecture is reliable and precise in determining the condition of the thin aluminium plate.

Conclusion

In this study, an MLP framework is applied to identify damage in a thin aluminium plate. The dataset, derived from both experimental and simulated sources, includes Lamb wave responses under healthy and damaged conditions for training and testing.

The problem of damage detection has been framed as a binary classification task. The proposed MLP model has attained a classification accuracy of 96% when evaluated on an unseen dataset, highlighting the efficacy of MLPs in classifying the state of the structure. The model's high accuracy illustrates its capability to distinguish between wave responses from damaged and undamaged structures, thereby indicating the potential of MLPs as reliable, data-driven instruments for real-time structural monitoring and fault detection.

References

[1] Farrar, C. R., Worden, K. An introduction to structural health monitoring. *Philos. Trans. Royal Soc. A Mathem. Phy. Engg. Sci.*, 2007;365(1851):303–315.

[2] Balageas, D., Fritzen, C.-P., Gu¨emes, A. Structural health monitoring. John Wiley & Sons, 2010;90:496 pages.

[3] Mitra, M., Gopalakrishnan, S. Guided wave based structural health monitoring: A review. *Smart Mater. Struct.*, 2016;25(5):053001.

[4] Staszewski, W. J. Ultrasonic/guided waves for structural health monitoring. *Key Engg. Mater.*, 2005;293:49–62.

[5] Raghavan, A. Guided-wave structural health monitoring. *PhD Thesis*, 2007.

[6] Farrar, C. R., Worden, K. Structural health monitoring: a machine learning perspective. John Wiley & Sons, 2012.

[7] Zinno, R., Haghshenas, S. S, Guido, G., Rashvand, K., Vitale, A., Sarhadi, A. The state of the art of artificial intelligence approaches and new technologies in structural health monitoring of bridges. *Appl. Sci.*, 2022;13(1):97.

[8] Azimi, M., Eslamlou, A. D., Pekcan, G. Data-driven structural health monitoring and damage detection through deep learning: State-of-the-art review. *Sensors*, 2020;20(10):2778.

[9] Aval, S. B. B., Ahmadian, V., Maldar, M., Darvishan, E. Damage detection of structures using signal processing and artificial neural networks. *Adv. Struct. Engg.*, 2020;23(5):884–897.

[10] Neves, A. C., Gonzalez, I., Leander, J., Karoumi, R. Structural health monitoring of bridges: A model-free ann-based approach to damage detection. *J. Civil Struct. Health Monit.*, 2017;7(5):689–702.

[11] Nick, W., Shelton, J., Asamene, K., Esterline, A. C. A study of supervised machine learning techniques for structural health monitoring. *MAICS*, 2015;1353:36.

[12] Gui, G., Pan, H., Lin, Z., Li, Y., Yuan, Z. Data-driven support vector machine with optimization techniques for structural health monitoring and damage detection. *KSCE J. Civil Engg.*, 2017;21:523–534.

[13] Tang, Y., Chang, Y., Li, K. Applications of k-nearest neighbor algorithm in intelligent diagnosis of wind turbine blades damage. *Renewable Energy*, 2023;212:855–864.

[14] Vitola, J., Pozo, F., Tibaduiza, D. A., Anaya, M. A sensor data fusion system based on k-nearest neighbor pattern classification for structural health monitoring applications. *Sensors*, 2017;17(2):417.

[15] Lu, Y., Ye, L., Su, Z., Zhou, L., Cheng, L. Artificial neural network (ann)-based crack identification in aluminium plates with lamb wave signals. *J. Intell. Mater. Sys. Struct.*, 2009;20(1):39–49.

[16] An, D., Kim, N. H., Choi, J.-H. Practical options for selecting data-driven or physics-based prognostics algorithms with reviews. *Reliab. Engg. Sys. Safety*, 2015;133:223–236.

[17] Atashipour, S. A., Mirdamadi, H. R., Hemasian-Etefagh, M. H., Amirfattahi, R., Ziaei-Rad, S. An effective damage identification approach in thick steel beams based on guided ultrasonic waves for structural health monitoring applications. *J. Intell. Mater. Sys. Struct.*, 2013;24(5):584–597.

[18] de Lautour, O. R., Omenzetter, P. Damage classification and estimation in experimental structures using time series analysis and pattern recognition. *Mech. Sys. Signal Proc.*, 2010;24(5):1556–1569.

[19] Harirchian, E., Hosseini, S. E. A., Jadhav, K., Kumari, V., Rasulzade, S., I¸sık, E., Wasif, M., Lahmer, T. A review on application of soft computing techniques for the rapid visual safety evaluation and damage classification of existing buildings. *J. Build. Engg.*, 2021;43:102536.

[20] Gomez-Cabrera, A., Escamilla-Ambrosio, P. J. Review of machine-learning techniques applied to structural health monitoring systems for building and bridge structures. *Appl. Sci.*, 2022;12(21):10754.

[21] Pandey, P. C., Barai, S. V. Multilayer perceptron in damage detection of bridge structures. *Comp. Struct.*, 1995;54(4):597–608.

[22] Saadatmorad, M., Jafari-Talookolaei, R.-A., Pashaei, M.-H., Khatir, S., Wahab, M. A. Application of multilayer perceptron neural network for damage detection in rectangular laminated composite plates based on vibrational analysis. *Proc. 2nd Internat. Conf. Struct. Damage Model. Assess. SDMA 2021,* 2021:163–178.

[23] Saadatmorad, M., Siavashi, M., Jafari-Talookolaei, R.-A., Pashaei, M. H., Khatir, S., Thanh, C.-L. Multilayer perceptron neural network for damage identification based on dynamic analysis. *Struct. Health Monitor. Engg. Struct. Select Proc. SHM&ES 2020*, 2021:29–48.

[24] Melville, J., Alguri, K. S., Deemer, C., Harley, J. B. Structural damage detection using deep learning of ultrasonic guided waves. *AIP Conf. Proc.*, 2018;1949:230004.

[25] Khan, S., Yairi, T. A review on the application of deep learning in system health management. *Mech. Sys. Signal Proc.*, 2018;107:241–265.

[26] Rai, A., Mitra, M. Lamb wave based damage detection in metallic plates using multi-headed 1-dimensional convolutional neural network. *Smart Mater. Struct.*, 2021;30(3):035010.

31 Geometrically non-linear bending of composite imperfect cylindrical shells under hygrothermal environment

Ashish Giri[a] *and Kaustav Bakshi*[b]

Department of Civil Engineering, Indian Institute of Technology, Indore, Madhya Pradesh, India

Abstract

The composite cylindrical panels are preferred by the engineers to construct wide variety of weight sensitive engineering structures. The roofs of railway stations, wings of an aircraft and submarine hulls are typical examples. The popularity of cylindrical panel is attributed to easy to construct and aesthetically appealing surface. The laminated composites are preferred due to extraordinary strength for constant material consumption and flexibility to orient the major load carrying capacity along the governing stresses. The industrial applications of composite shells often subjected to hygrothermal stresses. The effects of hygrothermal stresses are more pronounced for imperfect panels. Research studies on composite cylindrical panels are not very few in numbers but the bending performances under hygrothermal stresses including imperfections is yet to receive due attention. Hence, this paper aims to work on that area. The cylindrical panels are considered to have imperfect (longer/shorter) dimensions due to fabrication error. The initial stresses due to the hygrothermal environment are considered. The governing equations include constant shear strains with proper correct factors, von-Karman strains and minimising total potential energy. The bending performances in terms of maximum non-dimensional central deflections and mode shapes are critically studied for varying radius of curvature, ratios of plan dimensions and imperfection amplitudes.

Keywords: Laminated composite, shells, imperfect shell, hygrothermal environment, geometric non-linearity

Introduction

The laminated composites gained preferences for aerospace, civil and marine structures due to lightweight applications. The curved structures suit the requirements of greater rigidity and better aesthetics of modern engineering structures especially for long unsupported spans [1]. The natural ventilation and construction advantages of singly curved, developable, cylindrical surfaces are recognised by civil engineers to fabricate the roofs of car parking lots, aircraft hangers, airport terminal buildings, shopping malls and stadiums. The parallelogram shaped structures require skewed cylindrical roofs. The performances of skewed panels were studied by Liao and Lee [2], Singha and Daripa [3], Upadhyay and Shukla [4], Zhang and Xiao [5] and Sengar et al. [6]. The research was done on composite flats only. The geometrically non-linear von- Karman strains were involved. The studies included elastic stability [2], free vibration [3], flexure responses [4], dynamics of CNT – reinforced plates [5] and sandwiches [6]. Similar studies of cylindrical panels were done by Kandasamy and Singh [7], Haldar [8], Kiani et al. [9] and Mandal et al. [10] using linear finite elements. Although, the non-linear approach is essential for realistic performance predictions of moderately thin panels none worked on this area. There is a recent paper [11] which reported non-linear displacements and stresses of stiffened skewed cylindrical panels.

The skewed curved panels can have imperfect dimensions during fabrication. The deviations from the design values are geometric imperfections. The longer/shorter panels, when fabricated within fixed span between column supports, can alter the curvature of skewed cylindrical panels. Naturally, such imperfections can influence the rigidity and performance of skewed panels. Researchers worked on this subject. Godoy [12] modelled the initial imperfections as out of plane displacements of shells for the first time. Hui [13] reported that imperfections influence the free vibration. The dynamic performances of laminated cylindrical panels were studied by Nanda and Pradyumna [14] using von-Karman strains and constant transverse shear strains. Kochurov and Avramov [15] worked on frequencies of imperfect cylinders using von-Karman strains. Kitipornchai et

[a]phd2201204001@iiti.ac.in, [b]kaustav.bakshi@iiti.ac.in

DOI: 10.1201/9781003770442-31

al. [16] researched on vibration of imperfect functionally graded plates. The authors reported soft-spring vibrations for plates with free sides. The stability performances of imperfect panels also attracted the attention of researchers. Dey and Ramachandra [17], Garcea et al. [18] and Baciu et al. [19] used von-Karman non-linearity to predict the instability of cylindrical panels with imperfections.

The hygrothermal variations initiate initial stresses in skewed curved panels. The aircraft body parts experience high temperature variations during high-speed operations. The panels of marine structures experience variations of moisture concentration during services. The sloshing of sea waves initiates thermal variations on ship hulls. The initial stresses caused by hygrothermal variations lead to premature failure through delamination. Although Naidu and Sinha [20–22], Nanda et al. [23], Biswal et al. [24] worked on bending, free and forced vibrations, impact of composite panels for hygrothermal variations but they did not consider imperfections. Research studies on cylindrical panels for combined hygrothermal variations and initial imperfections are scanty. Nanda and Pradyumna [14], Foroutan et al. [25] and Ahmadi et al. [26] worked on this area and concluded that the combinations of hygrothermal variations and initial imperfections have greater influence on performance of panels. However, they [14, 25, 26] did not consider the skewness of panels. The shell panels can have compromised rigidity (soft-spring behaviour) due to the combination of imperfections and skewness. The deflections of such shells may exceed the limits of serviceability due to hygrothermal environment even before application of the external loads. Hence, bending studies of imperfect panels including the variations of skewness and hygrothermal environment is the pre-requisite for confident practical applications. Despite of the importance, the research studies in this area are absolutely missing in literature which the present paper aims to fill up. This paper adopts non-uniformly restrained cylindrical panels having imperfect dimensions. The shorter panels have negative imperfections while the longer panels have positive ones. The hygrothermal and skewness variations are included. Several design guidelines are proposed after critically discussing the results.

Numerical formulation

This paper adopts a composite cylindrical panel (Figure 31.1) of uniform thickness "*h*".

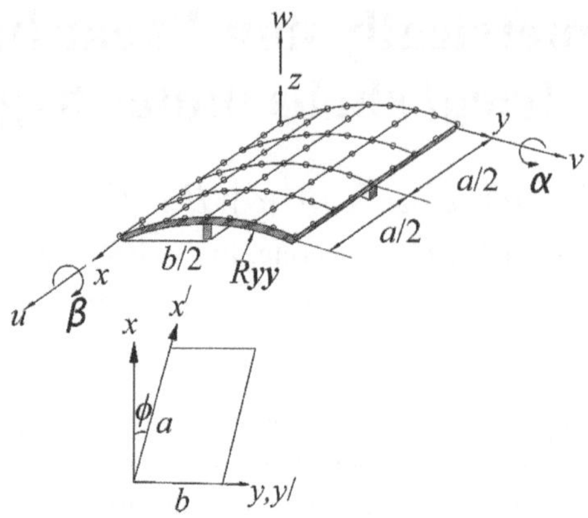

Figure 31.1 Skewed cylindrical panel
Source: Authors

The panel has two different system of reference which are (x', y') and (x, y). The axes are the following:

$$x = x'\cos\phi,\ y = y + x'\sin\phi,\ z = z' \tag{1}$$

ϕ = Skew angle. "*a*", "*b*"= sides along the x'- y' axes.

Constitutive relationship

This study considers the composite material as linearly elastic. The constitutive relations are as follows:

$$\{F\}' = [E]\{\varepsilon\}' - \{F^h\}' \tag{2}$$

The force vector, $\{F\}'$, and matrix $[E]$ are same as those were given by Bakshi and Giri [27].

$$\{F^h\}' = \{N^h_{x'} \quad N^h_{y'} \quad N^h_{x'y'} \quad M^h_{x'} \quad M^h_{y'} \quad M^h_{x'y'} \quad 0 \quad 0\}^T \tag{3}$$

The hygrothermal strains $\{\varepsilon\}_h$ are the following:

$$\{\varepsilon\}_h = \begin{bmatrix} m^2 & n^2 \\ n^2 & m^2 \\ -2mn & 2mn \end{bmatrix} \begin{Bmatrix} \beta_1 \\ \beta_2 \end{Bmatrix} (\Delta C) + \begin{bmatrix} m^2 & n^2 \\ n^2 & m^2 \\ -2mn & 2mn \end{bmatrix} \begin{Bmatrix} \alpha_1 \\ \alpha_2 \end{Bmatrix} (\Delta T) \tag{4}$$

where $m = \cos\theta$ and $n = \sin\theta$. θ = Fibre alignment from x' axis. β_1 and β_2 are moisture coefficients, α_1 and α_2 are thermal coefficients. ΔT and ΔC are changes in temperature and moisture concentration.

Geometric imperfection

The imperfect dimensions vary curvature of panels supported along all edges. The imperfections are considered as out-of-plane deflections, which follow Godoy [12]. The deflections follow the equation given below,

$$\bar{w}_0 = \eta h \sin\left(\frac{\pi x'}{a}\right)\sin\left(\frac{\pi y'}{b}\right) \quad (5)$$

η = Amplitude of imperfection.

Finite element formulation

Eight nodded curved elements are adopted to develop a C^0 isoparametric code. The degrees of freedom are depicted in Figure 31.1.

$$\{\delta'\} = \{u' \; v' \; w' \; \alpha' \; \beta'\}^T$$

The generalised displacement field is the following:

$$u' = u'_0 + z\alpha',\; v' = v'_0 + z\beta',\; w' = w'_0 \quad (6)$$

The displacements from (x', y') axes to (x, y) ones are related though the following:

$$\{\delta'\} = [T]\{\delta\} \quad (7)$$

The matrix $[T]$ is the same as it is given in Reference [11]. The mid-surface strains of a perfect panel, $\{\varepsilon'_p\}$, are given below which combines von-Karman non-linearity and constant transverse shear strains with appropriate correction factors.

$$\{\varepsilon'_p\} = \begin{Bmatrix} \frac{\partial u'_0}{\partial x'} + \frac{1}{2}\left\{\frac{\partial(w'_0+\bar{w}_0)}{\partial x'}\right\}^2 \\ \left\{\frac{\partial v'_0}{\partial y'} - \frac{(w'_0+\bar{w}_0)}{R'_{yy}}\right\} + \frac{1}{2}\left\{\frac{\partial(w'_0+\bar{w}_0)}{\partial y'} - \frac{v'_0}{R'_{yy}}\right\}^2 \\ \frac{\partial u'_0}{\partial y'} + \frac{\partial v'_0}{\partial x'} + \left\{\frac{\partial(w'_0+\bar{w}_0)}{\partial x'}\right\}\left\{\frac{\partial(w'_0+\bar{w}_0)}{\partial y'} - \frac{v'_0}{R'_{yy}}\right\} \\ \alpha' + \frac{\partial w'_0}{\partial x'} \\ \beta' + \frac{\partial w'_0}{\partial y'} \end{Bmatrix} \quad (8)$$

The net mid-surface strains of the imperfect shell panel,

$$\{\varepsilon^0\}' = \{\varepsilon'_p\} - \{\bar{\varepsilon}'_0\} \quad (9)$$

$$\text{where } \{\bar{\varepsilon}'_0\} = \begin{Bmatrix} \frac{1}{2}\left(\frac{\partial \bar{w}_0}{\partial x'}\right)^2 \\ -\frac{\bar{w}_0}{R'_{yy}} + \frac{1}{2}\left(\frac{\partial \bar{w}_0}{\partial y'}\right)^2 - \left(\frac{\partial \bar{w}_0}{\partial y'}\right)\left(\frac{v'_0}{R'_{yy}}\right) \\ \left\{\frac{\partial \bar{w}_0}{\partial x'}\right\}\left\{\frac{\partial(\bar{w}_0)}{\partial y'} - \frac{v'_0}{R'_{yy}}\right\} \\ 0 \\ 0 \end{Bmatrix}$$

The net strains are as follows:

$$\{\varepsilon^0\}' = \begin{Bmatrix} \frac{\partial u'_0}{\partial x'} \\ \frac{\partial v'_0}{\partial y'} - \frac{w_0}{R'_{yy}} \\ \frac{\partial u'_0}{\partial y'} + \frac{\partial v'_0}{\partial x'} \\ \alpha' + \frac{\partial w_0}{\partial x'} \\ \beta' + \frac{\partial w_0}{\partial y'} \end{Bmatrix} + \begin{Bmatrix} \frac{1}{2}\left(\frac{\partial w_0}{\partial x'}\right)^2 \\ \frac{1}{2}\left(\frac{\partial w_0}{\partial y'} - \frac{v'_0}{R'_{yy}}\right)^2 \\ \left(\frac{\partial w_0}{\partial x'}\right)\left(\frac{\partial w_0}{\partial y'} - \frac{v'_0}{R'_{yy}}\right) \\ 0 \\ 0 \end{Bmatrix} + \begin{Bmatrix} \frac{1}{2}\left(\frac{\partial w_0}{\partial x'}\right)^2 \\ \frac{1}{2}\left(\frac{\partial w_0}{\partial y'} - \frac{v'_0}{R'_{yy}}\right)^2 \\ \left(\frac{\partial w_0}{\partial x'}\right)\left(\frac{\partial w_0}{\partial y'} - \frac{v'_0}{R'_{yy}}\right) \\ 0 \\ 0 \end{Bmatrix} \quad (10)$$

$\{\varepsilon^0_L\}'$ = net linear strains. $\{\varepsilon^0_I\}'$ = net strains due to imperfection. $\{\varepsilon^0_{NL}\}'$ = net non-linear strains.

The net strains $\{\varepsilon^0\}'$ to the nodal degrees of freedom $\{\delta_i\}$ are related as given below,

$$\{\varepsilon^0\}' = \left([B']^L + [B']^I + \tfrac{1}{2}[B']^{NL}\right)\{\delta_i\}' \quad i = 1 \text{ to } 8 \quad (11)$$

where

$$\{\varepsilon^0\}' = \{\varepsilon^0_{x'} \; \varepsilon^0_{y'} \; \gamma^0_{xy'} \; k_{x'} \; k_{y'} \; k_{xy'} \; \gamma^0_{xz'} \; \gamma^0_{yz'}\}$$

and $\{k_{x'} \; k_{y'} \; k_{xy'} \; k_{xz'} \; k_{yz'}\}^T = \left\{\frac{\partial\alpha'}{\partial x'} \; \frac{\partial\beta'}{\partial y'} \; \frac{\partial\alpha'}{\partial y'} + \frac{\partial\beta'}{\partial x'} \; 0 \; 0\right\}^T$

Refer to reference [27] for $[B']^L$ and $[B']^{NL}$. $[B']^I$ can be obtained from Equation (10). By using Equations (10), Equation (2) takes the following form,

$$\{F\}' = [E]\left(\{\varepsilon^0_L\}' + \{\varepsilon^0_I\}' + \{\varepsilon^0_{NL}\}'\right) - \{F^h\}' \quad (12)$$

Governing equations

The total potential energy ($\pi = U + W$) is minimised with respect to the shell displacements (δ').

$$\left\{\frac{\partial}{\partial \delta'}(U + W)\right\} = \Delta \quad (13)$$

where $(U) = \frac{1}{2}\int_v \{\varepsilon\}'^T\{\sigma'\}dv',\ (W) = -\int_A \{\delta'\}^T\{q\}dA'$ and δ' = displacements for (x', y') axes. $\Delta \neq 0$ indicates residual forces. That is minimised following the Newton –Raphson iterations [28]. Equation (13) can be modified as the following:

$$\sum_{l=1}^{ne}\left(\int_{A/}[\bar{B}^{/}]^{T}\{F\}^{/}dA^{/}-\sum_{i=1}^{8}\int_{A}[N_i]^{T}\{q\}dA^{/}\right)_{l}=\sum_{l=1}^{ne}\Delta_i \quad (14)$$

$\{F\}^{/}$ is defined in Equation (2). ne = Number of elements.

and $\frac{\partial\{\varepsilon^0\}^{/T}}{\partial\{\delta^{/}\}}=[\bar{B}^{/}]^{T}$, $[\bar{B}^{/}]=[B^{/}]^{L}+[B^{/}]^{I}+[B^{/}]^{NL}$

$[N_i]$ = Shape functions, $\{q\}$ = external pressure intensities.

Equation (14) is given as the following:

$$\sum_{l=1}^{ne}\left(\int_{A/}[\bar{B}^{/}]^{T}\{F\}^{/}dA^{/}-\{R\}\right)_{l}=\sum_{l=1}^{ne}\Delta_i \quad (15)$$

where the external force is $\{R\}=\sum_{i=1}^{8}\int_{A}[N_i]^{T}\{q\}dA^{/}$

and the internal force is $\int_{A/}[\bar{B}^{/}]^{T}\{F\}^{/}dA^{/}$.

Solution process

Equation (15) is solved following Bakshi and Giri [27]. The final form is given below,

$$[K_s]\{\delta_i\}-\{Q_e\}=\Delta \quad (16)$$

where $[K_s]=\int_{A}\left\{[\bar{B}]^{T}[E]\left([B]^{L}+[B^{/}]^{I}+\frac{1}{2}[B]^{NL}\right)dA\right\}$

and $\{Q_e\}=\{R\}+\int_{A/}[\bar{B}^{/}]^{T}\{F^{h}\}^{/}dA^{/}$.

No external pressure is considered, hence $\{R\}$ = 0. The 2×2 Gauss quadrature numerical integration is done for bending and shear energies. The solutions at element level are assembled afterwards to get the global solutions and convergence of Newton-Raphson method is checked as was given in the ref. paper [27].

Numerical problems

The first set of problems are solved to validate the proposed approach. Section "Validation problems" reports those problems. The solutions of authors' own problems on composite cylindrical panels are reported in Section "Additional problems".

Validation problems

The first problem solves the defections of a composite $[0^0/0^0/30^0/-30^0]_2$ cylindrical panel under hygrothermal variations. The panels have simple support along the edges. The results are given in Figures 31.2 and 31.3 which depict close match with what was reported by Naidu and Sinha [20]. Therefore, the

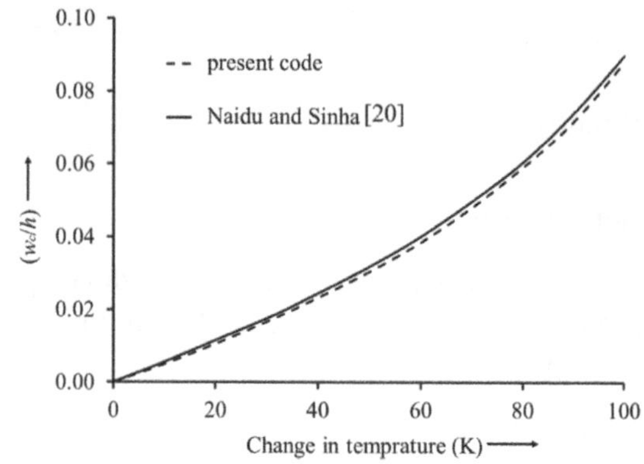

Figure 31.2 Comparison of deflections due to temperature variation
Source: Authors

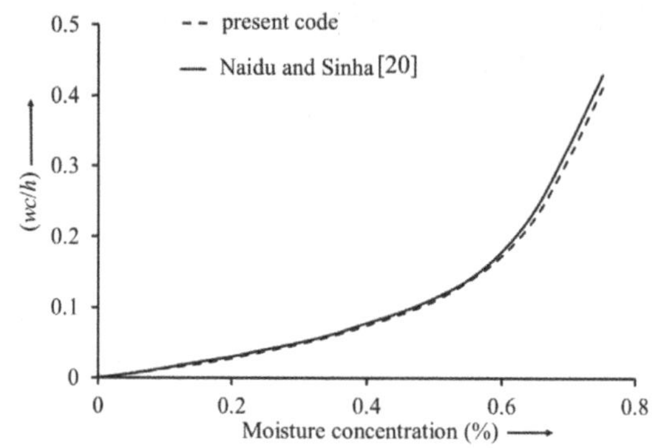

Figure 31.3 Comparison of deflections due to moisture variation
Source: Authors

correctness of proposed code to obtain non-linear deflections of cylindrical panels under hygrothermal variations is confirmed.

The non-linear deflections of isotropic cylindrical panels solved by present method are compared with CUF based results which were published by Wu et al. [28] and Carrera et al. [29]. The comparison is furnished in authors' own paper given in Reference [27] which depicted correctness of the present non-linear code. That paper reports the problem data also. The non-linear deflections at the centre of a square, isotropic, clamped plates are computed using the proposed code for skew angles varying from 0°, 15°, 30° and 45°. The results are compared with the values given by Buragohmt and Patodi [30] in authors' own paper given in Reference [11]. The close match of results confirmed correctness of the proposed formulation for skewed panels. For brevity, those results are not repeated here.

Additional problems

The corrected non-linear approach is used to compute deflections of laminated composite skewed cylindrical panels under varying temperatures and moisture concentrations. The following boundary condition is adopted:

CSCS shells = Curved ends ($x = 0$, $x = a$) are clamped and straight edges ($y = 0$, $y = b$) simple supported.

The material properties of the composite adopted in this paper are those which were reported by Nanda and Pradyumna [14]. The results are studied for varying skew angles measured from the x-axis. The imperfections are varied in percentage of shell thickness designated as imperfection amplitude here. Results are presented in tables and through diagrams. Dimensions of shells are reported with the tables.

Results and discussion

Table 31.1 reports the non-dimensional (central deflection/thickness) deflections of cylindrical panels subjected to temperature (T) varying from 300° Kelvin to 400° Kelvin with a constant step of 25° Kelvin. The panels at 300° Kelvin are considered as the reference from where the variations of deflections are measured. It is found that the deflections increase as temperature changes by 100° Kelvin and the maximum values are obtained at $T = 400°$ Kelvin. Table 31.2 reports the deflections of identical panels subjected to moisture concentration (C) varying from 0% to 1% following a constant step of 0.25%. This study considers panels at 0% moisture concentration as the baseline from where the variations of deflections are measured. Table 31.2 also reflects that deflections of panels increase due to exposure to moisture concentration and the values are maximum for $C = 1\%$. The stiffnesses of composite material decrease when the temperature and moisture concentration increase. That can lead to the flexibility of panels for exposure to elevation of temperature and moisture concentration. The panels studied in this paper have their in-plane movements locked along all four edges. Naturally, these panels show upward deflections when subjected to elevated temperature and moisture concentration.

Tables 31.1 and 31.2 report the deflections for antisymmetric and symmetric composites. For any given laminate, the panels are taken to be skewed by 0°–35° with constant step of 5°. Table 31.1 shows that at any given temperature, the deflections of panels monotonically increase as the skew angle varies from 0° to 35°. The panels skewed at 35° show the maximum deflections. The performances of skewed panels are

Table 31.1 Non-dimensional deflections for varying temperature and skew angles.

Lamination (degree)	ϕ	Temperature in (^{0}K)			
		325	350	375	400
0°/90°	0	0.0465	0.0867	0.1199	0.1339
	5	0.0468	0.0874	0.1210	0.1366
	10	0.0476	0.0890	0.1236	0.1409
	15	0.0485	0.0908	0.1266	0.1439
	20	0.0492	0.0925	0.1292	0.1456
	25	0.0498	0.0937	0.1313	0.1466
	30	0.0503	0.0947	0.1330	0.1473
	35	0.0507	0.0955	0.1343	0.1478
0°/90°/0°	0	0.0462	0.0866	0.1055	0.1302
	5	0.0462	0.0867	0.1060	0.1308
	10	0.0464	0.0871	0.1072	0.1322
	15	0.0467	0.0876	0.1085	0.1336
	20	0.0471	0.0881	0.1094	0.1348
	25	0.0474	0.0886	0.1101	0.1356
	30	0.0476	0.0891	0.1106	0.1362
	35	0.0479	0.0895	0.1110	0.1366
$(0°/90°)_2$	0	0.0434	0.0806	0.1047	0.1276
	5	0.0436	0.0810	0.1060	0.1292
	10	0.0441	0.0819	0.1084	0.1325
	15	0.0448	0.0832	0.1105	0.1352
	20	0.0454	0.0844	0.1118	0.1371
	25	0.0459	0.0855	0.1127	0.1383
	30	0.0464	0.0864	0.1133	0.1391
	35	0.0467	0.0871	0.1137	0.1396
$(0°/90°)_S$	0	0.0465	0.0870	0.1101	0.1355
	5	0.0466	0.0871	0.1108	0.1363
	10	0.0468	0.0876	0.1123	0.1381
	15	0.0472	0.0881	0.1138	0.1399
	20	0.0476	0.0888	0.1150	0.1413
	25	0.0479	0.0894	0.1158	0.1422
	30	0.0482	0.0899	0.1163	0.1429
	35	0.0484	0.0903	0.1167	0.1433

$a/b = 1$, $a/h = 100$, $R_{yy} = 750$ mm
Source: Authors

Table 31.2 Non-dimensional central deflections of CSCS panels for varying moisture concentration and skew angles.

Lamination (degree)	ϕ	Moisture concentration C (%)			
		0.25	**0.50**	**0.75**	**1.0**
0°/90°	0	0.1207	0.2423	0.3436	0.3716
	5	0.1153	0.2303	0.3314	0.3983
	10	0.1064	0.2113	0.3162	0.3911
	15	0.1002	0.1985	0.2955	0.3741
	20	0.0967	0.1910	0.2836	0.3622
	25	0.0946	0.1867	0.2768	0.3548
	30	0.0933	0.1840	0.2727	0.3502
	35	0.0925	0.1824	0.2700	0.3474
0°/90°/0°	0	0.0750	0.1471	0.2162	0.2822
	5	0.0755	0.1480	0.2176	0.2841
	10	0.0765	0.1501	0.2206	0.2881
	15	0.0775	0.1521	0.2236	0.2920
	20	0.0783	0.1536	0.2259	0.2950
	25	0.0789	0.1547	0.2274	0.2969
	30	0.0792	0.1554	0.2285	0.2983
	35	0.0795	0.1559	0.2292	0.2992
$(0°/90°)_2$	0	0.0918	0.1811	0.2682	0.3534
	5	0.0912	0.1798	0.2660	0.3502
	10	0.0900	0.1774	0.2620	0.3442
	15	0.0891	0.1753	0.2587	0.3394
	20	0.0884	0.1739	0.2565	0.3362
	25	0.0879	0.1730	0.2550	0.3342
	30	0.0877	0.1724	0.2541	0.3329
	35	0.0875	0.1720	0.2535	0.3329
$(0°/90°)_S$	0	0.0830	0.1632	0.2403	0.3142
	5	0.0837	0.1644	0.2422	0.3168
	10	0.0850	0.1672	0.2463	0.3222
	15	0.0863	0.1697	0.2501	0.3272
	20	0.0873	0.1716	0.2528	0.3308
	25	0.0879	0.1729	0.2547	0.3332
	30	0.0884	0.1737	0.2559	0.3348
	35	0.0887	0.1743	0.2567	0.3358

$a/b = 1$, $a/h = 100$, $R_{yy} = 750$ mm
Source: Authors

compromised because of their longer length along the relatively weaker side. This study aims to minimise the compromised performances of skewed panels by varying the laminations. Table 31.1 indicates

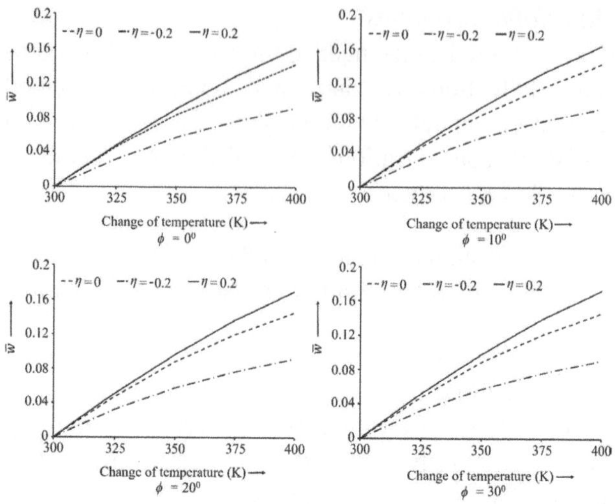

Figure 31.4 Variations of deflections for varying imperfection amplitudes (η), temperature and skew angle (ϕ)
Source: Authors

that at any given temperature, the compromises due to varying skewness are lower when the symmetric composites are adopted for fabrication. Among the symmetric ones studied in Table 31.1, the 0°/90°/0° option shows the minimum deflections of panels. This option also shows the minimum compromise with deflections for varying skewness.

Table 31.2 reports the deflections of similarly skewed panels for varying moisture concentrations. The deflections of anti-symmetric panels decrease as the skewness increases keeping the moisture concentration as constant. However, for symmetric laminates the deflections increase for similar variations skew angles and moisture concentration. This observation indicates that the lamination is crucial in determining the variations of deflections for varying skew angles. Among the anti-symmetric laminates, the 0°/90°/0°/90° option shows the minimum deflection for ϕ = 35° (Table 31.2). As Table 31.2 shows the 0°/90°/0° laminate shows the minimum deflection for ϕ = 0° when the comparison is limited within symmetric ones. It is noteworthy that the best performance of 0°/90°/0°/90° laminate is 10% higher than that of the 0°/90°/0° one. Naturally, the practical facilities must be covered by 0°/90°/0° panels to achieve the best performance.

The cylindrical 0°/90°/0° panels are selected for further investigations where deflections are studied for different imperfection amplitudes (η). Results are depicted in Figures 31.4 and 31.5.

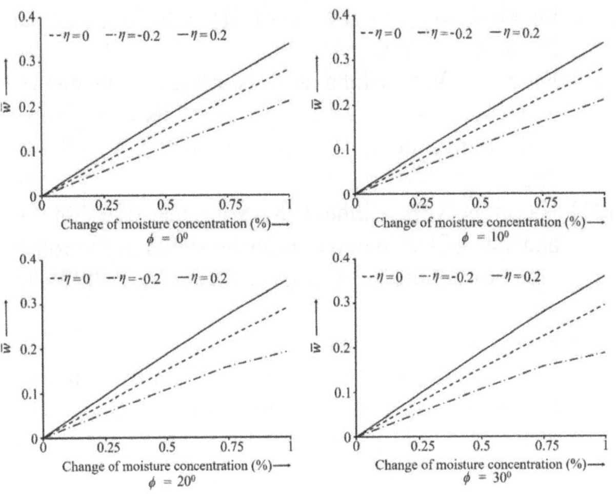

Figure 31.5 Variations of deflections for varying imperfection amplitudes (η), moisture concentration and skew angle (ϕ)
Source: Authors

Figure 31.4 depicts variations of deflections for varying temperature. The study of deflections for varying moisture concentration is given in Figure 31.5. It is found by comparing the deflections depicted in Figures 31.4 and 31.5 that at any constant skew angle, the variations of deflections for different moisture concentration are nearly double of what obtained for varying temperature. It indicates that panels are relatively more flexible when exposed to varying moisture concentration. The influences of imperfection amplitudes are studied also. The panels with positive imperfections (η = 0.2) indicate longer dimensions than designed ones. The negative imperfections (η = -0.2) represent shorter dimensions than intended. Hence, the panels with η = 0.2 have greater hogging curvature than perfect panels (η = 0). The negative amplitude refers to panels with minimum curvature. When the panels are under varying temperature and moisture concentration, the figures indicate that the patterns of load-deflection curves do not vary for variations of imperfection amplitudes. The figures reveal that the panels studied in this paper do not depict instability when amplitudes of imperfections are limited within 20% of panel thickness. However, for any constant temperature and moisture, the deflection values vary as the imperfection amplitude changes. Figures 31.4 and 31.5 show that for all skew angles, the panels with η = 0.2 show the maximum upward deflections. The maximum deflections are followed by those yielded by perfect panels (η = 0) and panels with η = -0.2. The influence of skewness is similar for all values of η considered in these figures. Figures 31.4 and 31.5 show that the deflections at any given temperature and moisture concentration monotonically increase as the skewness varies from 0° to 30° with a fixed step of 10°.

Conclusion

Correctness of the present approach to obtain deflections of skewed cylindrical panels under hygrothermal variations is confirmed through benchmark problems. It is recommended to avoid skewed panels for practical applications. However, if skewed panels are essential then the 0°/90°/0° composite is found to be the best choice to have among the options studied here. The shell panels studied here are found to yield upward deflections under hygrothermal stresses. Such deflections increase by exposure to elevated temperature and moisture. In cases where the amplitudes of imperfections do not exceed 20% of panel thickness, there are no instability issues of panels noted in this paper. However, the imperfectly fabricated longer panels may yield greater upward deflections than perfect ones.

References

[1] Bakshi, K., Chakravorty, D. First ply failure loads of composite conoidal shell roofs with varying lamination. *Mech. Adv. Mater. Struct.*, 2015;22(12):978–987.

[2] Liao, C. L., Lee, Z. Y. Elastic stability of skew laminated composite plates subjected to biaxial follower forces. *Int. J. Num. Meth. Eng.*, 1993;36:1825–1847.

[3] Singha, M. K., Daripa, R. Nonlinear vibration of symmetrically laminated composite skew plates by finite element method. *Int. J. Non-Linear Mech.*, 2007;42:1144–1152.

[4] Upadhyay, A. K., Shukla, K. K. Large deformation flexural behaviour of laminated composite skew plates: An analytical approach. *Compos. Struct.*, 2012;94:3722–3735.

[5] Zhang, L. W., Xiao, L. N. Mechanical behaviour of laminated CNT-reinforced composite skew plates subjected to dynamic loading. *Compos. Part B Engg.*, 2017;122:219–230.

[6] Sengar, V., Nynaru, M., Watts, G., Kumar, R., Singh, S. Postbuckled vibration behaviour of skew sandwich plates with metal foam core under arbitrary edge compressive loads using isogeometric approach. *Thin-Walled Struct.*, 2023;184:110524.

[7] Kandasamy, S., Singh, A. V. Free vibration analysis of skewed open circular cylindrical shells. *J. Sound Vib.*, 2006;290:1100–1118.

[8] Haldar, S. Free vibration of composite skewed cylindrical shell panel by finite element method. *J. Sound Vib.*, 2008;311:9–19.

[9] Kiani, Y., Dimitri, R., Tornabene, F. Free vibration of FG-CNT reinforced composite skew cylindrical shells using the Chebyshev-Ritz formulation. *Compos. Part B.*, 2018;147:169–177.

[10] Mandal, A., Ray, C., Haldar, S. Experimental and numerical studies on vibration characteristics of laminated composite skewed shells with cutout. *Compos. Part B.*, 2019;161:228–240.

[11] Giri, A., Bakshi, K. Geometrically nonlinear bending of stiffened composite skewed cylindrical shells under transverse pressure. *J. Strain Anal. Engg. Des.*, 2024;59(2):85–99.

[12] Godoy, L. A. On loads equivalent to geometrical imperfections in shells. *J. Eng. Mech.*, 1993;119:186–190.

[13] Hui, D. Large amplitude vibrations of geometrically imperfect shallow spherical shells with structural damping. *AIAA J.*, 1983;21:1736–1741.

[14] Nanda, N., Pradyumna, S. Nonlinear dynamic response of laminated shells with imperfections in hygrothermal environments. *J. Compos. Mater.*, 2011;45:2103–2112.

[15] Kochurov, R., Avramov, K. V. On effect of initial imperfections on parametric vibrations of cylindrical shells with geometrical non-linearity. *Int. J. Solids Struct.*, 2012;49:537–545.

[16] Kitipornchai, S., Yang, J., Liew, K. M. Semi-analytical solution for nonlinear vibration of laminated FGM plates with geometric imperfections. *Int. J. Solids Struct.*, 2004;41:2235–2257.

[17] Dey, T., Ramachandra, L. S. Static and dynamic instability analysis of composite cylindrical shell panels subjected to partial edge loading. *Int. J. Non-Linear Mech.*, 2014;64:46–56.

[18] Garcea, G., Liguori, F. S., Leonetti, L., Magisano, D., Madeo, A. Accurate and efficient a posteriori account of geometrical imperfections in Koiter finite element analysis. *Int. J. Num. Meth. Engg.*, 2017;112:1154–1174.

[19] Baciu, T. D., Franzoni, F., Degenhardt, R., Kalnins, K., Bisagni, C. Shape and loading imperfection sensitivity of the vibration-correlation technique for buckling analysis of cylindrical shells. *Engg. Struct.*, 2024;304:117605.

[20] Naidu, N. V. S., Sinha, P. K. Nonlinear finite element analysis of laminated composite shells in hygrothermal environments. *Compos. Struct.*, 2005;69:387–395.

[21] Naidu, N. V. S., Sinha, P. K. Nonlinear free vibration analysis of laminated composite shells in hygrothermal environments. *Compos. Struct.*, 2007;77:475–483.

[22] Naidu, N. V. S., Sinha, P. K. Nonlinear transient analysis of laminated composite shells in hygrothermal environments. *Compos. Struct.*, 2006;72:280–288.

[23] Nanda, N., Sahu, S. K., Bandyopadhyay, J. N. Effect of delamination on the nonlinear transient response of composite shells in hygrothermal environments. *Int. J. Struct. Integrity.*, 2013;1:259–279.

[24] Biswal, M., Sahu, S. K., Asha, A. V. Experimental and numerical studies on free vibration of laminated composite shallow shells in hygrothermal environment. *Compos. Struct.*, 2015;127:165–174.

[25] Foroutan, K., Ahmadi, H., Carrera, E. Nonlinear vibration of imperfect FG-CNTRC cylindrical panels under external pressure in the thermal environment. *Compos. Struct.*, 2019;227:111310.

[26] Ahmadi, H., Bayat, A., Duc, N. D. Nonlinear forced vibrations analysis of imperfect stiffened FG doubly curved shallow shell in thermal environment using multiple scales method. *Compos. Struct.*, 2021;256:113090.

[27] Bakshi, K., Giri, A. Application of von-Karman model for laminated composite singly curved shells with non-uniform boundary conditions. *Mech. Adv. Mater. Struct.*, 2024:1–11.

[28] Wu, B., Pagani, A., Chen, W. Q., Carrera, E. Geometrically nonlinear refined shell theories by Carrera Unified Formulation. *Mech. Adv. Mater. Struct.*, 2021;28:1721–1741.

[29] Carrera, E., Pagani, A., Augello, R., Wu, B. Popular benchmarks of nonlinear shell analysis solved by 1D and 2D CUF-based finite elements. *Mech. Adv. Mater. Struct.*, 2020;27(13):1098–1109.

[30] Buragohmt, D. N., Patodi, S. C. Large displacement analysis of skew plates by lumped triangular element formulation. *Comput. Struct.*, 1978;9:183–189.

32 Geometrically non-linear performances of laminated composite toroidal shells

Alok Sharma[a] *and Kaustav Bakshi*[b]
Indian Institute of Technology, Indore, Madhya Pradesh, India

Abstract

An efficient structural model combines an economic structural form with advanced, efficient material. The curved toroidal shells are aesthetically appealing, stiff structures that are preferred to cover long spans of stadiums, auditoriums, museums, and airports. Since the second half of the last century, laminated composites gained popularity among practicing civil engineers owing to the high strength-to-weight ratio, flexibility to orient the major load-carrying capacity along the directions of governing stresses, and superior fatigue resistance. The research in the field of laminated composite shells is quite vast. However, the same effort on the composite toroidal shells are relatively scanty. Hence this paper aims to address that gap in the literature to popularise the use of composite toroidal shells further among practicing engineers. A C^0 continuous, isoparametric, geometrically non-linear finite element code is proposed here which combines Reissner-Mindlin theory and Sander's non-linear shell theory applicable to moderately thin panels. The governing equation is derived by minimising the total potential energy with respect to the shell deformations and solved using the Newton-Raphson method. The results are so presented that practically viable design guidelines can be recommended for practical use.

Keywords: Toroidal shell, finite element method, geometric non-linearity

Introduction

Shell structures are popular choices among civil engineers due to the aesthetically appealing stiff geometry. Different shell forms can be seen in practical applications. However, the easy to fabricate, developable, singly curved cylindrical and doubly curved conoids are the most preferred ones as roofs of long span metro stations, railway stations, auditoriums, international airports and aircraft hangers. The cylindrical and conoidal panels also permit ambient ventilation through both ends which obviates the artificial measures. The curved panels are lighter in weight when fabricated using laminated composites owing to the greater rigidity of the advanced material. In addition to that, the lower probability of decay due to atmospheric variations, resistance to fire and directional dependence of design strength along major load transfer directions are the attributes for enhanced popularity of composite panels over the traditional reinforced concrete and metal structures. Several research studies worked on composite shell panels for wide variety of parameters.

The closed form deflections and fundamental frequencies of laminated composite cylindrical panels for varying thickness values were reported by Reddy [1]. The dynamic performances of composite panels with concentric and eccentric cut-outs were studied by Chakravorty et al. [2] using geometrically linear isoparametric finite elements. Dynamics of delaminated panels were researched by Acharyya et al. [3] using the similar approach. The functionally graded CNT reinforced composite cylindrical panels were the topic of research for Arani et al. [4]. The research studies on composite cylindrical panels mostly adopted linear formulation. A few authors dedicated their efforts to non-linear formulations. Tsai and Palazotto [5] and Chang and Chang [6] reported dynamic frequencies of composite cylindrical shells. A hybrid finite-element approach was combined with non-linear strains by Toorani [7] to study deep orthotropic cylindrical panels. The vibration performance of composite panels under hygrothermal stresses for varying radius of curvature were studied by Naidu and Sinha [8]. The free and forced vibrations of cylindrical panels with imperfections under hygrothermal variations were reported by Nanda and Pradyumna [9]. The combinations of non-linearity and R-functions method was adopted by Kurpa [10].

[a]phd2401104003@iiti.ac.in, [b]kaustav.bakshi@iiti@gmail.com

DOI: 10.1201/9781003770442-32

Recently, the von-Karman non-linearity is adopted for solutions of bending and vibration problems of cylindrical panels with non-uniform supports [11].

The research studies on conoidal panels also have received attention from researchers. The linear finite elements were opted by Nayak and Bandyopadhyay [12] and Das and Chakravorty [13, 14] to work on bending and vibration performances of composite panels of moderate thickness. Kumari and Chakravorty [15] used that isoparametric formulation which is given in [14] for vibration problems of delaminated conoids. The buckling problem of conoids made of functionally graded composites were reported by Pradyumna and Bandyopadhyay [16]. The failure initiation in composite conoids were researched by Bakshi and Chakravorty [17–20] where both linearity and non-linearity were adopted. The CNT reinforced sandwiched shells were studied by Rout and Hota [21] using von-Karman strains.

There are several industrial applications like roofs of stadiums, shopping malls and airports where the shell panels are curved in plan. Researchers worked on toroidal geometries but mostly have solved circular sections. Vick and Gramoll [22] developed a technique derived from membrane theory to model composite cylindrical toroids which accounts for variation in thickness due to fibre stacking. Sun [23] worked on circular cylinders of toroidal geometry and suggested that the bending theory should be used in analysis. The non-linear approach for bending analysis of underwater elastic toroidal containers was reported by Jiammeepreecha [24]. The vibration of circular toroids was reported by Redekop [25] and Leung and Kwok [26] but the authors worked on isotropic cylinders. Bidzard et al. [27] first worked on toroidal shaped cylindrical panels (panels curved in plan). The authors solved vibration problems of FG-GPLRC composites for edges restrained against rotation using linear approach. The von-Karman non-linearity was later adopted by Bidzard et al. [28] to solve free vibration of FG-GPLRC toroidal panels.

The literature clearly shows that even though research studies on cylindrical and conoidal panels are ample in numbers but those studied were done for rectangular plan areas only. The shell panels which are curved in plan received very limited attention from the researchers, especially using geometric non-linearity. The moderately thin panels may undergo large bending deformations and similar amplitudes of vibration which necessitates the use of non-linear approach. There is only one paper [28] which used non-linear approach for toroidal shaped (curved in plan) cylindrical panels. Hence, the research on toroidal cylindrical panels is far from being complete. Moreover, the studies on toroidal shaped conoidal panels are absolutely missing in literature. Hence, this paper aims to fill that gap. A C^0 continues non-linear finite element code is used here to solve the bending and vibration problems of toroidal shaped cylindrical and conoidal panels. The shell panels are considered as made of laminated composites and fully clamped along the edges. The downward deflections and fundamental frequencies are studied in non-dimensionalised forms for varying layups of cross-ply composites and sizes cut-outs. This study finally suggests a series of engineering guidelines which can optimise the use cut-outs for varying curvatures of plan area.

Bi

Numerical problems

The problems are selected in this paper to confirm correctness of the proposed approach and to study the bending and vibration performances of toroidal shaped cylindrical and conoidal panels.

The correctness of conoid formulation is confirmed by computing displacements of an isotropic panel and comparing those with what was given by Hadid [28]. Figure 32.1 depicts the comparison which confirms that the conoidal panel is correctly

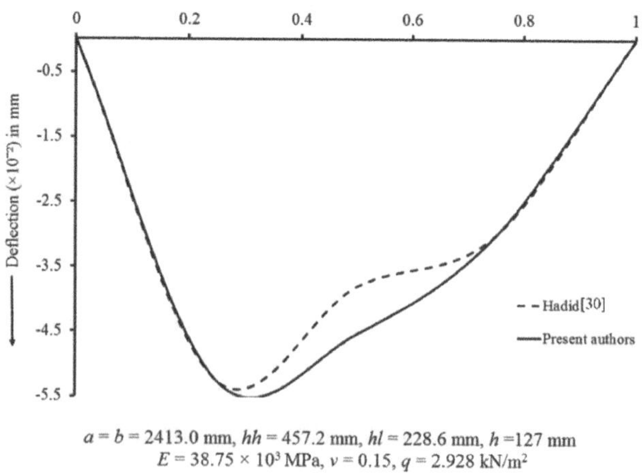

$a = b = 2413.0$ mm, $hh = 457.2$ mm, $hl = 228.6$ mm, $h = 127$ mm
$E = 38.75 \times 10^3$ MPa, $\nu = 0.15$, $q = 2.928$ kN/m^2

Figure 32.1 A profile of deflections along $\frac{\bar{y}}{b} = 0.5$
Source: Authors

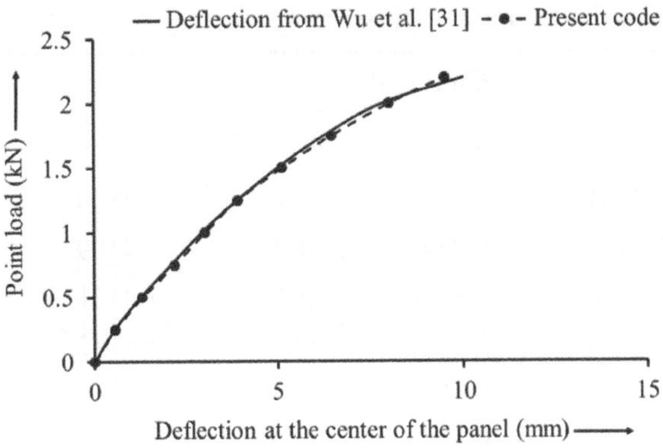

(a) Deflections of cylindrical panels with thickness (h) = 12.7 mm

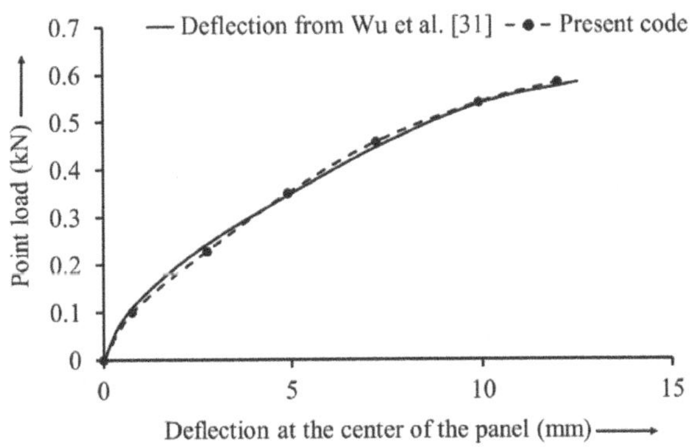

(b) Deflections of cylindrical panels with thickness (h) = 6.35 mm

Figure 32.2 Deflections of centrally loaded cylindrical panels
Source: Authors

formulated. The material properties and dimensions of the panel are given with the figure.

The correctness of cylindrical panel formulation is confirmed by Figure 32.2 where the displacements of an isotropic panel obtained using present approach are compared with those given by Wu et al. [29] and Carrera et al. [30]. The properties of cylindrical panels are: $E = 3.10275$ kN/mm^2, $v = 0.3$, length of 0.508 m, $R = 2.540$ m and included angle = 0.1 radian. The figure depicts the comparison till the first limit point only as the present code is developed for civil engineering shells only. The displacements at limit points are generally greater than the serviceable limits and hence, the solutions do not exceed that. Figure 32.2 confirms the correctness of cylindrical formulation.

Once it is confirmed that the code is correctly coded, it is used to solve non-linear deflections and fundamental frequencies of clamped composite cylindrical and conoidal panels on curved platforms. The maximum offset is "h" as shown in Figure 32.2. The results are obtained for panels with cut-outs. Tables 32.1–32.4 report the results for concentric and eccentric locations of cut-outs. The material constants and dimensions are given with the figures.

Results and discussion

The non-dimensional downward deflections ($\bar{w} = wE_{22}h^3/qa^4$) of conoidal and cylindrical laminated composite panels under uniform pressure are shown in Table 32.1 for both symmetric and antisymmetric cross-ply composite stackings. Table 32.2 provides the non-dimensional fundamental frequencies [$\varpi = \omega a^2 (\rho/E_{22}h^2)^{1/2}$] for panels made of identical composites. Both tables include panels which have curved plan areas where the maximum offset from the x-axis varies from 0 to 70 mm with a constant step of 10 mm. Table 32.1 shows that for all offset values and composites, conoidal panels offer at least 4% lower deflections than cylindrical panels for $a_1/a = 0.0$. The panels with 0°/90°/0° composite and $a_1/a = 0$ depicts an exception in performance, where cylindrical panels deflect 8% less than conoidal ones. For $a_1/a > 0.0$, the deflections of conoids at minimum are 35% less than what is obtained from cylindrical panels across all offsets and composites. Table 32.2 depicts that fundamental frequencies of conoidal panels are more than 30% higher than those of cylindrical panels for a fixed composite and offset value. The cylindrical geometry is singly curved. The conoids are doubly curved with cross-curvature. Thus, the conoidal panels are more rigid due to the additional curvature. Hence, to cover a given plan area (rectangular or curved), the engineers must prefer the conoidal panels to obtain superior bending and vibration performances within a fixed cost of construction.

Table 32.1 reveals that the deflections of cylindrical panels for all a_1/a ratios continue to increase when offset length varies from 0 to 70 mm. The minimum deflections are obtained when the base is rectangular (offset = 0 mm). The curved base of cylindrical panels leads to marginal increase of deflections. The highest deflections (offset = 70 mm) for all a_1/a ratios are within 2% of the corresponding minimum counterparts. The fundamental frequencies given in Table 32.2 across all a_1/a ratios show that these are minimum when the base of cylindrical panels is curved. For 90% of the solutions reported in Table 32.2, the minimum frequencies are obtained for offset of 70 mm. Hence, if the practical applications must have

Table 32.1 Non-dimensional downward deflections of clamped panels under uniform pressure.

Lamination (degree)	Maximum offset "*h*" (mm)	Cylindrical			Conoidal		
		$a_1/a=0$	$a_1/a=0.125$	$a_1/a=0.25$	$a_1/a=0$	$a_1/a=0.125$	$a_1/a=0.25$
0°/90°	0	0.6047	0.5530	1.2131	0.8547	2.7006	1.4109
	10	0.6049	0.5528	1.2134	0.8547	2.7015	1.4109
	20	0.6053	0.5527	1.2140	0.8550	2.7039	1.4121
	30	0.6060	0.5525	1.2152	0.8556	2.7083	1.4142
	40	0.6070	0.5523	1.2169	0.8564	2.7143	1.4175
	50	0.6083	0.5520	1.2188	0.8575	2.7218	1.4216
	60	0.6098	0.5516	1.2214	0.8588	2.7310	1.4269
	70	0.6116	0.5512	1.2242	0.8604	2.7417	1.4331
0°/90°/0°	0	0.4561	0.4963	1.0562	0.6943	1.8340	0.9360
	10	0.4562	0.4965	1.0562	0.6942	1.8344	0.9362
	20	0.4565	0.4966	1.0565	0.6944	1.8357	0.9371
	30	0.4570	0.4967	1.0571	0.6950	1.8379	0.9389
	40	0.4576	0.4968	1.0581	0.6960	1.8409	0.9415
	50	0.4587	0.4970	1.0595	0.6974	1.8448	0.9449
	60	0.4601	0.4972	1.0612	0.6991	1.8497	0.9490
	70	0.4616	0.4974	1.0632	0.7013	1.8551	0.9539
$(0°/90°)_2$	0	0.6062	0.4985	1.0567	0.6985	1.9840	1.0922
	10	0.6064	0.4985	1.0567	0.6985	1.9845	1.0924
	20	0.6068	0.4986	1.0570	0.6987	1.9861	1.0934
	30	0.6074	0.4987	1.0577	0.6993	1.9885	1.0953
	40	0.6084	0.4987	1.0587	0.7001	1.9921	1.0980
	50	0.6096	0.4989	1.0601	0.7012	1.9965	1.1015
	60	0.6111	0.4990	1.0619	0.7026	2.0019	1.1058
	70	0.6128	0.4992	1.0640	0.7043	2.0081	1.1109
$(0°/90°)_S$	0	0.5905	0.5747	1.1471	0.7442	2.0058	1.0611
	10	0.5907	0.5747	1.1471	0.7442	2.0063	1.0613
	20	0.5910	0.5747	1.1474	0.7445	2.0077	1.0623
	30	0.5917	0.5748	1.1482	0.7452	2.0101	1.0641
	40	0.5926	0.5749	1.1493	0.7462	2.0133	1.0669
	50	0.5940	0.5750	1.1508	0.7477	2.0177	1.0704
	60	0.5957	0.5752	1.1527	0.7495	2.0227	1.0748
	70	0.5977	0.5754	1.1549	0.7516	2.0287	1.0800

$E_{11} = 142.5$ GPa, $G_{12} = G_{13} = 4.72$ GPa, $E_{22} = 9.79$ GPa, $v_{12} = 0.27$, $G_{23} = 1.192$ GPa, $a/b = 1$, $h/a=0.01$
Source: Authors

a cylindrical panel, then that can be applied to cover rectangular areas only. For a given offset and a_1/a ratio, the 0°/90°/0° composite yields the best bending and vibration performances of cylindrical panels among the composites studied in Tables 32.1 and 32.2. For any given a_1/a ratio and composite option, the deflections of conoidal panels also increase when offset length varies from 0 to 70 mm as is evident from Table 32.1. The rectangular (offset = 0 mm) panels show minimum deflections.

Table 32.2 Non-dimensional fundamental frequency of clamped panels.

Lamination (degree)	H (mm)	Cylindrical			Conoidal		
		a_1/a =0	a_1/a = 0.125	a_1/a = 0.25	a_1/a =0	a_1/a =0.125	a_1/a =0.25
0°/90°	0	104.401	103.523	96.975	138.632	140.670	139.906
	10	104.404	103.520	96.900	138.613	140.606	139.833
	20	104.403	103.517	96.891	138.904	140.612	139.739
	30	104.398	103.509	96.873	139.926	140.685	139.920
	40	104.398	103.502	96.856	140.140	140.826	140.141
	50	104.392	103.494	96.834	140.656	141.029	140.400
	60	104.384	104.004	96.817	141.193	141.298	140.703
	70	104.378	103.472	96.964	141.277	141.627	141.035
0°/90°/0°	0	151.410	148.462	127.416	179.195	177.311	175.051
	10	151.276	148.463	127.412	179.166	177.243	174.992
	20	151.272	149.057	127.406	179.197	177.227	174.967
	30	151.267	149.049	127.397	179.288	177.265	174.979
	40	151.259	149.039	127.384	179.444	177.351	175.033
	50	151.250	149.026	127.372	179.653	177.489	175.121
	60	151.238	148.433	127.356	179.922	177.674	175.243
	70	151.225	148.420	127.334	180.246	177.904	175.401
(0°/90°)$_2$	0	123.749	122.9036	114.855	158.665	157.308	157.253
	10	123.749	122.5581	114.835	158.626	157.402	157.173
	20	123.919	122.8897	114.810	158.678	157.795	157.141
	30	123.742	122.8769	114.784	158.820	157.804	157.161
	40	123.737	122.8637	115.016	159.051	157.995	157.229
	50	123.729	122.8479	114.725	159.369	158.486	157.346
	60	123.722	122.8321	113.443	159.770	158.926	157.515
	70	123.711	122.8129	114.653	160.249	158.816	157.727
(0°/90°)$_S$	0	146.401	144.146	128.272	175.980	174.766	172.799
	10	146.402	144.755	128.264	176.105	174.685	172.724
	20	146.393	144.146	128.251	175.992	174.672	172.700
	30	146.393	144.141	128.237	176.130	174.480	172.723
	40	146.388	144.134	128.219	176.348	174.612	172.793
	50	146.372	144.126	128.202	176.650	174.812	172.915
	60	146.370	144.698	128.179	177.032	175.078	173.082
	70	146.359	144.678	128.156	177.489	175.406	174.279

E_{11} = 142.5 GPa, $G_{12} = G_{13}$ = 4.72 GPa, E_{22} = 9.79 GPa, v_{12} = 0.27, G_{23} = 1.192 GPa, a/b = 1, h/a=0.01
Source: Authors

It indicates that the bending rigidity of conoids with curved bases interacts in a complex manner with composite stackings and offset length values and thus, a unified conclusion may not be always achievable. For any a_1/a ratio, the difference between minimum deflections to their maximum equivalents (for offset = 70 mm) are lower for conoids. It shows that due to the curved base, the bending rigidity of conoids compromise to a lower extent than the cylindrical panels. In contrast to the vibration performances of cylindrical panels, the fundamental frequencies of conoids increase monotonically with cumulative offset values for all composites and a_1/a ratios. The maximum frequencies are obtained for panels with offset = 70 mm.

Hence, for practical situations where curved plan areas are to be covered, the conoids with offset of 70 mm should be preferred to achieve the best vibration performances and negligible compromises with deflections. For any given offset and a_1/a ratio, Tables 32.1 and 32.2 show that the performances of conoidal panels are the best for $0^\circ/90^\circ/0^\circ$ composite. The minimum deflections of $0^\circ/90^\circ/0^\circ$ conoids, considering all a_1/a ratios, is nearly 10% lower than the minimum of $0^\circ/90^\circ$, 0.5% lower than the minimum of $0^\circ/90^\circ/0^\circ/90^\circ$ and 12% lower than the minimum of $0^\circ/90^\circ/90^\circ/0^\circ$ composites. The fundamental frequencies follow a similar pattern in Table 32.2 where the $0^\circ/90^\circ/0^\circ$ conoids show atleast 24% higher frequency for $0^\circ/90^\circ$, 11% higher frequency for $0^\circ/90^\circ/0^\circ/90^\circ$ and 0.6% higher frequency for $0^\circ/90^\circ/90^\circ/0^\circ$ composites. Hence, the performances of $0^\circ/90^\circ/0^\circ$ composite cannot be replicated by the other cross-ply composites studied in this paper.

The performances of cylindrical and conoidal panels are studied in Tables 32.1 and 32.2 for varying sizes of cut-outs (a_1/a ratios) also. The panels with a_1/a = 0 are solid ones. Those with a_1/a = 0.25 have double the size of cutout than panels with a_1/a = 0.125. Tables 32.1 and 32.2 show that for all laminates, the bending and vibration performances of cylindrical panels are compromised for cut-outs, and solid panels show the best performances for all offset vales. The deflections of conoids also follow the similar pattern for varying a_1/a ratios. However, for fundamental frequencies, the $0^\circ/90^\circ$ conoids interestingly show that panels with a_1/a = 0.125 show superior performances than the solid ones. The cut-outs decreases both mass and stiffness of panels. The frequencies decrease when the stiffness of a panel reduces but that increases again due to decrease in mass. This is why the conoids with cut-outs show higher frequencies than the solid ones. Since it is noted that the performances of conoidal panels can be improved by cut-outs (a_1/a = 0.125) the authors are curious to investigate the bending and vibration performances of conoidal panels further for varying eccentric locations of cutout. The $0^\circ/90^\circ/0^\circ$ conoids are selected for this investigation as this option offered the best performances among the composites

Table 32.3 Non-dimensional deflections of $0^\circ/90^\circ/0^\circ$ clamped conoids panels for eccentric cut-outs.

e_y/b	e_x/a			
	0	**0.125**	**0.25**	**0.375**
0	0.6943	0.5584	0.4981	0.4978
0.125	0.6760	0.5531	0.4953	0.4961
0.25	0.5584	0.4990	0.4961	0.4965
0.375	0.5008	0.4983	0.4966	0.4966

Source: Authors

Table 32.4 Non-dimensional fundamental frequencies of $0^\circ/90^\circ/0^\circ$ clamped conoids panels for eccentric cut-outs.

e_y/b	e_x/a			
	0	**0.125**	**0.25**	**0.375**
0	177.904	179.506	179.295	178.444
0.125	181.329	181.889	181.047	180.059
0.25	179.020	180.257	180.247	180.189
0.375	176.714	178.325	179.210	179.831

Source: Authors

studied in Tables 32.1 and 32.2. For same reason, the conoids with curved bases of offset 70 mm are taken up. The deflections and fundamental frequencies of conoids are furnished in Table 32.3 and 32.4 for varying eccentric locations of cut-outs where the eccentricities are varied along both plan axes.

Table 32.3 shows that the deflections of conoids are lower when cut-outs are eccentrically located. The minimum values are obtained for e_x/a = 0.25 and e_y/b = 0.125. The minimum deflection is 29.24% lower than that obtained for the panel with concentric cutout (e_x/a = 0, e_y/b = 0). Table 32.4 also shows that the vibration performances can be improved if the cutout is located eccentrically, and the maximum values are obtained for e_x/a = 0.125 and e_y/b = 0.125 which is 0.92% higher than that having concentric cutout.

For practical applications, the engineers must have a fixed eccentricity of cut-outs which maximises both bending and vibration performances. To attain that Tables 32.3 and 32.4 are studied and it is found that the cutout can be located at e_x/a = 0.25 and e_y/b = 0.125 where the deflection is minimum, and the fundamental frequency is 0.5% lower than the maximum value.

Conclusion

The conclusions of the present study are the following:

1. The correctness of proposed approach to predict deflections and fundamental frequencies of composite cylindrical and conoidal panels is confirmed through solutions of benchmark problems.
2. The cylindrical panels, being singly curved, are found as less stiff than the doubly curved conoids for a given plan area and material consumption. The bending and vibration performances of cylindrical panels also compromises when the base area is curved. Hence, the cylindrical panels are at best suitable for rectangular plan areas. Those panels should be fabricated using $0^{\circ}/90^{\circ}/0^{\circ}$ composites to maximise the bending and vibration performances.
3. The conoidal panels can be adopted to cover plan areas which are curved. The vibration performances are found to be at the best and bending compromises are negligible if the maximum offset of conoids from the *x*-axis don't exceed 70 mm. The $0^{\circ}/90^{\circ}/0^{\circ}$ composite is found as the best choice to fabricate the conoidal panels with curved bases which maximises the fundamental frequencies and minimises the deflections.
4. The cut-outs are found to deprive the bending and vibration performance of cylindrical panels. Hence, for cylindrical panels, the solid ones are recommended. But for conoidal panels, the cut-out of $a_1/a = 0.125$ improves the vibration performance. That cutout improves both bending and vibration performances to notable extents if located eccentrically at $e_x/a = 0.25$ and $e_y/b = 0.125$.

References

[1] Reddy, J. N. Exact solution of moderately thick laminated shells. *J. Eng. Mech.*, 1984;110(5):794–809.

[2] Chakravorty, D., Bandyopadhyay, J. N., Sinha, P. K. Finite element free vibration analysis of point supported laminated composite cylindrical shells. *J. Sound Vib.*, 1995;181(1):43–52.

[3] Acharyya, A. K., Chakravorty, D., Karmakar, A. Natural frequencies and mode shapes of composite cylindrical delaminated shells by finite element. *J. Reinf. Plast. Compos.*, 2010;29(2):226–237.

[4] Arani, A. G., Kiani, F., Afshari, H. Free and forced vibration analysis of laminated functionally graded CNT reinforced composite cylindrical panels. *J. Sandw. Struct. Mater.*, 2019;23(1):255–278.

[5] Tsai, C. T., Palazotto, A. N. On the finite element analysis of non-linear vibration for cylindrical shells with high-order shear deformation theory. *Int. J. Non-Linear Mech.*, 1991;26(3–4):379–388.

[6] Chang, T. P., Chang, H. C. Nonlinear vibration analysis of geometrically nonlinear shell structures. *Mech. Res. Commun.*, 2000;27(2):173–180.

[7] Toorani, M. H. Dynamics of the geometrically non-linear analysis of anisotropic laminated cylindrical shells. *Int. J. Non-Linear Mech.*, 2003;38(9):1315–1335.

[8] Naidu, N. V. S., Sinha, P. K. Nonlinear free vibration analysis of laminated composite shells in hygrothermal environments. *Compos. Struct.*, 2007;77(4):475–483.

[9] Nanda, N., Pradyumna, S. Nonlinear dynamic response of laminated shells with imperfections in hygrothermal environments. *J. Compos. Mater.*, 2011;45(20):2103–2112.

[10] Kurpa, L. V. Nonlinear free vibrations of multilayer shallow shells with a symmetric structure and with a complicated form of the plan. *J. Math. Sci.*, 2009;162(1):85–98.

[11] Bakshi, K., Giri, A. Application of von-Karman model for laminated composite singly curved shells with non-uniform boundary conditions. *Mech. Adv. Mater. Struct.*, 2024:1–11.

[12] Nayak, A. N., Bandyopadhyay, J. N. Free vibration analysis and design aids of stiffened conoidal shells. *J. Eng. Mech.*, 2002;128:419–427.

[13] Das, H. S., Chakravorty, D. Finite element application in analysis and design of point supported composite conoidal shell roofs: Suggesting selection guidelines. *J. Strain Anal. Engg. Des.*, 2009;45:165–177.

[14] Das, H. S., Chakravorty, D. Natural frequencies and mode shapes of composite conoids with complicated boundary conditions. *J. Reinf. Plast. Compos.*, 2008;27(13):1397–1415.

[15] Kumari, S., Chakravorty, D. Bending of delaminated composite conoidal shells under uniformly distributed load. *J. Eng. Mech.*, 2011;137:660–668.

[16] Pradyumna, S., Bandyopadhyay, J. N. Free vibration and buckling of functionally graded shell panels in thermal environments. *Int. J. Struct. Stab. Dyn.*, 2010;10(05):1031–1053.

[17] Bakshi, K., Chakravorty, D. First ply failure study of composite conoidal shells used as roofing units in civil engineering. *J. Fail. Anal. Prev.*, 2013;13:624–633.

[18] Bakshi, K., Chakravorty, D. First ply failure study of thin composite conoidal shells subjected to uniformly distributed load. *Thin Walled Struct.*, 2014; 76: 1–7.

[19] Bakshi, K., Chakravorty, D. First ply failure loads of composite conoidal shell roofs with varying lamination. *Mech. Adv. Mater. Struct.*, 2015; 22: 978–987.

[20] Bakshi, K., Chakravorty, D. Numerical study on failure of thin composite conoidal shell roofs considering geometric nonlinearity. *KSCE J. Civ. Engg.*, 2020;24:913–921.

[21] Rout, M., Hota, S. S. Geometrically nonlinear free vibration of CNTs reinforced sandwich conoidal shell in thermal environment. *Acta Mech.*, 2023;234:2677–2694.

[22] Vick, M. J., Gramoll, K. Finite element study on the optimization of an orthotropic composite toroidal shell. *J. Pressure Vessel Technol.*, 2012;134(5):051201.

[23] Sun, B. Small symmetrical deformation of thin torus with circular cross-section. *Thin Walled Struct.*, 2021;163:107680.

[24] Jiammeepreecha, W., C., K. Nonlinear static response of an underwater elastic toroidal storage. *Int. J. Solids Struct.*, 2021;228:111134.

[25] Redecop, D. Dynamic response of a toroidal shell panel. *Comput. Struct.*, 1994:235–239.

[26] Leung, A. Y. T., Kwok, N. Dynamic stiffness analysis of toroidal shells. *Thin Walled Struct.*, 1994;21:43–64.

[27] Karimipour Bakshi, K., Chakravorty, D. Geometrically linear and nonlinear first-ply failure loads of composite cylindrical shells. *J. Eng. Mech.*, 2014;140(12):04014094.

[28] Hadid, H. A. An analytical and experimental investigation into the bending theory of elastic conoidal shells. *Doctoral Disserta.*, 1964.

[29] Wu, B., Pagani, A., Chen, W. Q., Carrera, E. Geometrically nonlinear refined shell theories by Carrera Unified Formulation. *Mech. Adv. Mater. Struct.*, 2021;28:1721–1741.

[30] Carrera, E., Pagani, A., Augello, R., Wu, B. Popular benchmarks of nonlinear shell analysis solved by 1D and 2D CUF-based finite elements. *Mech. Adv. Mater. Struct.*, 2020;27(13):1098–1109.

33 Numerical characterisation of auxetic behaviour and flexural response in novel hybrid bio-inspired sandwich panel

Bibhu Prasad Mahapatra[a], *Rahul Chauhan and Prasun Jana*

Department of Aerospace Engineering, Indian Institute of Technology, Kharagpur, India

Abstract

This paper presents a comprehensive flexural and modal analysis of a novel nature-inspired auxetic sandwich plate using a finite element (FE) framework within the ANSYS® platform. The sandwich structure features a hybrid auxetic core, combining a turtle-inspired cellular configuration with an integrated conventional re-entrant architecture and is attached to carbon fibre-reinforced polymer (CFRP) face sheets on both the top and bottom surfaces. The effective elastic properties of the core are calculated through a full-field homogenisation approach using the material designer tool in ANSYS, and these effective properties are utilised further for the flexural and modal response (free vibration) of the sandwich panel in ANSYS mechanical. The study explores the effect of key design constraints or parameters on the deformation behaviour of the structure, highlighting the effects of various configurations on its mechanical performance. First, to check the robustness of the present method, the RVE-assisted FE methodology has been validated against benchmark results from existing literature. Subsequently, a detailed flexural and modal analysis is carried out on the sandwich structures, both with and without an enclosing re-entrant structure within the turtle-inspired core unit cell. Two specific configurations are considered: (i) A turtle-inspired sandwich structure (TISS), and (ii) A turtle-inspired hybrid sandwich structure (TIHSS). Results indicate a noticeable increase in the overall stiffness of TIHSS when the re-entrant structure is enclosed within the turtle-inspired core, while TISS exhibits a higher natural frequency due to lower mass density. Several additional numerical examples are provided to explain the effects of varying influential design parameters, demonstrating the potential of this hybrid auxetic structure to improve the mechanical performance of sandwich plates.

Keywords: Auxeticity, FEM, sandwich structure, bio-mimicry, flexural analysis

Introduction

Auxetic structures are distinguished by their negative Poisson's ratio (NPR), meaning they widen in the transverse direction when pulled and shrink laterally under compression. This counterintuitive response originates from their carefully engineered geometric arrangements. This exceptional property enhances key mechanical characteristics such as impact resistance, energy absorption, and vibration control, making them invaluable in aerospace, automotive, and protective applications. Replacing conventional cores with auxetic cores in sandwich structures has demonstrated significant improvements in stiffness, strength, and dynamic performance. Early designs, including re-entrant honeycombs with inwardly slanted walls and chiral structures featuring rotating ligaments, showcased the potential of auxetic materials. Building on these foundations, modern innovations such as star-shaped and rotating unit cells have refined auxetic behaviour, offering tailored mechanical responses, increased energy dissipation, and structural stability. These advancements highlight the growing significance of auxetic structures in engineering, paving the way for versatile and high-performance material systems.

Recent research has increasingly focused on bio-inspired auxetic structures, leveraging the unique mechanical efficiencies found in natural systems such as plant cellular arrangements, animal skins, and fish scales, which inherently exhibit NPR properties. These nature-inspired designs are known for their ability to enhance adaptability, energy absorption, and mechanical performance, making them ideal for engineering applications requiring lightweight yet robust materials. For instance, Jiang et al. [1] introduced a concentric auxetic re-entrant honeycomb

[a]bibhuprasadmahapatra76@gmail.com

DOI: 10.1201/9781003770442-33

(CARH) model based on nature-inspired strategy. This design exhibited improved energy absorption and crashworthiness, attributed to its unique deformation mechanisms and tuneable parameters such as wall thickness and concentric spacing. Similarly, studies [2, 3] on honeycomb sandwich panels inspired by the beak structure of woodpeckers revealed significant enhancements in energy dissipation and impact resistance. These improvements were linked to increased core thickness and wave amplitude, which mimic the woodpecker's natural ability to withstand repetitive impacts.

Peng et al. [4] explored additively manufactured sandwich structures with an innovative minimal surface core. Their findings highlighted the critical role of core topology and relative density in governing the bending stiffness and energy absorption of the panel, demonstrating the versatility of bio-inspired designs in optimising mechanical responses. Furthermore, beetle-inspired auxetic structures [5], modelled after the resilient exoskeleton of beetles, showcased improved impact toughness and energy absorption, underscoring the potential of such designs for high-stress applications. A 3D voronoi-based composite mimicking nacre's multilayer structure was developed [6], validated through uniaxial tests, and assessed for blast resistance. Research on auxetic core sandwich plates with graphene-reinforced facings [7] showed that fibre content and helicoidal lay-ups significantly affect aerodynamic pressure resistance. Butterfly-inspired auxetic structures have attracted attention for their superior auxeticity and energy absorption [8], outperforming traditional honeycombs in sound transmission loss (STL) for acoustic applications [9]. A novel butterfly-shaped auxetic core sandwich plate also showed improved buckling load, with core-to-face sheet thickness being a crucial factor [10]. Zhang et al. [11] developed a turtle shell-inspired auxetic honeycomb that exhibited improved crushing load uniformity compared to traditional re-entrant honeycombs. In another work, Pei et al. [12] investigate the turtle carapace's chemical composition and mechanical properties to design a bionic sandwich structure with enhanced impact resistance, validated through finite element analysis and experimental test.

The literature indicates that research on bio-inspired auxetic structures, particularly turtle shell-inspired cores, has predominantly focused on crushing and energy absorption analyses, with limited exploration of their broader mechanical behaviour. In this work, we perform free vibration and bending analysis of a turtle-inspired hybrid auxetic sandwich structure, where the stiffness of the auxetic core unit cell is enhanced by integrating a conventional re-entrant structure within it (refer Figure 33.1). Using RVE-FE homogenisation, we predict the effective elastic properties of the auxetic core layer to ensure an accurate representation of its mechanical behaviour. The macro-mechanical analysis is conducted using ANSYS-FE software, with results validated against previously published literature for reliability.

Methodology

This section highlights the study's contribution and scope. Figure 33.2 illustrates the numerical investigation, which bridges the micro- and macro-scales used in the work. RVE-FE full-field homogenisation is performed in ANSYS to compute the effective elastic properties of the turtle-inspired hybrid auxetic

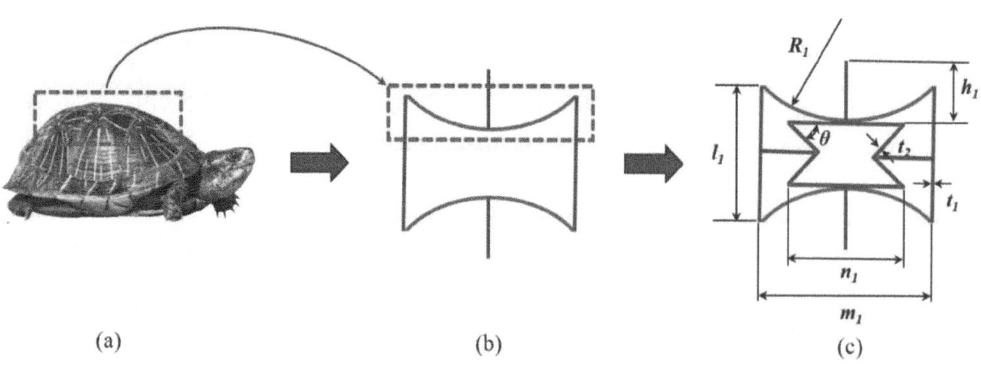

Figure 33.1 Evolution of proposed turtle-inspired hybrid auxetic core from natural turtle
Source: Authors

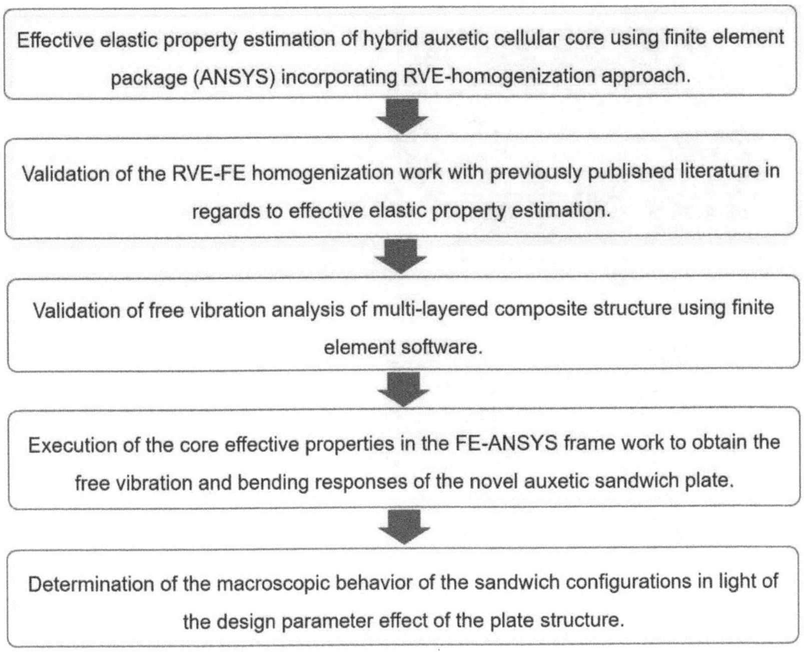

Figure 33.2 Sequential representation of frequency and bending analysis of proposed sandwich plate
Source: Authors

core. The effective elastic properties are incorporated into the macro-scale FE model to accurately simulate the free vibration and bending responses of the sandwich plate. SHELL281 elements, suitable for thin to moderately thick structures, are utilised for modelling, offering precise representation of bending and membrane behaviour. A structured (10×10) mesh is applied, ensuring accurate convergence and reliable predictions of deformation and vibration responses. This approach effectively captures the effect of the auxetic core design on the overall mechanical performance of the sandwich structure.

For CFFF boundary conditions, only the left edge at $x = 0$ is fixed, however for CFCF type end condition, the left edge at $x = 0$ and the right edge at $x = a$ are kept fixed keeping other edges free (refer Figure 33.3).

Results and discussion

Verification of numerical approach

We ensured the efficacy of the present RVE-based numerical approach by comparing our results with several key literature results. The RVE-FE homogenisation process has been validated using the results from Nguyen et al. (2020) [13]. For the free vibration of multi-layered laminated composites, our results are compared with those of Reddy (2003) [14].

RVE-FE full field homogenisation

In this comparison work, a re-entrant auxetic unit cell with inclined rib length l, vertical rib length d, rib thickness t and the outer inclined angle θ has been considered for full-field FE homogenisation. The effective NPR values (μ_{12}) of the auxetic RVE are calculated by varying d/l ratio and angle θ using the present homogenisation technique in ANSYS platform. The obtained results are then compared with the analytical solutions of Nguyen et al. (2020) [13], and illustrated in Table 33.1.

Free vibration comparison of multi-layered composite laminate

For validation, the FE results obtained in the present study for the free vibration behaviour of square composite plates ($a = b$) are compared with the numerical findings reported by Reddy (2003) [14]. Various configurations of multi-layered composite laminates are considered in this example for two different boundary conditions SS1 and SS2 as mentioned in the reference [14]. Table 33.2 shows a close match of the present dimensionless frequencies ($\varpi = \omega(b^2/h)\sqrt{\rho/E_2}$) with the reference results.

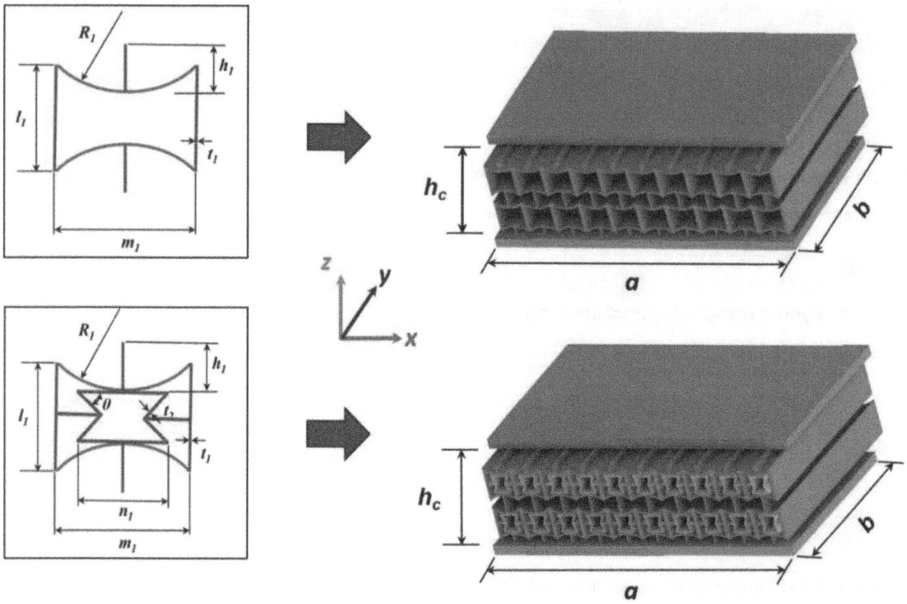

Figure 33.3 Schematic of turtle-inspired auxetic sandwich structures with and without integrating re-entrant structure within the core unit cell (TIHSS (bottom) and TISS (top))
Source: Authors

Table 33.1 Verification of the in-plane negative Poisson's ratio (μ_{12}) with varying d/l ratio.

θ (degree)	$d/l = 1.5$		$d/l = 2$		$d/l = 3$		$d/l = 4$	
	Present	Analytical [13]	Present	Analytical [13]	Present	Analytical [13]	Present	Analytical [13]
15	-0.339	-0.344	-0.478	-0.482	-0.755	-0.759	-1.031	-1.036
20	-0.442	-0.448	-0.635	-0.641	-1.022	-1.028	-1.408	-1.415

Source: Authors

Table 33.2 Validation of dimensionless frequency responses for simply supported square laminates ($E_1/E_2 = 40$, $G_{12} = G_{13} = 0.6\,E_2$, $G_{23} = 0.5\,E_2$, $\mu_{12} = 0.25$, $a/h = 10$).

	(0/90/0...) (SS1)			(-45/45/45...) (SS2)		
	$n = 2$	$n = 4$	$n = 8$	$n = 2$	$n = 4$	$n = 8$
Reddy [14]	11.186	17.101	18.279	14.504	23.294	25.052
Present	11.177	17.077	18.210	14.481	23.220	25.004

n = Number of plies in the laminate
Source: Authors

Free vibration responses of auxetic sandwich structures

In this section, the free vibration analysis has been carried out on auxetic sandwich structures having two different core topologies; (i) turtle-inspired core (TIC) and (ii) turtle-inspired hybrid core (TIHC). In later case, a conventional re-entrant structure has been incorporated inside the turtle-inspired core to form a stiffer hybrid auxetic core as shown in Figure 33.3. In this work, the two different sandwich structures are named as turtle-inspired sandwich structure (TISS) and turtle-inspired hybrid sandwich structure (TIHSS), respectively.

Effect of core height (h_c) on free vibration responses of auxetic sandwich structures

Free vibration analysis on both TISS and TIHSS has been carried out for three different core heights (h_c = 13 mm, 15 mm and 20 mm) in this section. Both the sandwich plates are having same edge length ($a = b = 400$ mm), and face-sheet thickness of 3 mm (each ply 1 mm). The auxetic cores are made of aluminium and face-sheets are considered to be carbon fibre reinforced polymer (CFRP). All the material properties considered for core and face-sheets are taken from Mahapatra et al. [15]. For TIC, the core design parameters; such as cell width m1, radius of curvature (R1) and vertical edge length (l_1) are set at 5 mm, 4 mm and 5.5 mm, respectively. Similarly, the cell-wall thickness (t_1) and vertical ligament length (h_1) are set at 0.4 mm and 2 mm, respectively. On the other hand, for TIHC, a re-entrant structure having wall thickness (t_2) 0.2 mm and internal angle (θ) 700 has been integrated inside the TIC to form a hybrid auxetic core for TIHSS.

`Now, both the core RVEs for TIC and TIHC have been homogenised in ANSYS platform to calculate the effective elastic properties and mass densities and tabulated in Table 33.3. TIHC exhibits better elastic properties than TIC due to its enhanced mass distribution in same volume of RVE. Also, TIHC possesses more mass density than TIC due to incorporation of re-entrant structure inside TIC. Now the effective elastic properties have been incorporated in ANSYS software to perform free vibration analysis of both TISS and TIHSS.

In this section, two different sandwich structures (i.e., TISS and TIHSS) are taken into consideration to calculate the frequency responses. Here, two different edge supports (CFFF and CFCF) are considered with varying core thickness (h_c = 13 mm, 15 mm and 20 mm). Side lengths ($a = b$) are kept constant at 400 mm each. The findings highlight that as the of core thickness (h_c) is increased the natural frequencies of both structures get increased in all three modes under both boundary conditions. Also, it is observed from Table 33.4 that, TISS exhibits more natural frequency than TIHSS due to lesser mass density. The higher density of TIHSS makes it bulkier thereby reducing the natural frequencies.

Effect of plate length (a) on frequency responses of auxetic sandwich panel

In this section, two different sandwich structures (i.e., TISS and TIHSS) are taken into consideration to calculate the frequency responses. Here, two different edge supports (CFFF and CFCF) are considered with varying side length (a = 200 mm, 400 mm

Table 33.3 Homogenised material properties for two different auxetic cores.

	E_1 (GPa)	E_2 (GPa)	E_3 (GPa)	G_{12} (GPa)	G_{23} (GPa)	G_{13} (GPa)	μ_{12}	μ_{23}	μ_{13}	Mass density (ρ) kg/m^3
TIC	3.04	18.7	0.540	3.38	1.88	0.097	0.053	0.009	-1.78	729.82
TIHC	3.97	24.4	0.751	4.21	2.09	0.125	0.053	0.010	-1.55	954.18

Source: Authors

Table 33.4 Natural frequencies (*Hz*) of TISS and TIHSS under various core height and boundary conditions ($a = b$ = 400 mm).

h_c (m)	**0.013**			**0.015**			**0.02**		
Modes	**1**	**2**	**3**	**1**	**2**	**3**	**1**	**2**	**3**
TISS (CFFF)	304.95	424.98	1019.06	318.00	448.92	1044.59	342.83	502.15	1094.40
TIHSS (CFFF)	296.24	420.18	1015.11	309.28	444.91	1040.11	334.39	497.21	1088.70
TISS (CFCF)	798.96	912.62	1596.58	809.59	945.83	1617.31	828.98	1024.53	1655.90
TIHSS (CFCF)	814.70	920.82	1629.64	823.45	950.65	1645.87	839.10	1029.0	1676.00

Source: Authors

and 800 mm) of plates keeping b fixed at 400 mm. Core thickness is kept constant at 13 mm. The findings highlight that as the length of the plate structure is increased the natural frequencies of both structures get reduced in all three initial modes under both boundary conditions. As we increase the length of the plate structure, the stiffness gets reduced, hence the natural frequency responses. Also, Table 33.5 shows that, TISS exhibits more natural frequency than TIHSS due to lesser mass density. The higher density of TIHSS makes it bulkier thereby reducing the natural frequencies. Figure 33.4 illustrates the mode shape plots of TIHSS for three initial modes under two different boundary conditions.

Flexural analysis of auxetic sandwich structures

Effect of core height (h_c) on flexural responses of auxetic sandwich structures

Flexural analysis on both TISS and TIHSS has been carried out for three different core heights (h_c = 13 mm, 15 mm and 20 mm) in this section. Both the sandwich plates are having same edge length ($a = b$ = 400 mm), and face-sheet thickness of 3 mm (each ply 1 mm). The auxetic cores are made of aluminium and face-sheets are considered to be carbon fibre reinforced polymer (CFRP). All material and core design parameters are same as mentioned in earlier section. A uniformly distributed load of magnitude 1 Pa has been applied on both the sandwich structures with 2 distinct edge supports (i.e., CFFF and CFCF) and the maximum deflection of the plates hence obtained are tabulated in Table 33.6. It has been found that TIHSS exhibits improved structural stiffness than TISS structure due to incorporation of re-entrant structure inside the turtle-inspired auxetic core unit cell, thereby undergoing less deformation under all boundary conditions. Also, it has been observed that, as we increase the core height (h_c), all the structures offer better resistance to deformation due to improved stiffness, thereby reducing the maximum deflection of the plate structure.

Table 33.5 Natural frequencies (*Hz*) of TISS and TIHSS under various plate length and boundary conditions (h_c = 13 mm).

***a* (mm)**	**200**			**400**			**800**		
Modes	1	2	3	1	2	3	1	2	3
TISS (CFFF)	305.01	347.24	740.48	97.76	143.95	413.60	26.62	61.79	142.64
TIHSS (CFFF)	296.31	340.92	715.50	91.63	139.20	406.06	24.55	59.48	135.21
TISS (CFCF)	798.98	829.98	1096.03	358.54	382.13	733.26	134.80	159.14	303.59
TIHSS (CFCF)	814.72	843.62	1088.02	354.11	375.12	727.46	129.12	154.14	297.44

Source: Authors

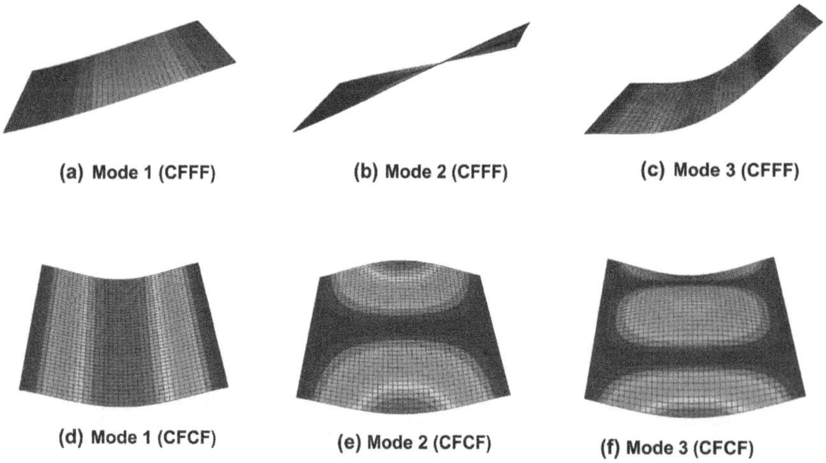

(a) Mode 1 (CFFF) (b) Mode 2 (CFFF) (c) Mode 3 (CFFF)

(d) Mode 1 (CFCF) (e) Mode 2 (CFCF) (f) Mode 3 (CFCF)

Figure 33.4 Mode shape plots for TIHSS for three initial modes and two edge supports (CFFF and CFCF)

Source: Authors

Table 33.6 Maximum deflection (mm) of TISS and TIHSS under various core height and boundary conditions ($a = b = 400$ mm, Load =).

h_c (mm)	13	15	20
TISS (CFFF)	0.2692	0.2229	0.1528
TIHSS (CFFF)	0.2401	0.1966	0.1321
TISS (CFCF)	0.0346	0.0305	0.2360
TIHSS (CFCF)	0.0276	0.2436	0.1870

Source: Authors

Table 33.7 Maximum deflection (mm) of TISS and TIHSS under various plate length and boundary conditions ($h_c = 13$ mm).

a (mm)	200	400	800
TISS (CFFF)	0.2701	2.788	38.492
TIHSS (CFFF)	0.2406	2.650	37.608
TISS (CFCF)	0.0346	0.1735	1.258
TIHSS (CFCF)	0.0276	0.1453	1.139

Source: Authors

Effect of plate length (a) on flexural responses of auxetic sandwich structures

In this section, flexural analysis has been performed on both TISS and TIHSS under two different boundary conditions (CFFF and CFCF) and varying side length ($a = 200$ mm, 400 mm and 800 mm) of plates keeping b fixed at 400 mm. Core thickness is kept constant at 13 mm. The findings highlight that as the length of the plate structure is increased the maximum deflection of both structures get increased under both boundary conditions.

As we increase the length of the plate structure, the stiffness gets reduced; hence the maximum deflection follows an increasing trend. Also, it is observed from Table 33.7 that, TIHSS exhibits better flexural stiffness than TISS. The higher density of TIHSS makes it bulkier thereby improving the load carrying capacity.

Conclusion

This study conducted a detailed flexural and frequency analysis of auxetic sandwich panels featuring nature-inspired hybrid core design strategies. The cores combine a turtle-inspired cellular architecture with an integrated conventional re-entrant structure to enhance mechanical properties. Using a finite element framework in ANSYS®, the effective elastic properties of the anisotropic cellular cores were determined through full-field homogenisation technique. These properties were then incorporated in a macro-scale to predict the free vibration and bending behaviour of the meta sandwich panel. The current methodology was validated against benchmark results. Numerical investigations explored the effect of core design variables on both flexural and dynamic behaviours. The findings revealed following conclusions.

- The RVE-assisted finite element methodology accurately predicts the mechanical behaviour of the turtle-inspired auxetic sandwich structures, as validated against benchmark literature results.
- TIHSS demonstrates significantly higher stiffness and improved mechanical strength compared to TISS due to incorporation of re-entrant structure inside the turtle-inspired unit cell.
- TISS shows higher natural frequencies than TIHSS, due to the lower mass density under same volume, indicating better performance in free vibration scenarios.

References

[1] Jiang, H., Ren, Y., Jin, Q., Zhu, G., Hu, Y., Cheng, F. Crashworthiness of novel concentric auxetic re-entrant honeycomb with negative Poisson's ratio biologically inspired by coconut palm. *Thin-Walled Struct.*, 2020;154:106911.

[2] Ha, N. S., Lu, G., Xiang, X. Energy absorption of a bio-inspired honeycomb sandwich panel. *J. Mater. Sci.*, 2019;54(8):6286–6300.

[3] Deng, X., Qin, S., Huang, J. Out of plane impact analysis for a bioinspired sinusoidal honeycomb. *Mech. Adv. Mater. Struct.*, 2021;29:1995547.

[4] Peng, C., Fox, K., Qian, M., Nguyen-Xuan, H., Tran, P. 3D printed sandwich beams with bioinspired cores: Mechanical performance and modelling. *Thin-Walled Struct.*, 2021;161:107471.

[5] Zhang, W., Xu, J., Yu, T. X. Dynamic behaviors of bio-inspired structures: Design, mechanisms, and models. *Engg. Struct.*, 2022;265:114490.

[6] Tran, P., Ngo, T. D., Ghazlan, A., Hui, D. Bimaterial 3D printing and numerical analysis of bio-inspired composite structures under in-plane and transverse loadings. *Comp. Part B Engg.*, 2017;108:210–223.

[7] Prajapati, V. K., Pitchaimani, J. Flutter behaviour of quadrilateral auxetic core sandwich plate with

bio-inspired three-phase composite facings numerical analysis and experimental verification. *Proc. Instit. Mech. Engrs. Part L J. Mat Des. Appl.*, 2024;239(3).

[8] Alomarah, A., Yuan, Y., Ruan, D. A bio-inspired auxetic metamaterial with two plateau regimes: Compressive properties and energy absorption. *Thin-Walled Struct.*, 2023;192:111175.

[9] Wang, X., Fu, T. Improvement of broadband low-frequency sound insulation of sandwich plates with negative Poisson's ratio butterfly-shaped auxetic cellular. *Eng. Anal. Bound Elem.*, 2024;166:105852.

[10] Xiaoquan, L., Zhong, Y., Liu, R., Shi, J. Buckling and post-buckling analysis of butterfly-shaped auxetic core sandwich plates based on variational asymptotic method. *Thin-Walled Struct.*, 2023;184:110464.

[11] Zhang, X., An, C., Shen, Z., Wu, H., Yang, W., Bai, J. Dynamic crushing responses of bio-inspired re-entrant auxetic honeycombs under in-plane impact loading. *Mater. Today Comm.*, 2020;23:100918.

[12] Pei, B., Guo, L., Wu, X., Hu, M., Wu, S., Wang, Y. Impact resistant structure design and optimization inspired by turtle carapace. *Materials*, 2022;15.

[13] Nguyen, N. V., Lee, J., Nguyen-Xuan, H. Active vibration control of GPLs-reinforced FG metal foam plates with piezoelectric sensor and actuator layers. *Compos. B Engg*, 2019;172.

[14] Reddy, J. N. Mechanics of composite plates and shells, 2nd edition, CRC Press, 2003.

[15] Mahapatra, B. P., Sinha, V., Maiti, D. K., Jana, P. Active vibration suppression of tetrachiral auxetic core sandwich panel with CFRP skin: An RVE homogenization-assisted finite element approach. *Eur. J. Mech. A/sol.*, 2024;106:105282.

34 A static behaviour of carbon-CNT-metal composite laminate: Role of interphase and waviness via micromechanics

Leeladhar Rajput[a] *and Prasun Jana*[b]
Department of Aerospace Engineering, Indian Institute of Technology, Kharagpur, West Bengal, India

Abstract

In this study, multiscale micromechanics-assisted finite element modelling (FEM) approach is utilised to investigate the influence of interphase properties and carbon nanotube (CNT) waviness on the static behaviour of CNT-grown carbon fibre-reinforced metal matrix (CGCMM) symmetric cross-ply hybrid composite laminates. A three-step approach is used to analyse the static behaviour. First, the elastic properties of a CNT-coated layer (CCL) surrounding carbon microfibers are predicted employing an extended Halpin-Tsai analytical model that incorporates the CNT/metal matrix interphase and CNT waviness. Next, the effective elastic properties of the CGCMM hybrid laminated composite are estimated using three-dimensional representative volume element (3D-RVE) modelling based on the mechanics of materials (MOM) approach. Finally, the effective properties calculated at the mesoscale are applied to examine the macroscopic static properties of cross-ply CGCMM hybrid composite laminated plates under various loading and boundary conditions through FEM. This analysis assesses the overall mechanical performance of the composites and provides insights into how different configurations and constraints influence their behaviour. A detailed bending analysis highlights how CNT waviness and the CNT/metal matrix interphase affect the laminate's static behaviour. Additionally, this study compares both the micromechanical analytical model and FEM methods from existing literature to validate the accuracy of the multiscale modelling and FEM formulation. The numerical examples evaluate the effectiveness of the FEM approach in analysing the bending behaviour of different laminate configurations and explore the effects of non-dimensional coordinate (x/a), edge-to-thickness ratio (a/h), and CNT volume fraction on cross-ply hybrid laminated composite plates.

Keywords: Carbon nanotube (CNT), finite element method (FEM), multi-scale micromechanics, CNT waviness, interphase

Introduction

Fibre-reinforced composites (FRCs) are favoured in structural engineering for their outstanding strength-to-weight ratio and superior stiffness [1]. However, they often show weaker mechanical characteristics in the transverse direction, primarily due to fibre/matrix interface interactions, which depend on factors like fibre type, matrix material, and interfacial characteristics [2–5]. In aerospace engineering, the demand for lightweight structures to enhance fuel efficiency and overall performance necessitates using materials that maintain high strength and stiffness [6]. During the fabrication of a metal matrix composite (MMC), the hemodynamic gradients between the matrix and reinforcement can form an interphase, a transitional layer that significantly influences the composite's structural properties [4, 7]. The interphase affects load transfer, resistance to degradation, and failure, and its thickness, often comparable to the diameter of carbon nanotubes (CNTs), is vital for the interaction between the matrix and nanofibre CNTs.

As one-dimensional nanomaterials, CNTs offer a much larger surface area for interaction with the matrix than traditional micro-scale fibres, resulting in a higher interphase volume fraction [8]. This enhanced surface-to-volume ratio improves mechanical bonding between the reinforcement and matrix. Samandari et al. developed a model for predicting the elastic modulus of nanoparticle/polymer nanocomposites, incorporating the interphase using the rule of mixtures [9]. Similarly, a multiscale approach is proposed by Macías et al. for nano-polymer composites, including representative fibres that show the impact of interphase on mechanical behaviour

[a]leelars2000@kgpian.iitkgp.ac.in, [b]prasun@aero.iitkgp.ac.in

DOI: 10.1201/9781003770442-34

[10]. These models emphasise the interphase's critical role in enhancing nanocomposite stiffness and elasticity.

Furthermore, the coating or grafting of CNTs on carbon fibres (CF) has proven highly effective in achieving higher CNT loadings, which can create structures with randomly oriented CNTs that enhance performance and functionality [11]. This method is particularly useful in CGCMM hybrid composites, which are essential for developing multifunctional polymer composites with excellent performance. By depositing or growing CNTs on the carbon fibre surface, the fibre's surface characteristics are significantly improved, thereby boosting composite performance [12]. CNT-coated carbon fibre hybridisation, an essential strategy for enhancing interfacial bonding between the fibre and matrix, involves methods like chemical vapour deposition (CVD) and electrophoresis, each offering distinct advantages and limitations. The structural and interfacial characteristics of CGCMM composites are largely determined by the CNTs grown on the fibre surface.

Building on these concepts, this study aims to determine the mechanical elastic properties and static bending behaviour of CGCMM hybrid composites by employing a multilevel (multi-scale) micromechanical model. A finite element modelling (FEM) technique, supported by multiscale micromechanics, is employed to investigate how interphase properties and CNT waviness affect the static response of CNT-grown carbon-reinforced metal matrix (CGCMM) symmetric cross-ply laminates. The methodology follows three steps: (1) determining the elastic properties of a CNT-coated layer surrounding carbon microfibers using a modified Halpin-Tsai model [7], (2) applying 3D-RVE model to estimate the effective stiffness of the CGCMM composite, and (3) using mesoscale effective properties to evaluate the macroscopic static response under various loading conditions through FEM. To validate the results, this study compares the micromechanical model with FEM approaches from the literature [7, 13].

Micromechanical and FEM analysis

This study employs a multiscale micromechanics methodology to analyse the effective mechanical properties of CGCMM hybrid composites. Figure 34.1a depicts the CGCMM hybrid composite and its RVE model. The macroscopic properties, including the bending analysis of the CGCMM, are examined to understand the impact of the interphase between the metal matrix and CNT nanofibres. First, the CNT-coated layer (CCL) surrounding the carbon fibre is analysed at the nanoscale to assess its elastic properties. Next, the effective properties obtained are then utilised to calculate the elastic properties of the CGCMM hybrid composite using the mechanics of materials (MOM) approach at the microscale. Finally, the static behaviour of the composite is simulated using the finite element method at the macroscale, incorporating the effective properties derived at the microscale. Further details are provided in Sections 2.1, 2.2, and 2.3.

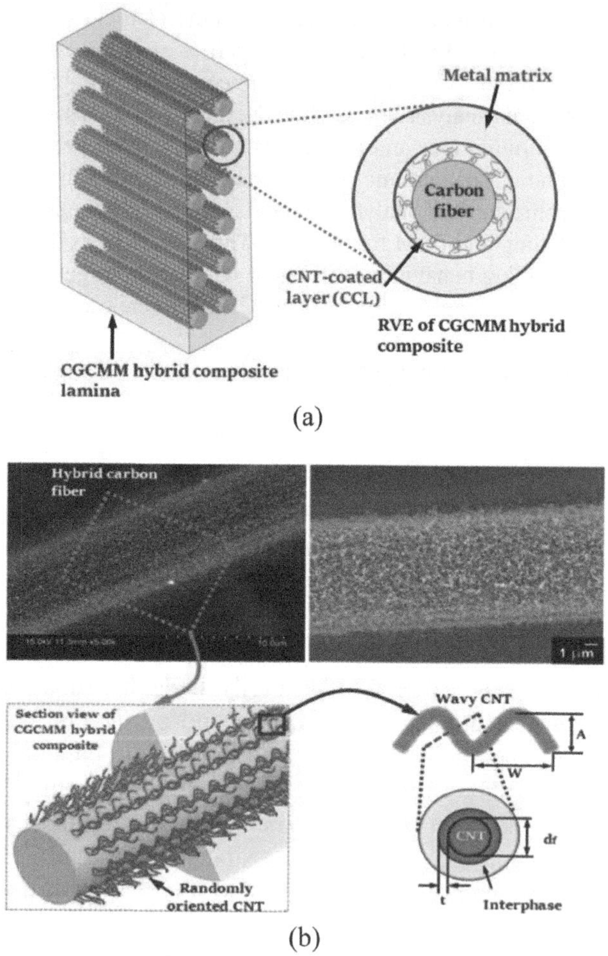

Figure 34.1 Schematic representation of the CGCMM hybrid composite and SEM images [14] of CNT-grown hybrid carbon fibre: (a) 3D illustration of the CGCMM composite lamina with the associated RVE model, and (b) SEM image of the CNT-grown hybrid carbon fibre with its corresponding RVE model
Source: Authors

Determination of effective modulus of CCL

In the present study, a micromechanical model known as modified Halpin-Tsai is employed to calculate the elastic modulus of the CCL, as illustrated in Figure 34.1b. The model accounts for the influence of CNT waviness and interfacial behaviour. The CNTs are randomly oriented and grown over the carbon fibre, as depicted in Figure 34.1. They are treated as isotropic materials, along with the metal matrix, which is also assumed to be isotropic. The elastic properties of the interphase, which describes the connection between the metal matrix and nanofibres, are calculated using empirical relations from [9, 15]. The elastic constant of the interphase is assumed to vary continuously between the CNT reinforcement and the matrix, adhering to the following conditions:

$$E_{in} = \begin{cases} E_f & \text{at } r = r_f \\ E_{in} & \text{at } r = r_{in} \end{cases}, \quad (1)$$

where, E_{in} represents the interphase's elastic properties at radius r, while E_m and E_f denote the moduli of the metal matrix and CNT at radii r_{in} and r_f, respectively, as shown in Figure 34.1b. Based on the equations above, the elastic constant at any given position within the interphase can be determined as follows:

$$E_{in} = E_m \frac{r_{in}}{r} + \left[\frac{(r_{in} - r)}{(r_{in} - r_f)}\right]^n \left[E_f - E_m \frac{r_{in}}{r_f}\right]. \quad (2)$$

The interfacial characteristic property, denoted as n, is a constant that quantifies the extent of interaction or bonding within the interphase region. The average elastic constant of the interphase is expressed using the following relation:

$$\bar{E}_{in} = \frac{1}{\pi\left(r_{in}^2 - r_f^2\right)} \int_{r_f}^{r_{in}} 2\pi r E_{in} dr = \frac{2}{\left(r_{in}^2 - r_f^2\right)} \int_{r_f}^{r_{in}} r E_{in} dr. \quad (3)$$

As presented in Figure 34.1b, both the interphase and the nanofibre (CNT) are considered as reinforcements. The effective property of CCL (E_{ccl}), accounting for the influence of the interphase, can be estimated using an adapted form of the Halpin-Tsai micromechanics model.

$$E_{ccl} = \frac{1 + k_f \zeta_f \alpha_r + k_{in} \zeta_{in} \alpha_{in}}{1 - \zeta_f \alpha_r - \zeta_{in} \alpha_{in}} E_m, \quad (4a)$$

$$\zeta_{in} = \frac{f_r f_w \frac{\bar{E}_{in}}{E_m} - 1}{f_r f_w \frac{\bar{E}_{in}}{E_m} + k_{in}}, \quad \zeta_f = \frac{f_r f_w \frac{E_f}{E_m} - 1}{f_r f_w \frac{\bar{E}_f}{E_m} + k_f}, \quad (4b)$$

$$k_{in} = \frac{l}{2(r_{in} + t)} \text{ and } k_f = \frac{l}{2(r_f + t)}. \quad (4c)$$

Where, α_r and α_{in} is used to define the volume fraction of CNT and interphase, respectively. Additionally, l, t, f_r and f_w denote the length of CNT fibre, interphase thickness, CNT orientation efficiency factor, and CNT-waviness efficiency factor, respectively. As illustrated in Figure 34.1b, the volume fraction of the interphase α_{in} can be calculated using the following formula:

$$\alpha_{in} = \frac{2r_f t + t^2}{r_f^2} \alpha_r \quad (5)$$

Prediction of elastic properties of CGCMM hybrid composite

The 3D-RVE model for the CGCMM hybrid laminated composite, depicted in Figure 34.1a is conceptualised using three-phase MOM. Within this model, the CCL and the metal matrix are treated as isotropic materials, whereas the carbon fibre is modelled as transversely isotropic. The constitutive relationship for the CGCMM hybrid composite is formulated accordingly:

$$\{\sigma^j\}_{6x1} = [C^j]_{6x6} \{\varepsilon^j\}_{6x1} \quad (6)$$

In this context, *j* represents *c, m, ccl, and hc*, where the superscripts *c, m, ccl,* and *hc* denote the carbon fibre, metal matrix, CNT-coated layer and CGCMM hybrid composite, respectively. The value of $[C^c]$ and $[C^m]$ and $[C^{ccl}]$ can be derived same as presented in [16].

By applying iso-strain and iso-stress assumptions in the axial and transverse directions, respectively, and utilising the rule of mixtures (ROM) as shown in the figure, the system can be effectively analysed.

$$\{\varepsilon_1^c\} = \{\varepsilon_1^{ccl}\} = \{\varepsilon_1^m\} = \{\varepsilon_1^{hc}\}, \quad (7)$$

$$v_c\{\sigma_1^c\} + v_{ccl}\{\sigma_1^{ccl}\} + v_m\{\sigma_1^m\} = \{\sigma_1^{hc}\} \quad (8)$$

Here, v_{ccl}, v_c and v_m are denoted as the volume fraction of CCL, carbon fibre, and metal matrix, respectively. By replacing Equation (6) with Equation (7), the strain and stress relationships for the CGCMM hybrid composite plate can be reformulated as follows:

$$\begin{aligned} \{\varepsilon^{hc}\}_{6x1} &= [V_1]_{6x6}\{\varepsilon^c\}_{6x1} + [V_2]_{6x6}\{\varepsilon^{ccl}\}_{6x1} + [V_3]_{6x6}\{\varepsilon^m\}_{6x1}, \\ \{\sigma^{hc}\}_{6x1} &= [C_1]_{6x6}\{\varepsilon^c\}_{6x1} + [C_2]_{6x6}\{\varepsilon^{ccl}\}_{6x1} + [C_3]_{6x6}\{\varepsilon^m\}_{6x1}. \end{aligned} \quad (9)$$

By applying Equations (7) through (9), we obtain Equations (10–13). Equation (13) provides the effective modulus for the CGCMM hybrid composite plate.

$$[C_4]\{\varepsilon^c\}-[C_5]\{\varepsilon^{ccl}\}=0\,,\ [C_5]\{\varepsilon^{ccl}\}-[C_6]\{\varepsilon^m\}=0 \tag{10}$$

$$[V_5]=[[V_4][C_6]^{-1}[C_4]+V_1]\,,\ [V_4]=[[V_2][C_5]^{-1}[C_6]+V_3]\,, \tag{11}$$

$$[C_7]=[C_2][C_5]^{-1}[C_6]+[C_3]\,,\ [V_6]=[[V_1][C_4]^{-1}[C_6]+V_4]\,, \tag{12}$$

$$[C^{hc}]=[C_7][V_6]^{-1}+[C_1][V_5]^{-1}. \tag{13}$$

Here, matrix C_1–C_4 and matrix V_1–V_3 can be written similar to the reference [16].

Transverse deformation analysis using FEM approach

The displacement components are defined using the first-order shear deformation theory (FSDT), as shown in the following relations:

$$u=u^0+z\varphi^x,\quad v=v^0+\varphi^y,\quad w=w^0 \tag{14}$$

The variables u^0, v^0, and w^0 denote the translational displacements in the *x, y*, and *z* coordinates, respectively, and rotational elements of displacement are denoted by φ^x and φ^y around the *x* and *y* axes. Next, the strain components can be determined by referring to Reddy [13].

The plate is discretised into a 4 × 4 mesh of a quarter plates with eight-nodded iso-parametric elements, with the polynomials in the natural coordinates ξ and η serving as the shape functions, which can be referred from [13]. The displacement and strain functions are interconnected in the following manner:

$$\{\varepsilon^b\}=\left[L^{tb}\right]\{d^t\}+z\left[L^{rb}\right]\{d^r\},\ \{\varepsilon^s\}=\left[L^{ts}\right]\{d^t\}+\{d^r\} \tag{15}$$

The nodal generalised displacement vectors represent the generalised displacement vectors at each location within the element.

$$\{d^t\}=\left[N^t\right]\{d_e^t\},\ \{d^r\}=\left[N^t\right]\{d_e^r\} \tag{16}$$

The elemental stiffness is obtained using the equation below:

$$K_{pq}^e=\int_{-1}^{1}\int_{-1}^{1}B_p^T D B_q\,|\,J\,|\,d\xi d\eta \tag{17}$$

In this context, *J* is the Jacobian matrix, *D* denotes the stiffness matrix, and *B* represents the strain-displacement matrix. The indices *p* and *q* range from 1 to 8, corresponding to the eight nodes per element. Once load vector and the stiffness matrix of the plate are computed, the relationship between displacement and load can be expressed as follows:

$$\left[K^g\right]\{u^g\}=\{F^g\} \tag{18}$$

The displacement vector by u^g, the global load vector is denoted by F^g and the global stiffness matrix by K^g.

The deformation of the laminate was analysed using finite element modelling, applying simply-supported boundary conditions as follows:

$$v^0=w^0=\varphi^x=0, N^{xy}=M^{yy}=0 \quad \text{at } y=0,\ b.$$
$$u^0=w^0=\varphi^y=0, N^{xy}=M^{xx}=0 \quad \text{at } x=0,\ a;$$

Results and discussion

This study investigates the effect of the interfacial behaviour between CNTs and the matrix on the bending characteristics of a cross-ply 0/90/0 CGCMM

Table 34.1 Key material properties of carbon nanotube-coated layers (CCL) applied to carbon fibres [7].

Parameters	CNT length (*l*) (μm)	CNT diameter (*d*) (nm)	CNT modulus (GPa)	CNT Poisson's ratio v_f	Matrix modulus (E_m) (GPa)	Matrix Poisson's ratio v_m	Interphase thickness (t) (nm)	Orientation efficiency factor (f_r)	Waviness efficiency factor (f_w)
Sample-1 (CNT/Al) [17]	5	10	500	0.3	0.3	70	8	1	0.75
Sample-1 (CNT/Al) [18]	0.6	12.5	500	0.3	0.3	73	8	1	0.6

Source: Authors

Table 34.2 Elastic properties of carbon fibre: Key mechanical parameters in different direction [16].

C_{11} (GPa)	C_{12} (GPa)	C_{22} (GPa)	C_{23} (GPa)	C_{44} (GPa)
236.4	10.6	24.8	10.7	7.0

Source: Authors

Table 34.3 Validation of the CCL zone's elastic constant using micromechanics model against analytical [7] and experimental [17].

CNT vol. fraction (α_r)	Experiment [17]	Analytical (n=1) [7]	Present (n=1)	Error (%)	Present (n = 5)	Error (%)	Present (n = 10)	Error (%)	Present (without interphase)	Error (%)
0.00	70.00	70.00	70.00	0.00	70.00	0.00	70.00	0.00	70.00	0.00
0.02	82.50	83.24	83.68	1.43	87.32	5.84	76.18	7.66	74.50	9.70
0.03	95.32	94.68	94.68	2.15	104.67	9.82	82.34	13.62	78.97	17.15
0.05	110.25	104.43	104.43	0.77	122.10	10.74	88.51	19.72	83.46	24.31

Source: Authors

hybrid composite laminated plate, considering sinusoidal loading and simply supported boundary conditions. The key material parameters of the CCL are arranged in Table 34.1, and the effective properties of carbon reinforcement are shown in Table 34.2. This study examines the random growth of wavy CNTs on the cylindrical surfaces of carbon fibres, with the analysis conducted in three steps and validated through comparisons with existing literature, and the results are presented and discussed in Sections 3.1, 3.2 and 3.3.

Validation through available experimental and analytical data

The data in Table 34.3 and Figure 34.2 present a comparison of the CCL elastic constant predicted by the present micromechanical model, analytical predictions [7], and experimental results [17]. The model with $n = 1$ demonstrates a strong correlation with the experimental data, with deviations typically less than 3%, confirming its accuracy in predicting the elastic constants.

Furthermore, the results without considering the interphase show a considerable error increase, indicating the importance of including the interphase in the micromechanical model for accurate predictions. Therefore, for further analysis of the CGCMM hybrid composite, the interphase elastic modulus is calculated using the value of $n = 1$ to ensure accurate predictions.

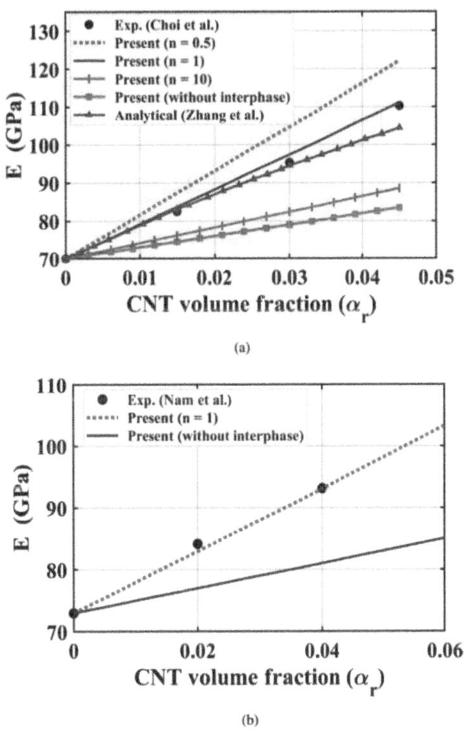

Figure 34.2 Validation of the CCL layer's elastic constant predicted by the micromechanical model for (a) Sample-1 (CNT/Al) and (b) Sample-2 (CNT/Al-cu) against analytical [7] and experimental data [17, 18]
Source: Authors

Mechanical elastic properties of CGCMM

Using a 3D micromechanics model that accounts for the interphase, the effective elastic constants of CNT/

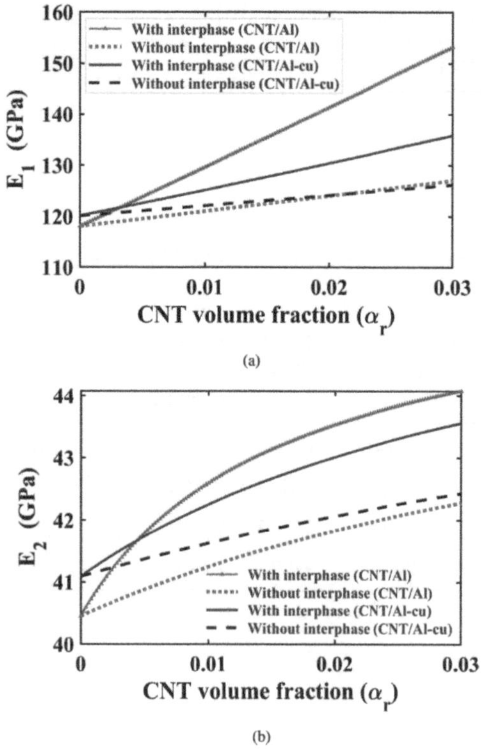

Figure 34.3 Prediction of the elastic constants: (a) along the longitudinal axis and (b) in the transverse direction for the CGCMM hybrid composite, using micromechanical models such as the Halpin-Tsai and MOM methods
Source: Authors

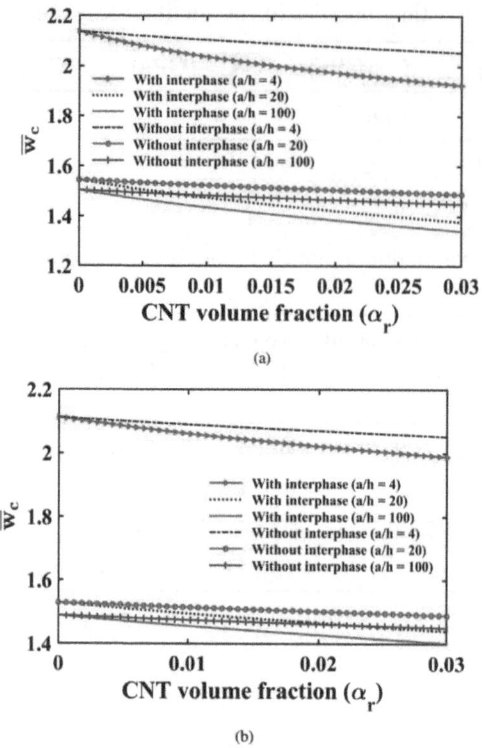

Figure 34.4 Maximum non-dimensional centre bending deflection of CGCMM hybrid composite with varying CNT volume fraction in the CCL – (a) Sample 1: CNT/Al and (b) Sample 2: CNT/Al-Cu-based CGCMM hybrid composite
Source: Authors

Al-based (sample-1) and CNT/Al-Cu-based (sample-2) CGCMM hybrid composites are calculated for both transverse and longitudinal directions.

As shown in Figures 34.3a,b, the inclusion of interfacial properties significantly improves the elastic properties in both directions. Interestingly, the CNT-Al-based composite (sample-1) is more significantly impacted by the omission of the interphase. The calculated effective properties for both samples are used at the macro-scale to analyse the static bending deformation of simply-supported cross-ply (0/90/0) CGCMM hybrid composite laminated plates under sinusoidal loading (SSL).

Bending of CGCMM hybrid composite

This section investigates the transverse deformation of CGCMM hybrid composite laminates under sinusoidal loading, considering various parameters such as CNT volume fraction α_r, *a/h* ratio, and non-dimensional coordinate (*x/a*). The validity and precision of the finite element modelling (FEM) approach have been confirmed in our previous work \cite{rajput-2024micromechanics}. The transverse deformation along the z-direction is non-dimensionalised and calculated using the following expression:

$$w = w_0\left(\frac{100E_2h^3}{b^4q_0}\right) \tag{19}$$

Figures 34.4a,b show the maximum non-dimensional centre bending deflection of CGCMM hybrid laminated composites with varying CNT volume fractions in the CCL: (a) Sample 1: CNT/Al and (b) Sample 2: CNT/Al-Cu based CGCMM hybrid composite. As the CNT volume fraction increases, both samples experience a reduction in central deflection, with the effect being more pronounced when the interphase is considered.

Figures 34.5a,b illustrate the maximum non-dimensional centre bending deflection of both the simply-supported CGCMM hybrid composites samples under sinusoidal loading (SSL), with varying

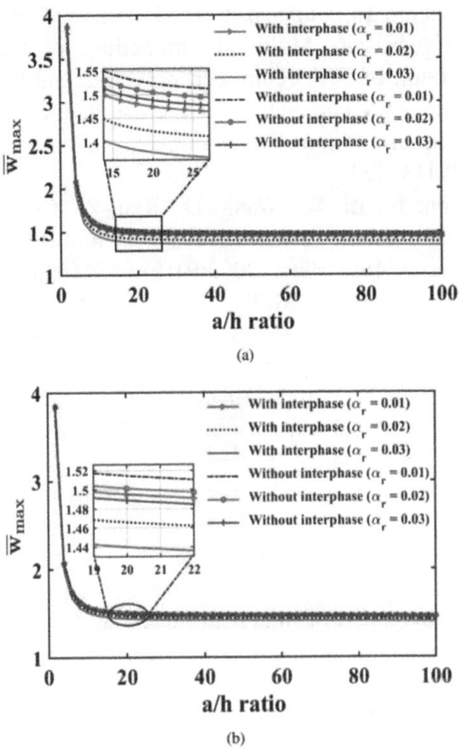

Figure 34.5 Maximum normalised centre deflection of CGCMM hybrid composite corresponding to varying length to thickness ratio (a/h): (a) sample 1: CNT/Al and (b) sample 2: CNT/Al-Cu based CGCMM hybrid composite
Source: Authors

length-to-thickness ratios (a/h). The results are provided for various CNT volume fractions, comparing cases that include and exclude the interphase effect.

Figure 34.5b demonstrates the convergence of the present FEM formulation for thin to moderately thick plates, showing that the transverse deflection stabilises for different CNT volume fractions once the *a/h* ratio reaches 20.

Furthermore, the non-dimensional centreline deflection of the simply-supported CGCMM hybrid composite laminate is plotted in Figures 34.6a,b for various a/h ratios, corresponding to the normalised coordinate (x/a), for both samples with and without considering the interphase effect. It is observed that the thick plate exhibits more deformation compared to the thin and moderately thick plates, for both cases (with and without interphase). Therefore, interfacial properties exert a greater influence on thin plates, where the interphase is vital in improving the composite's mechanical performance, especially by reducing deflection.

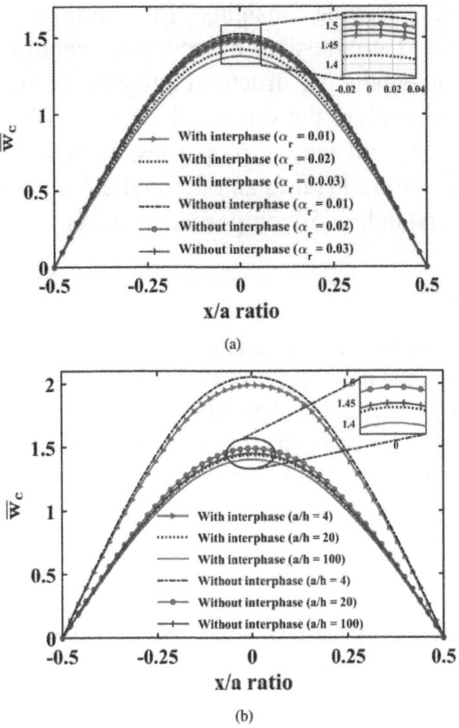

Figure 34.6 Non-dimensional bending deflection of CGCMM hybrid composite corresponding to normalised coordinate (x/a) for (a) Sample 1 (CNT/Al) and (b) Sample 2 (CNT/Al-Cu)
Source: Authors

Conclusion

The study highlights the significant impact of interfacial behaviour between CNTs and the matrix on the bending characteristics of CGCMM hybrid composite laminated plates under sinusoidal loading. The micromechanical model with $n = 1$ provides accurate predictions of the CNT-coated layers' elastic constants, with errors under 3%, showing good agreement with experimental and analytical data. Considering interfacial properties enhances the elastic constants in both longitudinal and transverse directions, particularly for CNT-Al-based hybrid composites. The interphase improves mechanical properties, especially for thinner plates, reducing deflection more effectively. The effective elastic properties for CNT/Al-cu and CNT/Al-based layer influence the bending behaviour, with increased CNT volume fraction leading to reduced centreline deflection, especially when the interphase is included. The FEM formulation's accuracy and convergence are validated, confirming its suitability for analysing bending deformation in thin to moderately thick plates.

The results provide insights for optimising CGCMM hybrid composites by reducing bending deflections and enhancing structural integrity. Future research could explore the effects of CNT distribution, orientation, and matrix properties, as well as incorporate dynamic loading and thermal effects for a deeper understanding of composite behaviour.

References

[1] Jones, D., Green, L. Sustainable approaches to the development of fibre-reinforced composites. *J. Compos. Mater.*, 2023;57(4):365–380.

[2] Rajput, L., Shrivastava, A. Structural analyses of nano- stitched composite laminates based on FSDT using finite element approach. *Mech. Mater.*, 2021;155:103771.

[3] Das, S., Jana, P. Buckling analysis of nonuniformly com- pressed rectangular FG-CNT reinforced laminated composite plate resting on elastic foundation: An analytical solution. *Mech. Adv. Mater. Struct.*, 2024:1–22.

[4] Lee, S., Brown, M. The role of fibre-matrix interface in the mechanical properties of fibre-reinforced composites. *Mater. Sci. Engg. A*, 2023;742:124–133.

[5] Smith, J., Clark, E. Recent advancements in fibre rein- forced polymer composites: A review of recent trends. *Compos. Sci. Technol.*, 2024;215:125–135.

[6] Davis, W., White, R. Fibre reinforced composites for aerospace applications: Performance and future directions. *Compos. Part B Engg.*, 2024;245:149–160.

[7] Zhang, X., Dong, P., et al. Effects of interphase, cnt waviness, and temperature on the Young's modulus of CNT/metal composites. *Compos. Struct.*, 2023;314:117182.

[8] Zhang, X., Li, W., Zheng, S., Zhao, Z., He, Y., Wang, S., et al. Temperature-dependent shear failure modes and tensile strength model of CNT/polymer nanocomposites. *Compos. Comm.*, 2021;25:100726.

[9] Saber-Samandari, S., Afaghi-Khatibi, A. Evaluation of elastic modulus of polymer matrix nanocomposites. *Poly. Compos.*, 2007;28(3):405–411.

[10] García-Macías, E., Guzmán, C. F., Flores, E. I. S., Castro-Triguero, R. Multiscale modelling of the elastic moduli of CNT-reinforced polymers and fitting of efficiency parameters for the use of the extended rule-of- mixtures. *Compos. Part B Engg.*, 2019;159:114–131.

[11] Thostenson, E., Li, W., Wang, D., Ren, Z., Chou, T. Carbon nanotube/carbon fibre hybrid multiscale composites. *J. Appl. Phys.*, 2002;91(9):6034–6037.

[12] Hassanzadeh-Aghdam, M. K., Ansari, R., Darvizeh, A. Micromechanical analysis of carbon nanotube-coated fibre-reinforced hybrid composites. *Internat. J. Engg. Sci.*, 2018;130:215–229.

[13] Reddy, J. N. Mechanics of laminated composite plates and shells: theory and analysis. 2nd Edition. Mechanical Engineering Series. Boca Raton, FL: CRC Press, 2003.

[14] Malekimoghadam, R., Icardi, U. Prediction of mechancal properties of carbon nanotube–carbon fibre rein- forced hybrid composites using multi-scale finite element modelling. *Compos. Part B Engg.*, 2019;177:107405.

[15] Maghsoudlou, M. A., Isfahani, R. B., Saber-Samandari, S., Sadighi, M. Effect of interphase, curvature and agglomeration of SWCNTs on mechanical properties of polymer-based nanocomposites: Experimental and numerical investigations. *Compos. Part B Engg.*, 2019;175:107119.

[16] Rajput, L., Jana, P. 3d micromechanics assisted finite element approach for bending analysis of straight/wavy CNT-reinforced multi-scale hybrid composite plate. *Mech. Based Design Struct. Mach.*, 2024:1–31.

[17] Choi, H., Shin, J., Min, B., Park, J., Bae, D. Reinforcing effects of carbon nanotubes in structural aluminum matrix nanocomposites. *J. Mater. Res.*, 2009;24(8):2610–2616.

[18] Nam, D. H., Cha, S. I., Lim, B. K., Park, H. M., Han, D. S., Hong, S. H. Synergistic strengthening by load transfer mechanism and grain refinement of cnt/al–cu composites. *Carbon*, 2012;50(7):2417–2423.

35 Enhancing volume reconstruction of spinal cord from medical images using Res-V-Net-XL and adaptive interpolation

Nikhil Y. N.[a], *Ramsha Fathima*[b], *Jamuna S. Murthy*[c] *and Muzammil Bagewadi*[d]

Department of Mechanical Engineering, Ramaiah Institute of Technology, Bengaluru, Karnataka, India

Abstract

Reconstructing a 3D volume of the spinal cord from medical images presents many challenges, including the preservation of fine structural details and handling variable slice orientations and gaps between slices. Conventional strategies, such as direct, cubic, nearest-neighbour, B-spline, Lanczos, and radial basis function (RBF) interpolation, find it hard to preserve spatial consistency and frequently come about in obscured or mutilated recreations. Deep learning models like U-Net variations fall flat to capture the complex highlights of the spinal cord structure completely, driving problematic execution in division and recreation assignments. To overcome these challenges, we propose a crossover strategy that coordinates an improved V-Net with versatile addition strategies, called Res-V-Net-XL, which stands for residual volume network with extended loss. The Res-V-Net-XL is developed for the exact division of spinal cord highlights, utilising deep residual connections with an added layer and spatial attention mechanisms to extricate anatomical information from medical images. This is then followed by adaptive interpolation, which dynamically adjusts the slice orientation and stitches them based on the size of the input to balance both the processing time as well as accuracy. The suggested hybrid approach works noticeably better than both conventional interpolation methods and the most advanced U-Net variations, according to our testing data. Res-V-Net-XL has a reconstruction accuracy of 98.89%, dice score of 98.98%, and precision of 99.2%. These findings outperform existing models, making our technique an important addition to the precise and trustworthy 3D reconstruction of the spinal cord.

Keywords: Spinal cord reconstruction, V-Net, Res-V-Net-XL, adaptive interpolation, MRI segmentation, 3D volume reconstruction

Introduction

Accurate segmentation and reconstruction of spine structures from medical images are of the utmost relevance for clinical applications such as surgery planning, diagnostics, and monitoring spinal health [1, 2]. Nonetheless, the complicated characteristics of the spinal cord anatomy and patient diversity provide major obstacles. Classical segmentation methods are insufficient to capture fine features, particularly when vertebrae are in close proximity or in case some regions, e.g., sacral vertebrae, have blurred boundaries [3, 4].

MRI and CT scans are commonly saved in formats such as .dcm (DICOM) and .nii (NIfTI), which store 2D slices as well as 3D volumetric data [5, 6]. The DICOM format is often preferred for clinical practice as it saves image data as well as important metadata, not the case with NIfTI, which is rather more suitable to volumetric analysis in research [7, 8]. Using these formats, 2D slices can be stacked into a 3D representation of anatomical structures [9].

However, a key challenge faced in visualising 3D volumes from the scanned data is that the outermost surfaces, i.e., either surrounding tissues in an MRI or thick bony structures in a CT scan, inevitably obstruct the view to the inner core of that volume [10]. For example, for a spine, exterior tissues or vertebrae obstruct the view to interior layers like the dura mater, arachnoid mater, and spinal cord [3, 5] so that they aren't visible.

Clinicians often have to browse through individual 2D slices to simulate interior views, be time-consuming and error-prone [2, 3]. To solve this, improved deep learning-based segmentation and reconstruction

[a]nikhilylkn@gmail.com, [b]ramshafathima95@gmail.com, [c]jamunamurthy.s@gmail.com, [d]bagewadi.muzammil212@gmail.com

DOI: 10.1201/9781003770442-35

algorithms are necessary. These approaches can digitally peel away surface layers, isolating and increasing the visibility of essential interior elements inside the 3D volume and thus assisting in more precise medical diagnosis and treatment planning [10–12].

Problem definition

Conventional approaches, such as linear, cubic, B-spline, and radial basis function (RBF) interpolation [13–16], cannot give the high degree of details necessary for successful spinal cord reconstruction. These approaches suffer from blurring effects, an inability to cope with tough anatomical regions, and losing important information at spots where vertebrae are closely spaced [3, 9]. Moreover, some of the existing deep learning algorithms, in particular U-Net-based techniques, might fail to distinguish microscopic properties of the spinal cord adequately, producing inaccuracies in 3D volume reconstruction [17, 18].

Objectives

This study aims to design a robust system that can accurately recreate 3D spinal cord volumes from medical images, given the tough restrictions imposed by anatomical complexity and the unpredictability of slice orientation. More specifically, this system must be able to:

- Minimise the disadvantages of traditional interpolations and deep learning systems.
- Improve the accuracy and structural consistency of spinal cord segmentation and reconstruction.
- Outperform the existing methods in terms of accuracy, dice score, and precision, ensuring reliable clinical results.

Proposed work

In order to reconstruct the spinal cord from medical images, this work presents a hybrid strategy that combines adaptive interpolation methods with an upgraded V-Net model (Res-V-Net-XL) that includes spatial attention processes and deep residue linkages along with a custom hybrid loss function. The extra feature merging layers, enhance gradient flow and keep gradients from fading. The suggested model outperforms the current methods U-Net implementations and classical interpolation. It outperforms previous algorithms in terms of segmentation accuracy and structural consistency, with a reconstruction accuracy of 98.98%, dice score of 98.98%, and precision of 99.2%. This discovery represents a significant step forward in spinal cord volume regeneration, with potential implications for improved clinical assessment and patient outcomes.

Literature review

In addressing the challenges of enhancing 3D volumes reconstructed from low-resolution 2D MRI slices, Zhang et al. (2021) introduced a GAN-based method to reduce noise and slice gaps, improving detail and outperforming traditional interpolation techniques by 97.8% [1]. Similarly, Jakhongir et al. (2021) utilised the visualisation toolkit (VTK) to create 3D spinal images from 2D MRI data, achieving 95.3% accuracy while maintaining geometric stability, noting that complex components like the spine require sequential approaches to avoid artefacts [2]. Toufani et al. (2021) used a deep-learning pipeline to detect spinal degeneration in MS and NMOSD patients with 97.9% accuracy, outperforming nearest-neighbour and B-spline interpolation [3]. Wu et al. (2022) achieved 98.1% accuracy in lumbar vertebrae segmentation using hybrid 2D visual projections, addressing bone concealment and avoiding U-Net reliance [4]. NC-PDNet, a non-Cartesian MRI organisation, reduced aliasing and achieved 98.6% accuracy, surpassing B-spline and Lanczos interpolation [5]. Deep learning and physics-based constraints further improved MRI quality for under-sampled data.

Ku et al. (2023) reconstructed 3D surgical fluoroscopic images from 2D MRI slices with 96.4% accuracy, outperforming nearest-neighbour and bicubic methods, improving real-time surgical assistance [6]. Vahedifard et al. (2023) developed a deep-learning system for detecting ventriculomegaly in neonatal MRI, achieving 98.5% accuracy while enhancing spatial precision in difficult brain regions [7]. Mittal et al. (2024) advanced 3D brain tumour modelling, achieving 98.2% accuracy using enhanced interpolation to improve therapeutic imaging [8]. Wang et al. (2024) designed a knowledge-driven deep-learning tool for MRI reconstruction, achieving 98.7% accuracy while reducing filter latency and surpassing B-spline and RBF enhancements for faster clinical applications [9]. Karimi and Hamghalam (2024) applied 2D CNNs to 3D new-born brain segmentation with 97.6% accuracy, enabling early detection of developmental abnormalities [10].

The proposed hybrid strategy, combining the Res-V-Net-XL architecture with adaptive interpolation, enhances 3D MRI reconstructions by addressing issues identified in prior methods. Res-V-Net-XL's 3D convolutional layers reduce reliance on post-processing interpolation, directly capturing volumetric structures, and ensuring robust handling of complex anatomical features like spinal vertebrae [1, 2, 4]. Adaptive interpolation refines reconstructed volumes, minimising artefacts and improving structural fidelity compared to conventional techniques like nearest-neighbour, cubic, and B-spline interpolation [11, 13–15]. Unlike Jakhongir et al.'s VTK-based method, the hybrid approach excels in detecting fine details in noisy or fragmented data, resolving slice misalignment and distortion issues [2, 4]. It ensures accurate vertebra segmentation through Res-V-Net-XL's spatial progression capabilities and adaptive interpolation's boundary fine-tuning [7].

Methodology

Dataset overview

The spine segmentation from the CT scans dataset contains high-resolution CT pictures of the spinal cord. This collection contains annotated pictures of the spine that are used to train and test segmentation algorithms. Each CT scan is well documented, making it suitable for operations such as spine identification and 3D modelling. The findings are especially beneficial for deep learning models that try to segment troublesome sections of the spine. This dataset is used to train the deep learning models described and compared in this paper [5]. The spine segmentation dataset contains MRI images of the spinal cord with pixel-level labelling for each vertebra. This dataset contains many imaging images and is intended to train machine learning algorithms for medical picture segmentation. Its detailed annotations help build effective segmentation algorithms, which are especially useful in clinical applications where identifying specific spinal components is critical for diagnosis and treatment planning. This was used to evaluate the inVesalius thresholding-based segmentation performance [6].

Deep learning approach for spinal segmentation

As shown in Figure 35.1, we use Res-V-Net-XL and Nested U-Net designs that have been updated for 3D and are appropriate for spinal segmentation [11, 12]. This architecture excels at processing 3D medical imaging data, resulting in improved feature extraction, segmentation accuracy, and adaptation to class imbalance in spinal data [11, 15].

Res-V-Net-XL architecture

Res-V-Net-XL builds on the V-Net's foundation while also offering critical enhancements, such as residual connections and a new feature-merging layer, as seen in Figure 35.1. These linkages ensure

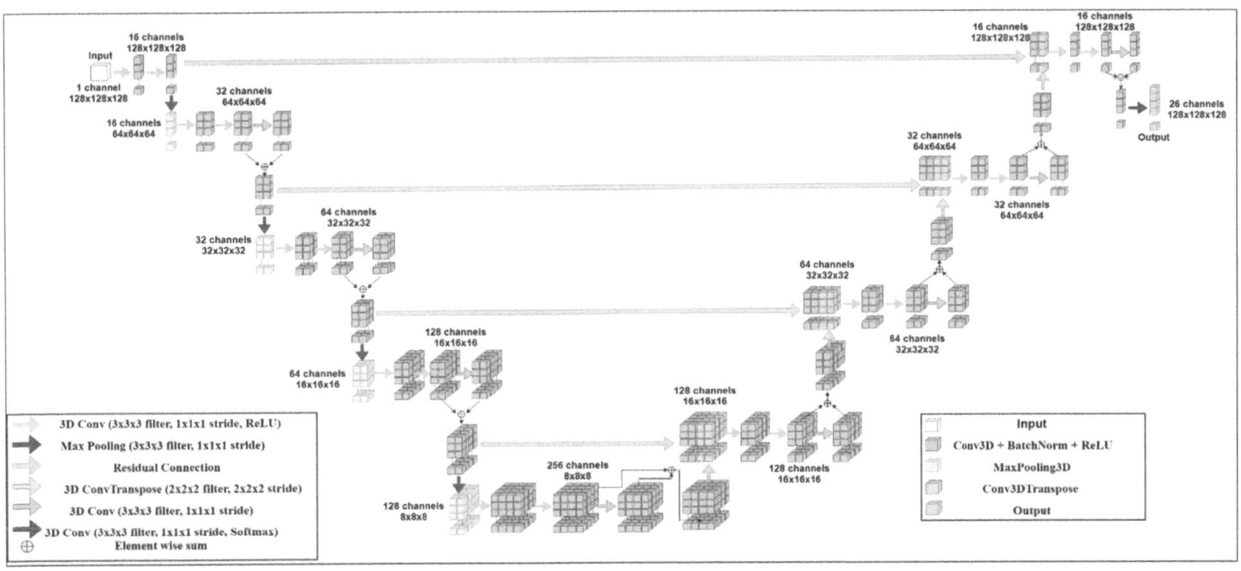

Figure 35.1 Res-V-Net-XL architecture diagram
Source: Authors

that information flows smoothly across the network, preventing the vanishing gradient issue. By mixing input from many layers, the feature-merging layer significantly improves the model's capacity to recognise complex spatial connections [11]. The system maintains a typical encoder-decoder structure, with the encoder compressing input data and extracting valuable information, and the decoder progressively restoring the original image's spatial dimensions. Skip connections link the encoder and decoder, allowing the model to keep low-level information from the input while increasing segmentation at each step. Res-V-Net-XL concentrates on microscopic features, along with the capacity to conduct 3D convolutions, enables it to effectively gather the complicated data essential for proper spinal cord segmentation. The hybrid loss function enhances its performance, assuring that the model achieves excellent accuracy in volume reconstruction, especially in clinical settings when precision is critical [11, 15].

Pseudocode of the training code

Start
1. Input: $\boldsymbol{X} \in \mathbb{R}^{1283} \times {}^{1}$, $\boldsymbol{Y} \in \mathbb{Z}^{1283} \times \boldsymbol{C}$, $\omega \in \mathbb{R}$^$\boldsymbol{C}$
2. Output: $\boldsymbol{Y} \in \mathbb{Z}^{1283}$

Procedure 1: $(\boldsymbol{X}, \boldsymbol{Y})$ – Pre-process data:
  1.1 Normalise: $\boldsymbol{X}$
  1.2 Resize: $\boldsymbol{X}, \boldsymbol{Y}$

Procedure 2: $\boldsymbol{V}$-Net$(\boldsymbol{X})$ – Build network:
  2.1 $\boldsymbol{S} \leftarrow \text{Conv}_{l}(\boldsymbol{X}) \oplus \boldsymbol{R}_{i}$
  2.2 $\boldsymbol{S} \rightarrow$ Down $\rightarrow$ Up ‖ Skip
  2.3 Final layer: $\boldsymbol{S} \in \mathbb{R}^{1283} \times \boldsymbol{C}$

Procedure 3: $\boldsymbol{L}\_(\boldsymbol{Y}, \boldsymbol{S}, \omega)$ – Loss computation:
  3.1 $\boldsymbol{L} \leftarrow f(\boldsymbol{L}, \boldsymbol{S}, \omega)$
  3.2 Return: $\boldsymbol{L}$

Procedure 4: Train$(\boldsymbol{X}, \boldsymbol{Y}, \omega)$ – Training phase:
  4.1 $\boldsymbol{V}$-Net $\leftarrow$ Compile$(\boldsymbol{L}_\omega)$
  4.2 Optimise$(\boldsymbol{X}, \boldsymbol{Y}, \omega)$

Procedure 5: $\boldsymbol{Y}(\boldsymbol{X})$ – Inference phase:
  5.1 $\boldsymbol{S} \leftarrow$ predict$(\boldsymbol{X})$
  5.2 $\boldsymbol{Y} \leftarrow$ argmax$(\boldsymbol{S})$
  5.3 Export: $\boldsymbol{Y}$

End

Nested U-Net architecture

The original U-Net architecture, the Nested U-Net, commonly referred to as U-Net++, increases segmentation performance [12] in medical imaging. It employs the same encoder-decoder architecture as U-Net. The usage of tiered skip routes is what makes it distinct. These algorithms, which enhance feature extraction at varied sizes, are crucial for detecting minute and intricate structures, such as those seen in medical images. Nested U-Net combines the high-level semantic features with the low-level semantic features, high-resolution features to achieve superior performance on challenging tasks and more comprehensive segmentation [12]. Nested U-Net model's depth and number of parameters are increased by several skip connections and stacking pathways, which may result in longer training times and higher memory use [17]. While Nested U-Net is very successful at combining features from several resolutions, it may not be excellent at capturing fine-grained 3D structures, particularly in volumetric data such as spinal cord segmentation. The proposed Res-V-Net-XL overcomes these constraints and delivers a more accurate and efficient solution for applications that need delicate anatomical features. It reduces training time and boosts the capacity to understand spatial relationships across 3D slices [11].

Thresholding-based segmentation

Thresholding-based segmentation is a standard medical imaging approach that separates an image based on intensity levels [13]. It is swift and simple to use, although it struggles with complicated or low-contrast photographs with blurred edges [13]. Its effectiveness falls when the intensity distributions overlap, which is prevalent in medical scans like MRI or CT [13]. Deep learning models, such as CNNs, automatically learn features from data, making them more adept at dealing with difficult, noisy, or low-contrast photographs [11]. Thresholding is rigid and rule-based, but deep learning adjusts to changing image situations, delivering higher accuracy. Despite larger CPU needs, deep learning models such as U-Net and Res-V-Net-XL outperform thresholding in terms of accuracy, particularly for 3D segmentation [15, 17].

Interpolation techniques

When working with multiple image resolutions or filling in gaps in 3D volumes, interpolation algorithms are vital for scaling or reconstructing medical images [13–16]. In order to infer unknown pixel or voxel values while keeping visual structure and spatial relationships, they leverage existing data points. In medical imaging applications such as segmentation, effective interpolation algorithms are critical for maintaining anatomical characteristics and lowering artefacts [19].

Nearest neighbour interpolation (order=0)

The most straightforward technique is nearest neighbour interpolation, where the value corresponding to the nearest pixel/voxel is mapped to the unknown position [16]. This step is especially important with mask images, since the targeted parts should not be smoothed or averaged throughout the scale. It is quick, although it frequently causes blocky, pixelated visuals, making it unsuitable for applications requiring fluidity [16].

$$I(x) = I_{nearest}(x) \tag{1}$$

Where: $I(x)$ is the interpolated value at the target voxel position x. - $I_{nearest}(x)$ is the value of the nearest voxel in the input volume to the position x.

Linear interpolation (order=1)

Linear interpolation computes the value of an unknown place by linearly mixing the values of its neighbours [16]. It produces smoother results than the nearest neighbour but may still exhibit some distortion, especially in high-detail locations. It is useful for scaling volumes with higher volume counts. It is unsuitable for occluded images due to its inherent inclination to smooth the data.

For a target voxel $(x,,)$ between eight surrounding voxels:

$$I(x, y, z) = (1-t_x)(1-t_Y)(1-tz)I_{000} + t_x(1-t_Y)(1-tz)I_{100} + (1-t_x)t_Y(1-tz)I_{010} + t_x t_Y(1-tz)I_{110} + (1-t_x)(1-t_Y)tzI_{001} + (2)\ t_x(1-t_Y)tzI_{101} + (1-t_x)t_Y tzI_{011} + t_x t_Y tzI_{111} \tag{2}$$

Where $t_X = (x-x_0)/(x_1-x_0)$, $t_Y = (y-y_0)/(y_1-y_0)$, and $tz = (z-z_0)/(z_1-z_0)$.

$I(x)$ is the interpolated value at position x.

Cubic interpolation (order=3)

By accounting for a larger number of surrounding points, cubic interpolation surpasses linear approaches. Suitable for medical imaging applications that need high accuracy but have a higher processing cost as it generates significantly cleaner and more exact findings [16]. Similar to linear interpolation, it cannot be applied in mask images. The following is the formula for cubic interpolation for a vector x:

$$I(x) = ax3 + bx2 + cx + d \tag{3}$$

where a, b, c, and are constants determined by the surrounding voxel values.

Lanczos interpolation

Lanczos interpolation uses *sinc* functions to interpolate across a larger area, producing the most precise and smooth results [19]. It is very effective for high-resolution image reconstruction, but it is the most computationally costly, rendering it unsuitable for real-time applications [19]. This may be used for smaller segmentation tasks since smaller segmentations are very tough, and applying it to larger datasets would be computationally expensive [19] due to Lanczos processing delays. This cannot be used for scaling mask images due to the same reasons stated in linear interpolation.

$$I(x) = \sum_{i=-a}^{a} \sum_{j=-a}^{a} V_{ij} sinc(x-i) sinc(y-j) \tag{4}$$

where V_{ij} are the neighbouring voxels (Table 35.1).

Loss functions in multi-class segmentation

To minimise class imbalance and increase border correctness, effective spinal vertebrae segmentation utilises rigorous loss functions. Res-V-Net-XL assesses performance in boosting segmentation accuracy and precision by using hybrid loss functions, including dice loss, Tversky loss, focal loss, and categorical cross-entropy [12–14]. Each lost function has a particular role in overcoming many hurdles during the spinal segmentation process.

Table 35.1 Interpolation technique.

Sl. No.	Interpolation technique	Pros	Cons
1.	Nearest Neighbour	Clear boundaries, fast	Aliasing, may need post-processing
2.	Linear	Smooth voxel transitions	Boundary displacement, less accurate
3.	Cubic	Smooth, better detail	Slower, overshoot (ringing)
4.	Lanczos	High quality, minimal artefacts	Slow, blurs fine details

Source: Authors

Dice loss

Dice loss is better suited for minute or irregular structures as it reduces the gap between expected and real segmentations [15]. By giving greater attention to neglected areas, it gets beyond class mismatch, a frequent obstacle in medical categorisation problems. It focuses on achieving a higher overlap, which helps the model better understand spatial dimensions. The formula for dice loss is:

$$1 - \frac{2|A \cap B|}{|A| + |B|} \tag{5}$$

where A is the set of predicted voxels and B is the ground truth.

Tversky loss

Tversky loss is made for boosting precision by emphasising and giving weights to false negatives and positives [16]. It is an extension of dice loss aiming to eliminate mismatches inside the overlap. It permits the model's sensitivity and accuracy to be fine-tuned, which is vital in the medical industry. Defined as:

$$1 - \frac{|A \cap B|}{|A \cap B| + \alpha|A - B| + \beta|B - A|} \tag{6}$$

where α and β control the weight of false positives and false negatives, respectively

Focal loss

It targets the underrepresented classes in the volume. This plays a key role as 3D volumes are primarily vacuum space [17], making it easier for the model to overfit them. Focal loss handles this by down-weighting simple instances. It helps the model focus on problematic places like lesions. Defined as:

$$-\alpha(1 - p_t)^{\gamma} \log \log (p_t) \tag{7}$$

where p_t is the probability estimate assigned by the model to a given class, α is a weighting factor, and γ is the focusing parameter.

Categorical cross-entropy

Categorical cross-entropy loss is recognised for being used for classification and to penalise the model for misclassification [18]. The sole negative of this is that it struggles with class imbalance, which is eliminated by adding pre-determined class weights.

Each loss function offers unique advantages. In our segmentation model, a combination of Dice, Tversky, Focal, and Categorical Cross-Entropy Loss was applied with pre-calculated dataset-based class weights, balancing the benefits of each to achieve optimal performance across all vertebrae [12–14].

Sliding window with overlap and padding

A sliding window approach (128×128×128 with 50% overlap) is used to reduce volumes into more manageable parts for examination. Padding ensures that volumes are continually divisible by the window size, limiting the loss of key features during segmentation, and overlapping segments minimise border artefacts [11, 12].

Morphological post-processing for spine isolation

After initial segmentation, binary opening eliminates noise like erroneously categorised tissues, whereas binary closure fills microscopic gaps in the segmented spine to build a totally connected structure [19]. These approaches increase segmentation by deleting extraneous components and preserving only the spine structure.

- Binary closing:

$$C(A) = (A \oplus B) \ominus B \tag{8}$$

- Binary opening:

$$O(A) = (A \ominus B) \oplus B \tag{9}$$

Where, O: binary opening operator, C: binary closing operator, A: input binary volume, B: structuring element, $\oplus$: morphological dilation operation, Θ: morphological erosion operation.

Advanced spine isolation using region properties

Spine being a long structure, it is determined by extension along the Z-axis [20]. Z-axis centres and related component marking further separate the spine by filtering out components that deviate from the central Z-position and omitting compact structures like ribs or hips.

- Volume and elongation filtering:

$$\frac{z_{extent}}{xy_{extent}} \tag{10}$$

where, z_{extent}: The extent of the bounding box along the z-axis and xy_{extent}: The maximum extent in the x and y dimensions.

• Z-axis centring: filters out components whose centroids deviate from this axis, helps improve capture the spine accurately.

Anisotropic diffusion for boundary preservation

By carefully changing voxel intensities and preserving critical features, anisotropic diffusion preserves spine boundaries [21]. This approach decreases noise without sacrificing the integrity of the borders between anatomical components and enhances contrast between the spine and surrounding tissues. Local gradients, the cornerstone of the diffusion process, conceal homogenous parts while preserving high-gradient places, such as the border of the spine. Anisotropic diffusion's update rule is provided by:

$$I^{t+1} = I^t + \Delta t(c_n \cdot \nabla I_{north} + c_s \cdot \nabla I_{south} + c_e \cdot \nabla I_{east} + c_w \cdot \nabla I_{west}) \quad (11)$$

Conductance calculation (Exponential):

$$c = e^{-(\frac{|\nabla I|}{\kappa})^2} \quad (12)$$

Where, I: Volume being diffused, I^t: Volume at iteration t, I^{t+1}: updated volume, Δt : time step, c_n, c_s, c_e, c_w the conductance coefficients for the north, south, east, and west gradients, respectively, ∇I: volume gradient (rate and direction of intensity change), κ: controls the sensitivity of conductance.

Edge detection for artefact removal

Edge detection approaches, such as the Sobel operator, are utilised to accurately identify the borders of non-spinal structures, further enhancing segmentation [22]. By identifying intensity gradients [23], Sobel filters capture both vertical and horizontal edges, helping to eliminate non-spinal features like ribs and sections of the hip. This step increases segmentation accuracy by ensuring that only spinal components are maintained while non-spinal parts are securely deleted without affecting the basic curve of the spine.

$$G_x = \begin{bmatrix} 1 & 0 & 1 \\ -2 & 0 & 2 \\ -1 & 0 & 1 \end{bmatrix}, G_y = \begin{bmatrix} 1 & 2 & 1 \\ 0 & 0 & 0 \\ -1 & -2 & -1 \end{bmatrix} \quad (13)$$

Where G_x is the gradient in the x-direction, detecting horizontal edges. It is represented by the kernel and G_y is the gradient in the y-direction, detecting vertical edges. It is represented by the kernel

Gaussian smoothing for visualisation

Gaussian smoothing is applied before analysing the segmentation to reduce voxel irregularity and soften edges while retaining the anatomical integrity of the spine [24]. A small sigma number is applied to achieve a superficially clean outcome without producing excessive distortion.

$$G(x, y) = \frac{1}{2\pi\sigma^2} e^{-\frac{x^2+y^2}{2\sigma^2}} \quad (14)$$

σ controls the Gaussian kernel's standard deviation, thereby setting the extent of smoothing applied.

Live rendering

After predicting segmentation using the Res-V-Net-XL model, anisotropic diffusion is used to reduce noise while preserving critical spinal properties [21]. To separate the spine, the approach searches for areas with considerable elongation in the z-axis, revealing the vertical architecture of the spinal cord. This ensures that the segmentation is restricted to the spine and does not include any non-spinal regions. The smoothed spinal volume is then processed using the marching cubes approach, yielding a surface mesh stored in STL format. Mayavi is used in real-time 3D rendering to provide complete, anatomically precise imaging of the spinal cord, which is essential for diagnostic and surgical planning [24].

Pseudocode for live rendering:

Start

1. Input: $\boldsymbol{X} \in \mathbb{R}$^n, Model: M, Parameters: *c_size*, *stride*, *n it*, Δt
2. Output: Segmented and Isolated Spine Volume: $\boldsymbol{V}$

Procedure 1: ($\boldsymbol{X}$) – Pre-process Volume:

- 1.1 Determine volume size and set interpolation order
- 1.2 Normalise $\boldsymbol{X}$, zoom with appropriate order

Procedure 2: ($\boldsymbol{X}$, *n_it*, Δt) – Anisotropic Diffusion:

- 2.1 Initialise diffused volume
- 2.2 For each iteration: Compute gradient (∇), update volume using diffusion

Procedure 3: ($\boldsymbol{X}$, M, *c_size*, *stride*) – Segment Volume:

- 3.1 Pre-process $\boldsymbol{X}$, pad volume to match *c_size*
- 3.2 For each sub-volume: Use M to predict segmentation
- 3.3 Apply morphological operations, aggregate segmented sub-volumes

Procedure 4: (V, min-size, elongation) – Isolate Spine:

4.1 Label volume regions, calculate elongation

4.2 Filter regions based on elongation and spatial position (z-centring)

4.3 Apply morphological closing/opening, isolate spine

Procedure 5: (V, σ) – Visualise Spine:

5.1 Apply Gaussian smoothing to V

5.2 Generate 3D iso-surface using scalar field

Procedure 6: Full Process:

6.1 Load input $\boldsymbol{X}$ from MRI

6.2 Segment volume using $\mathcal{S}$

6.3 Isolate spine with $\boldsymbol{I}$

6.4 Apply diffusion $\mathcal{A}$ to smooth spine volume

6.5 Visualise result $\boldsymbol{V}$

End

Results and discussion

Evaluation of metrics

To evaluate the segmentation model, we used multiple metrics to provide a comprehensive understanding of its performance. These metrics—Dice score and Precision—were chosen for their ability to capture both global segmentation performance and specify class-wise accuracy [15, 16].

Thresholding-based segmentation

Thresholding-based segmentation's effectiveness is reflected in its ability to highlight distinct structures, albeit at the expense of precision due to noise sensitivity (Table 35.2).

Res-V-Net-XL model

The Res-V-Net-XL model achieved consistently high metrics across both training and validation datasets. Detailed performance indicators are as follows:

As seen in Table 35.3, the loss appears to be high due to the usage of class weights and summation of multiple loss functions; it was not scaled between 0 and 1 [12]. This consistent performance, coupled with a low validation loss, underscores Res-V-Net-XL's robustness in segmentation tasks that demand high precision and spatial coherence [11, 15] (Table 35.4).

Nested U-Net model

The Nested U-Net model achieved consistently high metrics across both training and validation datasets. Detailed performance indicators are as follows:

As seen in Table 35.5, loss appears to be high due to the usage of class weights and summation of multiple loss functions; it was not scaled between 0 and 1. The custom loss was used in the Nested U-Net as well [12] (Figures 35.2 and 35.3).

Comparison of different methods by metrics

The dice score reflects spatial alignment between predicted and ground truth masks, where a higher score indicates better performance [15]. Precision measures the model's ability to control false positives

Table 35.2 Thresholding-based segmentation metrics.

Sl. No.	Metric	Value
1.	Accuracy	86%
2.	Dice score	80%
3.	Precision	70%
4.	Sensitivity to noise	High
5.	Manual intervention	Mod
6.	Over-segmentation rate	High

Source: Authors

Table 35.3 Res-V-Net-XL model metrics.

Sl. No.	Metric	Value
1.	Accuracy	98.98%
2.	Dice score	98.98%
3.	Precision	99.2%
4.	Validation accuracy	98.71%
5.	Validation dice score	99%
6.	Validation precision	98.88%
7.	Loss	0.9840

Source: Authors

Table 35.4 Comparison of metrics between the methods discussed.

Sl. No	Model	Accuracy	Dice score	Precision
1	Thresholding-based segmentation	86%	80%	70%
2	Res-V-Net-XL	98.98%	98.98%	99.2%
3	Nested U-Net	98%	96.68%	99.64%

Source: Authors

Table 35.5 Metrics of the Nested U-Net model.

Sl. No.	Metric	Value
1.	Accuracy	97%
2.	Dice score	96.68%
3.	Precision	99.64%
4.	Validation accuracy	98%
5.	Validation dice score	97%
6.	Validation precision	99.74%
7.	Loss	0.988

Source: Authors

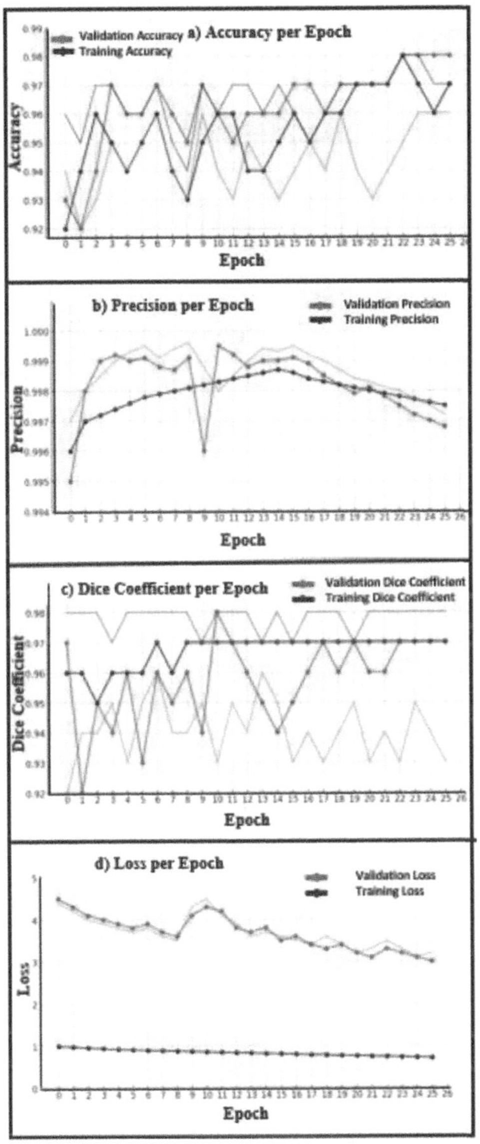

Figure 35.2 Metrics analysis of Res-V-Net-XL per epoch

Source: Authors

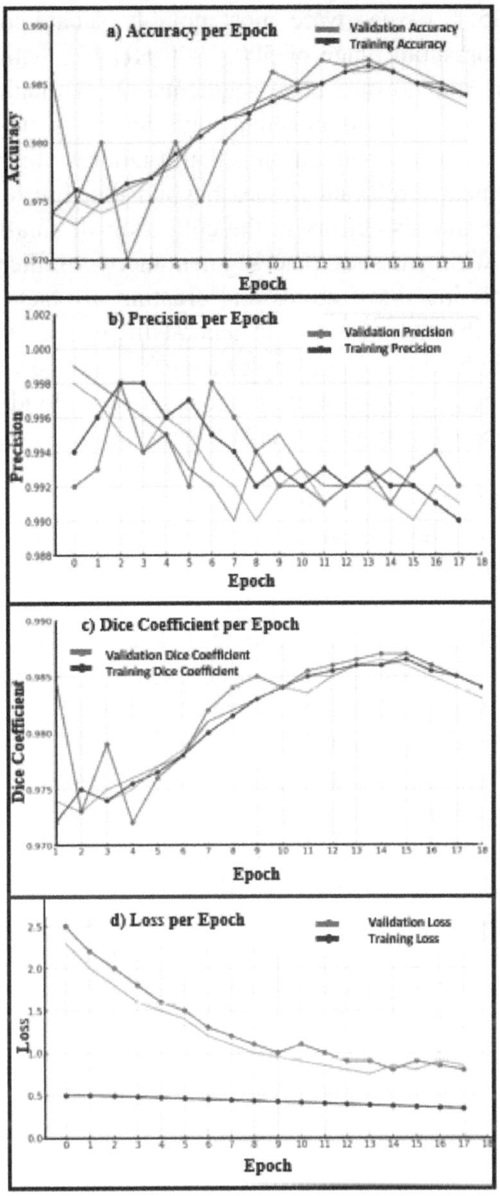

Figure 35.3 Metrics analysis of Nested U-Net per epoch

Source: Authors

and false negatives, and accuracy denotes the fraction of accurately classified instances (both true positives and true negatives) relative to the total number of predictions [16]. These three metrics were used to evaluate the methods tested in this research.

Evaluation of output

Thresholding-based segmentation (InVesalius)

The figures were produced using the open-source program InVesalius with DICOM data that has been processed using threshold techniques. As seen in

Figure 35.4, tissues were most notably separated using a threshold range of 500–1000 HU [6], with a specific aim toward bone segmentation. Though it works well in distinguishing high-density structures like bones, its limitation is that it ignores other more clinically relevant tissues beyond the threshold range and also errors in the collection of small features within tissues. The segmentation is rather crude, missing fine features and creating artefacts. Although the method assists in general diagnostics, it is not suitable for use in complex anatomy and tissue heterogeneity. Moreover, it fails to segment individual vertebrae due to similar HAU levels.

Res-V-Net-XL segmentation

The Res-V-Net-XL model achieves outstanding results in 3D segmentation and reconstruction as seen in Figure 35.5, especially for precise fine details like the anatomy of the vertebrae. It keeps individual bones and tissues for fine structure preservation, which is important for spinal use cases that require a clear visual representation of each vertebra. Such deep learning architecture retains minor variations in density, thereby improving diagnostics as well as usability.

In comparison with basic thresholding techniques, Res-V-Net-XL manages to segment rather complex anatomical shapes without sacrificing the finer details or introducing artificial boundaries. Some minor inaccuracies may appear in places with very high boundary ambiguity, but these are very infrequent and by and large lost in the noise of the model's general accuracy. Its flexibility guarantees intense, well-defined structures, so it finds ideal usage in medical applications that require very detailed segmentation of bones, cartilage, and soft tissues [12, 17].

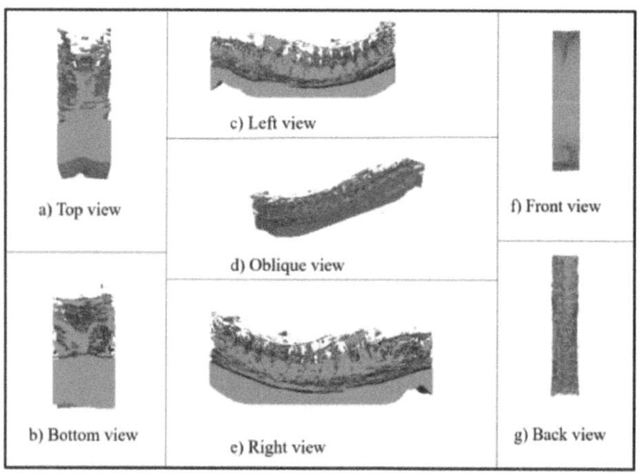

Figure 35.4 3D spinal cord extracted using thresholding-based segmentation
Source: Authors

Nested U-Net segmentation

Though the Nested U-Net is good for refining boundaries and collecting minor characteristics in complicated anatomical regions, it may struggle to segment some major, more homogenous components, especially in volumetric datasets. As seen in Figure 35.6 the additional lumps in between near the thoracic spine vertebrae. Skip connections with no sparseness help to control tissue heterogeneity, but

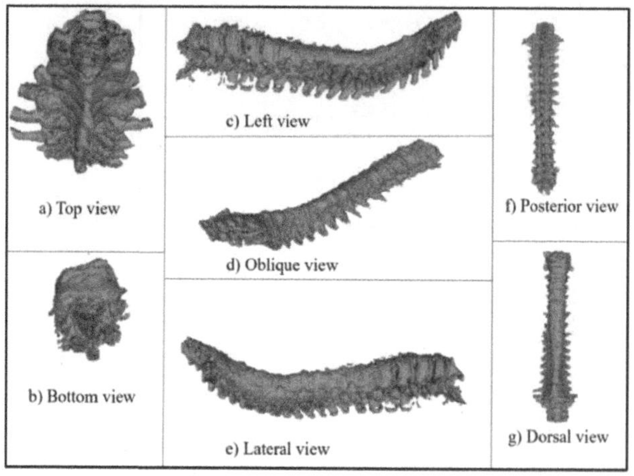

Figure 35.5 3D spinal cord extracted using Res-V-Net-XL
Source: Authors

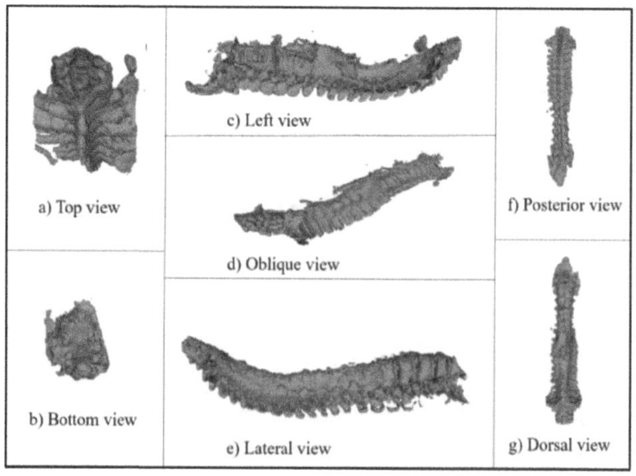

Figure 35.6 Spinal cord extracted using Nested U-Net
Source: Authors

they increase computational expense and may result in over-segmentation, which happens when tiny changes are overemphasised. It would be capable of distinguishing well-defined structures, such as vertebrae, which are suitable for volumetric segmentation. As observed it can compromise overall segmentation quality when fine-detailed separation of compact tissues such as bones is needed, making it less suited for medical applications that demand higher precision [12, 17].

Conclusion

This research presents a state-of-the-art solution for spinal cord segmentation, including Res-V-Net-XL, hybrid extended loss, and live rendering, and gives a comparative analysis with other methods that are commonly used. By concentrating on these advanced models, this research surmounted significant obstacles in spinal segmentation and shows how well suited these methods are to maintaining fine vertebral details. The introduction of adaptive interpolation improves segmentation, resulting in more accurate boundaries and anatomical feature continuity.

Live rendering, on the other hand, allows on-the-fly visualisation, further allowing monitoring of the segmented spine in real-time, culminating in a complete 3D model with anatomical accuracy. This enhances interaction with dynamic evaluation of the segmented structures – as well as their diagnosis and dynamically assess segmented structures, as well as their diagnostic and planning abilities.

Future approaches will require more adaptable, real-time feedback systems, as well as higher segmentation performance for underrepresented vertebral groups. The model's resilience will be examined as the dataset expands.

Acknowledgments

This research was supposed by M. S. Ramaiah Institute of Technology (MSRIT), Banglore–560054.

The datasets and thresholding-based segmentation code generated and/or analysed during the current study are available from the following sources:

- **Thresholding-based segmentation code**: The code used for thresholding-based segmentation can be accessed from the https://github.com/invesalius/invesalius3/tree/master/invesalius.
- **CT scans dataset**: The CT scan dataset used for spine segmentation is publicly available on Kaggle and can be accessed https://www.kaggle.com/datasets/pycadmk/spine-segmentation-from-ct-scans.
- **MRI dataset**: The MRI dataset for spine segmentation is also publicly available on Kaggle and can be accessed https://www.kaggle.com/datasets/trainingdatapro/spine-segmentation-dataset.

References

[1] Zhang, H., Shinomiya, Y., Yoshida, S. 3D MRI reconstruction based on 2D generative adversarial network super resolution. *Sensors*, 2021;21(9): 2978.

[2] Jakhongir, N., Abdusalomov, A., Whangbo, T. K. 3D volume reconstruction from MRI slices based on VTK. *2021 Internat. Conf. Inform. Comm. Technol. Converg. (ICTC)*, 2021:689–692.

[3] Toufani, H., Vard, A., Adibi, I. A pipeline to quantify spinal cord atrophy with deep learning: Application to differentiation of MS and NMOSD patients. *Physica Medica*, 2021;89:51–62.

[4] Wu, Z., Xia, G., Zhang, X., Zhou, F., Ling, J., Ni, X., Li, Y. A novel 3D lumbar vertebrae location and segmentation method based on the fusion envelope of 2D hybrid visual projection images. *Comp. Biol. Med.*, 2022;151:106190.

[5] Ramzi, Z., R, C. G., Starck, J., Ciuciu, P. NCPDNet: A density compensated unrolled network for 2D and 3D non Cartesian MRI reconstruction. *IEEE Trans. Med. Imag.*, 2022;41(7):1625–1638.

[6] Ku, P., Martin-Gomez, A., Gao, C., Grupp, R., Mears, S. C., Armand, M. Towards 2D/3D registration of the preoperative MRI to intraoperative fluoroscopic images for visualisation of bone defects. *Comp. Methods Biomech. Biomed. Engg. Imag. Visualiz.*, 2022;11(4):1096–1105.

[7] Vahedifard, F., Ai, H. A., Supanich, M. P., Marathu, K. K., Liu, X., Kocak, M., Ansari, S. M., Akyuz, M., Adepoju, J. O., Adler, S., Byrd, S. Automatic ventriculomegaly detection in fetal brain MRI: A step by step deep learning model for novel 2D3D linear measurements. *Diagnostics*, 2023;13(14):2355.

[8] Mittal, R., Malik, V., Singla, G., Kaur, A., Singh, M., Mittal, A. 3D reconstruction of brain tumors from 2D MRI scans: An improved marching cube algorithm. *Biomed. Signal Proc. Control*, 2024;91:105901.

[9] Wang, G., Luo, T., Nielsen, J., Noll, D. C., Fessler, J. A. B-spline parameterized joint optimization of reconstruction and Kspace trajectories (BJORK) for accelerated 2D MRI. *IEEE Trans. Med. Imag.*, 2022;41(9):2318–2330.

[10] Karimi, H., Hamghalam, M. Segmentation of 3D MRI using 2D convolutional neural networks in infants' brain. *Multim. Tools Appl.*, 2023.

[11] Milletari, F., Navab, N., Ahmadi, S. VNet: Fully convolutional neural networks for volumetric medical image segmentation. *2016 Fourth Internat. Conf. 3D Vision (3DV)*, 2016:565–571.

[12] Zhou, Z., Siddiquee, M. M. R., Tajbakhsh, N., Liang, J. UNet++: A nested UNet architecture for medical image segmentation. *Lec. Notes Comp. Sci.*, 2018:3–11.

[13] Getreuer, P. Linear methods for image interpolation. *Image Proc. Line*, 2011;1238–259.

[14] Keys, R. Cubic convolution interpolation for digital image processing. *IEEE Trans. Acoust. Speech Signal Proc.*, 1981;29(6):1153–1160.

[15] Duchon, C. E. Lanczos filtering in one and two dimensions. *J. Appl. Meteorol.*, 1979;18(8):1016–1022.

[16] He, K., Zhang, X., Ren, S., Sun, J. Deep residual learning for image recognition. *IEEE Conf. Comp. Vision Patt. Recog. (CVPR)*, 2016:770–778.

[17] Ronneberger, O., Fischer, P., Brox, T. UNet: Convolutional networks for biomedical image segmentation. *Lec. Notes Comp. Sci.*, 2015:234–241.

[18] Lin, T., Goyal, P., Girshick, R., He, K., Dollar, P. Focal loss for dense object detection. *IEEE Trans. Patt. Anal. Mac. Intel.*, 2018;42(2):318–327.

[19] Witkin, A. P. Scale-space filtering. *Elsevier eBooks*, 1987:329–332.

[20] Perona, P., Malik, J. Scale-space and edge detection using anisotropic diffusion. *IEEE Trans. Patt. Anal. Mac. Intel.*, 1990;12(7):629–639.

[21] Otsu, N. A threshold selection method from gray level histograms. *IEEE Trans. Sys. Man Cybernet.*, 1979;9(1):62–66.

[22] Rayed, M. E., Islam, S. S., Niha, S. I., Jim, J. R., Kabir, M. M., Mridha, M. Deep learning for medical image segmentation: State-of-the-art advancements and challenges. *Informat. Med. Unlock.*, 2024;47:101504.

[23] Niyas, S., Pawan, S., Kumar, M. A., Rajan, J. Medical image segmentation with 3D convolutional neural networks: A survey. *Neurocomputing*, 2022;493:397–413.

[24] Vania, M., Lee, D. Intervertebral disc instance segmentation using a multistage optimization mask-RCNN (MOM-RCNN). *J. Comput. Design Engg.*, 2021;8(4):1023–1036.

36 Relative performances of skewed hyperbolic paraboloidal stiffened shells – A geometrically non-linear approach

Madhu Trivedi[a], Natvar singh[b], Ravindra Solanki[c] and Kaustav Bakshi[d]

Department of Civil Engineering, Indian Institute of Technology Indore, Madhya Pradesh, India

Abstract

Laminated composites are lucrative construction materials for civil engineers. The material is lightweight, resistant to atmospheric decay and possesses high strength/stiffness to weight ratio. The doubly curved hyperbolic-paraboloidal shells are stiff and architecturally pleasing. Naturally, these shells made of laminated composites are popular choices among the practicing civil engineers to cover long column free spans. The skewed forms can satisfy the architectural requirements of industrial applications. But these are susceptible to instability which can be addressed by stiffeners. The research on hyperbolic paraboloids is ample in numbers but the same for panels with cut-outs are scanty. The literature shows that there are no research reports available on skewed stiffened hyperbolic-paraboloidal curved panels with cut-outs, especially using the geometrically non-linear approach. Therefore, an attempt has been made in this paper to study the bending and free vibration of simple supported skewed panels for different skew angles, laminations, cut-outs and stiffener orientations which can provide meaningful design guidelines to practicing civil engineers.

Keywords: Skewed panels, hyperbolic paraboloidal shells, geometric non-linearity, iso-parametric finite elements

Introduction

Laminated composite shells play crucial roles in aerospace, civil, and marine structures because of lightweight construction, defiance to atmospheric decay, and flexibility to modify rigidity along governing loads. Among different shells reported by researchers [1], the hyperbolic paraboloids are anticlastic, aesthetically pleasing and stiffer for roofs of long unsupported facilities. Researchers have worked on hyperbolic paraboloids using linear finite element approach. Chakravorty et al. [2] reported fundamental frequencies of point supported panels. Ortega and Robles [3] compared the mechanical behaviour of different geometrical parameters. Natural frequencies and buckling of simply supported shallow shells were computed by Matsunaga [4]. Kumar et al. [5] worked on vibration responses considering parabolic shear stresses. A Fractal-based model was reported by Rian [6] while functionally graded panels were researched by Li et al. [7] and Zhao et al. [8]. Since the moderately thin panels undergo finite deformations, the geometric non-linearity was preferred by Singh and Panda [9], Ghosh and Chakravorty [10], Sofiyev et al. [11] and Yusufoglu and Avey [12] to study the hyperbolic paraboloids. These papers included von-Karman strains to study free vibration [9] and first ply failure of laminated composites [10]. Similar approach was adopted for dynamics of functionally graded [11] and CNT reinforced panels [12].

The skewed shells are preferred to cover parallelogram shaped practical facilities. Research studies on skewed panels were concentrated on plates only using non-linear strains. The iso-parametric finite elements were applied for symmetric composites by Singha and Daripa [13]. Flexure of composite flats was studied by Upadhyay and Shukla [14]. Skewed plates made of CNT reinforced composites were researched by Zhang and Xiao [15]. The initial stresses due to hygrothermal variations were the focus of study for Kallannavar et al. [16]. Katariya et al. [17] worked on sandwiched skew flats while an isogeometric approach was preferred by Sengar et al. [18]. Recently, Kallannavar and Kattimani [19] and Jahanbazi and Kiani [20] studied free vibration of skew composite laminated and FG-GPLRC panels.

[a]mtphd2302104002@iiti.ac.in, [b]mtphd2302104007@iiti.ac.in, [c]mt2302104009@iiti.ac.in, [d]Kaustav.bakshi@iiti.ac.in

DOI: 10.1201/9781003770442-36

Giri and Bakshi [21] worked on bending of skewed shell panels using non-linear approach.

The finite deformations of moderately thin skewed panels can be arrested by stiffeners. The stiffeners are also preferred to control large amplitudes of vibrations. The composite stiffeners for singly curved cylindrical panels were adopted by Giri and Bakshi [21] and Bakshi [22]. Huang and Qiao [23] and Yang et al. [24] worked on stability and vibro-acoustic characteristics of stiffened curved forms. However, the research on stiffened hyperbolic paraboloids is relatively scanty. Although Sahoo and Chakravorty [25–27] worked in this area but these are limited to linear approach only. The skewed hyperbolic paraboloidal shells with composite stiffeners received limited attention from the researchers, especially using geometric non-linearity. Hence, this deficiency is aimed to be filled by this paper. A non-linear code is proposed which is isoparametric and follows C^0 continuity. It is developed using von-Karman strains and eight nodded quadratic elements. The code solves bending and free vibration of moderately thin panels, and the non-dimensionalised deflections and fundamental frequencies are reported for varying sizes and locations of cut-outs, laminations and orientations of central stiffeners. The cut-outs are preferred for atmospheric ventilations and ambient lighting. Moreover, the ducts for cooling and electrical cables can be passed through the cut-outs. These are advantageous to control the weight of panels also. This paper offers meaningful guidelines to civil engineers to achieve minimum compromises with deflections and maximum improvement of fundamental frequencies for an optimised shape and location of cut-outs. The performance studies include different orientations of central stiffeners.

Mathematical model

A laminated composite stiffened hyperbolic paraboloidal panel (Figure 36.1) with radii of curvatures R_{xx} and R_{yy} is taken up here. The shell is skewed at an angle "ϕ" from y-axis. The shell thickness "h" can be bonded by multiple laminae. Fibres are oriented at an angle "θ" from x'-axis. The $x' - y'$ axes are taken at mid-depth and are related to x - y system through the following:

$$x = x' + y'\sin\phi, y = y'\cos\phi, z = z' \qquad (1)$$

The sides of shell are "a" and "b."

Displacement field

The generalised displacements are assumed as the following:

$$u = u_0 + z\alpha, v = v_0 + z\beta, w = w_0 \qquad (2)$$

u_0, v_0 and w_0 = mid-surface translations. α and β are rotations (Figure 36.1). The displacements between x' - y' axes and x - y system are related through the following,

$$\begin{Bmatrix} u' \\ v' \\ w' \\ \alpha' \\ \beta' \end{Bmatrix} = \begin{bmatrix} 1 & -tan\phi & 0 & 0 & 0 \\ 0 & sec\phi & 0 & 0 & 0 \\ 0 & 0 & 1 & 0 & 0 \\ 0 & 0 & 0 & \sec\phi & 0 \\ 0 & 0 & 0 & -\tan\phi & 1 \end{bmatrix} \begin{Bmatrix} u \\ v \\ w \\ \alpha \\ \beta \end{Bmatrix} \qquad (3)$$

Finite element formulation

Present code adopts eight nodded quadratic elements. The mid-surface strains are assumed through von-Karman strains and first order shear deformation theory,

$$\{\varepsilon^0\} = \{\varepsilon_x^0 \quad \varepsilon_y^0 \quad \gamma_{xy}^0 \quad k_x \quad k_y \quad k_{xy} \quad \gamma_{xz}^0 \quad \gamma_{yz}^0\},$$
$$\{\varepsilon^0\} = \{\varepsilon^0\}^L + \{\varepsilon^0\}^{NL} \qquad (4)$$

$$\{\varepsilon^0\} = \{\varepsilon_x^0 \quad \varepsilon_y^0 \quad \gamma_{xy}^0 \quad k_x \quad k$$

$$\{\varepsilon^0\} = \{\varepsilon^0\}^L + \{\varepsilon^0\}^{NL}$$

$$\{\varepsilon^0\}^L = \begin{Bmatrix} \frac{\partial u_0}{\partial x} - \frac{w_0}{R_{xx}} \\ \frac{\partial v_0}{\partial y} - \frac{w_0}{R_{yy}} \\ \frac{\partial u_0}{\partial y} + \frac{\partial v_0}{\partial x} + \frac{w_0}{R_{xy}} \\ \alpha + \frac{\partial w_0}{\partial x} \\ \beta + \frac{\partial w_0}{\partial y} \end{Bmatrix},$$

$$\{\varepsilon^0\}^{NL} = \begin{Bmatrix} \frac{1}{2}\left(\frac{\partial w_0}{\partial x} - \frac{u}{R_{xx}}\right)^2 \\ \frac{1}{2}\left(\frac{\partial w_0}{\partial y} - \frac{v}{R_{yy}}\right)^2 \\ \left(\frac{\partial w_0}{\partial x} - \frac{u}{R_{xx}}\right)\left(\frac{\partial w_0}{\partial y} - \frac{v}{R_{yy}}\right) \\ 0 \\ 0 \end{Bmatrix}$$

$$\text{and} \begin{Bmatrix} k_x \\ k_y \\ k_{xy} \\ k_{xz} \\ k_{yz} \end{Bmatrix} = \begin{Bmatrix} \frac{\partial \alpha}{\partial x} \\ \frac{\partial \beta}{\partial y} \\ \frac{\partial \alpha}{\partial y} + \frac{\partial \beta}{\partial x} \\ 0 \\ 0 \end{Bmatrix}$$

The strains $\{\varepsilon^0\}$ and nodal degrees of freedom $\{\delta_i\}$ are related as following,

$$\{\varepsilon^0\} = \left([B]^L + \frac{1}{2}[B]^{NL}\right)\{\delta_i\}^{/}, \frac{\partial\{\varepsilon^0\}}{\partial\{\delta\}} = [\bar{B}], \qquad (5)$$

$$[\bar{B}] = [B]^L + [B]^{NL}$$

$[B]^L$ is the same as given by Sahoo and Chakravorty [25].

$$[B]^{NL} = \begin{bmatrix} \frac{\partial w_0}{\partial x} - \frac{u}{R_{xx}} & 0 \\ 0 & \frac{\partial w_0}{\partial y} - \frac{v}{R_{yy}} \\ \frac{\partial w_0}{\partial y} - \frac{v}{R_{yy}} & \frac{\partial w_0}{\partial x} - \frac{u}{R_{xx}} \end{bmatrix}$$

$$\begin{bmatrix} -\frac{N_i}{R_{xx}} & 0 & \frac{\partial N_i}{\partial x} & 0 & 0 \\ 0 & -\frac{N_i}{R_{yy}} & \frac{\partial N_i}{\partial y} & 0 & 0 \end{bmatrix}$$

$$\{\delta_i\}^{/} = \{u_i^{/} \quad v_i^{/} \quad w_i^{/} \quad \alpha_i^{/} \quad \beta_i^{/}\}^T \; i = 1 \text{ to } 8$$

The stiffener element

The stiffeners are aligned along $x^{/}$ - $y^{/}$ - axes only. Three noded elements formulate the stiffeners. The formulation of stiffeners adopted in this paper are same as it was given by Giri and Bakshi [21].

The displacements of $x^{/}$-stiffener between x' - y' axes and x - y system are related through the following,

$$\begin{Bmatrix} u'_{sx'} \\ w'_{sx'} \\ \alpha'_{sx'} \\ \beta'_{sx'} \end{Bmatrix} = \begin{bmatrix} 1 & 0 & 0 & 0 \\ 0 & 1 & 0 & 0 \\ 0 & 0 & 1 & 0 \\ 0 & 0 & 0 & 1 \end{bmatrix} \begin{Bmatrix} u_{sx} \\ w_{sx} \\ \alpha_{sx} \\ \beta_{sx} \end{Bmatrix} \qquad (6)$$

Strains of the $x^{/}$-stiffener are given below,

$$\{\varepsilon_{sx^{/}}\} = \begin{Bmatrix} \frac{\partial u'_{sx^{/}}}{\partial x^{/}} - \frac{w_{sx^{/}}}{R_{xx}} \\ \frac{\partial \alpha'_{sx^{/}}}{\partial x^{/}} \\ \frac{\partial \beta_{sx^{/}}}{\partial x^{/}} \\ \alpha'_{sx^{/}} + \frac{\partial w'_{sy^{/}}}{\partial x^{/}} \end{Bmatrix} + \begin{Bmatrix} \frac{1}{2}\left(\frac{\partial w'_{sx^{/}}}{\partial x^{/}} - \frac{u_{sx^{/}}}{R_{xx}}\right)^2 \\ 0 \\ 0 \\ 0 \end{Bmatrix} \qquad (7)$$

The above equation can be written as the following,

$$\{\varepsilon_{sx^{/}}\} = [B]_{x^{/}}\{\delta'_{sx^{/}j}\} \text{ where } [B]_{x^{/}} = [B_{sx^{/}}]^L + \frac{1}{2}[B_{sx^{/}}]^{NL}$$

$[B]_{x^{/}}$ is the same as given by Sahoo and Chakravorty [26].

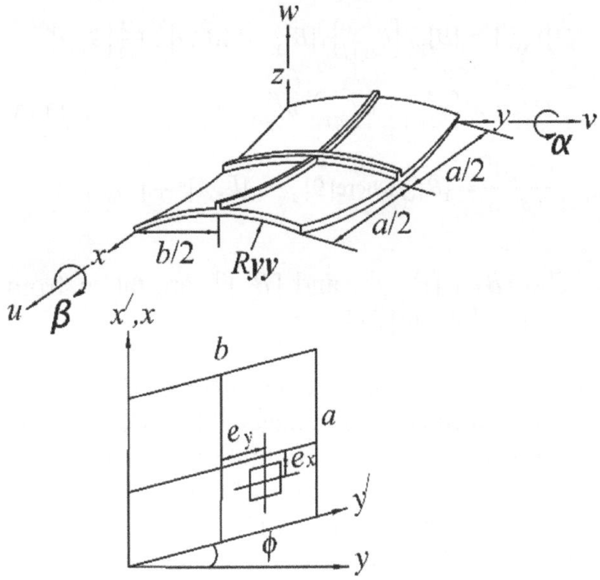

Figure 36.1 Skewed hyperbolic paraboloidal composite shell with orthogonal stiffeners
Source: Authors

$$[B_{sx^{/}}]^{NL} = \begin{bmatrix} \frac{\partial w'_{sx^{/}}}{\partial x} \end{bmatrix} \begin{bmatrix} -\frac{N_j}{R_{xx}} & \frac{\partial N_j^{sx}}{\partial x} & 0 & 0 \\ 0 & 0 & 0 & 0 \\ 0 & 0 & 0 & 0 \\ 0 & 0 & 0 & 0 \end{bmatrix}$$

$$\{\delta'_{sxj}\} = \{u'_{sx^{/}j} \quad w'_{sx^{/}j} \quad \alpha'_{sx^{/}j} \quad \beta'_{sx^{/}j}\}^T \; j = 1 \text{ to } 3 \qquad (8)$$

$$\frac{\partial\{\varepsilon_{sx^{/}}\}^T}{\partial\{\delta'_{sx^{/}}\}} = [\bar{B}]_{x^{/}}, [\bar{B}]_{x^{/}} = [B_{sx^{/}}]^L + [B_{sx^{/}}]^{NL}$$

The displacements of $y^{/}$-stiffener between x' - y' axes and x - y system are related through the following,

$$\begin{Bmatrix} v'_{sy'} \\ w'_{sy'} \\ \alpha'_{sy'} \\ \beta'_{sy'i} \end{Bmatrix} = \begin{bmatrix} sec\phi & 0 & 0 & 0 \\ 0 & 1 & 0 & 0 \\ 0 & 0 & sec\phi & 0 \\ 0 & 0 & -tan\phi & 1 \end{bmatrix} \begin{Bmatrix} v_{sy} \\ w_{sy} \\ \alpha_{sy} \\ \beta_{sy} \end{Bmatrix} \qquad (9)$$

Strains of the $y^{/}$-stiffener are given below,

$$\{\varepsilon_{sy^{/}}\} = \begin{Bmatrix} \frac{\partial v_{sy^{/}}}{\partial y^{/}} - \frac{w_{sy^{/}}}{R_{yy}} \\ \frac{\partial \beta_{sy^{/}}}{\partial y^{/}} \\ \frac{\partial \alpha_{sy^{/}}}{\partial y^{/}} \\ \beta_{sy^{/}} + \frac{\partial w_{sy^{/}}}{\partial y^{/}} \end{Bmatrix} + \begin{Bmatrix} \frac{1}{2}\left(\frac{\partial w_{sy^{/}}}{\partial y^{/}} - \frac{v_{sy^{/}}}{R_{yy}}\right)^2 \\ 0 \\ 0 \\ 0 \end{Bmatrix} \qquad (8)$$

$$\{\varepsilon_{sy/}\} = [B]_{y/}\{\delta'_{sy/j}\}, [B]_{y/} = [B_{sy/}]^L + \frac{1}{2}[B_{sy/}]^{NL}$$

$$\{\delta'_{syi}\} = \{u'_{sy/i} \quad w'_{sy/i} \quad \alpha'_{sy/i} \quad \beta'_{sy/i}\}^T \qquad j=1 \text{ to } 3$$

$$\frac{\partial\{\varepsilon_{sy/}\}^T}{\partial\{\delta'_{sy/}\}} = [\bar{B}]_{y/} \text{where} [\bar{B}]_{y/} = [B_{sy/}]^L + [B_{sy/}]^{NL}$$

The $[\bar{B}]_{y'}$, $[\bar{B}_{sy'}]^{NL}$, and $[B_{sy'}]^{NL}$ are taken from Giri and Bakshi [21].

Force-displacement relationship of shell

The laminated composite material is assumed to act as linearly elastic. The constitutive relationship for the shell and those for the stiffeners are taken up from Sahoo and Chakravorty [27]. The compatibility between shell and stiffeners is ensured through the paper by Bakshi [22].

Governing equations

The governing equation is formulated using Lagrange's equation of motion based on Hamilton's principle for shell panels. The steps to formulate the governing equations of composite shell are the same as given by Bakshi and Giri [28],

$$[M_s]\{\ddot{\delta}_i\} + \int_A [\bar{B}]^T[E]\left([B]^L + \frac{1}{2}[B]^{NL}\right) dA \ \{\delta_i\} = \{Q_e\} (9)$$

where $[M_s]$ = mass matrix of the stiffened shell [27]. From Equation (9), the equations of static bending and free vibration can be developed as it was given by reference [28].

The governing equation of static bending is,

$$[K_s]\{\delta_i\} - \{Q_e\} = \varphi \qquad (10)$$

$$\text{where} [K_s] = \int_A \left\{[\bar{B}]^T[D]\left([B]^L + \frac{1}{2}[B]^{NL}\right) dA\right\}$$

The governing equation of free vibration is given below,

$$[M_e]\ \{\ddot{\delta}_i\} + \int_A [\bar{B}]^T[E]\left([B]^L + \frac{1}{2}[B]^{NL}\right) dA \{\delta_i\} = 0 \qquad (11)$$

Equation (11) can be written as the following:

$$([K] - \omega^2[M])\{\delta_i\} = 0 \qquad (12)$$

Equations (10) and (12) are solved following reference [28]. As given in Bakshi [22], the matrices of stiffeners are obtained. The matrices for stiffened shells are coded following the same paper. The matrices are numerically integrated using 2×2 Gauss quadrature rule. It is adopted for both bending and shear energies. The solutions at element level are assembled for global ones.

Table 36.1 Non-dimensional deflections ($\bar{w} \times 10^4$) of simply supported hyper shell.

Stiffeners	Lamination (degree)	Present approach	Sahoo and Chakravorty [27]
n_x=1, n_y=0	0°/90°	21.567	21.967
	0°/90°/0°	13.589	13.970
n_x=0, n_y=1	0°/90°	21.987	22.096
	0°/90°/0°	25.835	25.935
n_x=1, n_y=1	0°/90°	15.246	15.334
	0°/90°/0°	12.058	12.431

a/b=1, c/a=0.2, h/a=0.01, $G_{12}=G_{13}=0.5E_{22}$, $E_{11}=25E_{22}$, $G_{23}=0.2E_{22}$, $\nu_{12}=\nu_{21}=0.25$

Source: Authors

Problems

The problems solved in this paper have two different objectives. The first set of problems have the purpose of validating the present approach. Those are given in Section 3.1. Apart from these problems, authors' own problems on simply supported, stiffened, skewed hyperbolic paraboloids made of composites are given in Section 3.2.

Validation problems

The first problem computes deflections of simple supported composite skewed hyper shells which have point load at the centre. The deflections are computed for three different orientations of central stiffeners. The results computed using the present code are reported in Table 36.1. The table compares these with the values of Sahoo and Chakravorty [27]. The earlier results were obtained using geometric linearity. The table confirms that the present approach has correctly coded the bending of stiffened composite shell. The footnotes report material properties and dimensions of the shell.

The fundamental frequencies of simply supported composite hyper panels with orthogonal stiffeners

Table 36.2 Non-dimensional fundamental frequencies (ϖ) of simply supported hyper shell.

Stiffeners	Lamination (degree)	Present approach	Sahoo and Chakravorty [26]
n_x=1, n_y=0	0°/90°	6.0987	6.10935
	0°/90°/0°	8.5365	8.73346
n_x=0, n_y=1	0°/90°	6.0578	6.0659
	0°/90°/0°	6.3859	6.4731
n_x=1, n_y=1	0°/90°	7.5897	7.6083
	0°/90°/0°	8.7256	8.7410

a/b=1, c/a=0.2, h/a=0.01, $G_{12}=G_{13}=0.5E_{22}$, $E_{11}=25E_{22}$, $G_{23}=0.2E_{22}$, $v_{12}=v_{21}=0.25$

Source: Authors

are solved as the second problem. Table 36.2 reports values computed by the present code and compare these with the ones given by Sahoo and Chakravorty [26]. The earlier results were obtained using geometric linearity. Thc tablc confirms that present code correctly computes fundamental frequencies of stiffened composite shells. The footnotes report material properties and dimensions of the shell.

The correctness of present non-linear code for bending and vibration problems is established earlier in the paper [28] where outputs of present code were compared with CUF based results reported by Wu et al. [29] and Carrera et al. [30]. The comparison depicted correctness of the present non-linear code. For stiffened skewed panels, the correctness of proposed code for non-linear bending solutions was reported in the paper [21]. For non-linear free vibration, the correctness is confirmed in paper [22]. For the sake of brevity, the graphs are not repeated here.

Additional problems

The corrected code is used to solve the bending and vibration of composite skewed hyperbolic paraboloids under uniform pressure. The panels are simply supported and stiffened along the central chords. The results are given in tabular forms.

Results and discussion

Table 36.3 furnishes the non-dimensional deflections ($\bar{w}$) of uniformly loaded hyperbolic paraboloidal panels skewed at 0°–35° with the constant step of 5°. The deflections are reported for two, three and four layered cross-ply composites. The performances are studied for varying orientations of central stiffeners. For any given lamination and stiffener orientation, the deflections generally increase as the skew angle (ϕ) varies from 0° to 15°. For ϕ > 15°, the deflections decrease as panels get skewed. The table shows that minimum deflections for any laminate depends on orientation of central stiffeners. For n_x=0, n_y=1 stiffeners, the deflections are minimum for ϕ = 35° which is obtained for all laminates. No unified trend can be obtained when panels have atleast one stiffener along x' – axis (n_x=1, n_y=0 and n_x=1, n_y=1). The deflections can be minimum for ϕ = 0° or ϕ = 35° depending on orientations of stiffeners and lamination. For panels which show the minimum deflections for ϕ = 0°, the deflections for ϕ = 35° are marginally higher than the corresponding minimum counterparts. Considering all laminates, the differences between minimum deflection and that for ϕ = 35° is maximum for n_x=1, n_y=1 stiffeners. That maximum difference is less than 7.5%.

If these minor differences are ignored, then panels for all cross-ply composites which are studied in Table 36.3 should have ϕ = 35° to get the preferred bending performances. The symmetric composites offer lower deflections than anti-symmetric ones due to absence of bending-stretching couplings. The 0°/90°/90°/0° laminate skewed at 35° can be preferred to have the minimum deflections. Table 36.4 reports the non-dimensional fundamental frequencies (ϖ) of similarly skewed hyperbolic paraboloidal panels. The variations of composites and stiffener orientations are the same as it is given in Table 36.3. For all laminations and stiffener orientations, the fundamental frequencies continuously increase as the skew angle varies from 0° to 35° and the maximum values are obtained for ϕ = 35°. The maximum frequencies are more than

Table 36.3 Non-dimensional deflections ($\bar{w} = wE_{22}\, h^3/qa^4$) of simply supported hyperbolic-paraboloidal shell under uniform pressure.

Lamination (degree)	ϕ	$a_1/a = 0.0$			$a_1/a = 0.125$			$a_1/a = 0.25$		
		n_x=1, n_y=0	n_x=0, n_y=1	n_x=1, n_y=1	n_x=1, n_y=0	n_x=0, n_y=1	n_x=1, n_y=1	n_x=1, n_y=0	n_x=0, n_y=1	n_x=1, n_y=1
0/90	0	0.9921	0.9864	0.7115	1.0945	1.1111	0.7606	11.1890	1.2436	0.7850
	5	1.1823	1.4009	1.0713	1.2408	1.4990	1.1262	11.1665	1.5885	1.1590
	10	1.3282	1.5286	1.2079	1.3865	1.6147	1.2582	11.1489	1.6892	1.2883
	15	1.3371	1.3967	1.1633	1.3941	1.4546	1.2067	11.1361	1.5116	1.2326
	20	1.2753	1.1970	1.0534	1.3293	1.2386	1.0901	11.1273	1.2887	1.1127
	25	1.1917	1.0166	0.9341	1.2426	1.0447	0.9669	11.1205	1.0965	0.9870
	30	1.1088	0.8635	0.8244	1.1570	0.8848	0.8558	11.1165	0.9360	0.8739
	35	1.0349	0.7384	0.7289	1.0921	0.7574	0.7609	11.1156	0.8065	0.7771
0/90/0	0	0.8598	1.0216	0.6537	0.9657	1.1449	0.6876	10.4097	1.3051	0.7123
	5	0.9546	1.1249	0.8059	1.0019	1.1922	0.8343	10.3862	1.3350	0.8568
	10	1.0244	1.1995	0.8952	1.0652	1.2558	0.9210	10.3696	1.3437	0.9404
	15	1.0493	1.1687	0.9149	1.0852	1.2109	0.9375	10.3559	1.2830	0.9538
	20	1.0376	1.0752	0.8877	1.0697	1.1049	0.9077	10.3461	1.1618	0.9218
	25	1.0013	0.9563	0.8356	1.0327	0.9764	0.8546	10.3402	1.0297	0.8659
	30	0.9526	0.8348	0.7719	0.9856	0.8506	0.7906	10.3363	0.9005	0.7999
	35	0.8997	0.7221	0.7039	0.9377	0.7359	0.7232	10.3353	0.7831	0.7320
0/90/0/90	0	0.9235	0.9211	0.6642	0.9826	0.9898	0.6959	11.1890	1.0906	0.7146
	5	1.0262	1.1121	0.8581	1.0754	1.1800	0.8914	11.1665	1.2420	0.9132
	10	1.1090	1.1999	0.9563	1.1562	1.2587	0.9875	11.1489	1.3147	1.0084
	15	1.1284	1.1492	0.9529	1.1698	1.1949	0.9811	11.1361	1.2412	0.9989
	20	1.0948	1.0321	0.8928	1.1344	1.0661	0.9183	11.1263	1.1029	0.9336
	25	1.0372	0.9008	0.8134	1.0794	0.9262	0.8373	11.1185	0.9559	0.8507
	30	0.9735	0.7785	0.7333	1.0198	0.7975	0.7563	11.1165	0.8284	0.7683
	35	0.9166	0.6721	0.6594	0.9653	0.6867	0.6823	11.1146	0.7190	0.6932
0/90/90/0	0	0.8367	1.0250	0.6467	0.9383	1.1003	0.6756	11.1880	1.2423	0.6943
	5	0.9326	1.1282	0.8063	0.9779	1.1972	0.8357	11.1655	1.3085	0.8532
	10	1.0028	1.1988	0.8922	1.0401	1.2570	0.9190	11.1479	1.3377	0.9336
	15	1.0195	1.1512	0.8986	1.0487	1.1954	0.9219	11.1351	1.2611	0.9328
	20	0.9933	1.0414	0.8564	1.0199	1.0730	0.8767	11.1263	1.1244	0.8855
	25	0.9454	0.9138	0.7930	0.9728	0.9356	0.8113	11.1195	0.9765	0.8190
	30	0.8900	0.7909	0.7236	0.9198	0.8056	0.7410	11.1156	0.8458	0.7478
	35	0.8349	0.6817	0.6555	0.8804	0.6941	0.6728	11.1146	0.7316	0.6791

E_{11} = 142.5 GPa, $G_{12} = G_{13}$ = 4.72 GPa, E_{22} = 9.79 GPa, v_{12} = 0.27, G_{23} = 1.192 GPa, R_{xx}/R_{yy} = -1, a/b = 1, h/a=0.01, $b_{st} = h$, d_{st}/b_{st} = 2

Source: Authors

Table 36.4 Non-dimensional fundamental frequencies [$\varpi = \omega a^2 (\rho/E_{22}h^2)^{1/2}$] of simple supported composite hyperbolic-paraboloidal shell.

Lamination (degree)	ϕ	a_1/a = 0.0			a_1/a = 0.125			a_1/a = 0.25		
		n_x=1, n_y=0	n_x=0, n_y=1	n_x=1, n_y=1	n_x=1, n_y=0	n_x=0, n_y=1	n_x=1, n_y=1	n_x=1, n_y=0	n_x=0, n_y=1	n_x=1, n_y=1
0/90	0	36.454	36.481	36.392	36.282	36.292	36.235	36.131	36.044	36.220
	5	37.640	37.731	37.902	37.469	**38.006**	37.563	37.326	37.373	37.879
	10	41.244	41.969	42.417	41.081	41.815	41.784	40.963	41.647	**42.623**
	15	46.653	48.136	48.858	46.482	47.988	47.817	46.309	47.748	47.640
	20	53.095	55.287	54.944	52.878	**56.399**	54.718	**53.706**	54.726	54.456
	25	60.042	64.467	63.887	59.730	64.060	61.914	**60.666**	63.395	59.164
	30	67.206	72.238	71.510	66.746	71.667	69.149	66.117	70.711	66.552
	35	76.442	78.210	78.941	75.583	77.701	76.548	72.915	77.814	**79.696**
0/90/0	0	46.900	46.799	46.496	46.744	46.643	41.659	46.574	46.436	41.505
	5	47.627	48.143	47.612	**47.669**	47.657	42.899	47.331	47.474	42.597
	10	50.090	51.292	50.155	**50.416**	50.454	46.264	49.780	50.243	46.172
	15	54.004	55.712	54.377	53.843	55.500	51.054	**54.284**	55.185	50.818
	20	58.988	60.142	59.529	**59.878**	**61.052**	56.713	**59.382**	**60.811**	57.160
	25	66.379	66.208	66.857	64.467	65.888	62.962	65.157	**66.757**	63.642
	30	73.250	72.960	73.778	72.634	72.411	69.709	69.882	**73.145**	70.574
	35	80.808	82.372	81.202	77.685	81.304	**81.797**	76.348	79.988	77.957
0/90/0/90	0	44.733	44.750	44.500	44.581	44.591	44.377	44.426	44.393	44.227
	5	45.782	45.784	45.764	45.641	**46.060**	45.485	45.519	45.616	45.314
	10	48.867	49.370	49.595	48.730	49.232	49.070	48.590	49.071	49.101
	15	53.643	54.836	55.310	53.490	54.690	54.422	53.289	54.463	54.242
	20	59.553	62.732	62.144	59.348	62.457	60.813	**59.954**	62.052	60.544
	25	66.171	70.417	69.533	**67.247**	70.014	69.068	65.368	69.207	65.211
	30	75.025	78.214	77.146	74.495	77.423	76.486	72.043	76.495	**77.516**
	35	80.789	86.174	84.898	80.072	85.303	82.384	78.884	83.816	83.408
0/90/90/0	0	46.996	46.931	46.655	46.839	46.770	46.522	46.676	46.569	46.370
	5	47.936	47.868	47.855	47.791	**47.924**	47.549	47.651	47.753	47.258
	10	50.869	51.149	51.388	50.729	**51.620**	50.849	50.568	**51.409**	50.758
	15	55.416	57.125	55.893	55.262	56.929	55.748	**55.648**	56.633	52.374
	20	61.076	63.735	61.824	60.859	63.477	61.609	**61.438**	63.020	58.723
	25	67.477	70.792	69.859	**68.533**	70.367	69.360	**67.782**	69.673	**70.149**
	30	76.285	78.286	77.193	73.985	77.403	74.938	73.072	75.061	75.859
	35	82.085	86.217	84.881	81.173	85.216	82.151	79.926	82.158	83.190

E_{11} = 142.5 GPa, $G_{12} = G_{13}$ = 4.72 GPa, E_{22} = 9.79 GPa, v_{12} = 0.27, G_{23} = 1.192 GPa, R_{xx}/R_{yy} = -1, a/b = 1, h/a=0.01, $b_{st} = h$, d_{st}/b_{st} = 2

Source: Authors

40% higher than those panels for ϕ = 0°, and for all laminations. The symmetric ones perform better than anti-symmetric ones and the 0°/90°/90°/0° composite offers the maximum frequencies among all composites studied in Table 36.4. Hence, the engineers may adopt the 0°/90°/90°/0° panels skewed at 35° for practical facilities. The skew angle of hyperbolic paraboloids is varied here by maintaining rise/span = 0.2 (shallow shell geometry). Since the skewed panels are longer along the y' – axis, for any given ϕ>0°, the panels have greater hogging curvature than the sagging one to maintain the constant rise/span ratio. This is why the skewed panels under downward forces are stiffer and show lower deflections. Due to the greater curvature, the frequencies are higher for skewed panels (ϕ>0°).

Table 36.3 varies the sizes (a_1/a) of concentric cut-outs for constant skew angle and lamination of panels. For practical applications, an optimum cut-out needs to be recommended. The panels with fixed lamination and orientation of stiffener show that the cut-outs yield higher deflections, and solid shells (a_1/a = 0) offer the maximum bending rigidity. For a_1/a = 0.125, the deflections of solid shells increase minimum by 6.45% considering all laminations and stiffeners reported in Table 36.3. Similar panels show a 9.36% increase of deflections for a_1/a = 0.25. Hence, the 0°/90°/90°/0° panels with ϕ = 35° can have a_1/a = 0.125. For fixed skew angle, Table 36.4 shows that several panels with cut-outs offer higher fundamental frequencies than the solid ones. Table 36.4 highlights these performances in bold letters. For panels a_1/a = 0.125, a total of 12 such performances are highlighted. For panels a_1/a = 0.25, these performances are 16 in numbers. Cut-outs decrease both mass and stiffness of a shell structure. The fundamental frequencies decrease with decrease of stiffness and increase with decrease of mass. This is why the panels with cut-outs can offer superior free vibration performances than the solid ones. Comparing the free vibration performances of shells with cut-outs, the panels with a_1/a = 0.125, 0°/90°/90°/0° composite and ϕ = 35° show the maximum fundamental frequencies. Naturally, if cut-outs are to be provided for practical applications, then a_1/a ratio should not exceed 0.125 considering the results furnished in Tables 36.3 and 36.4. The results given in these tables can be interesting to civil engineers as cut-outs can control the cost, weight and resonating frequency of shell structures.

Three different orientations of central stiffeners are taken up in Table 36.3 for all skewed panels. The table clearly confirms that for any given composite and a_1/a ratio, the n_x=1, n_y=1 stiffeners should be preferred to minimise the deflections. The panels with n_x = 1, n_y = 0 stiffeners should be avoided as these show the highest deflections for all laminations and a_1/a ratios. It is quite expected as the stiffeners along sagging curvature of hyperbolic paraboloids lead to relatively flexible panels for downward pressure. From the free vibration point of view, Table 36.4 shows that n_x=0, n_y=1 stiffeners generally offer maximum fundamental frequencies for all skew angles. There are a few exceptions where the n_x=1, n_y=1 stiffeners offer maximum frequencies but those are marginally higher (within 5%) than what was obtained for n_x=0, n_y=1 stiffeners. A further investigation from Table 36.3 reveals that if the engineers adopt n_x=0, n_y=1 stiffeners for all skewed forms then the deflections are to be compromised by less than 10%. Hence, civil engineers may choose the n_x=0, n_y=1 stiffeners to fabricate the skewed (ϕ = 35°) hyperbolic paraboloids with a_1/a = 0.125 and 0°/90°/90°/0° composite so that the fundamental frequencies are maximised and compromises with deflections are within 10%.

The variations of deflections and fundamental frequencies are studied further for varying eccentricities (refer to Figure 36.1) of the cut-out (a_1/a= 0.125). The eccentricities are varied along x'-axis (e_x) and y'-axis (e_y). The hyperbolic paraboloids with 0°/90°/90°/0° composite, n_x=0, n_y=1 stiffeners skewed at 35° are selected for this investigation. The deflections are reported in Table 36.5. The table clearly shows that bending performance obtained for concentric cut-outs can be lowered by varying the eccentricities. The best performance is obtained for e_x/a= 0.125 and

Table 36.5 Non-dimensional deflections of panels for a_1/a = 0.125, 0°/90°/90°/0° composite, ϕ = 35° and n_x=0, n_y=1 stiffeners.

e_y/b	e_x/a			
	0	**0.125**	**0.25**	**0.375**
0	0.6941	0.7184	0.8263	0.7956
0.125	0.6843	0.7335	0.7410	0.7040
0.25	0.6858	0.7463	0.6908	0.6809
0.375	0.8997	0.6872	0.7683	0.6924

Source: Authors

Table 36.6 Non-dimensional fundamental frequency of panels for $a_1/a = 0.125$, 0°/90°/90°/0° composite, $\phi = 35°$ and n_x=0, n_y=1 stiffeners.

e_y/b	e_x/a			
	0	**0.125**	**0.25**	**0.375**
0	85.2161	85.1590	85.7983	86.0291
0.125	85.1158	85.5344	87.0286	86.7262
0.25	85.3962	85.9091	86.5237	84.2571
0.375	79.1711	80.1075	80.9895	84.0038

Source: Authors

$e_y/b = 0.375$ which offers the minimum deflection. The best performance is an improvement over what obtained for panels with concentric cut-out. Table 36.6 reports the frequencies for varying eccentricities. The maximum value is obtained for $e_x/a = 0.25$ and $e_y/b = 0.125$ which is higher than that yielded by panels with concentric cut-out. Hence, the best bending and vibration performances cannot be obtained simultaneously. For practical situations, eccentricities are to be optimised which offers improvement over both performances simultaneously. Tables 36.5 and 36.6 indicate that for $e_x/a = 0$ and $e_y/b = 0.25$, the bending and vibration performances of hyperbolic paraboloids improve over those obtained for concentric cut-outs. Hence, the engineers may adopt these eccentricities for the cut-out.

Conclusion

The correctness of the present approach is confirmed through benchmark problems. The bending and vibration performances of hyperbolic paraboloids are found to be the best when 0°/90°/90°/0° composite is selected among the ones studied here. The performances can be improved further if n_x=0, n_y=1 stiffeners are adopted, and the panels are skewed to 35° from the y'-axis. In practical situations, the cut-outs are generally adopted for natural air, controlling the weight and resonating frequency. For such purposes, a cut-out size of $a_1/a = 0.125$ may be adopted which compromises minimum with deflections and improves the fundamental frequency compared to solid shells. The cut-out can be located eccentrically ($e_x/a = 0$ and $e_y/b = 0.25$) which leads to the best bending and vibration responses simultaneously.

Acknowledgement

The authors gratefully acknowledge the financial assistance by M/s Mazagaon Dock Shipbuilders to support this research.

References

[1] Bakshi, K., Chakravorty, D. First ply failure loads of composite conoidal shell roofs with varying lamination. *Mech. Adv. Mater. Struct.*, 2015;22(12):978–987.

[2] Chakravorty, D., Bandyopadhyay, J. N., Sinha, P. K. Free vibration analysis of point-supported laminated composite doubly curved shells—A finite element approach. *Comput. Struct.*, 1995;54(2):191–198.

[3] Ortega, N. F., Robles, S. I. The design of hyperbolic paraboloids based on their mechanical behaviour. *Thin-Walled Struct.*, 2003;41(8):769–784.

[4] Matsunaga, H. Vibration and stability of cross-ply laminated composite shallow shells subjected to in-plane stresses. *Compos. Struct.*, 2007;78(3):377–391.

[5] Kumar, A., Bhargava, P., Chakrabarti, A. Vibration of laminated composite skew hypar shells using higher order theory. *Thin-Walled Struct.*, 2013;63:82–90.

[6] Rian, I. M. Fractal-based computational modeling and shape transition of a hyperbolic paraboloid shell structure. *Nexus Netw. J.*, 2018;20(2):437–458.

[7] Li, H., Pang, F., Li, Y., Gao, C. Application of first-order shear deformation theory for the vibration analysis of functionally graded doubly-curved shells of revolution. *Compos. Struct.*, 2019;212:22–42.

[8] Zhao, J., Xie, F., Wang, A., Shuai, C., Tang, J., Wang, Q. Vibration behaviour of the functionally graded porous (FGP) doubly-curved panels and shells of revolution by using a semi-analytical method. *Compos. Part B-Eng.*, 2019;157:219–238.

[9] Singh, V. K., Panda, S. K. Nonlinear free vibration analysis of single/doubly curved composite shallow shell panels. *Thin-Walled Struct.*, 2014;85:341–349.

[10] Ghosh, A., Chakravorty, D. First ply failure analysis of laminated composite thin hypar shells using nonlinear finite element approach. *Thin-Walled Struct.*, 2018;131:736–745.

[11] Sofiyev, A. H., Turan, F., Zerin, Z. Large-amplitude vibration of functionally graded orthotropic double-curved shallow spherical and hyperboloidal shells. *Int. J. Press. Vessels Pip.*, 2020;188:104235.

[12] Yusufoglu, E., Avey, M. Nonlinear dynamic behaviour of hyperbolic paraboloidal shells reinforced by carbon nanotubes with various distributions. *J. Appl. Comput. Mech.*, 2021;7(2):913–921.

[13] Singha, M. K., Daripa, R. Nonlinear vibration of symmetrically laminated composite skew plates by finite element method. *Int. J. Non-Linear Mech.*, 2007;42:1144–1152.

[14] Upadhyay, A. K., Shukla, K. K. Large deformation flexural behaviour of laminated composite skew plates: An analytical approach. *Compos. Struct.*, 2012;94:3722–3735.

[15] Zhang, L. W., Xiao, L. N. Mechanical behaviour of laminated CNT-reinforced composite skew plates subjected to dynamic loading. *Compos. Part B Engg.*, 2017;122:219–230.

[16] Kallannavar, V., Kumaran, B., Kattimani, S. C. Effect of temperature and moisture on free vibration characteristics of skew laminated hybrid composite and sandwich plates. *Thin-Walled Struct.*, 2020;157:107113.

[17] Katariya, P. V., Mehar, K., Panda, S. K. Nonlinear dynamic responses of layered skew sandwich composite structure and experimental validation. *Int. J. Non-Linear Mech.*, 2020;125:103527.

[18] Sengar, V., Nynaru, M., Watts, G., Kumar, R., Singh, S. Postbuckled vibration behaviour of skew sandwich plates with metal foam core under arbitrary edge compressive loads using isogeometric approach. *Thin-Walled Struct.*, 2023;184:110524.

[19] Kallannavar, V., Kattimani, S. Effect of temperature and porosity on free vibration characteristics of a doubly-curved skew laminated sandwich composite structures with 3D printed PLA core. *Thin-Walled Struct.*, 2023;182:110263.

[20] Jahanbazi, R., Kiani, Y. Vibrations of skew cylindrical shells made of FG-GPLRCs under rapid surface heating. *Engg. Anal. Bound. Elem.*, 2024;165:105785.

[21] Giri, A., Bakshi, K. Geometrically nonlinear bending of stiffened composite skewed cylindrical shells under transverse pressure. *J. Strain Anal. Eng. Des.*, 2024;59(2):85–99.

[22] Bakshi, K. A numerical study on nonlinear vibrations of laminated composite singly curved stiffened shells. *Compos. Struct.*, 2021;278:114718.

[23] Huang, S., Qiao, P. A new semi-analytical method for nonlinear stability analysis of stiffened laminated composite doubly curved shallow shells. *Compos. Struct.*, 2020;251:112526.

[24] Yang, Y., Ye, T., Li, S., Jin, G., Chen, Y., Zhong, S. A semi-analytical method for vibro-acoustic characteristics of orthogonal stiffened laminated cylindrical shells. *J. Sound Vib.*, 2023;558:117770.

[25] Sahoo, S, Chakravorty, D. Finite element bending behaviour of composite hyperbolic paraboloidal shells with various edge conditions. *J. Strain Anal. Engg. Des.*, 2004;39(5):499–513.

[26] Sahoo, S, Chakravorty, D. Stiffened composite hypar shell roofs under free vibration: Behaviour and optimization aids. *J. Sound Vib.*, 2006;295(1–2):362–377.

[27] Sahoo, S, Chakravorty, D. Bending of composite stiffened hypar shell roofs under point load. *J. Engg. Mech.*, 2008;134(6):441–454.

[28] Bakshi, K., Giri, A. Application of von-Karman model for laminated composite singly curved shells with non-uniform boundary conditions. *Mech. Adv. Mater. Struct.*, 2024:1–11.

[29] Wu, B., Pagani, A., Chen, W. Q., Carrera, E. Geometrically nonlinear refined shell theories by Carrera Unified Formulation. *Mech. Adv. Mater. Struct.*, 2021;28:1721–1741.

[30] Carrera, E., Pagani, A., Augello, R., Wu, B. Popular benchmarks of nonlinear shell analysis solved by 1D and 2D CUF-based finite elements. *Mech. Adv. Mater. Struct.*, 2020;27(13):1098–1109.

37 Influence of constitutive relation on stress eigenvectors of homogeneous deep beam subjected to point loads

Gaurav Chandra Bridhani[a] *and U. Saravanan*[b]

Department of Civil Engineering, Indian Institute of Technology Madras, Chennai, India

Abstract

This paper aims to conduct a numerical study on a two-dimensional homogenous deep beam subjected to equally spaced four-point bending. The existing works study the behaviour of deep beams for linearised elasticity-based concrete models wherein Cauchy stress and linearised strain are explicitly related. The strut angle in the strut-and-tie method is used to analyse and design deep beams, ideally representing the stress eigenvectors. Further, the literature suggests that the constitutive relation used for modeling plain concrete has no bearing on strut angle. Recently, a non-linear, non-dissipative constitutive model has been proposed for plain concrete having linearised strain as a non-invertible function of Cauchy stress. Additionally, the analytical solution of the deep beam boundary value problem for the non-linear elastic plain concrete model is infeasible. Therefore, an existing numerical scheme with multifield formulation and appropriate element-free basis functions is employed to study deep beams. The existing multifield formulation assumes divergence-free *Airy's* stress potential function and displacement field as primary unknowns. The Airy's function is written as a linear combination of Hermite-type moving least square approximants. The displacement field is assumed using a linear combination of Lagrange Moving least squares approximants. The Cauchy stress is obtained from Airy's function, and linearised strain is computed from the assumed displacement field satisfying strain-displacement relationship point wise. The constitutive relation between stress and linearised strain is satisfied in a weak integral sense such that the assumed stress and displacement field satisfy traction and displacement boundary conditions, respectively. The unknown coefficients in the assumed stress function and displacement field are obtained by casting the boundary value problem into a minimisation problem with the Frobenius norm of the constitutive relation as the objective function, and the traction and displacement boundary conditions are enforced as constraints. Finally, the numerically computed stress eigenvectors are obtained and compared for linear elastic Hooke's model and non-linear elastic constitutive model obtained from an implicit constitutive theory framework.

Keywords: Deep beam, non-linear elastic models, element-free methods, moving least squares method (MLS), Hermite-type moving least squares method (HMLS), minimisation, Strut-and-tie method

Introduction

Concrete deep beams are critical structural elements that resist heavier concentrated or distributed loads. The deep beams are mainly used in corbels, transfer girders, and pile caps. Geometrically, deep beams differ from shallow beams in terms of span-to-depth ratio and shear span to overall depth ratio. The American Concrete Institute ACI 318:19 [1] defines a deep beam as a beam capable of transferring the load through a strut-tie action with a ratio of clear span to overall depth ≤ 4 and ratio of shear span to overall depth ≤ 2. Further, the mechanics of load transfer is different in deep beams than in shallow beams. The behaviour of deep beams under transverse loading is governed by flexure and shear. Due to significant shear stress generated in deep beams during bending, the plane section does not remain plane. High shear also causes non-linear variation of normal stress over the depth of the beam.

In order to analyse and design reinforced concrete deep beams, strut-and-tie model based on the truss analogy of beams exists in the literature [2, 3]. Ritter [4] and Mörsch [5] first presented the truss analogy of beams. Later, Leonhardt [6], Rüsch [7], Kupfer [8], and others contributed to improving the truss analogy of beams by creating the scientific basis for it. Extensive literature exists on applying the strut-and-tie model for designing deep beams. The "D" or disturbed region is subjected to high shear stress and causes non-linear strain distribution. The

[a]gbridhani@gmail.com, [b]saran@iitm.ac.in

DOI: 10.1201/9781003770442-37

disturbed region in the deep beam is identified by sudden changes in stress values and non-smooth stress trajectories [2]. Therefore, the strut-and-tie method is employed to analyse and design the disturbed region.

Struts are reinforced or unreinforced concrete regions subjected to pure compression and inclined at *strut angle* from the connecting ties, as shown in Figure *37.1*. Theoretically, the strut angle represents the Eigen direction of the stress tensor obtained at an appropriate location. The strut angle is an essential parameter in the design of deep beams as it is utilised to determine the load in concrete struts and steel ties. Therefore, determining the correct strut angle value is vital for the analysis and design of deep beams.

In the existing literature, the strut angle can be obtained either by elastic stress analysis or the *Load Path Method* [2]. The analytical stress analysis of deep beams for a newer class of elastic constitutive relations with a general loading is non-viable. Further, the load path method for a general loading condition assumes a linear stress distribution in the region adjoining the disturbed region. Moreover, in existing literature, the strut angle is assumed empirically and sometimes depends upon the choice of analyst [7].

Currently, the concrete struts are modelled using classical linearised elasticity-based constitutive models wherein the stress and strain are related explicitly. However, Rajagopal [10, 11] provided an essential framework to develop a general representation of implicit constitutive models between Cauchy stress and linearised strain for elastic bodies. In a recent work by Gokulnath et al. [10], a general

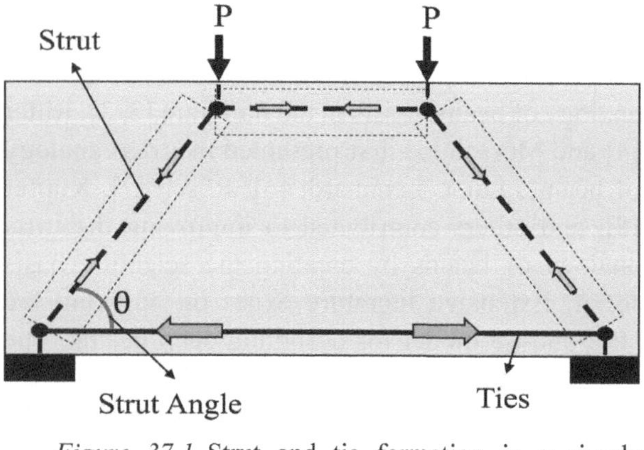

Figure 37.1 Strut and tie formation in a simply supported deep beam subjected to four-point bending
Source: Authors

representation of an implicit constitutive relation between Cauchy stress and Henky strain has been proposed, satisfying the second law of thermodynamics. A three-dimensional constitutive relation has been proposed to capture the non-linear non-dissipative response of plain concrete, wherein the linearised strain is a non-linear, non-invertible function of Cauchy stress [11]. Hence, there is a need to estimate and study the influence of constitutive relations on the strut angle in deep beams.

This paper aims to conduct a numerical study on a two-dimensional homogeneous deep beam subjected to four-point bending for linear elastic and non-linear elastic [11] constitutive models by assuming a divergence-free stress potential function satisfying traction boundary condition and kinematically admissible displacement field satisfying displacement boundary condition as primary unknowns [12]. The constitutive relation between stress and linearised strain is satisfied in a weak integral sense. Contrary to the finite element basis function employed in the work did by Shankar et al. [12], this study approximates stress potential and displacement field using Hermite-type MLS and Lagrangian MLS basis functions, respectively. The boundary value problem is cast into a minimisation problem with the Frobenius norm of the constitutive relation as the objective function, and the boundary conditions are enforced as constraints. Next, the stress eigenvectors are obtained and compared for linear and non-linear elastic constitutive models obtained from an implicit constitutive theory framework.

Material models

This work employs two constitutive models to solve the deep beam boundary value problem numerically. First, the linear elastic Hooke's law model is given as:

$$\mathbf{f}(\boldsymbol{\epsilon}, \boldsymbol{\sigma}) = \boldsymbol{\epsilon} + K_1 \frac{\nu}{E}\mathbf{1} - \frac{(1+\nu)}{E}\boldsymbol{\sigma} = \mathbf{0}, \tag{1}$$

where $K_1 = tr(\sigma)$, and ν and E are Poisson's ratio and Young's modulus, respectively. The value of material parameters is assumed as $\nu = 0.17$ and $E = 26{,}164$ MPa. The second model used to study the deep beam is a non-linear, non-dissipative model for plain concrete developed by Gokulnath et al. [11]. The general representation of the model is given as follows:

$$\mathbf{f}(\boldsymbol{\epsilon}, \boldsymbol{\sigma}) = \boldsymbol{\epsilon} - \alpha_0\mathbf{1} - \alpha_1\boldsymbol{\sigma} = \mathbf{0}, \tag{2}$$

where the coefficients are defined as $\alpha_0 = 2m_1K_4K_2^{1/2} + 2m_3K_4K_2^2 + 4m_5(K_4^2 - 1)K_4K_2^2$ and $\alpha_1 = 2m_2 + K_2^{3/2}(3m_3K_4^2 + 2m_4 + (K_4^4 - 6K_4^2 + 5)m_5)$. The stress invariants are defined as: $K_2 = tr(\boldsymbol{\sigma}^2)$ and $K_4 = K_1/\sqrt{K_2}$, and **1** and **0** represent unit and null matrices of appropriate dimension. The model presented in Equation (2) is curve fitted with uniaxial and equal biaxial compression test data for normal strength concrete (NSC) from [13] having 28 days characteristics cylinder strength of 38.1 MPa. The material parameters m_1 to m_5 for the normal strength grade concrete is provided in the Table 37.1.

The linear elastic model is tangent to the stress-strain curve of the NSC model at zero stress and strain levels.

Multifield formulation

Shankar et al. [12] present a multifield formulation for the mixed finite element method within the implicit constitutive framework by considering both displacement and stress fields as unknowns. The stress field is obtained by assuming divergence-free *Airy's* stress function [12]. Also, for stress to be continuous over the domain, Airy's function should be $\mathcal{C}^2$ continuous. Recent work by Bridhani and Saravanan [14] provides a general framework for applying the multifield formulation presented in the work by Shankar et al. [12] for the element-free approximation of field variables. A detailed discussion on multifield formulation is provided in the study by Bridhani and Saravanan [14]; however, for the sake of completeness, multifield formulation is briefly discussed.

Computation of stress field

This work assumes the Airy's stress function Φ as:

$$\Phi = \sum_{i=1}^{6N} \phi_i \Psi_s^i(\mathbf{x}), \tag{3}$$

Table 37.1 Material parameters of the normal strength grade concrete (NSC).

Concrete type	$m_1 \times 10^{-6}$ MPa^{-1}	$m_2 \times 10^{-5}$ MPa^{-1}	$m_3 \times 10^{-10}$ MPa^{-4}	$m_4 \times 10^{-10}$ MPa^{-4}	$m_5 \times 10^{-11}$ MPa^{-4}
NSC	-3.29	2.24	1.09	-0.26	1.12

Source: Authors

from which the Cauchy stress components for plane stress condition can be derived in terms of Airy's stress potential as:

$$\sigma_{xx} = \sum_{i=1}^{6N} \frac{\partial^2 \Psi_s^i}{\partial y^2} \phi_s^i, \sigma_{yy} = \sum_{i=1}^{6N} \frac{\partial^2 \Psi_s^i}{\partial x^2} \phi_s^i, \quad \sigma_{xy} = \sum_{i=1}^{6N} -\frac{\partial^2 \Psi_s^i}{\partial x \partial y} \phi_s^i, \tag{4}$$

where N is the total number of field nodes.

Computation of strain field

The current study assumes the displacement field **u** as:

$$\mathbf{u} = \sum_{i=1}^{N} \mathrm{d}_i^j \Psi_d^i(\mathbf{x}) \mathbf{e}_j, \tag{5}$$

where $j = \{x, y\}$. Therefore, the linearised strain components are obtained using the moving least squares approximation as:

$$\epsilon_{xx} = \sum_{i=1}^{N} \frac{\partial \Psi_d^i}{\partial x} d_i^x, \ \epsilon_{yy} = \sum_{i=1}^{N} \frac{\partial \Psi_d^i}{\partial y} d_i^y, \quad 2\epsilon_{xy} = \sum_{i=1}^{N} \left(\frac{\partial \Psi_d^i}{\partial x} d_i^y + \frac{\partial \Psi_d^i}{\partial y} d_i^x \right). \tag{6}$$

Governing equations

A detailed mathematical formulation of the governing equation, boundary conditions, and the resulting formulation is presented in the work by Bridhani and Saravanan [14]. The governing equations are briefly described here. For a body under static equilibrium

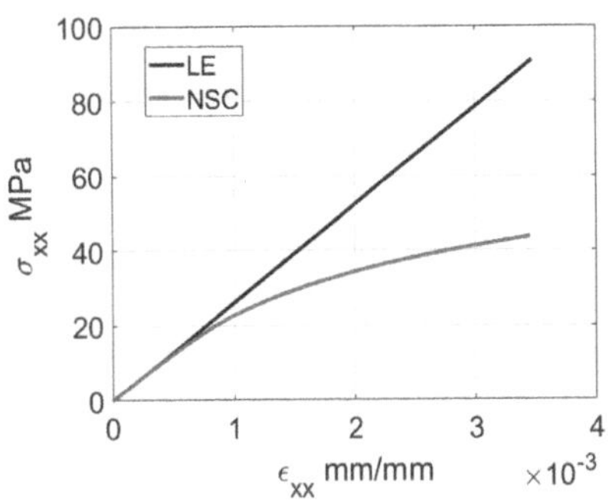

Figure 37.2 Plot showing stress-strain drawn in compression for linear elastic Hooke's model and non-linear elastic constitutive model for different grades of concrete

Source: Authors

with zero body force and subjected to mixed boundary conditions, the stress ($\boldsymbol{\sigma}$) and displacement ($\mathbf{u}$) are governed by

$$div(\boldsymbol{\sigma}) = \mathbf{0}, \forall\, \mathbf{x} \in \mathfrak{B}, \tag{7}$$

$$\boldsymbol{\sigma}\mathbf{n} = \tilde{\mathbf{t}}_{(\mathbf{n})}, \forall\, \mathbf{x} \in \partial\mathfrak{B}^{\sigma}, \tag{8}$$

$$\mathbf{u}(\mathbf{x}) = \tilde{\mathbf{u}}_b, \forall\, \mathbf{x} \in \partial\mathfrak{B}^{u}, \tag{9}$$

$$\mathbf{f}(\boldsymbol{\epsilon}, \boldsymbol{\sigma}) = \mathbf{0}, \forall\, \mathbf{x} \in \mathfrak{B} \tag{10}$$

where $\mathbf{f}$ is implicit constitutive relation, $\mathbf{n}$ is normal to the boundary at point $\mathbf{x}$, $\tilde{\mathbf{t}}_{(\mathbf{n})}$ denotes the specified boundary traction on part $\partial\mathfrak{B}$, $\tilde{\mathbf{u}}_b$ is the prescribed boundary displacement on the part $\partial\mathfrak{B}^u$ of the boundary $\partial\mathfrak{B}$. Figure 37.2.

The six stress coefficients ϕ^i's corresponding to each node are linearly related through traction boundary conditions. Further, wherever displacements are prescribed, the corresponding traction to enforce that displacement is considered unknown. Thus, the traction boundary conditions are prescribed over the entire boundary. Further, boundary over which the displacement is specified, there exists a linear relation between d_i's and the prescribed displacement, $\tilde{\mathbf{u}}_b$. Thus, the initial guess for the ϕ^i's and d_i's is chosen such that the linear constraint due to boundary conditions are met in a least square sense.

Then, for these guess values of ϕ^i's and d_i's the stress and strain are obtained and substituted in the implicit constitutive relation Equation (10). The error in satisfying the constitutive relation is obtained as:

$$\delta = \int_a \sqrt{\mathbf{f}(\boldsymbol{\epsilon}, \boldsymbol{\sigma}) \cdot \mathbf{f}(\boldsymbol{\epsilon}, \boldsymbol{\sigma})}\, da, \tag{11}$$

where δ represents the area integration of the Frobenius norm of the implicit constitutive model.

To arrive at the next guess value of the unknown coefficient, the Frobenius norm is minimised with respect to the unknowns subject to the satisfaction of boundary conditions. Due to the numerical approximation involved, δ does not approach precisely zero. However, to obtain the optimum solution within the function space of element-free basis functions, we seek to find within some specified tolerance. Therefore, the problem is cast into a non-linear constrained minimisation problem with Frobenius norm as the objective function and boundary conditions as constraints. A script is developed in MATLAB® and implemented using the inbuilt function *fmincon* to solve the problem numerically.

Element-free approximation

Some of the shortcomings of mesh-based methods, such as mesh-dependent generation of approximation function, remeshing involved in solving problems with changing domain, and inability to construct higher-order basis functions for an arbitrarily-shaped quadrilateral domain, gave impetus to element-free approximation. However, approximating function and its derivatives with high accuracy at a field node is still challenging. Recent work by Bridhani and Saravanan [15] suggests that the Hermite-type MLS method is able to approximate function and its derivatives reasonably well at a node. Therefore, following the discussion in the work by Bridhani and Saravanan [15], this work approximates the Airy's stress function using $\mathcal{C}^2$ continuous Hermite-type moving least squares method. The displacement field is approximated using $\mathcal{C}^1$ continuous Lagrange moving least squares (MLS) method and is discussed next.

Approximation of displacement field

The introduction of the MLS method in solid mechanics has been pioneered by Belytschko et al. [16] to approximate the displacement field within a Galerkin-based framework. In this work, a brief discussion on the derivation of MLS is conducted. A thorough derivation of the MLS method utilised in this work is done in the work by Bridhani and Saravanan [14, 15]. For any arbitrary scalar function u: $\Omega \rightarrow \mathbb{R}$ with its approximation u^h: $\Omega \rightarrow \mathbb{R}$ around a point $\mathbf{x}_0$ and evaluated at an arbitrary point $\mathbf{x}$ is given as:

$$u^h(\mathrm{x}, \mathrm{x}_0) = \mathbf{p}^T(\mathbf{x})\mathbf{a}(\mathbf{x}_0), \tag{12}$$

where $\mathbf{p}$ is a complete polynomial of order m, and $\mathbf{a}$ is the vector of unknown coefficients. This work uses a quadratic polynomial to approximate the displacement field. Further, the unknown vector $\mathbf{a}$ is obtained by minimising the discrete weighted L_2 norm. The quartic spline weighting function is used for approximation. The support domain size is taken as $d_u = \alpha_u d_{avg}$, where d_{avg} is the mean spacing between nodes and α_u is a dimensionless factor. Finally, the approximation u^h is derived as:

$$u^h(\mathbf{x}, \mathbf{x}_0) = \mathbf{p}^T(\mathbf{x})\mathbf{A}^{-1}(\mathbf{x}_0)\mathbf{B}(\mathbf{x}_0)\hat{\mathbf{u}} = \boldsymbol{\Psi}_d(\mathbf{x})\hat{\mathbf{u}}, \tag{13}$$

where $\mathbf{A}$ and $\mathbf{B}$ are matrices defined in terms of polynomial and weight function, $\boldsymbol{\Psi}_d$ is the displacement basis function, and $\hat{\mathbf{u}}$ is the vector of unknown nodal coefficients.

Approximation of Airy's stress function

In order to obtain the stress components from Airy's stress function, the function Φ and its first ($\Phi_{,x}$, $\Phi_{,y}$) and second-order derivatives ($\Phi_{,xx}$, $\Phi_{,yy}$, $\Phi_{,xy}$) are approximated using C^2 Hermite-type MLS method. Therefore, a total of six unknowns are approximated at each node. The mathematical background of Hermite-type MLS is presented in the work by Komargodski and Levin [17]. A detailed derivation of Hermite-type MLS is given in the work by Bridhani and Saravanan [14, 15].

Let a scalar-valued vector function Φ: $\Omega \rightarrow \mathbb{R}$ be approximated as Φ^h: $\Omega \rightarrow \mathbb{R}$ using the MLS method about an arbitrary point $\mathbf{x}_0$ and evaluated at as:

$$\Phi^h(\mathbf{x}) = \mathbf{p}^T(\mathbf{x})\tilde{\mathbf{a}}(\mathbf{x}_0), \tag{14}$$

where $\tilde{\mathbf{a}}$ is the vector of unknown coefficient, and $\mathbf{p}$ is a fifth-degree polynomial function having 21 terms. To obtain the unknown coefficients in the least square sense, the approximation in Equation (14) is evaluated at n nodes inside the support domain of point $\mathbf{x}_0$ such that $n > 21$ as:

$$\mathbf{\Phi}^h = \tilde{\mathbf{P}}\tilde{\mathbf{a}}, \tag{15}$$

where $\tilde{\mathbf{P}}$ is the matrix of polynomial function $\mathbf{p}$ evaluated at n nodes. Next, the first and second derivatives of Φ^h is obtained as:

$$\begin{aligned}\mathbf{\Phi}^h_{,x} = \tilde{\mathbf{P}}_{,x}\tilde{\mathbf{a}}, \mathbf{\Phi}^h_{,y} = \tilde{\mathbf{P}}_{,y}\tilde{\mathbf{a}}, \mathbf{\Phi}^h_{,xx} = \tilde{\mathbf{P}}_{,xx}\tilde{\mathbf{a}},\\ \mathbf{\Phi}^h_{,yy} = \tilde{\mathbf{P}}_{,yy}\tilde{\mathbf{a}}, \mathbf{\Phi}^h_{,xy} - \tilde{\mathbf{P}}_{,xy}\tilde{\mathbf{a}}\,,\end{aligned} \tag{16}$$

and combined in order to formulate the Hermite-type approximation as:

$$\begin{aligned}\hat{\mathbf{\Phi}}^h &= \hat{\mathbf{P}}\tilde{\mathbf{a}},\\ \hat{\mathbf{\Phi}}^h &= \left[\mathbf{\Phi}^h\ \mathbf{\Phi}^h_{,x}\ \mathbf{\Phi}^h_{,y}\ \mathbf{\Phi}^h_{,xx}\ \mathbf{\Phi}^h_{,yy}\ \ \mathbf{\Phi}^h_{,xy}\right]^T,\end{aligned} \tag{17}$$

and $\hat{\mathbf{P}}$ is matrix of dimension $6n \times m$ and is obtained using Equations (15) and (16). To find the unknowns $\tilde{\mathbf{a}}$, discrete weighted L_2 norm is minimised with respect to all six degrees of freedom at each node. A new fifth-order spline weighting function is developed and represented as:

$$\widehat{W}(r) = \begin{cases}1 - 10r^2 + 20r^3 - 15r^4 + 4r^5, 0 \le r \le 1\\ 0, r > 1\end{cases}, \tag{18}$$

where $r = \mathbf{x} - \mathbf{x}_i / d_s$, $d_s = \alpha_\sigma d_{avg}$ is the support domain size and α_σ is a dimensionless factor.

Finally, the approximation of Airy's stress function using Hermite-type moving least squares method is given as:

$$\Phi^h(\mathbf{x}_i) = \mathbf{p}^T(\mathbf{x}_i)\hat{\mathbf{A}}^{-1}\hat{\mathbf{B}}\boldsymbol{\phi}_i = \mathbf{\Psi}_s(\mathbf{x}_i)\boldsymbol{\phi}_i, \tag{19}$$

where matrices $\hat{\mathbf{A}}$ and $\hat{\mathbf{B}}$ are functions of weight function and polynomial function $\hat{\mathbf{P}}$, $\mathbf{\Psi}_s$ is the stress potential basis function and ϕ_i's are the six unknowns per node. The first and second derivatives of the element-free approximation in Equation (19) can be obtained by usual differentiation.

Boundary value problem

The deep beam boundary value problem studied in this work has traction and displacement boundary conditions. The applied point load is converted into an equivalent distributed load of triangular shape over a small distance with maximum value at the load application point. The displacement is restrained on the left and right edges of the beam, as shown in Figure 37.3, so that the simply supported boundary conditions are reproduced. A detailed derivation of the procedure to enforce boundary conditions is presented in Bridhani and Saravanan [14]. However, a brief discussion on boundary conditions is presented here for completeness. The traction boundary condition is enforced by satisfying the following three conditions on each boundary side of the body: (1) equivalence between the traction applied on each field node and the traction obtained from assumed Airy's function, (2) equivalence between the net boundary force due to applied traction and the net force due to assumed stress field, and, (3) equivalence between the net moment due to applied traction and the moment due to assumed stress field. The traction condition results in seven linear equations relating the unknown stress coefficients to the known applied traction.

Similarly, the assumed displacement field at the boundary field node is equated to the prescribed displacement to satisfy the displacement boundary condition. Unlike the existing Galerkin-based element-free methods, the assumed displacement can be equated to the applied displacement in the current method since the unknowns in the assumed displacement field do not represent nodal displacement. Finally, the traction and displacement boundary conditions relate the unknown stress and displacement coefficients to the applied traction and displacement. Hence, the boundary conditions provide a constraint over the values of the unknown coefficient ϕ^{i}'s and d_i's.

Results

A homogeneous deep beam of dimension L = 4 m, and D = 1.5 m is subjected to a four-point bending, as shown in Figure 37.3. The shear span is a = 1.5 m for both loads. The deep beam problem is solved numerically for linear elastic Hooke's law and non-linear elastic model for plain concrete [11] with normal strength grades of concrete, as reported in Table 37.1. The loading applied for linear elastic is equally proportioned to the 28-day characteristics cylinder strength of non-linear elastic models and is presented in Table 37.2.

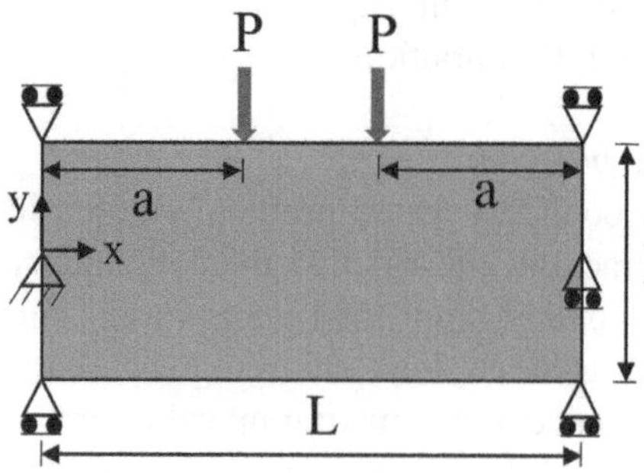

Figure 37.3 The simply supported deep beam under four-point bending studied in the current work
Source: Authors

Table 37.2 The applied load for different constitutive relations.

Linear elastic constitutive model (kN)	Non-linear elastic constitutive model NSC (kN)
8	8

Source: Authors

The domain is discretised into 33 × 19 field nodes with a total of 5055 unknowns. The dimensionless factors α_σ and α_u used for the approximation of Airy's stress function and displacement field, respectively, for different constitutive models is presented in Table 37.3. The value of dimensionless factors reported in Table 37.3 leads to the minimum objective function value for a given distribution of nodes and boundary conditions. The dimensionless parameters are fixed such that the approximation of field variables is localised. The number of nodes is fixed such that the objective function has the least value.

The numerical integration in Equation (11) is performed using composite Simpson's 1/3 rule. Table 37.4 presents the error in satisfying the constitutive relation for different constitutive models studied. The minimised objective value for the linear elastic model is the smallest among the studied models since the linear elastic model does not exhibit non-linearity.

Figure 37.5 shows the plot showing the variation of stress components over the deep beam subjected to four-point bending for the linear elastic Hooke's model. Figure 37.5a shows the normal stress σ_{xx} is compressive above the demarcated neutral axis and tensile below it as expected. In Figure 37.5b, the normal stress σ_{yy} is within the zero tolerance over the domain except at the loading points. Figure 37.5c

Table 37.3 Dimensionless factors used to approximate Airy's stress function and displacement field.

Parameters	Linear elastic constitutive model	Non-linear elastic constitutive model NSC
α_u	5	5
α_σ	17	17.5

Source: Authors

Table 37.4 Minimised value of Frobenius norm of the constitutive relation for different constitutive models.

Frobenius norm	Linear elastic constitutive model (× 10⁻⁴)	Non-linear elastic constitutive model (NSC)
δ	1.7598	8.8961

Source: Authors

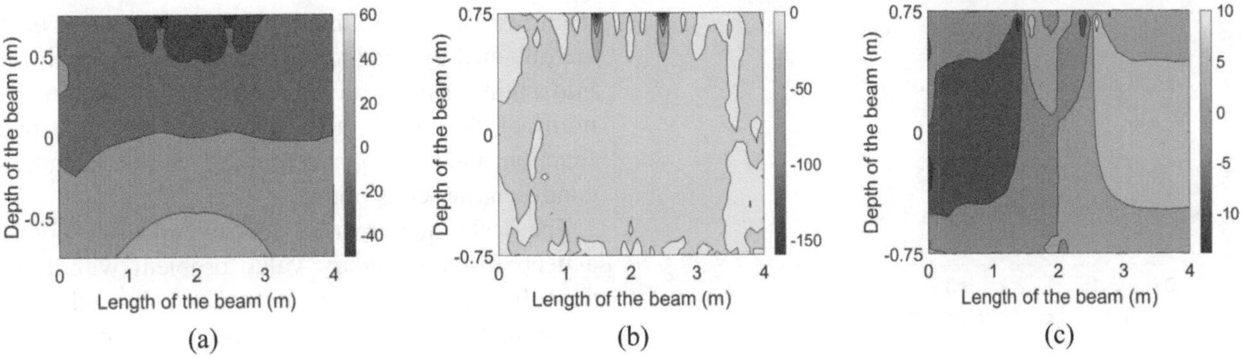

Figure 37.4 Plot showing variation of stress components (a) σ_{xx}, (b) σ_{yy}, and (c) σ_{xy} in MPa over the domain of the deep beam subjected to four-point bending for linear elastic constitutive model
Source: Authors

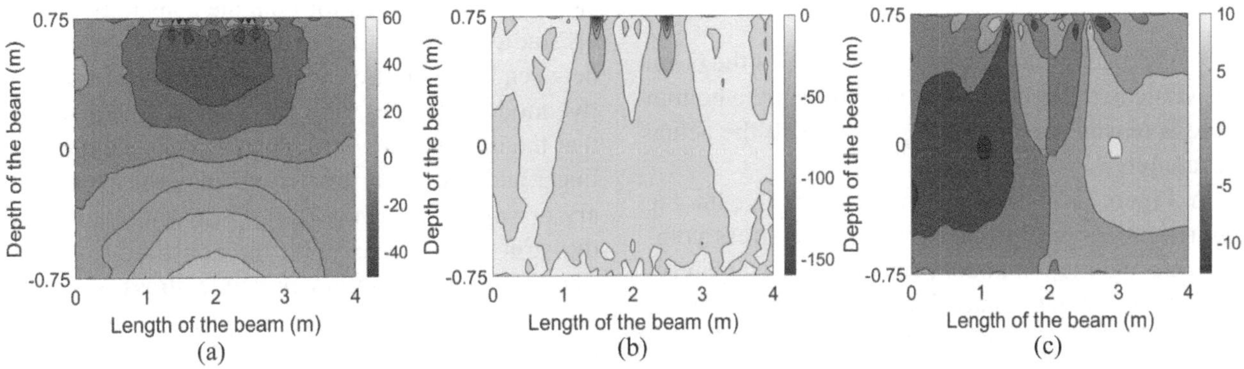

Figure 37.5 Plot showing variation of stress components (a) σ_{xx}, (b) σ_{yy}, and (c) σ_{xy} in MPa over the domain of the deep beam subjected to four-point bending for non-linear elastic constitutive model by [11] with normal strength grade concrete
Source: Authors

shows that zero shear stress variation over the depth of the beam at midspan, which is subjected to pure flexure. The nature of stresses developed in Figure 37.5 matches with the expected outcome.

The plots shown in Figure 37.4 depict the variation of stress components over the deep beam subjected to four-point bending for the non-linear elastic constitutive model of normal-strength concrete (NSC). Figure 37.5a shows the variation in normal stress σ_{xx} over the beam. The stress contours in Figure 37.4a are concentric and higher in numbers than in Figure 37.5a. Therefore, Figure 37.4a substantiates that the normal stress σ_{xx} varies non-linearly over the depth of the beam. The normal stress σ_{yy} in Figure 37.4b is as expected within tolerance over the domain except the boundary where point load is applied. Figure 37.4c represents the shear stress variation over the domain of the beam, with zero shear value along the depth at midspan. However, the region between the support and nearest point load, also known as the disturbed region, is subjected to shear in Figures 37.4c, 37.5c as the bending moment linearly varies.

Figure 37.6 shows the numerically obtained minimum principle stress eigenvectors over the domain of the deep beam for linear elastic Hooke's model and the non-linear elastic model for normal strength grade concrete. In Figure 37.6, eigenvectors for both models show the compression strut formation in the deep beam. The formation of two inclined struts between the supports and loading points and one horizontal strut between the load points is demonstrated. The strut angle represented by the eigenvectors for the two models differs near the support. However, Figure 37.6 depicts a significant difference in the strut angle for the two models along the strut direction and near the top surface of the beam. In Figure

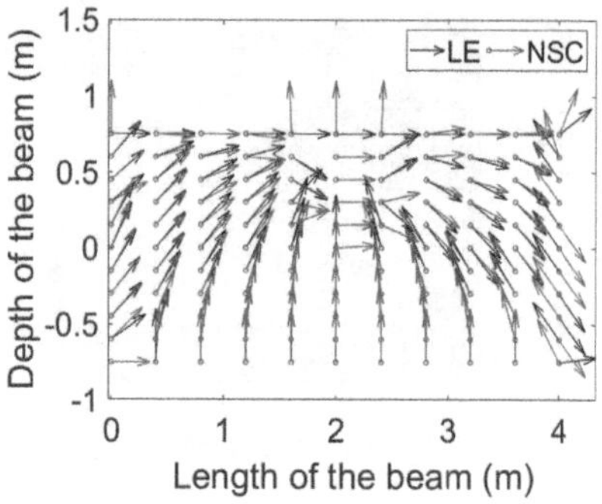

Figure 37.6 Plot showing minimum principle stress eigenvectors drawn over the domain of the deep beam for linear elastic Hooke's model and non-linear elastic constitutive model for normal strength concrete
Source: Authors

37.6, the eigenvectors at the bottom fibre of the beam are vertical, representing that the stress in the bottom fibre is tensile, which is comparable with the actual behaviour.

In Figure 37.4 through Figure 37.6 due to the numerical approximation involved, the stress components and the minimum principle stress eigenvectors at the boundary may deviate from the expected value. The difference in value is attributed to the fact that the element-free basis functions become spurious at boundaries, specifically at corners.

Conclusion

This study uses an existing alternate numerical technique based on multifield formulation and involves element-free basis functions for approximation of field variables to study the influence of constitutive relation on the strut angle in simply supported deep beams subjected to point loads. Airy's stress function and displacement field are considered as the primary unknowns. The Airy's stress function is approximated using C^2 Hermite-type moving least squares method. The displacement field is approximated using C^1 Moving least squares method. The Cauchy stress is obtained using Airy's function by exactly satisfying the equilibrium equation. The strain field is obtained from the assumed displacement field by exactly satisfying the strain-displacement relation. Finally, of all the stress fields satisfying traction boundary conditions and kinematically admissible displacement fields satisfying displacement boundary conditions, we sought the one that satisfies the constitutive relation in a weak integral sense. The unknowns in Airy's function and displacement field are obtained by casting the boundary value problem into a minimisation problem, such that the Frobenius norm of the constitutive relation is the objective function, and the displacement and traction boundary conditions are constraints.

The technique is used for numerically solving a deep beam boundary value problem with four-point bending for linear elastic Hooke's model and a non-linear, non-dissipative constitutive model for plain concrete. The stress response of the deep beam is obtained, and the eigenvectors corresponding to minimum principle stress are plotted. Firstly, the numerically obtained stress eigenvectors validate the formation of struts in deep beams. Inclined compression struts are formed for both the constitutive models studied. Further, it is observed that the strut angle between linear elastic and non-linear elastic constitutive models changes over the domain. It is observed that the difference in strut angles exists between the linear and non-linear elastic models near the boundary or support. However, away from the boundary, the strut angle varies with the constitutive model. Therefore, the minimum principle stress eigenvectors change for different constitutive models. Further, it would be interesting to find the combined influence of shear span to depth ratio and constitutive relation over the strut angle.

Acknowledgement

The first author gratefully acknowledges the research funding through the Prime Minister Research Fellowship (PMRF) and Half-Time Research Assistantship (HTRA) provided by the Ministry of Education (MoE), Government of India. Computational support was provided by the High-Performance Computing Environment (HPCE) at IIT Madras.

References

[1] A. C. Institute. Building code requirements for structural concrete (ACI 318-19) commentary on building code requirements for structural concrete (ACI 318R-19), 2019.

[2] Schlaich, J., Schaefer, K., Jennewein, M. Toward a consistent design of structural concrete. *PCI J.*, 1987;32(3):74–150.

[3] Schlaich, J., Schaefer, K. Design and detailing of structural concrete using strut-and-tie models. *Struct. Engg.*, 1991;69(6):113–125.

[4] Leonhardt, F. Reducing the shear reinforcement in reinforced concrete beams and slabs. *Mag. Concr. Res.*, 1965;17(53):187–198.

[5] Rüsch, H. On the limitations of applicability of the truss analogy for shear design of rc beams. 1964.

[6] Kupfer, H. Expansion of Morsch's truss analogy by application of the principle of minimum strain energy. *CEB Bull.*, 1964;40.

[7] El-zoughiby, M. E. Load-spread spectrum in Strut-and-Tie modeling of structural concrete. *ACI Struct. J.*, 2021;118:3–16.

[8] Rajagopal, K. R. On implicit constitutive theories. *Appl. Math.*, 2003;48(4):279–319.

[9] Rajagopal, K. R. The elasticity of elasticity. *Zeitschrift fur Angew. Math. und Phys.*, 2007;58(2):309–317.

[10] Gokulnath, C., Saravanan, U., Rajagopal, K. R. Representations for implicit constitutive relations describing non-dissipative response of isotropic materials. *Zeitschrift fur Angew. Math. und Phys.*, 2017;68(6):1–14.

[11] Gokulnath, C., Varaprasad, D., Saravanan, U. A three dimensional constitutive model for plain cement concrete. *Constr. Build. Mater.*, 2019;203:456–468.

[12] Shankar, L. S., Rajthilak, S., Saravanan, U. Numerical technique for solving truss and plane problems for a new class of elastic bodies. *Acta Mech.*, 2016;227(11):3147–3176.

[13] Hussein, A., Marzouk, H. Behavior of high-strength concrete under biaxial stresses. *ACI Struct. J.*, 2000;97(1):27–36.

[14] Bridhani, G. C., Saravanan, U. Multi-variable formulation using element-free basis functions for solving plane problems for a new class of constitutive relation. (Under review).

[15] Bridhani, G. C., Saravanan, U. Comparative study of methods to generate C^2 continuous mesh-free basis using the moving least square method. *VIII Internat. Conf. Particle-Based Methods PARTICLES 2023*, 2023:1–12.

[16] Belytschko, T., Lu, Y. Y., Gu, L. Element-free Galerkin methods. *Int. J. Numer. Methods Engg.*, 1994;37:229–256.

[17] Komargodski, Z., Levin, D. Hermite type moving-least-squares approximations. *Comput. Math. Appl.*, 2006;51(8):1223 1232.

For Product Safety Concerns and Information please contact our EU representative GPSR@taylorandfrancis.com Taylor & Francis Verlag GmbH, Kaufingerstraße 24, 80331 München, Germany